Lecture Notes in Computer Science   11639

Commenced Publication in 1973
Founding and Former Series Editors:
Gerhard Goos, Juris Hartmanis, and Jan van Leeuwen

More information about this series at http://www.springer.com/series/7407

Keren Censor-Hillel · Michele Flammini (Eds.)

Structural Information and Communication Complexity

26th International Colloquium, SIROCCO 2019
L'Aquila, Italy, July 1–4, 2019
Proceedings

 Springer

Editors
Keren Censor-Hillel
Technion
Haifa, Israel

Michele Flammini
University of L'Aquila
L'Aquila, Italy

ISSN 0302-9743   ISSN 1611-3349 (electronic)
Lecture Notes in Computer Science
ISBN 978-3-030-24921-2   ISBN 978-3-030-24922-9 (eBook)
https://doi.org/10.1007/978-3-030-24922-9

LNCS Sublibrary: SL1 – Theoretical Computer Science and General Issues

This Springer imprint is published by the registered company Springer Nature Switzerland AG
The registered company address is: Gewerbestrasse 11, 6330 Cham, Switzerland

Preface

This volume contains the papers presented at SIROCCO 2019: the 26th International Colloquium on Structural Information and Communication Complexity held during July 1–4, 2019, in L'Aquila. SIROCCO is devoted to the study of the interplay between structural knowledge, communication, and computing in decentralized systems of multiple communicating entities. Special emphasis is given to innovative approaches leading to better understanding of the relationship between computing and communication. SIROCCO has a tradition of interesting and productive scientific meetings in a relaxed and pleasant atmosphere, attracting leading researchers in a variety of fields in which communication and knowledge play a significant role. This year, SIROCCO was held in L'Aquila, a beautiful historical city on a mountain side, 100 km away from Rome.

For SIROCCO 2019 we received 39 submissions. Each submission was reviewed by three reviewers, either Program Committee (PC) members or external reviewers. The PC decided to accept 19 papers and to invite nine papers to be presented as brief announcements. The committee decided to give the SIROCCO 2019 Best Student Paper Award to Sebastian Brandt, Manuela Fischer and Jara Uitto, for their paper "Breaking the Linear-Memory Barrier in MPC: Fast MIS on Trees with Strongly Sublinear Memory."

Selected papers will also be invited to appear in a special issue of the *Theoretical Computer Science journal*, devoted to SIROCCO 2019.

In addition to the contributed talks, the conference program included invited talks by Susanne Albers, Pierre Fraigniaud, and Merav Parter, and a featured talk by Paola Flocchini – recipient of the 2019 Prize for Innovation in Distributed Computing. Before the start of the technical program, a 6-hour mini-course by Roger Watthenofer on blockchain was offered to the participants.

We would like to thank all of the authors for their high-quality submissions and all of the speakers for their excellent talks. We are grateful to the PC and all external reviewers for their efforts in putting together a great conference program, to the Steering Committee, chaired by Andrzej Pelc, for their help and support, and to everyone who was involved in the local organization. The EasyChair system was used to handle the submission of papers, manage the review process, and generate these proceedings.

July 2019

Keren Censor-Hillel
Michele Flammini

Organization

Program Committee

Dan Alistarh	IST, Austria
James Aspnes	Yale, USA
Alkida Balliu	Aalto University, Finland
Vittorio Bilò	University of Salento, Italy
Lelia Blin	LIP6, France
Ioannis Caragiannis	University of Patras, Greece
Keren Censor-Hillel	Technion, Israel
Michele Flammini	GSSI and University of L'Aquila, Italy
Luisa Gargano	Università di Salerno, Italy
Chryssis Georgiou	University of Cyprus, Cyprus
Magnus Halldorsson	Reykjavik University, Iceland
Tomasz Jurdzinski	University of Wroclaw, Poland
Fabian Kuhn	University of Freiburg, Germany
Toshimitsu Masuzawa	Osaka University, Japan
Alessia Milani	LaBRI, University of Bordeaux 1, France
Ivan Rapaport	Universidad de Chile, Chile
Dror Rawitz	Bar-Ilan University, Israel
Mordechai Shalom	Tel-Hai College, Israel
Jennifer Welch	Texas A&M University, USA
Prudence W. H. Wong	University of Liverpool, UK
Yukiko Yamauchi	Kyushu University, Japan

Steering Committee

Shantanu Das	Aix-Marseille University, France
Rastislav Kralovic	Comenius University, Slovakia
Zvi Lotker	Ben-Gurion University, Israel
Boaz Patt-Shamir	Tel Aviv University, Israel
Andrzej Pelc	University of Québec, Canada
Nicola Santoro	Carleton University, Canada
Sebastien Tixeuil	University of Paris 6, France
Jukka Suomela	Aalto University, Finland

Organizing Committee

Guido Proietti	University of L'Aquila, Italy
Gianlorenzo D'Angelo	GSSI, Italy
Mattia D'Emidio	University of L'Aquila, Italy
Michele Flammini	GSSI and University of L'Aquila, Italy

Luciano Gualà	University of Rome Tor Vergata, Italy
Ludovico Iovino	GSSI, Italy
Cosimo Vinci	University of L'Aquila, Italy

Additional Reviewers

Brandt, Sebastian
Böhnlein, Toni
Chalopin, Jérémie
Cordasco, Gennaro
Devismes, Stéphane
Dubois, Swan
Feuilloley, Laurent
Fraigniaud, Pierre
Gelles, Ran
Helgason, Sigurur
Hirvonen, Juho
Karakostas, George
Kim, Yonghwan
Klasing, Ralf
Korhonen, Janne H.
Kranakis, Evangelos
Kuszner, Lukasz
Labourel, Arnaud

Lotker, Zvi
Marcoullis, Ioannis
Markou, Euripides
Montealegre, Pedro
Nowicki, Krzysztof
Ooshita, Fukuhito
Pajak, Dominik
Rabie, Mikaël
Rescigno, Adele
Rybicki, Joel
Sudo, Yuichi
Tonoyan, Tigran
Uitto, Jara
Vaccaro, Ugo
Viglietta, Giovanni
Voudouris, Alexandros
Zaks, Shmuel

Laudatio: 2019 SIROCCO Prize for Innovation in Distributed Computing

It is a pleasure to award the 2019 SIROCCO Prize for Innovation in Distributed Computing to Paola Flocchini. Paola is a well-known member of the distributed computing community and her work is very closely related to SIROCCO's area of interest, i.e., the relationships between information and efficiency in decentralized computing. Most of her research can be divided into two parts. The first is the analysis of a property of labeled graphs called "sense of direction". This property which is a generalization of the orientation of rings or grids to arbitrary graphs, turns out to have major impacts on complexity and computational feasibility of communication systems. Paola's first publication on this topic [1] appeared in the first edition of SIROCCO and its journal version was later published in [2]. After that introductory paper, she wrote over 20 other papers focused on sense of direction, further exploring the impact of this property in distributed computations, all published in international conferences and top level international journals, such as the SICOMP paper [3].

The core of this part of Paola Flocchini's scientific work showed the dramatic effect that sense of direction has on the communication complexity of several important distributed problems, such as broadcast, depth-first traversal, election, spanning tree construction, in several classes of graphs. Before the seminal 1994 paper, an extensive body of evidence existed on the impact that specific edge labelings have on the communication complexity of distributed problems, suggesting that these (very different) labelings shared a common property. However, despite the obvious practical importance, a formal characterization of this property did not exist. Thus, the impact of this paper has been to provide a formal definition of sense of direction, defining those properties which make it possible to reduce the communication complexity in a distributed scenario. The final outcome of this work was to make explicit the very specific relationship between three factors: the labeling, the topological structure, and the local view that an entity has of the system.

Paola's work on sense of direction attracted a lot of attention in the distributed computing community, and generated a substantial amount of follow-up, producing a great body of scientific effort on this topic, as witnessed by over 150 citations (SCOPUS) of Paola's papers related to this matter.

The second important part of Paola's work concerns the theory of computation by autonomous mobile agents, i.e., distributed systems populated by moving and computing entities, that aim at solving various tasks. She contributed to the study of such entities both in the continuous setting (robots moving on a 2D plane), and in the discrete setting (agents acting in a network). In the continuous setting, she co-created the asynchronous oblivious mobile robots model called ASYNCH, and then studied several problems in this model, such as the gathering and the arbitrary pattern formation problem. She published over 20 papers on this topic, that collected more than 650 citations (SCOPUS), witnessing again the great interest of the distributed computing community in this area. One of the first of Paola's papers on mobile robots [4] was published in SIROCCO, and one of the most influential of her papers on asynchronous robots computing was the SICOMP paper [5].

Paola's body of work on mobile agents stimulated a lot of subsequent research, contributing to the development of this field, witnessed by numerous citations, and by establishing of the Moving and Computing international workshops, that periodically gather researchers from the distributed computing community interested in mobile agents computing.

We award the 2019 SIROCCO Prize for Innovation in Distributed Computing to Paola Flocchini for her contributions to the study of sense of direction in labeled graphs and to the analysis of asynchronous systems of mobile agents.

The 2019 Award committee[1]

Shantanu Das	Aix-Marseille Université
Pierre Fraigniaud	Université Paris Diderot, CNRS
Andrzej Pelc (Chair)	Université du Québec en Outaouais
Christian Scheideler	Paderborn University
Jukka Suomela	Aalto University
Sébastien Tixeuil	Sorbonne Université

Selected Publications Related to Paola Flocchini's Contribution:

1. P. Flocchini, B. Mans, N. Santoro,
   Sense of Direction: Formal Definitions and Properties,
   Proc. 1st Colloquium on Structural Information and Communication Complexity, (SIROCCO 94), 9–33, 1994.
2. P. Flocchini, B. Mans, N. Santoro
   Sense of Direction: Definitions, Properties, and Classes,
   Networks 32(3), 165–180, 1998.
3. P. Flocchini, A. Roncato, N. Santoro,
   Sense of Direction and Backward Consistency in Advanced Distributed Systems,
   SIAM Journal on Computing 32(2), 281–306, 2003.
4. P. Flocchini, G. Prencipe, N. Santoro, P. Widmayer,
   Pattern Formation by Anonymous Robots Without Chirality,
   Proc. 8th International Colloquium on Structural Information and Communication Complexity (SIROCCO 2001), 147–162, 2001.
5. M. Cieliebak, P. Flocchini, G. Prencipe, N. Santoro,
   Distributed Computing by Mobile Robots: Gathering,
   SIAM Journal on Computing, 41(4): 829–879, 2012.

---

[1] We wish to thank the nominators for the nomination and for contributing heavily to this text.

Abstracts of Invited Talks

On Energy Conservation in Data Centers

Susanne Albers

Department of Computer Science, Technical University of Munich
albers@in.tum.de

Abstract. We study algorithmic problems arising in data center operations with the objective to minimize the consumed energy. Specifically, we examine two settings that dynamically rightsize the pool of active servers depending on the current demand for computing capacity.

Data centers host a large number of power-heterogeneous servers. Each server has an active state and several standby/sleep states with individual power consumption rates. The demand for computing capacity varies over time. Idle servers may be transitioned to low-power modes so as to adjust the pool of active servers. The goal is to find a state transition schedule for the servers that minimizes the total energy consumed. For this power/capacity management problem, we present two main results. First, we investigate the scenario that each server has two states, i.e. an active state and a sleep state. We show that an optimal solution, minimizing energy consumption, can be computed in poly-nomial time by a combinatorial algorithm. The algorithm resorts to a single-commodity min-cost flow computation. Second, we study the general scenario that each server has an active state and multiple standby/sleep states. We devise a τ-approximation algorithm that relies on a two-commodity min-cost flow computation. Here τ is the number of different server types. A data center has a large collection of machines but only a relatively small number of different server architectures. Moreover, in the optimization one can assign servers with comparable energy consumption to the same class. Technically, both of our algorithms involve non-trivial flow modification procedures.

Additionally, we address an optimization problem introduced by Lin, Wier-man, Andrew and Thereska [3] that, over a time horizon, minimizes a combined objective function consisting of operating cost, modeled by a sequence of convex functions, and server switching cost. All prior work addresses a con-tinuous setting in which the number of active servers, at any time, may take a fractional value. We investigate for the first time the discrete data-center opti-mization problem where the number of active servers, at any time, must be integer valued. Thereby we seek truly feasible solutions. First, we show that the offline problem can be solved in polynomial time. Our algorithm relies on a new, yet intuitive graph theoretic model of the optimization problem and performs binary search in a layered graph. Second, we study the online problem and extend the algorithm *Lazy Capacity Provisioning* (LCP) by Lin et al. [3] to the discrete setting. We prove that LCP is 3-competitive. Moreover, we show that no deterministic online algorithm can achieve a competitive ratio smaller than 3. Hence, while LCP does not attain an optimal competitiveness in the continuous setting, it does so in the discrete problem examined here.

The presentation is based on our publications [1, 2].

References

1. Albers, S.: On energy conservation in data centers. In: Proceedings of 29th ACM Symposium on Parallelism in Algorithms and Architectures (SPAA), pp. 35–44 (2017)
2. Albers, S., Quedenfeld, J.: Optimal algorithms for right-sizing data centers. In: Proceedings of 30th ACM Symposium on Parallelism in Algorithms and Architectures (SPAA), pp. 363–372 (2018)
3. Lin, M., Wierman, A., Andrew, L.L.H., Thereska, E.: Dynamic right-sizing for power-proportional data centers. IEEE/ACM Trans. Netw. **21**(5),1378–1391 (2013)

A Topological Perspective on Distributed Network Algorithms

Armando Castañeda[1], Pierre Fraigniaud[2], Ami Paz[2],
Sergio Rajsbaum[1], Matthieu Roy[3], and Corentin Travers[4]

[1] UNAM, Mexico
{armando.castaneda,rajsbaum}@im.unam.mx
[2] CNRS and Université de Paris, France
{pierref,amipaz}@irif.fr
[3] CNRS, France
roy@laas.fr
[4] CNRS and University of Bordeaux, France
travers@labri.fr

Abstract. More than two decades ago, combinatorial topology was shown to be useful for analyzing distributed fault-tolerant algorithms in shared memory systems and in message passing systems. In this work, we show that combinatorial topology can also be useful for analyzing distributed algorithms in networks of arbitrary structure. To illustrate this, we analyze consensus, set-agreement, and approximate agreement in networks, and derive lower bounds for these problems under classical computational settings, such as the LOCAL model and dynamic networks.

On Sense of Direction and Mobile Agents

Paola Flocchini

University of Ottawa, Canada
paola.flocchini@uottawa.ca

Abstract. An edge-labeled graph is said to have *Sense of Direction* if the labeling satisfies a particular set of global consistency properties. When the graph represents a system of communicating entities, the presence of sense of direction has been shown to have a strong impact on computability and complexity.

Since its introduction, sense of direction has been investigated from various view points, revealing interesting graph theoretical properties and providing useful tools for the design of efficient distributed algorithms; furthermore, its presence allows to solve some otherwise unsolvable problems.

Far from being exhausted, the study of sense of direction and other consistency properties of edge-labeled graphs is still filled with interesting questions, open problems, and important new research directions.

In this paper, we revisit sense of direction reviewing the main results in the context of message passing point-to-point models, showing its impact in the more recent mobile agents models, and indicating directions for future study.

This work was supported in part by an NSERC Discovery Grant and by Dr. Flocchini's University Research Chair.

Secure Distributed Algorithms

Merav Parter

Weizmann Institute, Rehovot, Israel
merav.parter@weizmann.ac.il

Abstract. In the area of distributed graph algorithms a number of network's entities with local views solve some computational task by exchanging messages with their neighbors. Quite unfortunately, an inherent property of most existing distributed algorithms is that throughout the course of their execution, the nodes get to learn not only their own output but rather learn quite a lot on the inputs or outputs of many other entities. This leakage of information might be a major obstacle in settings where the output (or input) of network's individual is a private information (e.g., networks of selfish agents, decentralized digital currency such as Bitcoin).

While being quite an unfamiliar notion in the classical distributed setting, the notion of secure multi-party computation (MPC) is one of the main themes in the Cryptographic community. The existing secure MPC protocols do not quite fit the framework of classical distributed models in which only messages of bounded size are sent on graph edges in each round. In [1, 2], we present a new framework for *secure distributed graph algorithms* and provide the first *general compiler* that takes any "natural" non-secure distributed algorithm that runs in r rounds, and turns it into a secure algorithm that runs in $\widetilde{O}(r \cdot D \cdot \mathsf{poly}(\varDelta))$ rounds[1] where $\varDelta$ is the maximum degree in the graph and D is its diameter. A "natural" distributed algorithm is one where the local computation at each node can be performed in polynomial time. An interesting advantage of our approach is that it allows one to decouple between the price of locality and the price of *security* of a given graph function f. The security of the compiled algorithm is information-theoretic but holds only against a semi-honest adversary that controls a single node in the network.

The main technical part of our compiler is based on a new cycle cover theorem: We show that the edges of every bridgeless graph G of diameter D can be covered by a collection of cycles such that each cycle is of length $\widetilde{O}(D)$ and each edge of the graph G appears in $\widetilde{O}(1)$ many cycles. This is (existentially) optimal upto polylogarithmic terms.

In the second part of the talk, I will also discuss the notion of optimality in secure computation [3]. We will see how to adapt the existentially nearly optimal compiler into one that is nearly *optimal* (w.r.t. running time) for the given input graph G.

Keywords: CONGEST model · Cycle cover · Secure computation

Supported in part by grants from the Israel Science Foundation (no. 2084/18).

[1] The $\widetilde{O}(\cdot)$ notation hides poly-logarithmic terms in the number of vertices n.

References

1. Parter, M., Yogev, E.: Low congestion cycle covers and their applications. In: Proceedings of the Thirtieth Annual ACM-SIAM Symposium on Discrete Algorithms, SODA, pp. 1673–1692 (2019)
2. Parter, M., Yogev, E.: Distributed algorithms made secure: a graph theoretic approach. In: Proceedings of the Thirtieth Annual ACM-SIAM Symposium on Discrete Algorithms, SODA, pp. 1693–1710 (2019)
3. Parter, M., Yogev, E.: Secure distributed computing made optimal. In: Proceedings of the 2019 ACM Symposium on Principles of Distributed Computing, PODC (2019)

Contents

Brief Announcements

Invited Talks

A Topological Perspective on Distributed Network Algorithms

Armando Castañeda[1], Pierre Fraigniaud[2(✉)], Ami Paz[2], Sergio Rajsbaum[1],
Matthieu Roy[3], and Corentin Travers[4]

[1] UNAM, Mexico City, Mexico
{armando.castaneda,rajsbaum}@im.unam.mx
[2] CNRS and Université de Paris, Paris, France
{pierref,amipaz}@irif.fr
[3] CNRS, Toulouse, France
roy@laas.fr
[4] CNRS and University of Bordeaux, Bordeaux, France
travers@labri.fr

Abstract. More than two decades ago, combinatorial topology was shown to be useful for analyzing distributed fault-tolerant algorithms in shared memory systems and in message passing systems. In this work, we show that combinatorial topology can also be useful for analyzing distributed algorithms in networks of arbitrary structure. To illustrate this, we analyze consensus, set-agreement, and approximate agreement in networks, and derive lower bounds for these problems under classical computational settings, such as the LOCAL model and dynamic networks.

Keywords: Distributed computing · Distributed graph algorithms · Combinatorial topology

1 Introduction

1.1 Context and Objective

A breakthrough in distributed computing was obtained in the 1990's, when *combinatorial topology*, a branch of Mathematics extending graph theory to higher dimensional objects, was shown to provide a framework in which a large variety of models can be studied [29,41]. Combinatorial topology provides a powerful arsenal of tools, which considerably expended our understanding of the solvability and complexity of many distributed problems [2,9,10,30]. We refer to the book by Herlihy et al. [25] for an extended and detailed description of combinatorial topology applied to distributed computing, in a wide variety of settings.

In a nutshell, combinatorial topology allows us to represent all possible executions of a distributed algorithm, along with the relations between them, as a single mathematical object, whose properties reflect the solvability of a problem. Combinatorial topology was primarily used to study distributed computing

K. Censor-Hillel and M. Flammini (Eds.): SIROCCO 2019, LNCS 11639, pp. 3–18, 2019.
https://doi.org/10.1007/978-3-030-24922-9_1

in the context of shared memory and message passing systems, but not in the context of systems in which the presence of a network connecting the processing elements needs to be taken into account. On the other hand, a large portion of the study of distributed computing requires to take into account the structure of the network connecting the processors, e.g, when studying *locality*. This paper is a first attempt to approach distributed network computing through the lens of combinatorial topology.

The base of the topological approach for distributed computing consists of modeling all possible input (resp., output) configurations as a single object called input *complex* (resp., output complex), and specifying a task as a relation between the input and output complexes. Moreover, computation in a given model results in a topological deformation that modifies the input complex into another complex called the *protocol complex*. The fundamental result of combinatorial topology applied to distributed computing [25] is that a task is solvable in a computational model if and only if there exists a simplicial mapping, called *decision map*, from the protocol complex to the output complex, that agrees with the specification of the task. In other words, for every input configuration, (1) the execution of the algorithm should lead the system into one or many configurations, forming a subcomplex of the protocol complex, and (2) the decision map should map every configuration in this subcomplex (i.e., each of its *simplexes*) into a configuration of the output complex that is legal for the given input configuration, with respect to the specification of the task.

Understanding the power and limitation of a distributed computing model with respect to solving a given task requires to understand under which condition the decision map exists. This requires to understand the nature of topological deformations of the input complex resulting from the execution of an algorithm, and the outcome of this deformation, i.e., the protocol complex. That is, one needs to establish the connections between the distributed computing model at hand, and the topological deformations incurred by the input complex in the course of a computation under this model.

The connections between the computational models and the topological deformations are now well understood for several distributed computing models. For instance, in shared-memory wait-free systems, the protocol complex results from the input complex by a series of specific *subdivisions* of its simplexes. Note that the impossibility result for consensus in shared-memory wait-free systems is a direct consequence of this fact, as the input complex of consensus is connected, subdivisions maintain connectivity, but the output complex of consensus is not connected—this prevents the existence of a decision map, independently of how long the computation proceeds. Similarly, in shared-memory t-resilient systems, the protocol complex results from the input complex not only by a series of specific subdivisions, but also by the appearance of some *holes* in the course of the computation. This is because every process can wait for hearing from at least $n - t$ other processes in any n-node t-resilient system. These holes enable the existence of a decision map in the case of $(t + 1)$-set-agreement, but are not sufficient to enable the existence of a decision map for consensus, as long as $t \geq 1$. And indeed, the FLP result [19] implies that consensus is not solvable in asynchronous systems even in the presence of at most one failure.

This paper addresses the following issues: What is the nature of the topological deformations incurred by the input complex in the context of network computing, i.e., when nodes are bounded to interact only with nearby nodes according to some graph metric? And, what is the impact of these deformations on the ability to solve tasks efficiently (e.g., locally) in networks? As a first step towards answering these questions in general, we tackle them in the framework of synchronous failure-free computing, which is actually the framework in which most studies of distributed network computing are conducted [37].

1.2   Our Results

We place ourselves in the context of synchronous failure-free computing in networks [37]. As a first step towards understanding the nature of computation in this model from a topological perspective, we focus on lower bounds. We make a simplifying assumption which significantly strengthens the model, and therefore strengthens our lower bounds as well. We assume *structure awareness*. This assumption essentially asserts that each processing node is fully aware of the network it belongs to. More specifically, it assumes that all processes are given the same adjacency matrix of the network, and every process is given the index in the matrix of the vertex it occupies in the network. Structural awareness makes many tasks trivial. This is, for instance, the case of graph problems such as computing a vertex-coloring, an independent set, or a matching, which are among the main concerns of distributed network computing. Nevertheless, input-output tasks such as consensus and set-agreement, which are less studied in networks, yet important tasks as far as distributed computing and combinatorial topology are concerned [40], remain non-trivial.

The main contribution of this paper is in studying the topological model of distributed computing in networks, under the assumption of structure awareness. In particular, we show that the protocol complex involves deformations that were not observed before in the context of distributed computing, deformations which we call *scissor cuts*. These cuts appear between the facets of the input complex, and depend on the structure of the underlying network governing the way the information flows between nodes.

We show that this characterization is useful for deriving lower bounds on agreement tasks. For this purpose, we model the way information flows between nodes in the network by the so-called *information-flow graph*, and establish tight connections between structural properties of this graph, and the ability to solve agreement tasks in the network. This is achieved thanks to our understanding of the topology of the protocol complex. For instance, we show that if the domination number of the information-flow graph is at least $k + 1$, then the protocol complex is at least $(k - 1)$-connected, and if the protocol complex is at least $(k - 1)$-connected, then k-set agreement is not solvable.

Interestingly, our results connecting the structure of the information-flow graph with the topology of the protocol complex, imply lower bounds for solving agreement problems in the classical LOCAL model, as well as in dynamic networks. For instance, a consequence of our results is that, in the LOCAL model, solving k-set agreement in a network requires at least r rounds, where r is the smallest integer such that the r-th power of the network (two nodes are adjacent when their distance in the network is at most r) has domination number at most k. Similarly, we show that solving k-set agreement in a dynamic network $(H_t)_{t\geq1}$ requires at least r rounds, where r is the smallest integer such that $(H_t)_{1\leq t\leq r}$ has temporal dominating number at most k.

Applying the topological approach to network computing also enables to derive fine grained results. For instance, we show that in every n-node network where consensus is not solvable, ϵ-approximate agreement is also not solvable whenever $\epsilon < \frac{1}{n-1}$. This bound is tight, in the sense that there exists a network where consensus is impossible, while $\frac{1}{n-1}$-approximate agreement is solvable.

1.3   Related Work

The deep connections between combinatorial topology and distributed computing were concurrently and independently identified in [29] and [41]. Since then, numerous outstanding results were obtained using combinatorial topology for various types of tasks, including agreement tasks such as consensus and set-agreement [40], and symmetry breaking tasks such as renaming [2,9,10]. A recent work [1] provides evidence that topological arguments are sometimes necessary. All these contributions were obtained in the asynchronous shared memory model with crash failures, but combinatorial topology was shown to be applicable to Byzantine failures as well [36]. Note that the message passing model restricts itself to complete graphs [16,28]. Recent results showed that combinatorial topology can also be applied in the analysis of mobile computing [38], demonstrating the generality and flexibility of the topological framework applied to distributed computing. The book [25] provides an extensive introduction to combinatorial topology applied to distributed computing.

In contrast, distributed network computing has not been impacted by combinatorial topology. This domain of distributed computing is extremely active and productive this last decade, analyzing a large variety of network problems in the so-called LOCAL model [37], capturing the ability to solve task locally in networks[1]. We refer to [4,5,8,13,18,20,21,24,42] for a non exhaustive list of achievements in context. However, all these achievements were based on an operational approach, using sophisticated algorithmic techniques and tools solely from graph theory. Similarly, the existing lower bounds on the round-complexity of tasks in the LOCAL model [3,8,23,32,35] were obtained using graph theoretical and combinatorial arguments. The question of whether adopting a higher dimensional approach by using topology would help in the context of local computing, be it for a better conceptual understanding of the algorithms, or providing stronger technical tools for lower bounds, is, to our knowledge, entirely open.

---

[1] The CONGEST model has also been subject of tremendous progresses, but this model does not support full information protocols, and thus is out of the scope of our paper.

Similarly to (static) distributed network computing, the fundamental research on dynamic networks [6,11,12,34] has rarely been impacted by combinatorial topology. Relevant works in this framework study consensus [17,33], set-agreement [7,22] and approximate agreement [14]. We also refer to [15,31,39] which analyze distributed computation in a model where all processes broadcast messages at each round, but the recipients of these messages are defined by a graph which may change from round to round. The information-flow graph introduced and analyzed in this paper can be viewed as an abstraction of computation in dynamic networks, as this graph contains a summary of how information was transmitted among processes in the network during some interval of time.

2 Model and Definitions

In this section, we describe an abstract model of computation that captures various models of distributed computing, including the LOCAL model, and computing in dynamic graphs. This model is called KNOW-ALL, for reason that will soon be apparent.

2.1 The KNOW-ALL Model

We consider a set of n synchronous fault-free processes, with distinct names in $\{1, \ldots, n\}$, all running the same algorithm. The processes can model computing entities exchanging messages through a network, but also software agents or physical robots moving in space and exchanging messages whenever they meet, or computing entities in a dynamic network whose links evolve over time. The processes communicate using some communication medium, and the interactions are specified by a sequence $\mathcal{H}$ of n-node directed labeled graphs $(H_t)_{1 \leq t \leq T}$. The label of a node of H_t is a value in $\{1, \ldots, n\}$, different from the labels of all other nodes. The process with name $p \in \{1, \ldots, n\}$ occupies the node labeled p in each of the graphs H_t, $1 \leq t \leq T$. The arcs in H_t represent the interactions that can take place at the t-th rounds of an algorithm. The core property of the KNOW-ALL model is that every process is a priori given its name, and the sequence $\mathcal{H} = (H_t)_{1 \leq t \leq T}$, so every node is given the complete knowledge of the communication patterns occurring during the T rounds. The only uncertainty is about the inputs to the nodes.

The KNOW-ALL model is stronger than several classical distributed computing models. For example, the LOCAL model is also synchronous, fault-free model but with a fixed communication graph H, i.e., $H_t = H$ for every $t \geq 1$, and the nodes learn only some of the graph topology during an execution. A dynamic graph computation is defined by a sequence of graphs on the same set of nodes, and the nodes only gain partial information on the graph sequence during the execution. This is generalized by the KNOW-ALL model, where all the graph sequence is given in advance to the processes. Hence, in both cases the KNOW-ALL model is stronger than the classical model, and lower bounds proven for the KNOW-ALL model imply lower bounds for the other models as well.

By no means we claim the KNOW-ALL model to be practical. We make several simplifying assumptions that are typical in these settings: unbounded computational power, unbounded communication, failure-freeness, and also structural awareness, which is not a typical assumption. However, this strong model is sufficient for exhibiting lower bounds, and for establishing impossibility results for weaker, more realistic models. More important perhaps, it enables us to exhibit interesting phenomenon regarding the impact of the communication pattern on the topology of the protocol complex.

2.2   Input-Output Problems and the Information-Flow Graph

We focus on input-output problems, naturally defined as follows. A task (I, O, F) in the n-process KNOW-ALL model is described by a set I of input values, a set O of output values, and a mapping

$$F : I^n \to 2^{O^n}$$

specifying, for every n-tuple of input values, the set of possible legal n-tuple of output values. (In the topological sense, we focus on tasks for which the input complex is a pseudosphere, as explained below.) The input value of process p is denoted by $in(p) \in I$.

A distributed algorithm solving a task has two components: a communication protocol enabling each process to gather information about the inputs of other processes, and a decision function f that maps the gathered information to an output value. In the KNOW-ALL model, we can restrict our attention to considering only *flooding* protocols. At round t of such a protocol, every process p sends to all its out-neighbors in H_t all the name-input pairs it is aware of, that is, the pair $(p, in(p))$, and all the pairs it has received in the previous rounds. After T rounds, the process takes a decision based on the set of pairs it is aware of. Considering only flooding protocols does not reduce the computational power, as the structural awareness allows each process to simulate any other protocol.

Assuming flooding protocols, designing an algorithm boils down to designing a decision function f which allows each process, given the set of received input values, to compute an output value such that the collection of output values produced by the processes is consistent with the collection of input values. More specifically, for every vector of input values $(v_1, \ldots, v_n) \in I^n$, given to process $(p_1, \ldots, p_n)$, respectively, let w_i be the vector where for every $j \in \{1, \ldots, n\}$,

$$w_i[j] = \begin{cases} v_j \text{ if } j = i, \text{ or process } i \text{ receives the pair } (j, v_j) \text{ when flooding in } \mathcal{H}; \\ \bot \text{ otherwise.} \end{cases}$$

Then, every process $i \in \{1, \ldots, n\}$ must compute an output value

$$v_i' = f(i, w_i)$$

such that the resulting n-tuple $(v_1', \ldots, v_n')$ is in $F(v_1, \ldots, v_n)$.

In order to analyze flooding protocols, we define the *information-flow graph*, which describes the execution of a flooding protocol in the KNOW-ALL model.

Definition 1. *Let $\mathcal{H} = (H_t)_{1 \leq t \leq T}$ be an instance of the* KNOW-ALL *model. The information-flow graph associated with $\mathcal{H}$ is the directed graph G whose n nodes are labeled by $1, \ldots, n$, and there is an arc (p, q) from p to q in G if q receives the pair $(p, in(p))$ when flooding in $\mathcal{H}$.*

A crucial observation is that whenever two instances $\mathcal{H}$ and $\mathcal{H}'$ of the KNOW-ALL model yield the same information flow graph, then these two instances have the same computational power. The structure of the information-flow graph has a crucial impact on the ability to solve input-output problems in the KNOW-ALL model, an impact which we study in this paper. In order to clarify the impact of the structure of the information flow graph on the ability to solve problems, we apply techniques of combinatorial topology.

3   Topological Description of the KNOW-ALL Model

3.1   Basics Definitions

A *simplicial complex* is a finite set V along with a collection of nonempty subsets $\mathcal{K}$ of V closed under containment (i.e., if $A \in \mathcal{K}$ and $\emptyset \neq B \subset A$, then $B \in \mathcal{K}$). An element of V is called a *vertex* of $\mathcal{K}$, and the vertex set of $\mathcal{K}$ is denoted by $V(\mathcal{K}) = V$. Each set in $\mathcal{K}$ is called a *simplex*. A subset of a simplex is called a *face* of that simplex. The *dimension* dim σ of a simplex σ is one less than the number of elements of σ, i.e., $|\sigma| - 1$. We use "d-face" as shorthand for "d-dimensional face". A simplex σ in $\mathcal{K}$ is called a *facet* of $\mathcal{K}$ if σ is not contained in any other simplex. Note that a set of facets uniquely defines a simplicial complex. The dimension of a complex is the largest dimension of any of its facets. A complex is *pure* if all its facets have the same dimension.

Let $\mathcal{K}$ and $\mathcal{L}$ be complexes. A *vertex map* is a function $h : V(\mathcal{K}) \rightarrow V(\mathcal{L})$. If h also carries simplexes of $\mathcal{K}$ to simplexes of $\mathcal{L}$, it is called a *simplicial map*. We add one or more *labels* to the vertices, $\lambda : V \rightarrow D$, where D is an arbitrary domain. In particular, we have the set $\{1, \ldots, n\}$ of process names, and a label associating each vertex with a name. Typically, each simplex is *properly colored* by these names: if u and v are distinct vertices of a simplex σ, then name$(u) \neq$ name(v). A simplicial map h is *chromatic* if it preserves names, i.e., name$(h(v)) =$ name(v) for any vertex v. In this paper, all simplicial maps between colored complexes will be chromatic. Given two complexes $\mathcal{K}$ and $\mathcal{L}$, a *carrier map* Φ maps each simplex $\sigma \in \mathcal{K}$ to a sub-complex $\Phi(\sigma)$ of $\mathcal{L}$, such that for every two simplexes τ and τ' in $\mathcal{K}$ that satisfy $\tau \subseteq \tau'$, we have $\Phi(\tau) \subseteq \Phi(\tau')$.

Roughly speaking, a *geometric realization* $|\mathcal{K}|$ of a simplicial complex $\mathcal{K}$ is a geometric object defined as follows. Each vertex in $V(\mathcal{K})$ is mapped to a point in a Euclidean space, such that the images of the vertices are affinely independent. Each simplex is represented by a polyhedron, which is the convex hull of points representing its vertices. Figure 1 displays the geometric representations of several simplicial complexes that are detailed later.

Let k be a positive integer. We say that a complex has *a hole in dimension k* if the k-sphere S^k embedded in a geometric realization of the complex cannot be continuously contracted to a single point within that realization. Informally, a complex is *k-connected* if it has no holes in dimension k. A complex $\mathcal{K}$ is *k-connected* if every continuous map $h : S^k \to |\mathcal{K}|$ can be extended to a continuous map $h' : D^{k+1} \to |\mathcal{K}|$ where D^{k+1} denotes the $(k+1)$-disk. In dimension 0, this property simply states that any two points can be linked by a path, i.e., the complex is path-connected. In dimension 1, it states that any loop can be filled into a disk, i.e., the complex is simply connected. By convention, a (-1)-connected complex is just a non-empty complex.

Finally, given a set I, a *pseudosphere* $\Psi(\{1,\ldots,n\},I)$ is the complex defined as follows: (1) every pair (i,v) is a vertex, where $v \in I$, and (2) for every index set $J \subseteq \{1,\ldots,n\}$, and every multi-set $\{v_j : j \in J\}$ of values, the set $\{(j,v_j) : j \in J\}$ is a simplex. Pseudospheres offer a convenient way to describe all possible initial configurations where each process input is an arbitrary value from I.

3.2 The Topology of Computing in the KNOW-ALL Model

Given a distributed computing task (I,O,F) to be solved in the KNOW-ALL model, two complexes play a major role in this framework, the *input complex*, denoted by $\mathcal{I}$, and the *output complex*, denoted by $\mathcal{O}$. Let us fix an information flow graph G. The input complex $\mathcal{I}$ is the pseudosphere $\Psi(\{1,\ldots,n\},I)$, also defined by its set of facets

$$\big\{\{(1,v_1),\ldots,(n,v_n)\} : v_i \in I\big\}.$$

The set of all facets of the output complex $\mathcal{O}$ is

$$\big\{\{(1,v_1'),\ldots,(n,v_n')\} : v_i' \in O, \text{ and } \exists v \in I^n, (v_1',\ldots,v_n') \in F(v)\big\}.$$

Note that the output complex includes only combinations of output values that are legal with respect to the problem at hand. Note also that the input and output complexes do not depend on the communication medium considered, and that both complexes are pure—all their facets have the same dimension.

For instance, in the case of binary consensus in an n-process system (see Fig. 1), the set of facets of the input complex is

$$\big\{\{(1,v_1),\ldots,(n,v_n)\} : v_i \in \{0,1\}\big\}.$$

This complex is homeomorphic to the $(n-1)$-dimensional sphere S^{n-1}. For the same example, the output complex is composed of two disjoints $(n-1)$-facets, τ_0 and τ_1:

$$\tau_0 = \{(1,0),\ldots,(n,0)\}, \text{ and } \tau_1 = \{(1,1),\ldots,(n,1)\}.$$

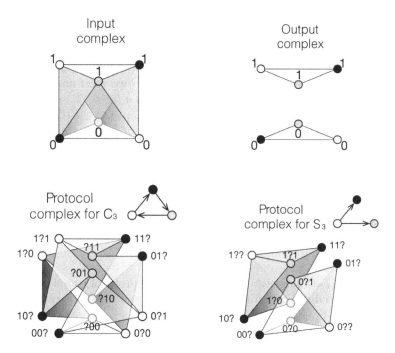

Fig. 1. Impact of the information flow graph on the protocol complex for binary consensus with three processes. Labels next to vertices are input and output values, in the input and output complexes respectively, or views in protocol complexes. A view "xyz" labeling a vertex means that the process corresponding to this vertex knows the input values x from process $\circ$, y from process $\bullet$, and z from process $\bullet$. A question mark in a label indicates that the process does not know the corresponding input value.

One can rephrase the operational definition (I, O, F) of *task* in Sect. 2.2 in the framework of combinatorial topology as follows: a task $(\mathcal{I}, \mathcal{O}, \Delta)$ is described by a carrier map Δ from $\mathcal{I}$ to $\mathcal{O}$. Note that, in absence of failures and asynchrony, a task can be described merely by a mapping Δ from the facets of $\mathcal{I}$ to subsets of facets of $\mathcal{O}$. For a given facet $\sigma = \{(1, v_1), \ldots, (n, v_n)\} \in \mathcal{I}$, the set of facets of $\Delta(\sigma)$ is defined by

$$\{(1, v'_1), \ldots, (n, v'_n)\} \in \Delta(\sigma) \iff (v'_1, \ldots, v'_n) \in F(v_1, \ldots, v_n). \tag{1}$$

The carrier map Δ of binary consensus maps every input facet σ containing both input values 0 and 1 to the two $(n-1)$-facets τ_0 and τ_1, and maps each $(n-1)$-facet σ_b with a unique input value $b \in \{0, 1\}$ to the output $(n-1)$-facet τ_b.

In any distributed computing model, in each point in time during the execution of an algorithm, one can define a complex whose vertices are pairs (p, w) where w is the state of process p, i.e., its view of the computation. A set of vertices with distinct process names forms a *protocol simplex* if there is a protocol execution where those processes collect those views. All possible protocol simplexes make up the *protocol complex*. The following fact is a direct consequence of the definition of the information flow graph.

Fact 1. *Given an information flow graph G, and a task $(\mathcal{I}, \mathcal{O}, \Delta)$, the protocol complex $\mathcal{P}$ associated with G and $\mathcal{I}$ is the complex whose facet are all the sets of the form $\{(1, w_1), \ldots, (n, w_n)\}$ for which there exists a facet $\{(1, v_1), \ldots, (n, v_n)\}$ of $\mathcal{I}$ such that, for $i = 1, \ldots, n$, $w_i = \{(j, v_j) : i = j \text{ or } (j, i) \in E(G)\}$. We define a carrier map $\Xi : \mathcal{I} \to \mathcal{P}$ which carries each facet of $\mathcal{I}$ to a single facet of $\mathcal{P}$, satisfying*

$$\Xi(\{(1, v_1), \ldots, (n, v_n)\}) = \{(1, w_1), \ldots, (n, w_n)\}.$$

An important observation is that the facets of the input complex are preserved in the protocol complex, i.e., there is a one-to-one correspondence between the facets of these two complexes. This is because the computation is synchronous and failure-free, from which it follows that each input configuration yields a single configuration in the protocol complex.

Example. Figure 1 displays two illustrations of the protocol complex for binary consensus, for two different information flow graphs on three processes: the consistently directed cycle C_3, and the directed star S_3 whose center has two out-neighbors. Process names are omitted, and instead are represented by the colors of the circles ($\circ$, $\bullet$, and $\bullet$). The number of vertices in the protocol complexes depends on the information flow graph.

Let us focus first on process $\circ$. A vertex $(\circ, v)$ in the input complex yields two vertices in the protocol complex for C_3, depending on the input value received from process $\bullet$. Instead, a vertex $(\circ, v)$ in the input complex yields a single vertex in the protocol complex for S_3 because, according to this information flow graph, process $\circ$ receives no inputs from other processes. On the other hand, every vertex $(\bullet, v)$ in the input complex yields two vertices in both protocol complexes. This is because, in both information flow graphs, C_3 and S_3, process $\bullet$ receives the input from process $\circ$. Similarly, every vertex $(\bullet, v)$ in the input complex yields two vertices in both protocol complexes, because in both information flow graphs process $\bullet$ receives the input from another process, from process $\bullet$ in C_3 and from process $\circ$ in S_3.

3.3 Topological Characterization of Task Solvability

So far, we have proceeded in two parallel paths. The first, *operational* path, was about algorithms in the KNOW-ALL model, where information propagates between processes according to some information flow pattern G. The second, *topological* path, relates the inputs of processes defined by an input complex, their views modeled in the protocol complex, and their desired outputs, appearing in the output complex. The connections between these paths is established in the next fact, which directly follows from the definitions.

Fact 2. *A task (I, O, F) is solvable in the KNOW-ALL model with information flow graph G if and only if, for the topological formulation $(\mathcal{I}, \mathcal{O}, \Delta)$ of the task,*

there exists a chromatic simplicial map $\delta : \mathcal{P} \to \mathcal{O}$ *satisfying* $\delta(\Xi(\sigma)) \in \Delta(\sigma)$
for every facet $\sigma \in \mathcal{I}$, *where* $\mathcal{P}$ *is the protocol complex associated with* G *and* $\mathcal{I}$.

The simplicial map $\delta : \mathcal{P} \to \mathcal{O}$ is called *decision map*. If $\delta(i, w_i) = (i, v_i')$,
then the corresponding algorithm specifies that process i with view w_i outputs
$f(i, w_i) = v_i'$.

Example. Let us consider Fig. 1 again. The protocol complex for S_3 is disconnected, while for C_3 it is 0-connected (i.e., path connected). The presence of a
universal node ○ (dominating all other nodes) in the information flow graph S_3
results in all processes becoming aware of the input of the process corresponding to that node. Therefore, the protocol complex for S_3 is split into two sub-
complexes, the one corresponding to process ○ with input 0, and the other corresponding to process ○ with input 1. Similarly, the protocol complex for the
complete graph K_3 with bidirectional edges is entirely disconnected, i.e., composed of eight pairwise non-intersecting facets, because there is no uncertainty
under the complete information flow graph, as every process receives the input
of every other process.

Since the protocol complex for S_3 is disconnected, consensus is solvable in
this graph. To see why, consider δ that maps every vertex $(p, 0**)$ of the protocol
complex to vertex $(p, 0)$ of the output complex, and every vertex $(p, 1**)$ of the
protocol complex to vertex $(p, 1)$ of the output complex. This is a chromatic
simplicial map, and thus, by Fact 2 consensus is solvable. In contrast, there is
no such mapping $\delta : \mathcal{P} \to \mathcal{O}$ for the protocol complex $\mathcal{P}$ corresponding to C_3,
because $\mathcal{P}$ is 0-connected. Let us consider the path $((○, 1?1), (\bullet, ?01), (\bullet, 00?))$ in
the protocol complex for C_3. Vertex $(○, 1?1)$ must be mapped to vertex $(○, 1)$
in the output complex because $(○, 1?1)$ belongs to a facet with all processes
having input value 1. Similarly, vertex $(\bullet, 00?)$ must be mapped to vertex $(\bullet, 0)$
because $(○, 00?)$ belongs to a facet with all processes having input value 0. If a
mapping δ maps $(\bullet, ?01)$ to $(\bullet, 1)$, then the simplex $\{(\bullet, ?01), (\bullet, 00?)\}$ is mapped
to $\{(\bullet, 1), (\bullet, 0)\}$, which is not a simplex of $\mathcal{O}$. The same occurs if $(\bullet, ?01)$ is
mapped to $(\bullet, 0)$, as $\{(○, 1), (\bullet, 0)\}$ is not a simplex of $\mathcal{O}$. Thus, there is no
simplicial map δ, and, by Fact 2, consensus is not solvable. We generalize this
result to every information flow graph G, and to k-set agreement, for every $k \geq 1$.

4   Applications to Agreement Tasks

In this section, we illustrate the power of using topology for analyzing the KNOW-
ALL model, and its implications on standard models such as LOCAL and dynamic
networks. First, we establish a connection between the structure of the information flow graph resulting from some instance of the KNOW-ALL model on the
one hand, and the topology of the protocol complex induced by this instance on
the other hand. Recall that the *domination number* $\gamma(G)$ of a graph G is the
number of vertices in a smallest dominating set for G, where, in directed graphs,
a vertex u dominates a vertex v if $(u, v) \in E(G)$.

Theorem 1. *Let $\mathcal{H}$ be an instance of the* KNOW-ALL *model, and G be the information flow graph associated with it. If $\gamma(G) > k$, then the protocol complex $\mathcal{P}$ for $\mathcal{H}$ is at least $(k-1)$-connected.*

Recall that, in the k-set agreement task, the processes must agree on at most k of the input values. In the context of asynchronous shared memory computing, the level of connectivity of the protocol complex is closely related to the ability to solve k-set agreement [26, 27, 30]. Using a similar connection, Theorem 1 implies the following.

Theorem 2. *Let $\mathcal{H}$ be an instance of the* KNOW-ALL *model, and G be the information flow graph associated with it. If $\gamma(G) > k$, then k-set agreement is not solvable in $\mathcal{H}$.*

To establish Theorem 2, we show that if the protocol complex $\mathcal{P}$ for $\mathcal{H}$ is at least $(k-1)$-connected, then k-set agreement is not solvable in $\mathcal{H}$, and then we apply Theorem 1. Observe that the converse of Theorem 2 also holds, i.e., if $\gamma(G) \leq k$ then k-set agreement is solvable in $\mathcal{H}$. The algorithm performs as follows. Let D be a dominating set for G, with $|D| \leq k$. Since D is dominating, every process p receives the input value of at least one process in D, and can decide on such a value as an output. In total, at most $|D| \leq k$ values are decided.

Theorem 2 implies that, in particular, consensus solvability requires the information flow graph to contain a universal node, i.e., a node that dominates all the other nodes. This theorem has implications for more traditional computational models, including the LOCAL model. Given a graph H, and $r \geq 1$, let H^r denote the graph on the same set of nodes as H, but in which two nodes are adjacent if their distance in H is at most r.

Corollary 1. *In the* LOCAL *model, solving k-set agreement in a network H requires at least r rounds, where r is the smallest integer such that $\gamma(H^r) \leq k$.*

Theorem 2 also applies to dynamic networks, in which edges appear and disappear over time. A dynamic network is a sequence $\mathcal{G} = (G_t)_{t \geq 1}$ of graphs on the same set of nodes V, where G_t is the actual network at round t. A set $D \subseteq V$ is a *temporal* dominating set for $(G_t)_{1 \leq t \leq r}$ if, for every node $v \notin D$, there is a temporal path from some node $u \in D$ to v, i.e., a sequence $(u_0, \ldots, u_s)$ of nodes with $u_0 = u$ and $u_s = v$, and a sequence $1 \leq t_0 < t_1 < \cdots < t_s \leq r$ of rounds such that $\{u_i, u_{i+1}\} \in E(G_{t_i})$ for every $i = 0, \ldots, s-1$.

Corollary 2. *Solving k-set agreement in dynamic network $\mathcal{G} = (G_t)_{t \geq 1}$ requires at least r rounds, where r is the smallest integer such that $(G_t)_{1 \leq t \leq r}$ has a temporal dominating set D with $|D| \leq k$.*

Finally, recall that, for $\epsilon \in [0, 1]$, binary ϵ-approximate agreement requires the processes to output values that are not more than ϵ apart, under the condition that if all the processes have the same input value $v \in \{0, 1\}$, then they all should output v. Using topological arguments applied to the information flow graph associated with the given instance $\mathcal{H}$ of the KNOW-ALL model, we show the following.

Theorem 3. *Let $\mathcal{H}$ be an instance of the* KNOW-ALL *model. If consensus is impossible under $\mathcal{H}$, then, for every $\epsilon < \frac{1}{n-1}$, ϵ-approximate agreement is also not solvable under $\mathcal{H}$. This bound is tight in the sense that there exists an instance $\mathcal{H}$ of the* KNOW-ALL *model for which consensus is impossible, while $\frac{1}{n-1}$-approximate agreement is solvable.*

The same way Theorem 2 has consequences on the complexity of solving k-set agreement in concrete computational models such as the LOCAL model and dynamic networks, Theorem 3 has similar consequences on the complexity of solving approximate agreement in these latter models.

5   Conclusion and Further Work

We demonstrate that combinatorial topology is applicable to distributed network computing. Of course, this is just a first step, and further work will require incorporating features of every distributed network model, in order to capture the specific characteristics of each of them. For instance, fully capturing the popular LOCAL model requires removing the structure awareness assumption, and studying the details of how the protocol complex evolves round after round.

Incorporating asynchrony and failures into network computing, from a topological perspective, requires understanding the topological impact of simultaneously stretching the facets, introducing holes resulting from t-resiliency, and introducing scissor cuts resulting from the presence of a network. This is definitely technically challenging, but our paper shows that there are no conceptual obstacles preventing us from addressing these questions.

Acknowledgments. Pierre Fraigniaud and Corentin Travers are supported by ANR projects DESCARTES and FREDA; Pierre Fraigniaud receives additional support from INRIA project GANG; Ami Paz is supported by the Fondation Sciences Mathématiques de Paris (FSMP); Sergio Rajsbaum is supported by project unam-papiit IN109917.

References

1. Alistarh, D., Aspnes, J., Ellen, F., Gelashvili, R., Zhu, L.: Why extension-based proofs fail. CoRR abs/1811.01421 http://arxiv.org/abs/1811.01421 (2018). To appear in STOC 2019
2. Attiya, H., Castañeda, A., Herlihy, M., Paz, A.: Bounds on the step and namespace complexity of renaming. SIAM J. Comput. **48**(1), 1–32 (2019). https://doi.org/10.1137/16M1081439
3. Balliu, A., Brandt, S., Hirvonen, J., Olivetti, D., Rabie, M., Suomela, J.: Lower bounds for maximal matchings and maximal independent sets. CoRR abs/1901.02441 http://arxiv.org/abs/1901.02441 (2019)
4. Barenboim, L., Elkin, M., Goldenberg, U.: Locally-iterative distributed $(\delta + 1)$-coloring below szegedy-vishwanathan barrier, and applications to self-stabilization and to restricted-bandwidth models. In: Proceedings of the 2018 ACM Symposium on Principles of Distributed Computing, (PODC), pp. 437–446 (2018). https://dl.acm.org/citation.cfm?id=3212769

5. Barenboim, L., Elkin, M., Pettie, S., Schneider, J.: The locality of distributed symmetry breaking. In: 53rd IEEE Symposium on Foundations of Computer Science (FOCS), pp. 321–330 (2012). https://doi.org/10.1109/FOCS.2012.60

6. Bhadra, S., Ferreira, A.: Computing multicast trees in dynamic networks and the complexity of connected components in evolving graphs. J. Internet Services Appl. **3**(3), 269–275 (2012). https://doi.org/10.1007/s13174-012-0073-z

7. Biely, M., Robinson, P., Schmid, U., Schwarz, M., Winkler, K.: Gracefully degrading consensus and k-set agreement in directed dynamic networks. Theor. Comput. Sci. **726**, 41–77 (2018). https://doi.org/10.1016/j.tcs.2018.02.019

8. Brandt, S., et al.: A lower bound for the distributed Lovász local lemma. In: 48th ACM Symposium on Theory of Computing (STOC), pp. 479–488 (2016). https://doi.org/10.1145/2897518.2897570

9. Castañeda, A., Rajsbaum, S.: New combinatorial topology bounds for renaming: the lower bound. Distrib. Comput. **22**(5–6), 287–301 (2010). https://doi.org/10.1007/s00446-010-0108-2

10. Castañeda, A., Rajsbaum, S.: New combinatorial topology bounds for renaming: the upper bound. J. ACM **59**(1), 3:1–3:49 (2012). https://doi.org/10.1145/2108242.2108245

11. Casteigts, A., Flocchini, P., Godard, E., Santoro, N., Yamashita, M.: On the expressivity of time-varying graphs. Theor. Comput. Sci. **590**, 27–37 (2015). https://doi.org/10.1016/j.tcs.2015.04.004

12. Casteigts, A., Flocchini, P., Quattrociocchi, W., Santoro, N.: Time-varying graphs and dynamic networks. Int. J. Parallel Emergent Distrib. Syst. **27**(5), 387–408 (2012). https://doi.org/10.1080/17445760.2012.668546

13. Chang, Y., Li, W., Pettie, S.: An optimal distributed $(\Delta + 1)$-coloring algorithm? In: 50th ACM Symposium on Theory of Computing (STOC), pp. 445–456 (2018). https://doi.org/10.1145/3188745.3188964

14. Charron-Bost, B., Függer, M., Nowak, T.: Approximate consensus in highly dynamic networks: the role of averaging algorithms. In: Halldórsson, M.M., Iwama, K., Kobayashi, N., Speckmann, B. (eds.) ICALP 2015. LNCS, vol. 9135, pp. 528–539. Springer, Heidelberg (2015). https://doi.org/10.1007/978-3-662-47666-6_42

15. Charron-Bost, B., Schiper, A.: The heard-of model: computing in distributed systems with benign faults. Distrib. Comput. **22**(1), 49–71 (2009). https://doi.org/10.1007/s00446-009-0084-6

16. Chaudhuri, S., Herlihy, M., Lynch, N.A., Tuttle, M.R.: Tight bounds for k-set agreement. J. ACM **47**(5), 912–943 (2000). https://doi.org/10.1145/355483.355489

17. Coulouma, E., Godard, E., Peters, J.G.: A characterization of oblivious message adversaries for which consensus is solvable. Theor. Comput. Sci. **584**, 80–90 (2015). https://doi.org/10.1016/j.tcs.2015.01.024

18. Fischer, M., Ghaffari, M., Kuhn, F.: Deterministic distributed edge-coloring via hypergraph maximal matching. In: 58th IEEE Annual Symposium on Foundations of Computer Science (FOCS), pp. 180–191 (2017). https://doi.org/10.1109/FOCS.2017.25

19. Fischer, M.J., Lynch, N.A., Paterson, M.: Impossibility of distributed consensus with one faulty process. J. ACM **32**(2), 374–382 (1985). https://doi.org/10.1145/3149.214121

20. Ghaffari, M.: An improved distributed algorithm for maximal independent set. In: 27th ACM-SIAM Symposium on Discrete Algorithms (SODA), pp. 270–277 (2016). https://doi.org/10.1137/1.9781611974331.ch20

21. Ghaffari, M., Kuhn, F., Maus, Y.: On the complexity of local distributed graph problems. In: 49th ACM Symposium on Theory of Computing (STOC), pp. 784–797 (2017). https://doi.org/10.1145/3055399.3055471
22. Godard, E., Perdereau, E.: k-set agreement in communication networks with omission faults. In: 20th International Conference on Principles of Distributed Systems (OPODIS), pp. 8:1–8:17 (2016). https://doi.org/10.4230/LIPIcs.OPODIS.2016.8
23. Göös, M., Hirvonen, J., Suomela, J.: Linear-in-Δ lowerbounds in the LOCAL model. Distrib. Comput. **30**(5), 325–338 (2017). https://doi.org/10.1007/s00446-015-0245-8
24. Harris, D.G., Schneider, J., Su, H.: Distributed $(\Delta + 1)$-coloring in sublogarithmic rounds. In: 48th ACM Symposium on Theory of Computing (STOC), pp. 465–478 (2016). https://doi.org/10.1145/2897518.2897533
25. Herlihy, M., Kozlov, D., Rajsbaum, S.: Distributed Computing Through Combinatorial Topology. Morgan Kaufmann, San Francisco (2013)
26. Herlihy, M., Rajsbaum, S.: Set consensus using arbitrary objects. In: Proceedings of the Thirteenth Annual ACM Symposium on Principles of Distributed Computing (PODC), pp. 324–333 (1994). https://doi.org/10.1145/197917.198119
27. Herlihy, M., Rajsbaum, S.: Algebraic spans. Math. Struct. Comput. Sci. **10**(4), 549–573 (2000). http://journals.cambridge.org/action/displayAbstract?aid=54601
28. Herlihy, M., Rajsbaum, S., Tuttle, M.R.: An axiomatic approach to computing the connectivity of synchronous and asynchronous systems. Electr. Notes Theor. Comput. Sci. **230**, 79–102 (2009). https://doi.org/10.1016/j.entcs.2009.02.018
29. Herlihy, M., Shavit, N.: The asynchronous computability theorem for t-resilient tasks. In: 25th ACM Symposium on Theory of Computing (STOC), pp. 111–120 (1993). https://doi.org/10.1145/167088.167125
30. Herlihy, M., Shavit, N.: The topological structure of asynchronous computability. J. ACM **46**(6), 858–923 (1999). https://doi.org/10.1145/331524.331529
31. Kuhn, F., Lynch, N.A., Oshman, R.: Distributed computation in dynamic networks. In: 42nd ACM Symposium on Theory of Computing (STOC), pp. 513–522 (2010). https://doi.org/10.1145/1806689.1806760
32. Kuhn, F., Moscibroda, T., Wattenhofer, R.: Local computation: lower and upper bounds. J. ACM **63**(2), 17:1–17:44 (2016). https://doi.org/10.1145/2742012
33. Kuhn, F., Moses, Y., Oshman, R.: Coordinated consensus in dynamic networks. In: 30th ACM Symposium on Principles of Distributed Computing (PODC), pp. 1–10 (2011). https://doi.org/10.1145/1993806.1993808
34. Kuhn, F., Oshman, R.: Dynamic networks: models and algorithms. SIGACT News **42**(1), 82–96 (2011). https://doi.org/10.1145/1959045.1959064
35. Linial, N.: Locality in distributed graph algorithms. SIAM J. Comput. **21**(1), 193–201 (1992). https://doi.org/10.1137/0221015
36. Mendes, H., Tasson, C., Herlihy, M.: Distributed computability in Byzantine asynchronous systems. In: 46th Symposium on Theory of Computing (STOC), pp. 704–713 (2014). https://doi.org/10.1145/2591796.2591853
37. Peleg, D.: Distributed Computing: A Locality-Sensitive Approach. SIAM, Philadelphia (2000)
38. Rajsbaum, S., Castañeda, A., Flores-Peñaloza, D., Alcantara, M.: Fault-tolerant robot gathering problems on graphs with arbitrary appearing times. In: 2017 IEEE International Parallel and Distributed Processing Symposium (IPDPS), pp. 493–502 (2017). https://doi.org/10.1109/IPDPS.2017.70
39. Rajsbaum, S., Raynal, M., Travers, C.: The iterated restricted immediate snapshot model. In: 14th International Conference on Computing and Combinatorics (COCOON), pp. 487–497 (2008). https://doi.org/10.1007/978-3-540-69733-6_48

40. Sakavalas, D., Tseng, L.: Network topology and fault-tolerant consensus. Synth. Lect. Distrib. Comput. Theory **9**, 1–151 (2019)
41. Saks, M.E., Zaharoglou, F.: Wait-free k-set agreement is impossible: the topology of public knowledge. In: 25th ACM Symposium on Theory of Computing (STOC), pp. 101–110 (1993). https://doi.org/10.1145/167088.167122
42. Suomela, J.: Survey of local algorithms. ACM Comput. Surv. **45**(2), 24:1–24:40 (2013). https://doi.org/10.1145/2431211.2431223

On Sense of Direction
and Mobile Agents

Paola Flocchini[(✉)]

University of Ottawa, Ottawa, Canada
paola.flocchini@uottawa.ca

Abstract. An edge-labeled graph is said to have *Sense of Direction* if the labeling satisfies a particular set of global consistency properties. When the graph represents a system of communicating entities, the presence of sense of direction has been shown to have a strong impact on computability and complexity.

Since its introduction, sense of direction has been investigated from various view points, revealing interesting graph theoretical properties and providing useful tools for the design of efficient distributed algorithms; furthermore, its presence allows to solve some otherwise unsolvable problems.

Far from being exhausted, the study of sense of direction and other consistency properties of edge-labeled graphs is still filled with interesting questions, open problems, and important new research directions.

In this paper, we revisit sense of direction reviewing the main results in the context of message passing point-to-point models, showing its impact in the more recent mobile agents models, and indicating directions for future study.

1 Introduction

Sense of direction ($\mathcal{SD}$) is a property of edge-labeled graphs that plays a special role in distributed computing. In fact, its presence has been shown to have a strong impact on computability of problems, as well as on their complexity. The power of sense of direction has been originally observed in message passing point-to-point systems and, more recently, in systems of mobile agents operating and moving on networks.

Given a simple undirected graph $G = (V, E)$ and a set Σ of labels, let λ_x be a function which associates a label to each edge incident on node $x \in V$. Let (G, λ) be the edge-labeled graph where $\lambda = \{\lambda_x\}$. In a distributed system with point-to-point communication, G would correspond to the communication topology of the system, and Σ to the set of possible communication link labels, called port numbers.

This work was supported in part by an NSERC Discovery Grant and by Dr. Flocchini's University Research Chair.

K. Censor-Hillel and M. Flammini (Eds.): SIROCCO 2019, LNCS 11639, pp. 19–33, 2019.
https://doi.org/10.1007/978-3-030-24922-9_2

The choice of an appropriate labeling $\lambda = \{\lambda_x : x \in V\}$ is very important, because it can be exploited when designing algorithms or can even have an impact on the solvability of problems. One of the basic properties of λ, underlying all point-to-point models, is *Local Orientation*: having distinct labels for distinct edges incident on the same node; another is *Edge Symmetry*, when the label $\lambda_y(y, x)$ can be derived from $\lambda_x(x, y)$ (this is the case, for example, of the left/right labeling of a ring or the "compass" labeling in a torus); a particular edge-symmetric labeling is *Coloring*, where $\lambda_x(x, y) = \lambda_y(y, x)$ (this is the case, for example, of the "dimensional" labeling in a hypercube).

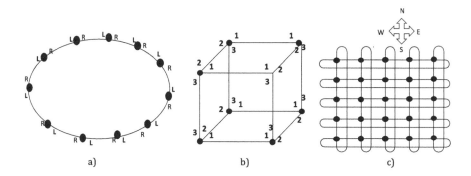

Fig. 1. Edge-labeled graphs.

Special edge labelings for specific topologies have been often exploited in the design of distributed algorithms. This has been the case, for example, of leader election in complete graphs, in hypercubes, in chordal rings, etc. (see [21] and references therein). In each of those topologies, the specific edge labeling played a crucial role; what all these different techniques had in common, if anything, was however quite elusive.

It turns out that there is indeed a particular property at the core of all those results and a commonality among all those labeled graphs: the fact that, from a given starting node in the graph, by following any two "labeled walks", it is possible to determine whether the walks terminate in the same node or not. Even more interestingly, the characteristics of those specific labelings, which seemed to be tightly linked to the topologies where they were defined, can be easily generalized to construct labelings with this property in any arbitrary topology. This observation gave rise to the formal definition of *Sense of Direction* [20].

In this paper, we revisit sense of direction, reviewing what is known in the context of message passing distributed models, showing its impact in the more recent mobile agents models, and indicating directions for future study.

2   Definitions

2.1   Labeled Graphs

Let $G = (V, E)$ be a simple undirected graph; let $E(x)$ denote the set of edges incident to node $x \in V$, and $d(x) = |E(x)|$ the degree of x.

Given $G = (V, E)$ and a set Σ of labels, a *local orientation* of $x \in V$ is any injective function $\lambda_x : E(x) \to \Sigma$ which associates a distinct label to each edge. A set $\lambda = \{\lambda_x : x \in V\}$ of local orientations will be called a *labeling* of G, and by (G, λ) we shall denote the corresponding *(edge-)labeled graph*. In the following, we indicate with n and m, respectively, the number of nodes and edges of G.

A labeling λ is *minimal* if it uses $\Delta_G = max\{d(x) : x \in V\}$ labels. It is *symmetric* if there exists a bijection $\psi : \Sigma \to \Sigma$ such that for each $(x, y) \in E$, $\lambda_y((y, x)) = \psi(\lambda_x((x, y)))$; ψ will be called the edge-symmetry function. A symmetric labeling is a *coloring* if the edge symmetry function is the Identity.

A *walk* π in G is a sequence of edges in which the endpoint of one edge is the starting point of the next edge. Let $P[x]$ denote the set of all the non empty walks starting from $x \in V$, $P[x, y]$ the set of walks starting from $x \in V$ and ending in $y \in V$. Let $\Lambda_x : P[x] \to \Sigma^+$ and $\Lambda = \{\Lambda_x : x \in V\}$ denote the extension of λ_x and λ, respectively, from edges to walks; let $\Lambda[x] = \{\Lambda_x(\pi) : \pi \in P[x]\}$, and $\Lambda[x, y] = \{\Lambda_x(\pi) : \pi \in P[x, y]\}$.

2.2   Sense of Direction

Sense of Direction is defined by the existence of a consistent coding and a consistent decoding function [20].

Definition 1. *Given (G, λ), a consistent coding function (or, simply, coding function) $\mathbf{c}$ for λ is any function with domain Σ^+ such that walks originating from the same node are mapped to the same value (called* local name*) if and only if they end in the same node; that is,*

$$\forall x, y, z \in V, \forall \pi_1 \in P[x, y], \pi_2 \in P[x, z], \mathbf{c}(\Lambda_x(\pi_1)) = \mathbf{c}(\Lambda_x(\pi_2)) \quad \Leftrightarrow \quad y = z.$$

A system (G, λ), has *weak sense of direction* $(\mathcal{WSD})$ iff there exists a coding function $\mathbf{c}$ for λ. We shall denote by $\mathcal{N}$ the codomain of $\mathbf{c}$.

Definition 2. *Given a coding function $\mathbf{c}$, a decoding function $\mathbf{d}$ for $\mathbf{c}$ is any function $\mathbf{d} : \Sigma \times \mathcal{N} \to \mathcal{N}$ such that $\forall x, y, z \in V$, with $(x, y) \in E(x)$ and $\pi \in P[y, z]$:*

$$\mathbf{d}(\lambda_x((x, y)), \mathbf{c}(\Lambda_y(\pi)) = \mathbf{c}(\lambda_x((x, y)) \cdot \Lambda_y(\pi)),$$

where $\cdot$ is the concatenation operator.

A system (G, λ), has a *sense of direction* $(\mathcal{SD})$ iff there exists both a coding function $\mathbf{c}$ for λ and a decoding function $\mathbf{d}$ for $\mathbf{c}$. We shall also say that $(\mathbf{c}, \mathbf{d})$ is a sense of direction in (G, λ).

Note that $\mathcal{SD}$ is a stronger notion than $\mathcal{WSD}$; in fact, it is easy to construct labeled graphs with $\mathcal{WSD}$ but without $\mathcal{SD}$.

Example 1: Chordal $\mathcal{SD}$. Given any graph $G = (V, E)$, with $|V| = n$, a chordal labeling can be obtained by fixing an arbitrary cyclic ordering of the nodes and labeling an edge (x, y) by the "distance" (modulo n) between y and x in the predefined ordering (see Fig. 2a). In this case the set of labels Σ is the set of positive integers modulo n.

With this labeling, (G, λ) has sense of direction (called *chordal*). The coding function $\mathbf{c} : \Sigma^+ \to \Sigma$ is the function that maps a sequence of labels into their sum modulo n: for any sequence of labels $a_1, a_2, \dots a_k$ with $a_i \in \Sigma$, $c(a_1, \dots a_k) = \sum_{i=1}^{k} a_i \bmod n$. The corresponding decoding function is defined as follows: $\forall a, b \in \Sigma$, $\mathbf{d}(a, b) = (a + b) \bmod n$. Note that this labeling is edge symmetric; in fact, $\lambda_x(x, y) = n - \lambda_y(y, x)$.

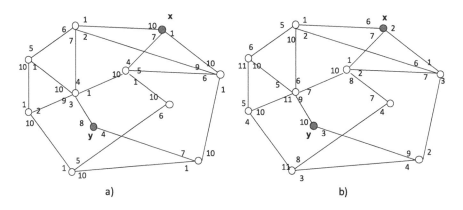

Fig. 2. (*a*) Chordal of $\mathcal{SD}$; (*b*) Neighbouring $\mathcal{SD}$.

Example 2: Neighboring $\mathcal{SD}$. A graph (G, λ) has *neighboring sense of direction* if λ is defined as follows: $\forall (x, y) \in E[x]$, $(z, w) \in E[z]$, $\lambda_x(x, y) = \lambda_z(z, w)$ iff $y = w$. That is, in a neighboring sense of direction, all the links ending in the same node are labeled with the same label (see Fig. 2b).

Let Σ be the set of labels; the coding function $\mathbf{c} : \Sigma^+ \to \Sigma$ is the following: for any sequence of labels $a_1, a_2, \dots a_k$ with $a_i \in \Sigma$, $\mathbf{c}(a_1, \dots a_k) = a_k$. The corresponding decoding function is the following: $\forall a, b \in \Sigma$, $\mathbf{d}(a, b) = b$. Note that this labeling is not edge symmetric.

A property of neighboring sense of direction that makes it particularly strong and sets it apart from most classes of sense of directions is that, when it is present in an *anonymous network* (i.e., where nodes are totally indistinguishable), it always destroys anonymity.

3   Sense of Direction: Properties

The study of the many interesting properties of edge-labeled graphs with sense of direction is fascinating on its own, even without considering implications on computability and complexity in distributed computing.

3.1 Symmetries and Minimal Sense of Direction in Regular Graphs

A *minimal SD* in a *d*-regular graph is a *SD* that uses only *d* labels. An interesting question is to determine under what conditions a *d*-regular graph admits a minimal *SD*.

Before describing the main results, we need to introduce some symmetry notions. A graph *G* is vertex transitive if, $\forall x, y \in V$, there exists an automorphism ρ such that $(\rho(x), \rho(y)) \in E$ iff $(x, y) \in E$. Consider now a labeled graph (G, λ).

The *surrounding N(u)* of a node *u* in (G, λ) is a labeled graph isomorphic to (G, λ) through a labeled graph isomorphism χ, together with the image of node u, $\chi(u)$ [22]. A graph is *surrounding symmetric* if every node has the same surrounding. Intuitively, surrounding symmetry is a generalization to edge-labeled graphs, of the well known notion of vertex transitivity. The hypercube of Fig. 1(b) is both vertex transitive and surrounding symmetric; the graph of Fig. 3, instead, is vertex transitive but there exists no labeling that makes it surrounding symmetric.

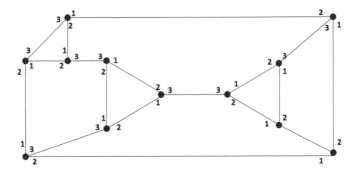

Fig. 3. A vertex symmetric graph *G*. The labeled graph (G, λ) is not surrounding symmetric.

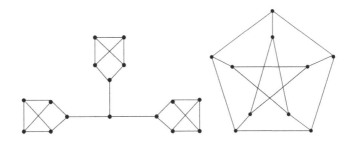

Fig. 4. Graphs that do not admit a Minimal *SD*.

Vertex transitivity of *G* is necessary for minimal *SD* in (G, λ) but it is not sufficient [15]. Figure 4 shows two regular graphs: the one on the left is not vertex symmetric, the one on the right is, but neither graph admits a minimal $\mathcal{SD}$, i.e., any $\mathcal{SD}$ requires at least fours labels in both graphs.

A regular graph with edge symmetric labeling has a minimal $\mathcal{SD}$ iff it is surrounding symmetric [22]. It can be shown that a labeled graph (G, λ) is surrounding symmetric iff it is a Cayley graph with a Cayley labeling[1]. It then follows that regular graphs with symmetric labeling admitting minimal sense of direction are all and only Cayley graphs (with Cayley labeling) [22]. This result holds also for directed graphs [4].

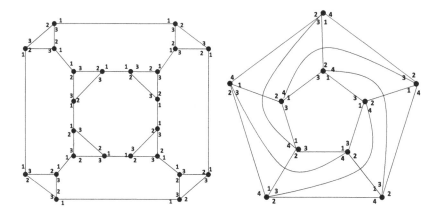

Fig. 5. Minimal $\mathcal{SD}$: (a) with edge symmetry; (b) without edge symmetry

When the labeling is not edge symmetric, the characterization is more complex: a regular labeled graph (G, λ) has a minimal $\mathcal{SD}$ iff G is the graph of a semigroup S which is the direct product of a group and a (particular type of semigroup called) left-zero semigroup, *and* λ corresponds to the generators of S. The graph of a semigroup is the graph where nodes correspond to the elements of the semigroup and the edges correspond to the action of the generators [26]. Examples of minimal sense of directions in edge symmetric and non edge symmetric graphs are shown in Fig. 5.

In [30] the class of d-regular graphs that admit a minimal Chordal $\mathcal{SD}$ has been studied, showing that this class is equivalent to that of circulant graphs, and presenting a polynomial-time algorithm for recognizing it, when the degree d is fixed.

---

[1] A Cayley graph is a graph where nodes correspond to the elements of a group and edges correspond to the action of the generators; a Cayley graph has a Cayley labeling when labels on the edges correspond to the generators of the group.

3.2 Minimum Sense of Direction

Let Δ_G be the maximum degree of G. When a minimal sense of direction does not exist in G, a natural question is to design one that uses the minimum possible number of labels.

It was conjectures that $\Delta_G + 1$ labels might be always sufficient. However, this is not the case; in fact, for sufficiently large n, there are graphs requiring $\Delta_G + \Omega(n \log \log n / \log n)$ labels for WSD, and this is true also for regular graphs. Moreover $\Omega(\Delta_G \log \log \Delta_G)$ labels are necessary in any graph G, and $\Omega(d \sqrt{\log \log d})$ in regular graphs of degree d [6,7].

3.3 Testing for Sense of Direction

Given a labeled graph (G, λ), how can we verify (decide) whether there is sense of direction? In [5], this question has been answered by showing that there exist polynomial algorithms for testing both $\mathcal{WSD}$ and $\mathcal{SD}$.

The time complexity of the algorithm for testing $\mathcal{WSD}$ is $O(n^{4.752} \log n)$. On the other hand, the algorithm for testing $\mathcal{SD}$ is significantly more complicated and its time complexity is $O(n^{14.256} \log n)$. Deciding $\mathcal{WSD}$ can be done efficiently in parallel. In fact, considering as model of computation a CRCW PRAM, $\mathcal{WSD}$ is in AC^1 for all graphs using n^6 processors. Unfortunately, this is not the case for $\mathcal{SD}$ which is in AC^1 only for some classes of graphs.

A consequence of these results is a polynomial time algorithm to test whether a given labeled graph is a Cayley graph (with a Cayley labeling).

A related question is the following: given a graph G and a coding function c: does G admit a sense of direction that uses c? In [10] it has been shown that the problem is NP-complete even for simple coding functions; this question has been answered finding interesting connections between weak sense of direction and graph embeddings.

3.4 Topological Constraints for Sense of Direction

Since the algorithm for testing sense of direction has a high polynomial complexity, an interesting research direction is the study of how to exploit the topological properties of G to find simpler testing algorithms.

The interplay between the topology of the system and the properties that a labeling must satisfy for having sense of direction has been investigated in [25] where a characterization of graph classes in which *every* labeling with basic properties (local orientation, edge-symmetric, coloring) guarantees the existence of sense of direction has been provided. Not surprisingly, an arbitrary labeling suffices for having sense of direction only in few trivial graphs; the simultaneous presence of local orientation and edge symmetry guarantees sense of direction in a larger class of graphs which includes trees and rings; finally, a coloring suffices for having sense of direction in an even larger class of graphs which includes particular types of "spiked rings". As a consequence, testing for sense of direction becomes very easy for graphs in those classes; for example, if G is a tree or a ring, the test consists in checking whether λ is symmetric and has local orientation which can be done in $O(n)$.

4   Sense of Direction in Message Passing Systems

4.1   Impact on Complexity

As mentioned in the introduction, some specific labelings in specific topologies, subsequently identified as particular classes of sense of directions, have been exploited to reduce the message complexity of solutions for LEADER ELECTION.

For example, in *complete graphs* with chordal $\mathcal{SD}$ the complexity of election has been reduced from $\Theta(n \log n)$ with arbitrary labelings to $O(n)$ [31,35]; in *chordal rings* with chordal $\mathcal{SD}$ various improvements have been achieved depending on the chord structures [2,33,37]; in *hypercubes*, the traditional dimensional sense of direction has been used to devise $O(n)$ algorithms [18,34,36]. The same achievement has been later obtained also using *any* $\mathcal{SD}$ [12]. The WAKE-UP problems, on the other hand, has been shown to be insensitive to $\mathcal{SD}$, at least in some topologies [14].

The most salient results on the impact of sense of direction hold for universal protocols (i.e., protocols for arbitrary topologies, without any additional topological information) [19]. BROADCAST and DEPTH-FIRST TRAVERSAL can be performed with $\Theta(n)$ messages with *any* sense of direction even if the system is anonymous, a dramatic improvement on the $\Omega(m)$ lower bound for these problems for arbitrary labelings; note that this lower bound holds even if the entities have distinct identities.

Similarly strong improvement can be achieved for the problems of LEADER ELECTION, SPANNING TREE CONSTRUCTION, and MINIMUM FINDING, which can be solved with $\Theta(n \log n)$ messages with *any* sense of direction, improving on the $\Omega(m + n \log n)$ lower bound for these problems in the case of arbitrary labelings.

4.2   Impact on Computability

A fundamental area of distributed computing is the study of computability in *anonymous systems*; i.e., the study of what distributed problems can be solved when nodes do not have distinct identifiers (e.g., [1,38]). Clearly, which problems can be solved depends on many factors including the properties of the system as well as the amount and type of knowledge available to the nodes.

The computational power of sense of direction in anonymous systems has been studied in [24] focusing on different levels of a-priori *structural knowledge*: (1) *no information*; (2) *upper bound on network size*; (3) *knowledge of network size*; (4) *topological awareness*; and (5) *complete topological knowledge*, the highest knowledge level[2].

---

[2] Topological awareness is the knowledge of the adjacency matrix of G. Complete topological awareness by node x includes knowledge of x's own position in the adjacency matrix representation.

The characterization of what is computable in presence of $\mathcal{SD}$, depending on the level of structural information available to the entities, has been shown to be linked to the notion of surrounding, introduced in [22] and already mentioned in Sect. 3. The surrounding of a node has been shown to be the maximum information that an entity can obtain by message transmissions in anonymous distributed systems with sense of direction; in particular, what is computable in these systems depends on the number of distinct surroundings, as well as on their multiplicity [24].

The main impact of the presence of sense of direction in a graph is that *with weak sense of direction, no additional knowledge is needed.* More precisely, given an arbitrary graph G and any structural knowledge $\mathcal{K}$, if a problem P is solvable in G with knowledge $\mathcal{K}$, then P is solvable in (G, λ) where λ is a weak sense of direction [24].

This result is based on the fact that, in a labeled graph with weak sense of direction (and without any other information except the coding function), every node can construct its surrounding even if the system is anonymous (and, thus, no leader can be elected). Note that, with an arbitrary labeling, the problem of constructing the surrounding of a node in anonymous networks is *unsolvable*.

A powerful implication of this result is that, with weak sense of direction, it is possible to do shortest path routing in anonymous networks, i.e., even if there are no global identifiers for neither source nor destination (nor for any other node in the graph).

5 Sense of Direction in Systems of Mobile Entities

Appropriate edge labelings have a positive impact also in systems of mobile entities, called *agents*, moving on graphs. Special labelings have been employed, for example, for EXPLORATION [27,28], GRAPH DECONTAMINATION [16,17], GRAPH SEARCH [8], RENDEZVOUS [3,29], BLACK HOLE SEARCH [9,13]. However, sense of direction in this context is largely unexplored, with the exception of two results, described in this Section, that provide a significant evidence of its impact.

5.1 Mobile Agents

A variety of mobile agents models have been defined (for a recent survey, see [11]). In the following we briefly describe the model used in the next two sections. Let (G, λ) be an *anonymous* edge-labeled network; that is, the nodes of G are unlabeled and thus undistinguishable. A set of autonomous mobile agents operates in (G, λ), each starting from a *homebase*. The agents have computing capabilities and bounded storage, execute the same protocol, and can move from node to neighbouring node in G. After moving from u to v, an agent has available the label $\lambda_u(u, v)$ of the edge from which it departed, as well as the label $\lambda_v(v, u)$ of the edge from which it arrived.

The agents are *asynchronous*, in the sense that every action they perform (computing, moving, etc.) takes a finite but otherwise unpredictable amount of time, and they are *anonymous*, thus undistinguishable from each other.

Each node is provided with a *whiteboard*, a local storage where agents can write and read (and erase) information; access to a whiteboard is done in fair mutual exclusion. Initially, all whiteboards contain the labels of the incident ports and indicate whether the node is a homebase of some agent.

The behavior of an agent can be described as follows: The agent *computes* based on its current state and on the content of the witheboard of the node currently visited. The computation is undivisible and, upon completion, the agent *changes its state* and then possibly *departs* through an exit port determined during the computation.

5.2   Black Hole Search

BLACK HOLE SEARCH in arbitrary topologies, and in presence of $\mathcal{SD}$, has been studied in [13]. The results are briefly summarized in this Section.

The Problem and the Setting. A *black hole* (BH) is a node that contains a stationary process that destroys any agent arriving at that node, leaving no observable trace of disappearance to the other agents. The location of the black hole is unknown. The BLACK HOLE SEARCH (BHS) problem consists of having a team of agents determining the location of the black hole. More precisely, BHS is solved if at least one agent survives and all surviving agents know the location of the BH. The main complexity measures of a solution $\mathcal{P}$ are: the number of agents used by $\mathcal{P}$ (*size*) and total number of moves performed by the agents (*cost*).

The problem has been studied in different topologies under a variety of assumptions on the activation scheduler, the communication mechanisms employed by the agents, the presence of a single or multiple BH, etc. For a recent survey, see [32].

Consider an arbitrary 2-connected graph[3] G with a team of agents starting on the same homebase under an asynchronous activation scheduler. Let Δ, the maximum degree of G, and n be known to the agents[4]. Three settings with availability of increasing information have been considered. *Topological ignorance*: when no additional topological information is available; *Sense of direction*: when a $\mathcal{SD}$ $(\mathbf{c}, \mathbf{d})$ is available; *Complete topological knowledge*: when the agents have: knowledge of (G, λ); correspondence between port labels and link labels of (G, λ); location of their homebase in (G, λ).

The Results. With *topological ignorance* $\Delta + 1$ agents are necessary and they cannot avoid performing $\Omega(n^2)$ moves in the worst case. An optimal algorithm exists that matches both these bounds.

With *topological ignorance* but with *sense of direction*: the size of the solution can be reduced: *two agents* suffice to locate the black hole; however, $\Omega(n^2)$ moves are still necessary in the worst case.

---

[3] If the graph is not 2-connected, the problem is unsolvable.
[4] If they are not known, the problem is unsolvable.

Finally, with *complete topological knowledge*: also the cost can be reduced and the problem can be solved with $\Theta(n \log n)$ moves.

The impact of $\mathcal{SD}$: Danger Awareness. $\mathcal{SD}$ is used by the agents to create a form of cautiousness that allows a small team of two agents to solve the problem.

First of all, observe that any BHS solution by mobile agents is based on a cooperative exploration of the graph where, unavoidably, some agent(s) will be trapped into the BH. The exploration is performed by having the agents visit the nodes to verify their status (possibly remaining trapped) following some traversal strategy.

If there is $\mathcal{SD}$, two agents (a and b) can devise some form of coordination to visit the nodes. The exploration of a node consists in the traversal of all its incident edges and it is called *expansion*. The crucial and difficult task during an expansion is to prevent both agents entering the BH from different ports. In particular, let u be the node currently being expanded. To expand u, agent a successively traverses the incident unexplored links. However, if some neighbouring node v is currently under exploration by the other agent (from a different link), the link (u, v) should be avoided because, if v is the black hole, it would kill both agents. Without $\mathcal{SD}$ this avoidance task is impossible. With $\mathcal{SD}$, on the other hand, coding and decoding functions can be used to identify the ports leading to the node currently under exploration by the other agent (and thus, potentially dangerous). If a dangerous node v is identified by agent a, from this moment on, until told otherwise, a will avoid entering v. The information that v is dangerous can be modified only by the other agent b. If v was not the BH, b will complete the exploration of u and it will reach a following it in its traversal. Once b reaches the node being expanded by a, it leaves a message on the whiteboard for a notifying it of its presence (and, thus, that v is no longer dangerous), and joins in the expansion.

It is very interesting to note that $\mathcal{SD}$ can be exploited to drastically reduce the size of the team of agents, but the solution incurs in the same $O(n^2)$ cost. Indeed, this cost has been shown to be also a lower bound in the worst case, by proving that there exists a $\mathcal{SD}$ with which any two agents algorithm for locating a black hole in arbitrary networks must perform $\Omega(n^2)$ moves in the worst-case. The cost can instead be reduced to $\Theta(n \log n)$ if there is full topological awareness, indicating that, in the case of BHS, full topological awareness has stronger implication than $\mathcal{SD}$ on the complexity of the problem.

5.3   Leader Election and Rendezvous

LEADER ELECTION and RENDEZVOUS in arbitrary topologies, and in presence of $\mathcal{SD}$, have been studied in [3]. The results are briefly summarized in this Section.

The Problem and the Setting. A group of k anonymous mobile agents operates asynchronously on an anonymous graph G. The only difference with the setting considered in Sect. 5.2 is that here the agents start from different home-bases. Initially, all agents have a predefined state variable set to *available*. The LEADER ELECTION problem consists of having one agent terminate with its state variable set to *leader*, while the others terminate setting theirs to *follower*. RENDEZVOUS consists of having all the agents gather at the same node (not necessarily chosen in advance) and terminate there. In this setting, the two problems are equivalent; that is, any solution for one can be easily modified to solve the other, so the rest of the section focuses on election.

The Results. The results depend on the relationship between k, the number of agents, and n, the number of nodes. If (G, λ) has an arbitrary edge labelings, the rendezvous problem is *unsolvable* (even if restricted to the class of inputs for which $\gcd(k, n) = 1$). If λ is a sense of direction, the election problem is still *unsolvable* when $\gcd(k, n) = d > 1$. On the other hand, it becomes *solvable* if $\gcd(k, n) = 1$, and the result holds for *any* sense of direction. In other words, *sense of direction overcomes anonymity if* $\gcd(k, n) = 1$.

The Impact of $\mathcal{SD}$: Overcoming Anonymity. The protocol uses a special mechanism, *dynamic name mutation*, that allows the anonymous agents in the anonymous network to distinguish agents and nodes in spite of anonymity.

In presence of $\mathcal{SD}$, even in a totally anonymous system, an agent can locally (i.e., privately) assign a unique "name" to itself and to the other agents. However, since all agents are behaviorally identical and start with the same initial values, there is no guarantee that such a name would be unique. In fact, it is possible that they all choose the same name for themselves, creating an *homonymous* universe. A mechanisms, called *dynamic name mutation*, is devised exploiting the presence of sense of direction, to allow the agents to operate in spite of these limitations, including homonimity.

In this mechanism, initially every agent chooses its private name based on the labels of the edges incident to its homebase. The private name is then modified whenever the agent moves on the graph. The name will be always relative to the current position of the agent. The main difficulty is to modify the names in such a way that, at any location v, two names will be different if and only if they refer to different agents. This will ensure that messages written on v's whiteboard by different agents will have different signatures. Another related difficulty is to ensure that an agent is capable of recognizing as its own any message it has written in previous visits. These difficulties are overcome by the use of the coding and decoding functions $(\mathbf{c}, \mathbf{d})$ by the mobile agents.

Strategy DNM

(1) To determine its initial name, an agent a with homebase u, chooses an arbitrary neighboring node $v \in E(u)$ and determines the label $\lambda_v(v, u)$ (e.g., by moving to v and coming back). Then, its name is $MyN := \mathbf{c}(\lambda_u(u, v) \cdot \lambda_v(v, u))$.

(2) When an agent with name MyN at node u moves to the neighboring node v, it modifies its name as follows: $MyN := \mathbf{d}(\lambda_v(v, u), MyN)$.

The election algorithm proceeds in a sequence of electoral phases with the anonymous agents following the dynamic name mutation mechanism. The agents that are candidates in each phase perform: (1) territory acquisition to acquire as many nodes as possible, and (2) a sequence of partitioning and pairing rounds to eliminate as many agents as possible so to decrease the number of candidates. At the end of a phase, at least half of the candidate agents which entered the phase become *passive*, and the number of those which will start the next phase is still co-prime with n. The total number of moves by the agents during the execution of the protocol is $O(kn)$.

6   Open Problems

Sense of direction is a global consistency property of local edge-labellings that has been shown to have an impact in distributed computing. Some of its benefits are due to the fact that it helps overcoming anonymity, breaking symmetries, and reducing redundancy of messages.

The study of $\mathcal{SD}$ has brought to light interesting graph theoretic properties of edge-labeled graphs, some of which can be seen as generalizations, to the edge-labeled realm, of very well known graph properties (for example, with the notion of labeled isomorphisms, surrounding symmetry is the labeled-analogous of vertex transitivity). Only the surface of this research area has been scratched, and a systematic study of *labeled graph classes* would be a very interesting and important research direction.

$\mathcal{SD}$ is just one of the several global consistency properties of edge-labeled graphs that could be defined. Only very few other label consistencies have been studied: the weaker labelings with local orientation, edge symmetry, edge coloring; backward sense of direction, where local orientation is not necessary. The study of other classes of labelings carrying some forms of consistency and of their computational relationships is wide open.

About its usefulness in message-passing distributed systems, by definition $\mathcal{SD}$ can exist only with point-to-point communications. However, in many message-passing distributed systems, communication is achieved through broadcast primitives where specific recipients cannot be selected (e.g., *wireless* networks, *optical* networks, ...). The natural question is whether a property analogous to $\mathcal{SD}$ can exist in such systems. The first steps in such direction are the results on a type of global consistency called *Backward $\mathcal{SD}$*, which can exist in broadcast-communication models and has been shown to be computationally equivalent to $\mathcal{SD}$ [23]. Apart from this, nothing is known.

The study of $\mathcal{SD}$ in mobile agents models is still largely unexplored. We have seen that it allows to overcome anonymity providing a private consistent naming scheme in a totally anonymous environment [3], and that it facilitates cooperation between agents in the case of the black hole search problem [13]. Nothing else is known and, in particular, no results exist where $\mathcal{SD}$ is shown to be useful to decrease the complexity of solutions in mobile agents' settings.

Another environment where sense of direction has never been studied is the one of *time varying graphs*, which are now intensively investigated from many perspectives. How can a $\mathcal{SD}$ be maintained in a network that changes in time? How can it be exploited in the design of algorithms? The investigation of its impact in dynamic environments (whether in message passing systems or in mobile agents settings) is a very interesting research avenue.

References

1. Angluin, D.: Local and global properties in networks of processors. In: Proceedings of 12th ACM Symposium on Theory of Computing, pp. 82–93 (1980)
2. Attiya, H., van Leeuwen, J., Santoro, N., Zaks, S.: Efficient elections in chordal ring networks. Algorithmica **4**, 437–446 (1989)
3. Barrière, L., Flocchini, P., Fraigniaud, P., Santoro, N.: Rendezvous and election of mobile agents: impact of sense of direction. Theory Comput. Syst. **40**(2), 143–162 (2007)
4. Boldi, P., Vigna, S.: Minimal sense of direction and decision problems for Cayley graphs. Inf. Process. Lett. **64**, 299–303 (1997)
5. Boldi, P., Vigna, S.: On the complexity of deciding sense of direction. SIAM J. Comput. **29**(3), 779–789 (2000)
6. Boldi, P., Vigna, S.: Lower bounds for sense of direction in regular graphs. Distrib. Comput. **16**(4), 279–286 (2003)
7. Boldi, P., Vigna, S.: Lower bounds for weak sense of direction. J. Discret. Algorithms **1**(2), 119–128 (2003)
8. Borowiecki, P., Dereniowski, D., Kuszner, L.: Distributed graph searching with a sense of direction. Distrib. Comput. **155**(3), 155–170 (2015)
9. Chalopin, J., Das, S., Labourel, A., Markou, E.: Black hole search with finite automata scattered in a synchronous torus. In: Peleg, D. (ed.) DISC 2011. LNCS, vol. 6950, pp. 432–446. Springer, Heidelberg (2011). https://doi.org/10.1007/978-3-642-24100-0_41
10. Cheng, C., Suzuki, I.: Weak sense of direction labelings and graph embeddings. Discret. Appl. Math. **159**(5), 303–310 (2011)
11. Das, S., Santoro, N.: Moving and computing models: agents. In: Flocchini, P., Prencipe, G., Santoro, N. (eds.) Distributed Computing by Mobile Entities, pp. 15–34. Springer, Cham (2019). https://doi.org/10.1007/978-3-030-11072-7_2
12. Dobrev, S.: Leader election using any sense of direction. In: Proceedings of 6th International Colloquium on Structural Information and Communication Complexity (SIROCCO), pp. 93–104 (1999)
13. Dobrev, S., Flocchini, P., Prencipe, G., Santoro, N.: Searching for a black hole in arbitrary networks: optimal mobile agent protocols. Distrib. Comput. **19**(1), 1–19 (2006)
14. Dobrev, S., Královic, R., Santoro, N.: On the cost of waking up. Int. J. Netw. Comput. **7**(2), 336–348 (2017)
15. Flocchini, P.: Minimal sense of direction in regular networks. Inf. Process. Lett. **61**, 331–338 (1997)
16. Flocchini, P., Huang, M., Luccio, F.L.: Decontamination of hypercubes by mobile agents. Networks **52**(3), 167–178 (2008)
17. Flocchini, P., Luccio, F., Pagli, L., Santoro, N.: Network decontamination under m-immunity. Discret. Appl. Math. **201**, 114–129 (2016)
18. Flocchini, P., Mans, B.: Optimal election in labeled hypercubes. J. Parallel Distrib. Comput. **33**(1), 76–83 (1996)
19. Flocchini, P., Mans, B., Santoro, N.: On the impact of sense of direction on message complexity. Inf. Process. Lett. **63**(1), 23–31 (1997)

20. Flocchini, P., Mans, B., Santoro, N.: Sense of direction: definition, properties and classes. Networks **32**(3), 165–180 (1998)
21. Flocchini, P., Mans, B., Santoro, N.: Sense of direction in distributed computing. Theor. Comput. Sci. **291**(1), 29–53 (2003)
22. Flocchini, P., Roncato, A., Santoro, N.: Symmetries and sense of direction in labeled graphs. Discret. Appl. Math. **87**, 99–115 (1998)
23. Flocchini, P., Roncato, A., Santoro, N.: Backward consistency and sense of direction in advanced distributed systems. SIAM J. Comput. **32**(2), 281–306 (2003)
24. Flocchini, P., Roncato, A., Santoro, N.: Computing on anonymous networks with sense of direction. Theor. Comput. Sci. **301**(1–3), 355–379 (2003)
25. Flocchini, P., Santoro, N.: Topological constraints for sense of direction. Int. J. Found. Comput. Sci. **9**(2), 179–198 (1998)
26. Foldes, S., Urrutia, J.: Sense of direction, semigroups, and Cayley graphs. Manuscript (1998)
27. Hanusse, N., Ilcinkas, D., Kosowski, A., Nisse, N.: Locating a target with an agent guided by unreliable local advice: how to beat the random walk when you have a clock? In: Proceedings of 29th ACM Symposium on Principles of Distributed Computing (PODC), pp. 355–364 (2010)
28. Ilcinkas, D.: Setting port numbers for fast graph exploration. Theor. Comput. Sci **401**(1–3), 236–242 (2008)
29. Kranakis, E., Krizanc, D., Markou, E.: Deterministic symmetric rendezvous with tokens in a synchronous torus. Discret. Appl. Math. **159**(9), 896–923 (2011)
30. Leão, R.S.C., Barbosa, V.C.: Minimal chordal sense of direction and circulant graphs. In: Královič, R., Urzyczyn, P. (eds.) MFCS 2006. LNCS, vol. 4162, pp. 670–680. Springer, Heidelberg (2006). https://doi.org/10.1007/11821069_58
31. Loui, M., Matsushita, T., West, D.: Election in complete networks with a sense of direction. Inf. Process. Lett. **22**, 185–187 (1986)
32. Markou, E., Shi, W.: Dangerous graphs. In: Flocchini, P., Prencipe, G., Santoro, N. (eds.) Distributed Computing by Mobile Entities, vol. 11340, pp. 455–515. Springer, Cham (2019). https://doi.org/10.1007/978-3-030-11072-7_18
33. Pan, Y.: A near-optimal multi-stage distributed algorithm for finding leaders in clustered chordal rings. Inf. Sci. **76**(1–2), 131–140 (1994)
34. Robbins, S., Robbins, K.: Choosing a leader on a hypercube. In: Proceedings of International Conference on Databases, Parallel Architectures and their Applications, pp. 469–471 (1990)
35. Singh, G.: Efficient leader election using sense of direction. Distrib. Comput. **10**, 159–165 (1997)
36. Tel, G.: Linear election in hypercubes. Parallel Proc. Lett. **5**(1), 357–366 (1995)
37. Tel, G.: Sense of direction in processor networks. In: Bartosek, M., Staudek, J., Wiedermann, J. (eds.) SOFSEM 1995. LNCS, vol. 1012, pp. 50–82. Springer, Heidelberg (1995). https://doi.org/10.1007/3-540-60609-2_3
38. Yamashita, M., Kameda, T.: Computing on anonymous networks, part I: characterizing the solvable cases. IEEE Trans. Parallel Distrib. Comput. **7**(1), 69–89 (1996)

Regular Papers

Locality of Not-so-Weak Coloring

Alkida Balliu[1(✉)], Juho Hirvonen[1], Christoph Lenzen[2], Dennis Olivetti[1],
and Jukka Suomela[1]

[1] Aalto University, Espoo, Finland
alkida.balliu@aalto.fi
[2] Max Planck Institute for Informatics, Saarbrücken, Germany

Abstract. Many graph problems are locally checkable: a solution is globally feasible if it looks valid in all constant-radius neighborhoods. This idea is formalized in the concept of *locally checkable labelings* (LCLs), introduced by Naor and Stockmeyer (1995). Recently, Chang et al. (2016) showed that in bounded-degree graphs, every LCL problem belongs to one of the following classes:
- *"Easy"*: solvable in $O(\log^* n)$ rounds with both deterministic and randomized distributed algorithms.
- *"Hard"*: requires at least $\Omega(\log n)$ rounds with deterministic and $\Omega(\log \log n)$ rounds with randomized distributed algorithms.

Hence for any parameterized LCL problem, when we move from local problems towards global problems, there is some point at which complexity suddenly jumps from easy to hard. For example, for vertex coloring in d-regular graphs it is now known that this jump is at precisely d colors: coloring with $d + 1$ colors is easy, while coloring with d colors is hard.

However, it is currently poorly understood where this jump takes place when one looks at *defective* colorings. To study this question, we define *k-partial c-coloring* as follows: nodes are labeled with numbers between 1 and c, and every node is incident to at least k properly colored edges.

It is known that 1-partial 2-coloring (a.k.a. weak 2-coloring) is easy for any $d \geq 1$. As our main result, we show that k-partial 2-coloring becomes hard as soon as $k \geq 2$, no matter how large a d we have.

We also show that this is fundamentally different from k-partial 3-coloring: no matter which $k \geq 3$ we choose, the problem is always hard for $d = k$ but it becomes easy when $d \gg k$. The same was known previously for partial c-coloring with $c \geq 4$, but the case of $c < 4$ was open.

1 Introduction

There is a broad family of graph problems—so-called *locally checkable labelings* or LCLs [20]—that exhibits the following dichotomy [9]: either the problem can be solved in $O(\log^* n)$ rounds with deterministic distributed algorithms, or any such algorithm requires at least $\Omega(\log n)$ rounds.

This work was supported in part by the Academy of Finland, Grants 285721 and 314888.

K. Censor-Hillel and M. Flammini (Eds.): SIROCCO 2019, LNCS 11639, pp. 37–51, 2019.
https://doi.org/10.1007/978-3-030-24922-9_3

Table 1. An overview of k-partial c-coloring in d-regular graphs: for each k and c, the table shows what is the smallest d such that the problem is easy. For example, "$4 \ldots 5$" means that for these parameters the problem is known to be easy in 5-regular graphs, while the case of 4-regular graphs is unknown. The new results are highlighted with a frame. Our main contributions are the new lower bounds for $c = 2$ (Theorem 2) and upper bounds for $c = 3$ (Theorem 1), which were previously completely open. We also obtain stronger lower bounds for e.g. $c = 3$, $k \geq 5$ (Theorem 3) and stronger upper bounds for $c = k \geq 4$ (Theorem 1).

(a) Before this work:

	$c = 2$	$c = 3$	$c = 4$	$c = 5$
$k = 1$:	1	1	1	1
2:	$3 \ldots \infty$	2	2	2
3:	$4 \ldots \infty$	$4 \ldots \infty$	3	3
4:	$5 \ldots \infty$	$5 \ldots \infty$	$5 \ldots 7$	4
5:	$6 \ldots \infty$	$6 \ldots \infty$	$6 \ldots 9$	$6 \ldots 9$
6:	$7 \ldots \infty$	$7 \ldots \infty$	$7 \ldots 11$	$7 \ldots 11$
7:	$8 \ldots \infty$	$8 \ldots \infty$	$8 \ldots 13$	$8 \ldots 13$

(b) After this work:

	$c = 2$	$c = 3$	$c = 4$	$c = 5$
$k = 1$:	1	1	1	1
2:	∞	2	2	2
3:	∞	$4 \ldots 5$	3	3
4:	∞	$5 \ldots 8$	$5 \ldots 6$	4
5:	∞	$7 \ldots 11$	$6 \ldots 9$	$6 \ldots 7$
6:	∞	$8 \ldots 14$	$7 \ldots 11$	$7 \ldots 11$
7:	∞	$10 \ldots 17$	$9 \ldots 13$	$8 \ldots 13$

Hence, for any parameterized LCL problem there is a sudden jump in its deterministic complexity, from $O(\log^* n)$, which is a very slowly-growing function of n, to $\Omega(\log n)$, which can be already as much as the diameter of the network. We will call these two cases *"easy"* and *"hard"* from now on.

If we look at d-regular graphs for constant $d = O(1)$, then by prior work the following thresholds are known [6,9,12,21]:

- Proper vertex coloring with c colors: easy for $c \geq d + 1$, hard for $c \leq d$.
- Proper edge coloring with c colors: easy for $c \geq 2d - 1$, hard for $c \leq 2d - 2$.

Here the easy cases are exactly those cases that can be solved with a greedy algorithm that picks the colors of the nodes or edges one by one; a straightforward parallelization of this idea then gives an $O(\log^* n)$-round distributed algorithm.

In this work, we study colorings that are not necessarily proper:

Definition 1. *Let $G = (V, E)$ be a graph. Mapping $f \colon V \to \{1, 2, \ldots, c\}$ is a k-partial c-coloring if for each node $v \in V$, there are at least k neighbors u of v with $f(u) \neq f(v)$.*

By prior work on defective colorings, it is known that e.g. k-partial 4-coloring is hard if $d = k \geq 4$ and easy if $d \gg k$. However, very little was known about partial 2-coloring and 3-coloring. In this work we complete the picture and show that the case of $c = 2$ is very different from the case $c \geq 3$:

- k-partial 2-coloring for any $k \geq 2$ is always hard, no matter how large a d we have,

– k-partial 3-coloring for any $k \geq 3$ is hard for $d = k$ but it becomes easy when $d \gg k$.

We summarize our contributions in Table 1.

2 Preliminaries and Related Work

2.1 LOCAL Model of Computing

We work in the usual LOCAL model of distributed computing [18,22]. Each node of the input graph $G = (V, E)$ is a computer and each edge is a communication link. Computation proceeds in rounds: in one round each node can exchange a message (of any size) with each of its neighbors. Initially each node knows only $n = |V|$, and when a node stops, it has to produce its own part of the output—in our case, its own color from $\{1, 2, \ldots, c\}$. We say that an algorithm runs in time T if after T rounds all nodes stop and announce their local outputs.

When we study *deterministic* algorithms, we assume that each node is labeled with a unique identifier from $\{1, 2, \ldots, n^{O(1)}\}$. When we study *randomized* algorithms, we assume that each node has an unlimited source of random bits. For a randomized algorithm, we require that it solves the problem correctly *with high probability*, i.e., with probability at least $1 - n^{-c}$ for an arbitrary, but predetermined constant $c > 0$.

Note that if a problem is solvable in T rounds in the LOCAL model, it also means that each node can produce its own part of the solution based on the information that is available in its radius-T neighborhood.

2.2 LCL Problems and Gap Theorems

LCL problems were introduced by Naor and Stockmeyer [20] in 1995. In an LCL problem, the input is a graph $G = (V, E)$ of maximum degree $\Delta = O(1)$, possibly labeled with some node labels from a constant-size set X. The task is to find a labeling $f: V \rightarrow Y$, for some constant-size set Y, that satisfies some local constraints—a labeling is globally feasible if it is feasible in all constant-radius neighborhoods.

For our purposes it is enough to note that k-partial c-coloring in d-regular graphs is an LCL problem, for any choice of constants $k, c, d = O(1)$. Hence also everything that we know about LCLs applies here.

In the past four years, we have seen a lot of progress in understanding the computational complexity of LCL problems in the LOCAL model [1,2,6–10,13–16]. For us, the most relevant result is the gap theorem by Chang et al. [9]. They show that every LCL problems belongs to one of the following classes, which we will here informally call "easy" and "hard"

"Easy": solvable in $O(\log^* n)$ rounds with both deterministic and randomized algorithms.

"Hard": requires $\Omega(\log n)$ rounds with deterministic algorithms and $\Omega(\log \log n)$ rounds with randomized algorithms.

In this work, our main goal is to understand for what values of k, c, d the problem of finding k-partial c-coloring in d-regular graphs is "hard" and when it is "easy". While we will focus on the case of d-regular graphs, most of the results directly generalize to the case of graphs of minimum degree d (and maximum degree some constant Δ).

2.3   Prior Work Related to Partial Colorings

Notes on Terminology. In d-regular graphs, a k-partial c-coloring is exactly the same thing as a $(d-k)$-*defective* c-*coloring* [3, Sect. 6]. While defective colorings are more commonly discussed in prior work, for our purposes the concept of a partial coloring is much more convenient, as we will often fix c and k and see what happens when d increases.

Our definition is in essence equal to k-*partially proper colorings* used by Kuhn [17]; for brevity, we call these partial colorings.

Weak Coloring, $k = 1$. In graphs without isolated nodes, a 1-partial c-coloring is identical to a *weak c-coloring* [20]. Weak 2-coloring can be solved in $O(\log^* n)$ rounds: find a maximal independent set $X \subseteq V$ using e.g. [12,21]; then color all nodes of X with color 1 and all other nodes with color 2. Naturally, this also gives a solution for weak c-coloring for any $c \geq 2$. Furthermore, this upper bound is tight: a weak 2-coloring breaks symmetry everywhere in a regular grid, and the usual lower bounds [18–20] apply.

Weak coloring corresponds to the first row of Table 1.

Partial Coloring for $k < c$. Above we have seen that we can find a 1-partial 2-coloring by simply finding a maximal independent set (MIS). The same idea can be generalized to $(c-1)$-partial c-coloring for $k = c - 1$: Find an MIS X_1, label X_1 with color 1, and remove X_1. Find an MIS X_2, label X_2 with color 2, and remove X_2, etc. We continue this for $c - 1$ steps and finally label all remaining nodes with color c.

The region where this simple (folklore?) strategy works is indicated with green color in Table 1.

Proper Vertex Coloring, $k = d$. In d-regular graphs, a d-partial c-coloring is a proper c-coloring. Recall that proper coloring with $d + 1$ colors is easy [12,21], while proper coloring with d colors is known to be hard [6,9].

Hardness of proper d-coloring implies the lower bounds in the blue and gray regions of Table 1a.

Partial Coloring for $c \geq 4$. Barenboim et al. [4] gave an algorithm that computes a $\lfloor d/p \rfloor$-defective p^2-coloring in time $O(\log^* n)$, which is essentially a defective variant of Linial's $O(\Delta^2)$-coloring algorithm [18]. This algorithm requires at least 4 colors, and for the case $c = 4$ it translates to a $\lceil d/2 \rceil$-partial 4-coloring. For example, 4-partial 4-coloring is therefore easy in 7-regular graphs, and 5-partial 4-coloring is easy in 9-regular graphs.

This algorithm gives the upper bounds in the blue region of Table 1a.

Partial Coloring for $c \leq 3$. To our knowledge, no $O(\log^* n)$-time algorithms are known for k-partial c-coloring for $c \leq 3$, $k \geq c$. In particular, it is not known if the problem becomes easy in d-regular graphs for sufficiently large values of $d \gg k$.

This unknown region is indicated with a gray shading in Table 1a.

Algorithms Based on Lovász Local Lemma. Chung et al. [11], Fischer and Ghaffari [13], and Ghaffari et al. [14, full version] present algorithms for defective coloring (and hence for partial coloring) that are based on the following idea: formulate a defective coloring as an instance of the Lovász local lemma (LLL), and then apply efficient distributed algorithms for LLL.

Unfortunately, this approach is unlikely to lead to an $O(\log^* n)$-time algorithm; LLL is known to be a hard problem for a wide range of parameters [6].

Other Algorithms. Bonamy et al. [5] show that there is an $O(\log n)$-round algorithm for trees that finds an MIS such that every component induced by non-MIS nodes is of size one or two. This can be interpreted as an algorithm for partial 2-coloring.

However, this approach cannot lead to an $O(\log^* n)$-time algorithm, either: if we color the MIS-nodes with color 1 and the non-MIS nodes with colors 2 and 3, we obtain a proper 3-coloring, and finding a 3-coloring in 3-regular trees is known to be a hard problem [6].

3   Our Contributions

To recap, by prior work, we have a good qualitative understanding of k-partial c-coloring for $c \geq 4$:

- $k < c$: easy for all $d \geq k$.
- $k \geq c$: hard for $d = k$ but easy for $d \gg k$.

We complete the picture for $c \leq 3$. For $c = 3$, we have precisely the same situation as above:

- $k < c$: easy for all $d \geq k$.
- $k \geq c$: hard for $d = k$ but easy for $d \gg k$.

However, the case of $c = 2$ is fundamentally different:

- $k < c$: easy for all $d \geq k$.
- $k \geq c$: hard for all values of d.

3.1   Corollary: Locally Optimal Cuts

Any partial 2-coloring can be interpreted as a *cut*; the properly colored edges are *cut edges*, and the *size* of the cut is the number of cut edges. Let us look at the problem of maximizing the size of a cut with a simple greedy strategy: start with any cut and change the color of a node if it increases the size of the cut. The process will converge to a *locally optimal cut*, in which changing the color of any single node does not help.

Now a locally optimal cut in d-regular graphs is precisely the same thing as a $\lceil d/2 \rceil$-partial 2-coloring. For example, in 3-regular graphs, any 2-partial 2-coloring is also a locally optimal cut, and vice versa.

Locally optimal cuts are easy to find in a centralized, sequential setting. However, previously it was not known if locally optimal cuts can be found efficiently in a distributed setting. As a corollary of our work, we now know that this is a hard problem.

3.2   Key Techniques

Upper Bound for 3-coloring. Prior algorithms for e.g. partial 4-coloring are based on the idea of organizing nodes in layers and doing two sweeps [4]: top to bottom, using colors from palette $A = \{1, 2\}$, and bottom to top using colors from palette $B = \{1, 2\}$. This way we eventually have a 4-coloring with colors from $A \times B = \{(1, 1), (1, 2), (2, 1), (2, 2)\}$. This idea generalizes easily to e.g. $6, 8, 9, \ldots$ colors, but it is not possible to use this idea to find a useful coloring with less than 4 colors.

We show how to do two sweeps so that the end result is only 3 colors. In brief, the first sweep uses *tentative* colors from palette $\{1, 2\}$, and the second sweep *finalizes* the colors, depending on the tentative colors that we chose in the first step. Here the second sweep depends on the result of the first sweep, while in prior algorithms the two sweeps are independent.

Lower Bound for 2-coloring. We show that 2-partial 2-coloring in d-regular graphs for any constant d is at least as hard to solve as *sinkless orientation*, which is known to be hard [6]. The key obstacle here is that sinkless orientation is known to be hard even if we are given a *proper* 2-coloring of the graph, so how could a *partial* 2-coloring help with it?

The basic idea is as follows: Assume we have a fast algorithm A_1 that finds a 2-partial 2-coloring in d_1-regular graphs. Then we can construct algorithm A_2 that finds a sinkless orientation in d_2-regular graphs, for a certain constant $d_2 \gg d_1$ that depends on the exact running time of A_1. Given a d_2-regular graph G_2, algorithm A_2 first replaces all nodes with appropriate gadgets to obtain a d_1-regular graph G_1, applies A_1 to G_1, and extracts enough information from the partial coloring so that it can find a sinkless orientation. But sinkless orientation is hard also in d_2-regular graphs, no matter how large a constant d_2 is.

4   Partial Colorings with More Than Two Colors

In this section we analyze the distributed complexity of k-partial c-coloring in the case where c is at least 3. More formally, we will prove the following theorem.

Theorem 1. *There exists an algorithm running in $O(\log^* n)$ that is able to compute:*

- *A k-partial 3-coloring, if $d \geq 3k - 4$ and $k \geq 3$;*
- *A k-partial k-coloring, if $d \geq k + 2$ and $k \geq 4$.*

In order to prove the theorem, we start by providing an algorithm, and then we will analyze it for the two cases separately.

The Algorithm. The algorithm that we propose is inspired by the procedure *Refine* in [3, Sect. 6]. This procedure starts by first finding an acyclic partial orientation, and then assigns two colors for each node by exploiting the two possible orders given by the orientation. It finally combines the two colors to determine the output color. Our algorithm starts in the same way, but it does *not* compute two independent colors, allowing us to be slightly more efficient in some cases.

We start by finding an acyclic partial orientation of the edges. That is, we first compute an $O(d^2)$ coloring in $O(\log^* n)$ rounds. Then, we assign a total order to the colors, and we orient the edges from the node with the smaller color to the node with the bigger one. The obtained directed graph is clearly acyclic, and all directed paths are of length at most $O(d^2)$. Nodes reachable from v through outgoing edges are considered to be *above* v, while the others are considered to be *below* v.

Now, we do two "sweeps" on the obtained acyclic graph, that is, we first process the nodes from above to below, and then we process them in the reverse order. More precisely, we start by processing the sinks, and then we continue by processing all nodes such that all of the nodes above them have already been processed. This is iterated until all nodes have been processed. Then, we repeat the same procedure in reverse order, i.e., from below to above. Each sweep takes $O(d^2)$ rounds.

During the first sweep, we assign to each node v a temporary color, by choosing the color that is the least used one among the neighbors above v. Crucially, during this phase, the choice is not among the full palette, but only colors from 1 to $c - 1$ are allowed. We call color c the *special* color. During the second sweep, each node v has three options:

- Keep the current color.
- Choose to switch to color c.
- Choose to switch to a color from 1 to $c - 1$. This option is allowed only if no neighbor below v is currently using that color.

Different choices give different guarantees. For example, by choosing to not change the color, or by choosing to switch to a color from 1 to $c - 1$, node v is ensured that the number of properly colored neighbors does not decrease when the nodes above it will be processed. This property is guaranteed by the fact that a node can switch to a color in $\{1, \ldots, c - 1\}$ only if no node below it is using that color. On the other hand, a node may switch to color c even if some

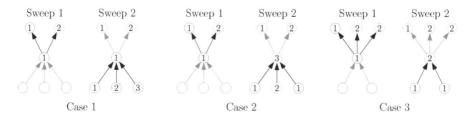

Fig. 1. Examples of the output of the 3-partial 3-coloring algorithm, running at the central node v, in a graph where the degree is 5. The figure shows 3 cases: v keeps the color chosen during the first sweep, v switches to the special color in the second sweep, v switches to color 2 in the second sweep.

neighbor below it is using color c as well, but then it loses any guarantee about the nodes above it, which may all decide to switch to color c. See Fig. 1 for an example of the execution of the algorithm.

The algorithm will make a choice that guarantees that k neighbors have different color, regardless of whether the nodes above change their color (subject to the above rules) or not. Accordingly, our task is to prove that such a choice always exists, provided that d is large enough.

k-partial 3-coloring. We now show that the above described algorithm is able to compute a k-partial 3-coloring if $d \geq 3k - 4$ and $k \geq 3$. In order to analyze the algorithm running on node v, assume without loss of generality that during the first sweep, v picks color 1 and t nodes above v chose color 2. Note that there are no more than t other nodes above v, as v picked the color out of $\{1, 2\}$ that was least used by the nodes above it in the first sweep. Denote by $x, y, z \geq 0$ the numbers of nodes below v that are colored 1, 2, and 3, respectively, after making their final choice in the second sweep. Thus, the number of nodes above v that chose 1 in their first sweep equals $d - t - x - y - z \leq t$.

We make a case distinction.

1. $t + y + z \geq k$. Thus, v can keep color 1, as the t nodes above it that have color 2 must then choose a color different from 1.
2. $x + y \geq k$. Then v can safely choose color 3.
3. $y = 0$ and none of the other cases apply. Thus, v is free to switch to color 2. If it does so, it has $x + z$ nodes of different color below, and $d - t - x - z$ nodes above that choose a different color than 2. As the first case does not apply and $k \geq 3$, these are at least $d - t \geq d - (k - 1) \geq 2k - 3 \geq k$ nodes. Hence, switching to color 2 is indeed a valid choice.

Hence, it suffices to show that one of the cases must apply. Assume for contradiction that this is false. Thus,

$$t + y + z \leq k - 1, \quad x + y \leq k - 1, \quad y \geq 1, \quad \text{and} \quad d \leq 2t + x + y + z,$$

yielding the contradiction that

$$d \leq 2t + x + y + z \leq 2(k - 1) - y - z + x \leq 3(k - 1) - 2y - z \leq 3k - 5.$$

k-*partial* k-*coloring.* We now show that the above described algorithm is able to compute a k-partial k-coloring if $d \geq k + 2$ and $k \geq 4$. We analyze this case similarly to the case before. Let t be the number of nodes above v of color different from v after the first sweep. Without loss of generality, assume that the color of v is 1, and the special color is k. Let x, y, and z be the number of nodes below v colored with 1, with color c such that $2 \leq c \leq k - 1$, and with color k, respectively. Recall that only colors from 1 to $k - 1$ are allowed during the first sweep. As v chooses a minority color among its above neighbors' choices, we have $r := d - t - x - y - z \leq t/(k - 2)$ remaining nodes above v that choose color 1 in the first sweep.

Let us analyze the second sweep. We make a case distinction.

1. $t + y + z \geq k$. Then v can keep color 1.
2. $x + y \geq k$. Then v can safely choose color k.
3. There are $f > 0$ "free" colors from $\{2, \ldots, k - 1\}$ that no neighbor below v chose and none of the other cases applies. If a free color was picked by at most $d - k$ above neighbors of v, it may select it with the guarantee that its other k neighbors end up with a different color.

   Assuming for contradiction that there is no such free color, observe that the least used free color was picked by at most $\lfloor t/f \rfloor$ neighbors above v in the first sweep. Accordingly, $t \geq f(d - k + 1) \geq d - k + f$. Moreover, clearly $f \geq k - 2 - y$ and, because the first case does not apply, $x + r = d - t - y - z \geq d - (k - 1) \geq 3$. Thus, we can lower bound the total number of neighbors of v by
   $$d = r + t + x + y + z \geq d - k + f + y + 3 \geq d + 1,$$
   a contradiction. Therefore, one of the free colors is a valid choice for v.

Hence, there is indeed always a valid choice if we can show that the above case distinction is exhaustive. Assuming otherwise, collecting inequalities from the cases and the earlier bound on r we obtain that

$$t + y + z \leq k - 1, \quad x + y \leq k - 1, \quad y \geq k - 2, \quad \text{and} \quad r \leq \frac{t}{k - 2}.$$

Together, this implies

$$(k - 2)d \leq (k - 1)(t + x + y + z) - y \leq (k - 1)^2 + (k - 1)(x + y) - ky$$
$$\leq 2(k - 1)^2 - k(k - 2) = k^2 - 2k + 2,$$

yielding the contradiction that $d \leq k + 2/(k - 2) < k + 2$ (using that $k \geq 4$).

5   Two-Partial Two-Coloring

In this section, we show that, in the LOCAL model, 2-partial 2-coloring requires $\Omega(\log n)$ deterministic time and $\Omega(\log \log n)$ randomized time in any d-regular tree, where $d \geq 2$. We show the result by reducing from the sinkless orientation problem, for which we know that its distributed deterministic complexity is $\omega(\log^* n)$ rounds in the LOCAL model.

Informally, the proof proceeds in two steps. We first show that, if we can solve 2-partial 2-coloring in constant time in a slightly modified version of the LOCAL model, which we call DC-LOCAL model, then we can solve sinkless orientation in the LOCAL model in $O(\log^* n)$ rounds. Subsequently, we show that, if we can solve an LCL problem P in $O(\log^* n)$ rounds in the LOCAL model, then we can solve P in the DC-LOCAL model in constant time using a simulation similar to that of Chang et al. [9]. Therefore no $O(\log^* n)$-round algorithm exists, i.e., the problem is not easy and hence it has to be hard, i.e., it requires $\Omega(\log n)$ deterministic time and $\Omega(\log \log n)$ randomized time.

Theorem 2. *Computing a 2-partial 2-coloring in d-regular trees in the* LOCAL *model requires* $\Omega(\log n)$ *deterministic time and* $\Omega(\log \log n)$ *randomized time, for any* $d \geq 2$.

DC-LOCAL *Model.* Consider the usual LOCAL model with the following modification. Instead of having unique identifiers, nodes are given as input a color from a palette of c colors, and this coloring of the nodes guarantees a distance-k coloring of the graph. In other words, each node sees different colors in its distance-k ball, but it may see repeated colors in its distance-$(k + 1)$ ball. We call this model DC-LOCAL(k, c) (where DC stands for distance coloring).

5.1   A Lower Bound for the **DC-LOCAL** Model

In this section, we show that 2-partial 2-coloring in d-regular trees is not solvable in $k = O(1)$ rounds in the DC-LOCAL$(k + 1, d^{2(k+1)})$ model. We show this by reducing from the sinkless orientation problem in 2-colored trees in the LOCAL model. More precisely, we show that, if there is an algorithm A solving 2-partial 2-coloring in time $k = O(1)$ in the DC-LOCAL model, then we can use it to design an $O(\log^* n)$-round algorithm that solves sinkless orientation in 2-colored trees in the LOCAL model. This would give an $\omega(1)$ lower bound for 2-partial 2-coloring in the DC-LOCAL model, since we know that sinkless orientation requires $\Omega(\log n)$ rounds, even in 2-colored trees, in the LOCAL model [8].

Gadgets. Let A be the algorithm that solves 2-partial 2-coloring in time $k = O(1)$ rounds in d-regular trees in the DC-LOCAL$(k + 1, d^{2(k+1)})$ model. We introduce two gadgets that we will use later for proving the lower bound. Let T be an arbitrarily distance-$(k + 1)$ colored d-regular tree of depth $k + 3$, and let u be the root of T. We run algorithm A on u and on each of its neighbors $v \in N(u)$. We denote with $A(v)$ the output of algorithm A on a node v. Notice that $A(v)$ is well-defined on these nodes, since their k-radius ball is properly distance-k colored and fully contained in T. Since algorithm A solves 2-partial 2-coloring, we are sure that there are two nodes $v, z \in N(u)$ such that $A(u) \neq A(v) = A(z)$. Let $b \in \{u, v, z\}$ be a node for which A outputs "black" and $w \in \{u, v, z\}$ be a node for which A outputs "white". Let T_w and T_b be the subtrees of depth k rooted at w and b respectively: these are our gadgets (see Fig. 2 for an example). The gadgets T_w and T_b satisfy the following property.

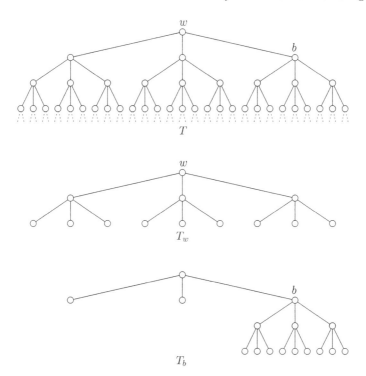

Fig. 2. In this example $k = 2$. Algorithm A outputs "white" on the root of T and "black" on the leftmost child of the root. The gadgets T_w and T_b are trees with depth 2 rooted at w and b respectively.

Property 1. Let $c \in \{\text{black, white}\}$ be the color of the root of the gadget and let $\bar{c}$ be the opposite color. Among all nodes at distance $2t$ from the root of the gadget, there must be at least one node for which the algorithm outputs color c.

Proof. Assume that A outputs $\bar{c}$ at all nodes at distance $2t$ from the root. Then all nodes at distance $2t - 1$ must have color c in order to guarantee a 2-proper 2-coloring. This would imply that all nodes at distance $2t - 2$ have color $\bar{c}$. By applying this reasoning recursively, we conclude that the root must have color $\bar{c}$, which is a contradiction.

Reduction. We now show that if there exists an algorithm A that solves 2-partial 2-coloring in time $k = O(1)$ rounds (where k is even) in d-regular trees in the DC-LOCAL$(k+1, d^{2(k+1)})$ model, then we can design an algorithm A' that solves sinkless orientation on trees of degree $\Delta = d^{2k}$ in which a 2-coloring of the tree is given, in $O(\log^* n)$ rounds in the LOCAL model.

Consider a Δ-regular 2-colored tree $B = (V \cup U, E)$, where V and U represent the set of nodes belonging to the two color classes. We construct a virtual tree in the following way. Each node $x \in V \cup U$ pretends to be the root of a d-regular tree of depth $2k$. We call this tree $T_{\text{virt}}(x)$. Then, each node $v \in V$ (resp. $u \in U$)

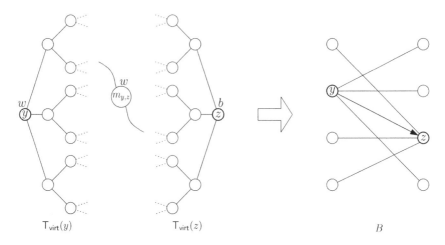

Fig. 3. In this example, the merged node of $\mathsf{T}_{\mathsf{virt}}(y)$ and $\mathsf{T}_{\mathsf{virt}}(z)$ has the same color as node y, hence the edge $\{y, z\}$ in B is oriented from y to z.

labels the nodes at distance at most k with the same colors of the nodes of the gadget T_w (resp. T_b). Note that this is possible since T_w and T_b are isomorphic to the subgraph induced by the nodes at distance at most k from v and u. Then, we merge the i-th leaf of $\mathsf{T}_{\mathsf{virt}}(v)$ with the j-th leaf of $\mathsf{T}_{\mathsf{virt}}(u)$ if and only if there is an edge $\{v, u\}$ connecting v and u through port i of v and port j of u. We call this node $m_{u,v}$ (merged node). In order to make the graph d-regular, we attach additional $d - 2$ virtual nodes to each merged node. Note that, since the original tree B is Δ-regular, and since each virtual tree $\mathsf{T}_{\mathsf{virt}}$ has Δ leaves, then all the leaves of $\mathsf{T}_{\mathsf{virt}}$ are merged nodes (except for the case in which the original node is a leaf, where just one leaf of $\mathsf{T}_{\mathsf{virt}}$ is merged). We then color the nodes that are still uncolored, that is, nodes at distance more than k from the root of the virtual trees, using a distance-k coloring algorithm. Since the already colored parts are far enough apart, this can be done efficiently, in $O(\log^* n)$ rounds.

Now, each node v in B simulates the k-round algorithm A on all nodes of $\mathsf{T}_{\mathsf{virt}}(v)$ and gets a color $c \in \{\text{black, white}\}$ for each node. This requires constant time. Algorithm A outputs "white" at all nodes $v \in V$, since they have the same view as the root of T_w up to distance k. Similarly, algorithm A outputs "black" at all nodes $u \in U$. We then orient an edge of B from node y to node z if and only if $m_{y,z}$ has the same color of y (see Fig. 3 for an example). By Property 1, we know that at least one such a leaf exists for each node in B, meaning that each node is guaranteed to have at least one outgoing edge. This would solve sinkless orientation on 2-colored trees in the LOCAL model in $O(\log^* n)$ rounds. Putting all together, we get the following lemma.

Lemma 1. *The complexity of the 2-partial 2-coloring problem in the* DC-LOCAL *model is* $\omega(1)$.

5.2   From DC-LOCAL to LOCAL

We now show that all LCLs solvable in $O(\log^* n)$ rounds can be solved in a standard manner, that is, first find a distance-k coloring, for some constant k, and then apply a constant time algorithm running in at most k rounds. In particular, we will prove the following lemma.

Lemma 2. *Any* LCL *problem that can be solved in* $O(\log^* n)$ *rounds in the* LOCAL *model can be solved in* $O(1)$ *rounds in the* DC-LOCAL *model.*

We prove the lemma by simulation. Let P be an LCL problem checkable in r rounds, where r is some constant. Assume that we have an algorithm A for the LOCAL model that solves P in $f(n) = O(\log^* n)$ rounds on graphs of size n in which nodes have unique identifiers in $\{1, \ldots, n\}$. Let $\Delta = O(1)$ be the maximum degree of the graph.

Fix N to be the smallest integer such that $t = f(N)+1$ and $\Delta^{2(t+r)} < N$. We show that we can construct an algorithm A' running in t rounds that solves P in the DC-LOCAL$\left(t+r, \Delta^{2(t+r)}\right)$ model. Note that t is constant. In other words, we design an algorithm that solves P in constant time given the promise that nodes are labeled with a $\Delta^{2(t+r)}$-coloring of distance $(t + r)$. We assume that the diameter of the graph is at least $2t$, otherwise nodes can gather the entire graph in constant time and solve P by brute force.

Algorithm A' executed by a node v is defined as follows. At first, node v gathers its distance-t neighborhood $B_v(t)$. Then, node v creates a virtual instance of P by renaming the nodes in $B_v(t)$ and setting their identifiers as their assigned colors. Now, node v simulates algorithm A on $B_v(t)$, by lying about the size of the graph and setting it to be N. Finally, the output of A' is defined to be the same as the output of algorithm A. Notice that this simulation is clearly possible, since A, running on instances of size N, terminates in strictly less than t rounds.

We still need to show that the output is valid for the original LCL. For this purpose, we show that, if the algorithm fails in some neighborhood, then we can construct an instance in which the original algorithm fails as well. Let G be the graph in which, given a $\Delta^{2(t+r)}$-coloring of distance $(t + r)$, there is a node v such that the verifier executed on v rejects (after running for r rounds). Consider $G' = B_v(t+r)$, the subgraph of radius $r+t$ centered at v. All nodes in G' have different colors and the number of nodes is at most N, since N satisfies $\Delta^{2(t+r)} < N$. We now modify G' in order to make it a graph of size exactly N. For this purpose, we pick an arbitrary node at distance $t+r$ from v (that exists by the diameter assumption), and we connect to it a path of as many nodes as needed. We then complete the coloring of these nodes in some consistent manner.

The identifiers of nodes in G' are set to be equal to their colors. The ID space in G' is in $1, \ldots, N$. At this point, we run algorithm A on G'. Consider the set S of nodes at distance at most r from v. For every node $u \in S$, the t-neighborhood of u is the same on G and G', hence the output of A on these nodes must be the same as the output of A'. Thus, the failure of A' on G implies the failure of A on G'. Theorem 2 follows by combining Lemmas 1 and 2.

6 Additional Hardness Results

Theorem 3. *Computing a k-partial c-coloring in d-regular graphs, for $k \geq \frac{c-1}{c}d+1$, requires $\Omega(\log n)$ deterministic time and $\Omega(\log \log n)$ randomized time.*

Proof. Assume the problem is easy to solve. Each monochromatic subgraph has a maximum degree $x = d - k \leq \frac{d}{c} - 1$, and hence it is easy to color with $x+1 \leq \frac{d}{c}$ colors. Hence overall we can easily find a proper coloring of a d-regular graph with at most $c \cdot \frac{d}{c} = d$ colors, but this is known to be hard [6,9]. ∎

Acknowledgments. We would like to thank anonymous reviewers for their helpful feedback on prior versions of this work.

References

1. Balliu, A., Brandt, S., Olivetti, D., Suomela, J.: Almost global problems in the LOCAL model. In: Proceedings of 32nd International Symposium on Distributed Computing (DISC 2018). Leibniz International Proceedings in Informatics (LIPIcs), Schloss Dagstuhl-Leibniz-Zentrum für Informatik (2018). https://doi.org/10.4230/LIPIcs.DISC.2018.9
2. Balliu, A., Hirvonen, J., Korhonen, J.H., Lempiäinen, T., Olivetti, D., Suomela, J.: New classes of distributed time complexity. In: Proceedings of 50th ACM Symposium on Theory of Computing (STOC 2018), pp. 1307–1318. ACM Press (2018). https://doi.org/10.1145/3188745.3188860
3. Barenboim, L., Elkin, M.: Distributed graph coloring: fundamentals and recent developments, vol. 4 (2013). https://doi.org/10.2200/S00520ED1V01Y201307DCT011
4. Barenboim, L., Elkin, M., Kuhn, F.: Distributed ($\Delta+1$)-coloring in linear (in Δ) time. SIAM J. Comput. **43**(1), 72–95 (2014). https://doi.org/10.1137/12088848X
5. Bonamy, M., Ouvrard, P., Rabie, M., Suomela, J., Uitto, J.: Distributed recoloring. In: Proceedings of 32nd International Symposium on Distributed Computing (DISC 2018). Leibniz International Proceedings in Informatics (LIPIcs), Schloss Dagstuhl-Leibniz-Zentrum für Informatik (2018). https://doi.org/10.4230/LIPIcs.DISC.2018.12
6. Brandt, S., et al.: A lower bound for the distributed Lovász local lemma. In: Proceedings of 48th ACM Symposium on Theory of Computing (STOC 2016), pp. 479–488. ACM Press (2016). https://doi.org/10.1145/2897518.2897570
7. Brandt, S., et al.: LCL problems on grids. In: Proceedings of 36th ACM Symposium on Principles of Distributed Computing (PODC 2017), pp. 101–110. ACM Press (2017). https://doi.org/10.1145/3087801.3087833
8. Chang, Y.J., He, Q., Li, W., Pettie, S., Uitto, J.: The complexity of distributed edge coloring with small palettes. In: Proceedings of 29th ACM-SIAM Symposium on Discrete Algorithms (SODA 2018), pp. 2633–2652. Society for Industrial and Applied Mathematics (2018). https://doi.org/10.1137/1.9781611975031.168
9. Chang, Y.J., Kopelowitz, T., Pettie, S.: An exponential separation between randomized and deterministic complexity in the LOCAL model. In: Proceedings of 57th IEEE Symposium on Foundations of Computer Science (FOCS 2016), pp. 615–624. IEEE (2016). https://doi.org/10.1109/FOCS.2016.72

10. Chang, Y.J., Pettie, S.: A time hierarchy theorem for the LOCAL model. In: Proceedings of 58th IEEE Symposium on Foundations of Computer Science (FOCS 2017), pp. 156–167. IEEE (2017). https://doi.org/10.1109/FOCS.2017.23
11. Chung, K.M., Pettie, S., Su, H.H.: Distributed algorithms for the Lovász local lemma and graph coloring. Distrib. Comput. **30**(4), 261–280 (2017). https://doi.org/10.1007/s00446-016-0287-6
12. Cole, R., Vishkin, U.: Deterministic coin tossing with applications to optimal parallel list ranking. Inf. Control **70**(1), 32–53 (1986). https://doi.org/10.1016/S0019-9958(86)80023-7
13. Fischer, M., Ghaffari, M.: Sublogarithmic distributed algorithms for Lovász local lemma, and the complexity hierarchy. In: Proceedings of 31st International Symposium on Distributed Computing (DISC 2017), pp. 18:1–18:16 (2017). https://doi.org/10.4230/LIPIcs.DISC.2017.18
14. Ghaffari, M., Harris, D.G., Kuhn, F.: On Derandomizing local distributed algorithms. In: Proceedings of 59th IEEE Symposium on Foundations of Computer Science (FOCS 2018) (2018). https://doi.org/10.1109/FOCS.2018.00069, http://arxiv.org/abs/1711.02194
15. Ghaffari, M., Hirvonen, J., Kuhn, F., Maus, Y.: Improved distributed Δ-coloring. In: Proceedings of 37th ACM Symposium on Principles of Distributed Computing (PODC 2018), pp. 427–436. ACM (2018). https://doi.org/10.1145/3212734.3212764
16. Ghaffari, M., Su, H.H.: Distributed degree splitting, edge coloring, and orientations. In: Proceedings of 28th ACM-SIAM Symposium on Discrete Algorithms (SODA 2017), pp. 2505–2523. Society for Industrial and Applied Mathematics (2017). https://doi.org/10.1137/1.9781611974782.166
17. Kuhn, F.: Weak graph colorings. In: Proceedings of 21st ACM Symposium on Parallelism in Algorithms and Architectures (SPAA 2009), pp. 138–144. ACM Press, New York (2009). https://doi.org/10.1145/1583991.1584032
18. Linial, N.: Locality in distributed graph algorithms. SIAM J. Comput. **21**(1), 193–201 (1992). https://doi.org/10.1137/0221015
19. Naor, M.: A lower bound on probabilistic algorithms for distributive ring coloring. SIAM J. Discret. Math. **4**(3), 409–412 (1991). https://doi.org/10.1137/0404036
20. Naor, M., Stockmeyer, L.: What can be computed locally? SIAM J. Comput. **24**(6), 1259–1277 (1995). https://doi.org/10.1137/S0097539793254571
21. Panconesi, A., Rizzi, R.: Some simple distributed algorithms for sparse networks. Distrib. Comput. **14**(2), 97–100 (2001). https://doi.org/10.1007/PL00008932
22. Peleg, D.: Distributed computing: a locality-sensitive approach. Society for Industrial and Applied Mathematics (2000). https://doi.org/10.1137/1.9780898719772

Near-Gathering of Energy-Constrained Mobile Agents

Andreas Bärtschi[1]([✉]), Evangelos Bampas[2], Jérémie Chalopin[3], Shantanu Das[3], Christina Karousatou[4], and Matúš Mihalák[5]

[1] Center for Nonlinear Studies,
Los Alamos National Laboratory, Los Alamos, NM, USA
baertschi@lanl.gov

[2] Laboratoire de Recherche en Informatique,
Université Paris-Sud, Orsay Cedex, France
evangelos.bampas@lri.fr

[3] CNRS, Aix-Marseille Université and Université de Toulon, LIS, Toulon, France
{jeremie.chalopin,shantanu.das}@lis-lab.fr

[4] Department of Mathematics, TU Darmstadt, Darmstadt, Germany
karousatou@mathematik.tu-darmstadt.de

[5] Department of Data Science and Knowledge Engineering,
Maastricht University, Maastricht, The Netherlands
matus.mihalak@maastrichtuniversity.nl

Abstract. We study the task of gathering k energy-constrained mobile agents in an undirected edge-weighted graph. Each agent is initially placed on an arbitrary node and has a limited amount of energy, which constrains the distance it can move. Since this may render gathering at a single point impossible, we study three variants of *near-gathering*:

The goal is to move the agents into a configuration that minimizes either (i) the radius of a ball containing all agents, (ii) the maximum distance between any two agents, or (iii) the average distance between the agents. We prove that (i) is polynomial-time solvable, (ii) has a polynomial-time 2-approximation with a matching NP-hardness lower bound, while (iii) admits a polynomial-time $2(1 - \frac{1}{k})$-approximation, but no FPTAS, unless P = NP. We extend some of our results to additive approximation.

Keywords: Mobile agents · Power-aware robots · Limited battery · Gathering · Graph algorithms · Approximation · Computational complexity

This work was partially supported by the SNF (project 200021L_156620) and by the ANR (project ANCOR anr-14-CE36-0002-01), while A. Bärtschi was working at ETH Zürich, and E. Bampas and C. Karousatou were working at Aix-Marseille Université. The Los Alamos National Laboratory report number is LA-UR-19-23906.

K. Censor-Hillel and M. Flammini (Eds.): SIROCCO 2019, LNCS 11639, pp. 52–65, 2019.
https://doi.org/10.1007/978-3-030-24922-9_4

1   Introduction

The problem of *gathering* is one of the fundamental problems in distributed computing with mobile entities, which includes mobile agents moving in a graph or robots moving in a continuous geometric space. In both cases, the objective is to bring together multiple autonomous agents to a single point (not predetermined). Gathering helps in coordination between the mobile agents, sharing of information between the entities, reassignment of duties among the entities, and even for protection of the agents (a group of robots gathered together is easier to protect than those dispersed in large area). Moreover, there are also theoretical reasons for studying gathering as the problem of selecting a gathering point is akin to problems of leader election and consensus in distributed systems. However, in some cases, it may be impossible to solve the problem of gathering, e.g. due to limitations in the capabilities of the agents, or due to symmetries in their perception of the environment. In some cases it may be desirable for the agents to get close to each other without actually meeting [27].

In this paper, we consider mobile agents moving on a graph, with severe limitations on their movements. We assume that the agents have limited energy resources and traversing any edge of the graph consumes some of this energy which can not be replenished. In other words, each agent has an initial energy budget which limits the total distance it can move in the graph. Under such constraints, it is not always possible to gather the agents at a single point. Thus, we consider the problem of moving the agents as close as possible to each other while respecting the movement constraints, defined below as the *near-gathering* problem.

Near-Gathering. A collection of k mobile agents is initially located at an arbitrary set of nodes of an undirected edge-weighted graph $G = (V, E, \omega)$. Each agent i, $i = 1, \ldots, k$, has an energy capacity b_i, which represents the maximum distance the agent can move in the graph. The agents have *global knowledge* of the graph and are controlled by a *central entity*. The goal is to move the agents to a configuration where they are as close to each other as possible under their respective limitations of movement. Closeness criteria can be measured, e.g., as the size of the smallest region enclosing all the agents, or as the maximum or average pairwise distance between the agents. We look at each of these criteria and give a more precise definition of the problem below.

Our Model. We consider an undirected graph $G = (V, E, \omega)$, where each edge $e \in E$ has a positive weight $\omega(e) > 0$. As usual, the length of a path is the sum of the weights of its edges. We think of every edge $e = \{u, v\}$ as a segment of infinitely many points, where every point in the edge is uniquely characterized by its distance from u, which is between 0 and $w(e)$. We consider every such point to subdivide the edge $\{u, v\}$ into two edges of lengths proportional to the position of the point on the edge. Thus, the distance $d(p, q)$ between two points p and q (nodes or points inside edges) is the length of a shortest path from p to q in G (with edges subdivided by p, q, respectively). For a point p inside an edge $e \in E$ we write $p \in G$ and $p \in seg(e)$.

A collection of k mobile agents is initially located at an arbitrary set of nodes $p_1, \ldots, p_k \in V$. Each agent i is equipped with an energy budget $b_i > 0$ and can move along edges of the graph, for a distance of at most b_i. In the *Near-Gathering* problem, the goal is to relocate every agent into a new position such that the resulting configuration minimizes one of the following objectives: (i) the radius of a smallest ball containing all agents, (ii) the maximum distance between any two agents, or (iii) the average distance between the agents (or, equivalently, the sum of all distances). We are further interested in two variants of the problem, where agents can: (I) only be relocated to *reachable nodes of the graph*, or (II) in a more general scenario, where the agents are allowed to be relocated to *reachable points* (i.e., nodes or points inside edges).

Definition 1 (Near-Gathering).
Instance: $I = \langle G, k, (p_i)_{i=1,\ldots,k}, (b_i)_{i=1,\ldots,k} \rangle$, *where* $G = (V, E, \omega)$ *is an undirected edge-weighted graph,* k *denotes the total number of agents,* p_i *denotes the initial positions of the agents and* b_i *denotes the total amount of energy each agent initially has at its disposal.*
Feasible solution: Any configuration $\mathbf{C} = (c_1, \ldots, c_k)$ *of agent end positions* c_i, *in which for each agent* i, $1 \leq i \leq k$, *we have* $d(p_i, c_i) \leq b_i$. *In the node-stop variant, we additionally require* $c_i \in V$.
Goals: (i) MINBALL*: Minimize* $Radius(\mathbf{C}, \mathbf{c})$ *of a smallest ball containing* $\mathbf{C}$ *around an optimally chosen center* $\mathbf{c}$, *where* $Radius(\mathbf{C}, \mathbf{c}) = \max_i d(\mathbf{c}, c_i)$. *We consider both the scenario with node centers only, and the scenario with arbitrary point centers.*
(ii) MINDIAM*: Minimize* $Diam(\mathbf{C})$, *where* $Diam(\mathbf{C}) = \max_{i,j} d(c_i, c_j)$.
(iii) MINSUM*: Minimize* $Sum(\mathbf{C})$, *where* $Sum(\mathbf{C}) = \sum_i \sum_j d(c_i, c_j)$.

Related Work. The gathering problem has been studied in two very different scenarios (i) Gathering of mobile agents in a connected (finite or infinite) graph, and (ii) Gathering of mobile robots in a (bounded or unbounded) plane or other geometric spaces. In the context of distributed robotics or swarm robotics [23], the problem of gathering many robots at a single point has been studied as an agreement problem, where the main issue is feasibility of gathering starting from arbitrary configurations [12] or gathering without full knowledge of the configuration [24, 26]. The problem of *convergence* requires the robots to converge towards a point [13], without actually arriving at the gathering point. When the robots are not allowed to collide, the problem of moving the robots closer avoiding any collisions has been studied by Pagli et al. [27]. In all these studies, the robots can move freely in any direction. For mobile agents on the graph that are restricted to move along the edges, gathering has been studied under different models (see e.g. [15, 28]). In particular, the gathering of two agents, often called rendezvous, has attracted a lot of attention, well documented in [1]. The problem of gathering with the objective of minimizing movements has been studied in [11]. However to the best of our knowledge, there have no previous studies on gathering with fixed constraints (budgets) on energy required for movements.

The model of energy-constrained agents was introduced in [3,7] for single agent exploration of graphs. Duncan et al. [20] consider a similar model where the agent is tied with a rope of length b to the starting location. Multi-agent exploration under uniform energy constraint of b, has been studied for trees [21,25] with the objective of minimizing the energy budget per agent [22] or the number k of agents [16] required for exploration, while time optimal exploration was studied by Dereniowski et al. [19] under the same model. Demaine et al. [17, 18] studied problems of optimizing the total or maximum energy consumption of the agents when the agents need to place themselves in desired configurations (e.g. connected or independent configurations); they provided approximation algorithms and inapproximability results. Similar problems have been studied for agents moving in the visibility graphs of simple polygons [8].

For the model studied in this paper, where each agent has a distinct energy budget, the problem of *Broadcast* and *Convergecast* was studied in [2] who provided hardness results for trees and approximation algorithms for arbitrary graphs. The problem of delivering packages by multiple agents having energy constraints was studied in [5,6,9,10]. All of these problems were shown to be NP-hard for general graphs even if the agents are allowed to exchange energy when they meet [4,14].

Our Contribution and Paper Organization. In Sect. 2, we establish a few preliminaries and prove that MINBALL is solvable in polynomial-time. In Sect. 3 we give a 2-approximation algorithm for MINDIAM, together with a matching NP-hardness lower bound; additionally we show that MINDIAM is polynomial-time solvable on tree graphs. In Sect. 4, we prove that MINSUM admits a $2(1-\frac{1}{k})$-approximation algorithm but no FPTAS, unless P = NP. We show that the analysis of the approximation ratio of the provided algorithm is tight.

We conclude with remarks on future research opportunities, including preliminary approximation hardness results for additive approximation of MINDIAM, in Sect. 5. All our results – with the exception of additive approximation – hold for both node-stop as well as arbitrary-stop scenarios. Omitted proofs are deferred to the full version of the paper.

2   Preliminaries and Minimizing the Radius

Preliminaries. We first point out some differences in the two scenarios we consider throughout this paper and our general approach on how to tackle and distinguish those. In the node stop scenario, where each agent i is only allowed to move to nodes v with distance $d(p_i, v) \leq b_i$, there is a finite number of feasible configurations **C**. For the scenario with arbitrary final positions, where agents are also allowed to move to points p inside edges (as long as $d(p_i, p) \leq b_i$), we discretize the set of configurations. In the MINBALL variant of Near-Gathering, the discretization turns out to contain at least one optimum solution, for MINDIAM and MINSUM it will at least contain a configuration approximating an optimum solution within a factor of 2 or $2(1 - \frac{1}{k})$, respectively. To this end, we define sets of reachable nodes and "maximally reachable" in-edge points as follows:

Algorithm 1. MINBALL (node centers)

Input: An instance $\langle G, k, (p_i)_{i \in 1, \ldots, k}, (b_i)_{i \in 1, \ldots, k} \rangle$.
Output: Configuration $\mathbf{C}$ and center $\mathbf{c} \in V$ with minimum radius $Radius(\mathbf{C}, \mathbf{c})$.
  1: **for** each $v \in V$ **do**
  2:     Compute $\mathbf{C}^v := (c_1^v, \ldots, c_k^v)$, where $c_i^v \in \arg\min\{d(v, c_i) \mid c_i \in B(i) \cup S(i)\}$
  3:         is a point in $B(i) \cup S(i)$ minimizing the distance to v, breaking ties arbitrarily.
  4:     Compute $Radius(\mathbf{C}^v, v)$.
  5: **end for**
  6: Return $\underset{\mathbf{C}^v, v:\, v \in V}{\arg\min}\ Radius(\mathbf{C}^v, v)$.

Definition 2 (Balls, Spheres). *For an instance $I = \langle G, k, (p_i), (b_i) \rangle$ with $i \in 1, \ldots, k$, we define*

- $B(i) := \{v \in V \mid d(p_i, v) \leq b_i\}$, *i.e. the ball containing all nodes that agent i can reach from its initial position p_i, and*
- $S(i) := \emptyset$ *for node stops, and* $S(i) := \{p \in G \mid d(p_i, p) = b_i\} \backslash B(i)$ *for arbitrary stops, i.e. the sphere of all in-edge points that agent i can reach from its initial position p_i only by spending its whole budget b_i.*

In the same spirit, we can study MINBALL-Gathering for centers $\mathbf{c}$ being restricted to nodes in V, or for the continuous set of center points being allowed to be placed both on nodes as well as the inside of edges of G. To discretize this set, it will be useful to define a set of midpoints, intuitively consisting of "points m lying in the middle of a trail between points p and q":

Definition 3 (Midpoints). *Given a set S of points in G, denote by $G' = (V', E', \omega')$ the graph we get from $G = (V, E, \omega)$ by subdividing the edges in E with points from S, i.e. $V' = V \cup S$. We define the midpoint set $M(S)$ of points in G' – and by bijection also of G – as:*

$$M(S) := \{m \in V' \mid \exists\, p, q \in S \colon d(p, m) = d(m, q)\}$$
$$\cup\, \{m \in seg(e) \mid e = \{u, v\} \in E', \exists\, p, q \in S \colon$$
$$d(p, u) + d(u, m) = d(m, v) + d(v, q)\}.$$

Lemma 1. *The sets $B(i)$, $S(i)$ and $M(S)$ can be computed in time polynomial in $|V|, k$ and $|V|, |S|$, respectively.*

MinBall for Node Centers. Having defined balls and spheres of reachable points for the agents, we can immediately give an exhaustive search algorithm for MINBALL for *centers restricted to nodes*. The main idea of Algorithm 1 is to fix a node in graph G as a *gathering point* and then for each agent i compute the minimum distance to this fixed center it can reach, given its starting position p_i and its energy budget b_i. Iterating over all possible center nodes, we find an optimal solution:

Algorithm 2. MINBALL (arbitrary centers), MINDIAM (2–apx / on Trees)

Input: An instance $\langle G, k, (p_i)_{i \in 1, \ldots, k}, (b_i)_{i \in 1, \ldots, k} \rangle$.
Output: Configuration $\mathbf{C}$ and center $\mathbf{c} \in G$ with minimum radius $Radius(\mathbf{C}, \mathbf{c})$.
1: **for** each $p \in M\left(V \cup \bigcup_i S(i)\right)$ **do**
2:     Compute $\mathbf{C}^p := (c_1^p, \ldots, c_k^p)$, where either $c_i^p = p$ if $d(p_i, p) \leq b_i$, or
3:     $c_i^p \in \arg\min\{d(p, c_i) \mid c_i \in B(i) \cup S(i)\}$ (breaking ties arbitrarily) otherwise.
4:     Compute $Radius(\mathbf{C}^p, p)$.
5: **end for**
6: Return $\underset{\mathbf{C}^p, p: \, p \in M\left(V \cup \bigcup_i S(i)\right)}{\arg\min} Radius(\mathbf{C}^p, p)$.

Theorem 1 (MinBall, node centers). *Algorithm 1 is a polynomial-time algorithm for* MINBALL *with node centers.*

The polynomial running time follows immediately from the fact that $B(i), S(i)$ can be computed in polynomial time and have polynomial size by Lemma 1. As the algorithm iterates over all possible center nodes, we can establish correctness by characterizing optimum stopping positions:

Lemma 2. *There exists an optimum solution* $(\mathbf{C}_{\mathrm{OPT}}, \mathbf{c}_{\mathrm{OPT}})$ *for* MINBALL *where every agent i either stops on* $\mathbf{c}_{\mathrm{OPT}}$ *or on a point in* $B(i) \cup S(i)$, *independent of whether* $\mathbf{c}_{\mathrm{OPT}}$ *is contained in* $\bigcup_i(B(i) \cup S(i))$ *or not.*

MinBall for Arbitrary Centers. As can be seen from Lemma 2, when testing for a fixed center $\mathbf{c}$, in addition to checking the points in $B(i) \cup S(i)$ we should also consider whether agent i can reach $\mathbf{c}$ itself. As candidates for the center $\mathbf{c}$ we take all points in the midpoint set $M(V \cup \bigcup_i S(i))$:

Theorem 2 (MinBall, arbitrary centers). *Algorithm 2 is a poly-time algorithm for* MINBALL *with arbitrary centers.*

As before, polynomial running time follows from the polynomial size of the candidate set $M(V \cup \bigcup_i S(i))$. Building upon Algorithm 1 and Theorem 1, it remains to show that this set contains an optimum center:

Lemma 3. *There exists an optimum solution* $(\mathbf{C}_{\mathrm{OPT}}, \mathbf{c}_{\mathrm{OPT}})$ *for* MINBALL *where* $\mathbf{c}_{\mathrm{OPT}}$ *is contained in* $M(V \cup \bigcup_i S(i))$.

3   Minimizing the Diameter

In this Section, we prove that Algorithm 2, which computes an optimum solution for MINBALL, also computes a 2-approximation for MINDIAM. As we will show, this is likely best-possible, as there is no polynomial-time $(2 - o(1))$-approximation for MINDIAM, unless P = NP. Nonetheless, for the special case of tree graphs, Algorithm 2 even computes an optimum solution for MINDIAM. We start with the positive results:

Theorem 3 (MinDiam, 2-apx). *Algorithm 2 is a polynomial-time 2-approximation algorithm for* MinDiam.

Proof. Let configuration $\mathbf{C}^* = (c_1^*, \ldots, c_k^*)$ with center $\mathbf{c}^*$ be the MinBall solution computed by Algorithm 2. We denote the radius of $(\mathbf{C}^*, \mathbf{c}^*)$ by $r^* = Radius(\mathbf{C}^*, \mathbf{c}^*) = \max_j d(\mathbf{c}^*, c_j^*)$ and the diameter of $\mathbf{C}^*$ by $d^* := Diam(\mathbf{C}^*) = \max_{i,j} d(c_i^*, c_j^*)$. Using the triangle inequality, we have for all configuration points c_i^*, c_j^* that $d(c_i^*, c_j^*) \leq d(c_i^*, \mathbf{c}^*) + d(c_j^*, \mathbf{c}^*)$ and thus $d^* \leq 2 \cdot r^*$. Now let $\mathbf{C}_{\mathrm{OPT}} = (o_1, \ldots, o_k)$ be an optimum configuration for MinDiam with diameter $d_{\mathrm{OPT}} := Diam(\mathbf{C}_{\mathrm{OPT}}) = \max_{i,j} d(o_i, o_j)$. We choose an arbitrary point $o \in \mathbf{C}_{\mathrm{OPT}}$ and compute the radius of a smallest ball around o containing $\mathbf{C}_{\mathrm{OPT}}$, $r_o = Radius(\mathbf{C}_{\mathrm{OPT}}, o) = \max_j d(o, o_j) \leq d_{\mathrm{OPT}}$. By Theorem 2, we have $r^* \leq r_o$ (even though o might not have been considered as a center candidate, see e.g. Fig. 1(left)). Combining all inequalities, we get $d^* \leq 2 \cdot r^* \leq 2 \cdot r_o \leq 2 \cdot d_{\mathrm{OPT}}$, hence $\mathbf{C}^*$ is a 2-approximation for MinDiam.   □

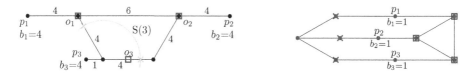

Fig. 1. (left) MinDiam-instance where agent 3's final position in the (unique) optimum solution $\mathbf{C}_{\mathrm{OPT}} = (o_1, o_2, o_3)$ is not in $B(3) \cup S(3)$. (right) Replacing $Radius(\mathbf{C}^p, p)$ in Lines 4&6 of Algorithm 2 with $Diam(\mathbf{C}^p)$ (yielding configurations depicted by × vs □) improves the quality of a MinDiam solution for certain instances by a factor of 2.

Theorem 4 (MinDiam, on Trees). *Algorithm 2 is a polynomial-time algorithm for* MinDiam *on trees.*

Proof. First note that if there is a configuration $\mathbf{C}_{\mathrm{OPT}}$ with maximum distance $Diam(\mathbf{C}_{\mathrm{OPT}}) = 0$, it also has radius $Radius(\mathbf{C}_{\mathrm{OPT}}, \mathbf{c}) = 0$ for some center $\mathbf{c}$, and thus will be found by Algorithm 2 as proven in Theorem 2. Otherwise the diameter $Diam(\mathbf{C}_{\mathrm{OPT}})$ of an optimum solution $\mathbf{C}_{\mathrm{OPT}}$ is lower bounded by the largest diameter among all optimal solutions of the instance reduced to pairs of agents i, j:

$$d^* := \begin{cases} \max_{i,j} \min_{q_i \in B(i),\, q_j \in B(j)} d(q_i, q_j) & \text{for the node stop scenario,} \\ \max_{i,j} d(p_i, p_j) - b_i - b_j & \text{for arbitrary final positions.} \end{cases}$$

We show that, indeed, Algorithm 2 computes a configuration $\mathbf{C}^*$ with $Diam(\mathbf{C}^*) = d^*$. To this end, denote by a, b two agents giving rise to d^*, and let $q_a \in B(a) \cup S(a)$, $q_b \in B(b) \cup S(b)$ be two points with $d(q_a, q_b) = d^*$. Since we consider tree graphs here, there is a unique shortest path from q_a to q_b and thus a unique midpoint $\mathbf{c}^* \in G$ with $d(\mathbf{c}^*, q_a) = d(\mathbf{c}^*, q_b) := \frac{d^*}{2}$. As $\mathbf{c}^*$ is contained in $M(V \cup \bigcup_i S(i))$, Algorithm 2 will use $\mathbf{c}^*$ as a center point candidate for which it computes a configuration $\mathbf{C}^* = (c_1^*, \ldots, c_k^*)$. By definition, we have $d(\mathbf{c}^*, c_a^*) = d(\mathbf{c}^*, q_a) = \frac{d^*}{2} = d(\mathbf{c}^*, q_b) = d(\mathbf{c}^*, c_b^*)$.

It is enough to show that for all other agents i we have $d(\mathbf{c}^*, c_i^*) \leq \frac{d^*}{2}$, too. Assume for the sake of contradiction that this is not the case and that there is an agent i with $d(\mathbf{c}^*, c_i^*) > \frac{d^*}{2}$. Consider the shortest c_i^*-$\mathbf{c}^*$-path P_i, the shortest c_a^*-$\mathbf{c}^*$-path P_a and the shortest c_b^*-$\mathbf{c}^*$-path P_b. By definition of d^* and c^*, the paths P_a and P_b must be interiorly disjoint, $P_a \cap P_b = \{\mathbf{c}^*\}$. Since P_i is a path on a tree ending in the same node $\mathbf{c}^*$, it must be interiorly disjoint with at least one of the two paths P_a, P_b, without loss of generality with P_a. Because any two points in a tree are connected by a unique path, we have $d(c_i^*, c_a^*) = d(c_i^*, \mathbf{c}^*) + d(\mathbf{c}^*, c_a^*) > d^*$ and thus also $\min_{q_i \in B(i) \cup S(i), \, q_a \in B(a) \cup S(a)} d(q_i, q_a) > d^*$, contradicting the maximality of d^*. Hence we have $Diam(C^*) \leq \max_{i,j} d(c_i^*, \mathbf{c}^*) + d(\mathbf{c}^*, c_j^*) = d^*$.

$\square$

Replacing the computation of $Radius(\mathbf{C}^p, p)$ in Lines 4 and 6 of Algorithm 2 by a computation of $Diam(\mathbf{C}^p)$ can improve the quality of a MINDIAM solution by a factor of up to 2 for some instances, see for example Fig. 1(right). However, this does not translate to the worst-case approximation guarantee, as one can see in the instance constructed in the following matching approximation hardness result.

Theorem 5. *There exists no deterministic polynomial-time $\left(2 - o(1)\right)$-approximation algorithm for MINDIAM, unless $P = NP$. This holds even in unweighted graphs with uniform budgets $b_i = 1$, $i = 1, \ldots, k$.*

Proof (Sketch). We prove Theorem 5 by a reduction from 3SAT to MINDIAM: Let ϕ be an arbitrary boolean formula in conjunctive normal form, where each clause contains 3 different literals, and let $x_1, \ldots, x_n$ be the n many variables and $C_1, \ldots, C_m$ be the m many clauses of ϕ. We show that any polynomial-time $(2 - o(1))$-approximation algorithm for MINDIAM can be used to decide whether ϕ is satisfiable. From ϕ, we construct an instance $I = \langle G, k, (p_i)_{i \in 1, \ldots, k}, (b)_{i \in 1, \ldots, k} \rangle$ with k agents of uniform budget $b = 1$ and a graph $G = (V, E, \omega)$ with uniform edge weights $\omega = 1$ in the following manner.

Set of Nodes V: Using T $=$ *true* and F $=$ *false*, we first define the set of all possible truth assignments of a clause C containing 3 literals, $L :=$ $\{TTT, TTF, TFT, TFF, FTT, FTF, FFT, FFF\}$. Note that every clause C is satisfiable by exactly 7 of the 8 possible truth assignments in L (e.g. $x_1 \vee x_2 \vee \bar{x}_n$ is satisfied by $x_1, x_2, x_n \in L \setminus \{FFT\}$). Now, let $V := V_x \cup V_\ell \cup V_C \cup V_L$, where

- $V_x = \{v_i \mid 1 \leq i \leq n\}$ are nodes corresponding to *variables* $x_1, \ldots, x_n$,
- $V_\ell = \{v_i^T \mid 1 \leq i \leq n\} \cup \{v_i^F \mid 1 \leq i \leq n\}$ are nodes corresponding to *literals*, i.e. true-value and false-value assignments of the variables x_i,
- $V_C = \{c_j \mid 1 \leq j \leq m\}$ are nodes corresponding to *clauses* $C_1, \ldots, C_m$,
- $V_L = \{c_j^l \mid 1 \leq j \leq m, \forall l \in L\}$ are nodes corresponding to all possible truth assignments of each clause C_i.

Agents and Reduction Idea: On each of the nodes in $V_x \cup V_C$ we place one agent with a budget of $b = 1$, for a total of $n + m$ agents. The main idea is to initially space the agents by a pairwise distance of 3. We then let agents on V_x "pick the value assignment of the variables x_i" by walking to their respective node in V_ℓ, whereas we let agents on V_C "pick the truth assignment of the clauses C_j" by walking to their respective node in V_L. Then a satisfiable assignment of ϕ exists, if and only if the variable agents and the clause agents "agree in their choice" which corresponds to an optimum MINDIAM configuration $\mathbf{C}_{\mathrm{OPT}}$ of diameter 1. Furthermore, any other configuration should have diameter ≥ 2. This gives rise to the

Set of edges $E := E_{x\ell} \cup E_{\ell L} \cup E_{CL} \cup E_{\ell\ell} \cup E_{LL}$, where:

- $E_{x\ell} = \{\{v_i, v_i^{\mathrm{T}}\}, \{v_i, v_i^{\mathrm{F}}\} \mid 1 \leq i \leq n \colon v_i \in V_x,\ v_i^{\mathrm{T}}, v_i^{\mathrm{F}} \in V_\ell\}$ are edges connecting each variable node x_i to its two literal nodes,
- $E_{CL} = \{\{c_j, c_j^l\} \mid 1 \leq j \leq m \colon c_j \in V_C,\ c_j^l \in V_L,\ c_j^l$ satisfies $C_j\}$ are edges connecting each clause node c_j with all nodes representing satisfying assignments for clause C_j,
- $E_{\ell L} = \{\{v_i, c_j^l\} \mid i \leq n,\ j \leq m \colon v_i \in \{v_i^{\mathrm{T}}, v_i^{\mathrm{F}}\} \subset V_x,\ c_j^l \in V_l$, such that
  - either x_i does not appear in C_j, or
  - x_i appears in C_j and v_i agrees with $c_j^l\}$
  are edges connecting unrelated literals and clause truth-assignments, as well as matching literals and clause truth-assignments.
- $E_{\ell\ell} = \{\{u, v\} \mid u, v \in V_\ell\}$ and $E_{LL} = \{\{u, v\} \mid u, v \in V_L\}$ are edges pairwise connecting nodes in V_ℓ, and nodes in V_L, respectively.

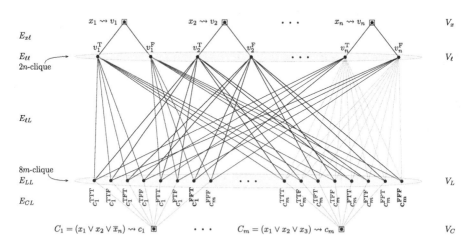

Fig. 2. A part of an instance of MINDIAM, constructed from the 3-SAT instance $C_1 \wedge \cdots \wedge C_m$ with variables $x_1, \ldots, x_n$, displaying the connections between nodes v_1, v_2, v_n, c_1 and c_m. Notice that nodes c_1^{FFT} and c_m^{FFF} are not connected to nodes c_1 and c_m, respectively. The location of mobile agents is denoted by squares ($\square$).

Figure 2 shows a part of an instance of MINDIAM which is constructed from an instance of 3SAT as described above. Before continuing with our proof we need to argue that no agent would stop in the middle of an edge:

Lemma 4. *For any configuration* $\mathbf{C}' = (c_1', \ldots, c_k')$ *with an agent* i *for which* $c_i' \notin \{V_\ell \cup V_L\}$, *there exists another configuration* $\mathbf{C}'' = (c_1'', \ldots, c_k'')$ *with diameter* $Diam(\mathbf{C}'') \leq Diam(\mathbf{C}')$ *for which* $\forall i : c_i'' \in \{V_\ell \cup V_L\}$.

$\Rightarrow$ Continuing with our proof of Theorem 5, we first show that if ϕ is satisfiable then there exists a configuration $\mathbf{C}$ of diameter $Diam(\mathbf{C}) = 1$. Since ϕ is satisfiable we have a truth assignment to the variables which satisfies every clause of ϕ. For each variable x_i, we let agent $a(v_i)$ move to node v_i^T if $x_i = true$ and to node v_i^F otherwise. Next, for each clause C_j, we let agent $a(c_j)$ move to the node c_i^l, which corresponds to the correct true/false-assignment picked by the three agents of the variables in C_j. Note that both types of moves can be done with an energy of $b = 1$. By construction of the instance, the maximum distance of any two agents in this final configuration is 1.

$\Leftarrow$ We now show that if ϕ is not satisfiable then every solution to MINDIAM is of size greater than or equal to 2.

By Lemma 4, we may assume that every agent starting on some node $v_i \in V_x$ moves to one of the nodes v_i^T, v_i^F, and every agent starting on some node $c_j \in V_C$ moves to one of the nodes c_j^l, $l \in L$ (otherwise, if an agent does not move, its distance is clearly at least 2 from any other agent). Therefore, by inspection of the final positions of agents starting in V_x, every MINDIAM solution corresponds to a truth assignment. Since ϕ is not satisfiable, this truth assignment must leave at least one clause C_y, involving variables x_r, x_s, x_t, unsatisfied. By construction of the instance, and in particular in view of the fact that the edge $\{c_y, c_y^{l^*}\}$ is missing (where l^* is the assignment to x_r, x_s, x_t falsifying C_y), the agent that started on c_y cannot move to $c_y^{l^*}$, and thus it will have a distance of 2 in the final configuration from at least one of the agents starting on v_r, v_s, v_t.

Since a polynomial-time $(2 - o(1))$-approximation algorithm for MINDIAM could distinguish between instances with an optimum solution with diameter 1 and instances with an optimum solution with diameter 2, it would also be able to decide whether ϕ is satisfiable of not. $\square$

Algorithm 3. MINSUM $(2(1 - \frac{1}{k})$–apx$)$

Input: An instance $\langle G, k, (p_i)_{i \in 1, \ldots, k}, (b_i)_{i \in 1, \ldots, k} \rangle$.
Output: Configuration $\mathbf{C}$ with $Sum(\mathbf{C}) \leq 2(1 - \frac{1}{k}) \cdot \min_{\text{feasible } \mathbf{C}'} Sum(\mathbf{C}')$.
1: **for** each $p \in V \cup \bigcup_i S(i)$ **do**
2:    Compute $\mathbf{C}^p := (c_1^p, \ldots, c_k^p)$, where either $c_i^p = p$ if $d(p_i, p) \leq b_i$, or
3:    $c_i^p \in \arg\min\{d(p, c_i) \mid c_i \in B(i) \cup S(i)\}$ (breaking ties arbitrarily) otherwise.
4:    Compute $Sum(\mathbf{C}^p)$.
5: **end for**
6: Return $\underset{\mathbf{C}^p \,:\, p \in V \cup \bigcup_i S(i)}{\arg\min} Sum(\mathbf{C}^p)$.

---

4   Minimizing the Average Distance

In this Section we describe and analyze an algorithm for minimizing the average pairwise distance between agents. We complement its approximation ratio of $2(1-\frac{1}{k})$ with a tight analysis and rule out an FPTAS for MinSum. The main idea of the presented Algorithm 3 for MinSum is similar to the idea of Algorithm 2 for MinDiam. We fix a point p in the graph G as a gathering point and move each agent i as close as possible to p with respect to its energy constraint, breaking ties arbitrarily. Algorithm 3 exhaustively tests all points in $V \cup \bigcup_i S(i)$ as possible gathering points and selects the point p for with a configuration $\mathbf{C} = (c_1, \ldots, c_k)$ of minimum sum of pairwise distances between the agents, $Sum(\mathbf{C}) = \sum_i \sum_j d(c_i, c_j)$. The choice of the search space for gathering points is based on a characterization of optimum solutions, similar in look to Lemmata 2 and 3:

Lemma 5. *There exists an optimum solution* $\mathbf{C}_{\mathrm{OPT}}$ *for* MinSum *where every agent stops on a point in* $V \cup \bigcup_i S(i)$.

Theorem 6 (MinSum, $2(1-\frac{1}{k})$-apx). *Algorithm 3 is a polynomial-time $2(1 - \frac{1}{k})$-approximation algorithm (and the approximation ratio is tight).*

Proof (Upper bound only). Let $\mathbf{C}^* = (c_1^*, \ldots, c_k^*)$ denote the configuration computed by Algorithm 3. We denote with $s^* := Sum(\mathbf{C}^*)$ the sum of all pairwise agent distances in $\mathbf{C}^*$. Furthermore, let $\mathbf{C}_{\mathrm{OPT}} = (o_1, \ldots, o_k)$ be an optimum MinSum solution in which each agent j stops on a point $o_j \in V \cup \bigcup_i S(i)$ and let $s_{\mathrm{OPT}} = Sum(\mathbf{C}_{\mathrm{OPT}}) = \sum_i \sum_j d(o_i, o_j)$. Choosing a point $o \in \arg\min_{o_i \in \mathbf{C}_{\mathrm{OPT}}} \sum_j d(o_i, o_j)$ we get

$$\sum_j d(o, o_j) = \tfrac{1}{k} \cdot k \sum_j d(o, o_j) \leq \tfrac{1}{k} \cdot \sum_i \sum_j d(o_i, o_j) = \tfrac{1}{k} \cdot s_{\mathrm{OPT}}.$$

Consider now the configuration $\mathbf{C}^o = (c_1^o, \ldots, c_k^o)$ which Algorithm 3 computed for point o in Step 2 and let $s^o := Sum(\mathbf{C}^o) = \sum_i \sum_j d(c_i^o, c_j^o)$. Clearly, we have $s^* \leq s^o$. Furthermore, o is reachable by at least one agent a, thus by Step 2 we also have $c_a^o = o$. Finally, as Step 2 moves agents as close to o as possible, we have $d(o, c_j^o) \leq d(o, o_j)$. Using the triangle inequality, we rewrite s^o to get

$$s^* \leq s^o = \sum_i \sum_j d(c_i^o, c_j^o) \leq 2 \sum_j d(c_a^o, c_j^o) + \sum_{i \neq a} \sum_{\substack{j \neq a \\ j \neq i}} d(c_i^o, o) + d(o, c_j^o)$$

$$= 2 \sum_j d(o, c_j^o) + (k-2) \sum_{i \neq a} d(c_i^o, o) + (k-2) \sum_{j \neq a} d(o, c_j^o)$$

$$= (2k-2) \sum_j d(o, c_j^o) \leq (2k-2) \sum_j d(o, o_j) \leq 2(1 - \tfrac{1}{k}) s_{\mathrm{OPT}}. \qquad \square$$

Theorem 7. *There is no FPTAS for* MINSUM, *unless P = NP.*

Proof. Assume for the sake of contradiction that there is a polynomial-time approximation scheme for MINSUM which for all $\varepsilon > 0$ computes a $(1 + \varepsilon)$-approximation in time $poly(k, \frac{1}{\varepsilon})$. We reuse the reduction to 3SAT already given in Theorem 5. Recall from its proof that (i) the underlying 3SAT-formula ϕ is satisfiable if and only if there is a Near-Gathering solution $\mathbf{C}^*$ in which all agents have pairwise distance 1, and that (ii) any other solution $\mathbf{C}$ has at least one pair of agents with distance 2.

Summing up the pairwise distances we get for (i) that $Sum(\mathbf{C}^*) = k(k-1)$, while for (ii) we have $Sum(\mathbf{C}) \geq k(k-1)+1$. The existence of an FPTAS, using $\varepsilon \leq \frac{1}{k^2}$, means that we can approximate $Sum(\mathbf{C}^*)$ to within $(1+\frac{1}{k^2}) \cdot k(k-1) = k^2 - k + 1 - \frac{1}{k} < k(k-1)+1 \leq Sum(\mathbf{C})$. Hence we could distinguish the existence of a solution $\mathbf{C}^*$ from any other solution and thus decide satisfiability of ϕ in time $poly(k, \frac{1}{1/k^2}) = poly(k)$, in contradiction to the assumption P $\neq$ NP.   □

5   Additive Approximation and Conclusion

In this paper, we explored the task of *Near-Gathering* a group of energy-constrained agents, whose movements are restricted by their energy budget. We showed how to compute, in polynomial time, an optimum solution for MINBALL (minimizing the radius of a smallest ball containing all agents), a 2-approximation for MINDIAM (minimizing the maximum distance between any two agents), and a $2(1 - \frac{1}{k})$-approximation for MINSUM (minimizing the average distance between any two agents). For MINDIAM, we provided a matching hardness result, while for MINSUM, we ruled out the existence of an FPTAS, unless P = NP. Hence for future work, a major open problem is to improve upon the (in)approximability of MINSUM.

A second possible research direction for Near-Gathering is an analysis of additive approximation. For this, we briefly review how we can reuse our hardness construction of multiplicative approximation of MINDIAM:

Theorem 8. *Unless P = NP, there is no deterministic polynomial-time additive* $+(2 \max_i b_i - o(1))$-*approximation algorithm for* MINDIAM *with node stops, and no deterministic polynomial-time additive* $+(\frac{4}{3} \max_i b_i - o(1))$-*approximation algorithm for* MINDIAM *with arbitrary stops.*

This is surprising for two reasons. On the one hand, *not moving the agents at all* is already an additive $+(2 \max_i b_i)$-approximation. On the other hand, this is the only result in this paper, in which the *two scenarios* of (I) node stops and (II) arbitrary stops *differ*. The difference in the hardness result boils down to the loss of Lemma 4 in the adaption of the proof of Theorem 5, which we can only fully salvage for the case of node stops. Does this mean that there is a polynomial-time $+(2 \max_i b_i - o(1))$-approximation for the scenario with arbitrary final positions? This remains completely open.

Finally, we aim to study the reverse problem of *Spreading* energy-constrained mobile agents, with the respective goals of (i) maximizing the radius of a smallest ball containing all agents, (ii) maximizing the minimum distance between any two agents, and (iii) maximizing the average distance between any two agents.

References

1. Alpern, S., Gal, S.: The Theory of Search Games and Rendezvous. Kluwer, Boston (2003)
2. Anaya, J., Chalopin, J., Czyzowicz, J., Labourel, A., Pelc, A., Vaxès, Y.: Convergecast and broadcast by power-aware mobile agents. Algorithmica **74**(1), 117–155 (2016)
3. Awerbuch, B., Betke, M., Rivest, R.L., Singh, M.: Piecemeal graph exploration by a mobile robot. Inf. Comput. **152**(2), 155–172 (1999)
4. Bampas, E., Das, S., Dereniowski, D., Karousatou, C.: Collaborative delivery by energy-sharing low-power mobile robots. In: ALGOSENSORS 2017, pp. 1–12 (2017)
5. Bärtschi, A., et al.: Collaborative delivery with energy-constrained mobile robots. In: SIROCCO 2016, pp. 258–274 (2016)
6. Bärtschi, A., et al.: Collaborative delivery with energy-constrained mobile robots. In: Theoretical Computer Science, SIROCCO 2016 (2017)
7. Betke, M., Rivest, R.L., Singh, M.: Piecemeal learning of an unknown environment. Mach. Learn. **18**(2), 231–254 (1995)
8. Bilò, D., Disser, Y., Gualà, L., Mihalák, M., Proietti, G., Widmayer, P.: Polygon-constrained motion planning problems. In: ALGOSENSORS 2013, pp. 67–82 (2013)
9. Chalopin, J., Das, S., Mihalák, M., Penna, P., Widmayer, P.: Data delivery by energy-constrained mobile agents. In: ALGOSENSORS 2013, pp. 111–122 (2013)
10. Chalopin, J., Jacob, R., Mihalák, M., Widmayer, P.: Data delivery by energy-constrained mobile agents on a line. In: ICALP 2014, pp. 423–434 (2014)
11. Cicerone, S., Stefano, G.D., Navarra, A.: Gathering of robots on meeting-points: feasibility and optimal resolution algorithms. Distrib. Comput. **31**(1), 1–50 (2018)
12. Cieliebak, M., Flocchini, P., Prencipe, G., Santoro, N.: Distributed computing by mobile robots: gathering. SIAM J. Comput. **41**(4), 829–879 (2012)
13. Cohen, R., Peleg, D.: Convergence properties of the gravitational algorithm in asynchronous robot systems. SIAM J. Comput. **34**, 1516–1528 (2005)
14. Czyzowicz, J., Diks, K., Moussi, J., Rytter, W.: Communication problems for mobile agents exchanging energy. In: SIROCCO 2016 (2016)
15. Czyzowicz, J., Labourel, A., Pelc, A.: How to meet asynchronously (almost) everywhere. ACM Trans. Algorithms **8**(4), 37:1–37:14 (2012)
16. Das, S., Dereniowski, D., Karousatou, C.: Collaborative exploration by energy-constrained mobile robots. In: SIROCCO 2015, pp. 357–369 (2015)
17. Demaine, E.D., Hajiaghayi, M., Mahini, H., Sayedi-Roshkhar, A.S., Oveisgharan, S., Zadimoghaddam, M.: Minimizing movement. ACM Trans. Algorithms **5**(3), 30:1–30:30 (2009)
18. Demaine, E.D., Hajiaghayi, M., Marx, D.: Minimizing movement: fixed-parameter tractability. ACM Trans. Algorithms **11**(2), 14:1–14:29 (2014)
19. Dereniowski, D., Disser, Y., Kosowski, A., Pająk, D., Uznański, P.: Fast collaborative graph exploration. Inf. Comput. **243**, 37–49 (2015)

20. Duncan, C.A., Kobourov, S.G., Kumar, V.S.A.: Optimal constrained graph exploration. ACM Trans. Algorithms **2**(3), 380–402 (2006)
21. Dynia, M., Korzeniowski, M., Schindelhauer, C.: Power-aware collective tree exploration. In: ARCS 2006, pp. 341–351 (2006)
22. Dynia, M., Łopuszański, J., Schindelhauer, C.: Why robots need maps. In: SIROCCO 2007, pp. 41–50 (2007)
23. Flocchini, P., Prencipe, G., Santoro, N. (eds.): Distributed Computing by Mobile Entities, Current Research in Moving and Computing. Lecture Notes in Computer Science, vol. 11340. Springer, Switzerland (2019). https://doi.org/10.1007/978-3-030-11072-7
24. Flocchini, P., Prencipe, G., Santoro, N., Widmayer, P.: Gathering of asynchronous robots with limited visibility. Theoret. Comput. Sci. **337**(1–3), 147–168 (2005)
25. Fraigniaud, P., Gasieniec, L., Kowalski, D.R., Pelc, A.: Collective tree exploration. Networks **48**(3), 166–177 (2006)
26. Lin, J., Morse, A.S., Anderson, B.D.O.: The multi-agent rendezvous problem (parts 1 and 2). SIAM J. Control Optim. **46**(6), 2096–2147 (2007)
27. Pagli, L., Prencipe, G., Viglietta, G.: Getting close without touching: near-gathering for autonomous mobile robots. Distrib. Comput. **28**(5), 333–349 (2015)
28. Pelc, A.: Deterministic rendezvous in networks: a comprehensive survey. Networks **59**(3), 331–347 (2012)

Optimal Multi-broadcast with Beeps
Using Group Testing

Joffroy Beauquier[1], Janna Burman[1], Peter Davies[2], and Fabien Dufoulon[1(✉)]

[1] LRI, CNRS UMR 8623,
Université Paris-Sud, Université Paris-Saclay, Orsay, France
{beauquier,burman,dufoulon}@lri.fr
[2] University of Warwick, Coventry, UK
Peter.Davies.4@warwick.ac.uk

Abstract. The *beeping model* is an extremely restrictive broadcast communication model that relies only on carrier sensing. In this model, we obtain *time-optimal and deterministic* solutions for the fundamental communication task of multi-broadcast. The proposed solutions are completely uniform, i.e., independent of the network and problem parameters.

We improve on previous results for multi-broadcast by giving *efficiently constructible* solutions, that is, with local computation cost polynomial in the identifiers' range. The originality of our approach lies in the use of *(combinatorial) group testing strategies*, originally developed in the centralized context.

Keywords: Beeping model · Group testing · Multi-broadcast

1 Introduction

Wireless networks with weak communication capabilities have received a great deal of interest recently. In particular, new models assuming very severe restrictions on communication capabilities have been proposed. One of them is the *discrete beeping model* ($\mathcal{BEEP}$), introduced by Cornejo and Kuhn [7]. Due to its weak assumptions, $\mathcal{BEEP}$ has broad applicability to many different communication networks. It has strong connections with the ad-hoc radio network model, and has been used to obtain optimal results in radio networks with collision detection [13,15]. In $\mathcal{BEEP}$, the wireless network is modeled by a *static communication graph* of diameter D, in which the n nodes represent devices and the edges represent reachability via direct transmission. Time is divided into synchronous steps (i.e., *rounds*), and in each step a node can either listen or transmit a unary signal (beep) to all its neighbors. As a beep is merely a detectable burst of energy, a listening node is not notified about the *identifiers*

Supported by the Centre for Discrete Mathematics and its Applications (DIMAP) and by EPSRC award EP/N011163/1. The full version of this paper can be found at https://hal.archives-ouvertes.fr/hal-02140017.

K. Censor-Hillel and M. Flammini (Eds.): SIROCCO 2019, LNCS 11639, pp. 66–80, 2019.
https://doi.org/10.1007/978-3-030-24922-9_5

(IDs) of its beeping neighbors. Even more critically, a beeping node receives no feedback, while a silent (listening) one can only detect whether at least one of its neighbors beeped or all of them were silent. Although algorithms can take advantage of the synchronous nature of the rounds to transmit information using beeps, doing so impacts the time complexity in a quantifiable manner.

Efficient solutions to fundamental communication primitives provide convenient and efficient abstractions of the actual communication mechanisms and serve as algorithmic building blocks, resulting in simpler algorithm design. Such primitives are even more important in weak communication models, such as $\mathcal{BEEP}$. In this model, simultaneous communications produce (non-destructive) interferences (i.e., two or more beeps cannot be distinguished from a single beep but can be distinguished from zero beeps), making it difficult for nodes to communicate on a global scale.

In the present paper we propose such communication primitives for the multi-broadcast task. In multi-broadcast, each *source* node in a subset of at most k (for some integer $k \leq n$) nodes (called *sources*) communicates its message m in $\{1, \ldots, M\}$ and its identifier id in $\{1, \ldots, L\}$ to the whole network (referred to as *multi-broadcast with provenance* in [8]). We present optimal and nearly optimal uniform solutions. Contrary to previous results, these solutions are *constructible*. It is important to emphasize that these results come from an entirely original approach based on *(combinatorial) group testing theory*. Group testing is a method coming from statistics, initially introduced during the Second World War to quickly detect an infection among a group of people [11]. In its original formulation (i.e., as *probabilistic group testing*), the defects were assumed to follow some probability distribution, and the goal was to design a strategy identifying all defects using a small expected number of tests. Probabilistic group testing has been used for local neighbor discovery tasks in some distributed settings [19]. In the *combinatorial* context [16,18], no assumptions are made about the distribution of the defects and the goal is to design a strategy with a small maximum number of tests (i.e., a worst-case scenario). Results from combinatorial group testing are crucial to the current work. They are used to efficiently detect all broadcasting sources, since these can be arbitrary, i.e. cannot be assumed to follow some known probability distribution.

Related Work for Multi-broadcast. In [8], an $O(D \cdot \log L + k \log \frac{LM}{k})$ round deterministic, completely uniform (in L, D and k) algorithm for k-source multi-broadcast is presented, and the lower bound of $\Omega(D + k \log \frac{LM}{k})$ rounds is given. In [13], a time-optimal leader election algorithm is given and is used to slightly improve these results: $O(D \cdot \log L)$ factors are reduced to $O(D \cdot \min\{k, \log L\})$ (by executing k consecutive leader elections). Finally, in [10], the lower bound for multi-broadcast given in [8] is extended to also apply to randomized algorithms and a time-optimal $O(D + k \log \frac{LM}{k})$ deterministic and uniform solution to multi-broadcast is proposed. However, this solution relies on a non-constructive existence proof of a complex combinatorial structure, meaning that it must be pre-computed for each possible set of network parameters, and provided to the network nodes in advance (see discussion below).

Importantly, by considering the $\mathcal{BEEP}$ model, the focus is on how non-destructive interferences impact the multi-broadcast problem. This improves the understanding of this problem even for stronger models. For example, in the related and well-established radio network model assuming $O(\log n)$ bit messages and collision detection (a strictly stronger model than $\mathcal{BEEP}$, to which $\mathcal{BEEP}$ algorithms can be straightforwardly translated), the fastest known (non-explicit) algorithms were designed in $\mathcal{BEEP}$ [10]. Somewhat counter-intuitively, efficient solutions for the stronger model do not use the $O(\log n)$ bits contained in the messages, but simply rely on collision detection. In comparison, in radio networks assuming $O(\log n)$ bit messages without collision detection (in which solutions cannot leverage the non-destructive interference for communication), the best multi-broadcasting randomized algorithm [2] requires $O(n \log n)$ time while the best deterministic algorithm [5] requires $O(n \log^4 n)$ time.

Explicitness. Algorithms in $\mathcal{BEEP}$ (and related models such as ad-hoc radio networks) generally seek to minimize the number of rounds required to complete communication tasks. As a result, the cost of local computations is often ignored. Indeed, the fastest deterministic communication algorithms in $\mathcal{BEEP}$, and in radio networks, are often *non-explicit*: they rely upon the use of combinatorial objects whose existence is only proven existentially (see e.g. [9,10]). Although the existence proofs of the combinatorial objects involved are 'non-constructive', they do imply a naive construction: one can simply generate candidate objects randomly, if shared randomness is available, or in lexicographical order otherwise, and test if they actually satisfy the conditions of the object. However, there are exponentially many possible candidates, and testing naively whether these candidates objects are the required combinatorial objects necessitates an exponential number of computations. Such an approach thus results in an impractically high computation cost.

In some settings an argument can be made that an exponential computation cost may still be acceptable, since the construction of suitable combinatorial objects only *ever* needs to be performed once, and henceforth the object can be stored and provided whenever needed to wireless devices. However, in $\mathcal{BEEP}$ this approach poses a problem: the combinatorial objects that we need depend on the parameters of the network which are not known in advance. Hence, network nodes would have to be pre-loaded with objects for every possible set of parameters. This is again impractical, especially since our aim is to model networks of weak devices which would generally have very limited space.

Consequently, we are only concerned by computationally tractable solutions. In $\mathcal{BEEP}$, *explicit solutions* correspond to algorithms with computation time polynomial in L and k (for the nodes), and *weakly explicit solutions* to algorithms with computation time polynomial in L and exponential in k. The latter can still be computationally feasible if $k << L$ when performing multi-broadcast, and thus of practical interest.

Contributions. First, group testing strategies based on *list disjunct matrices* (see Definition 2) are shown to give efficient solutions for multi-broadcast.

Then, several constructions of list disjunct matrices are presented, some novel and some from existing group testing literature, resulting in several algorithms for the multi-broadcasting task:

- An optimal $O(D + k \log \frac{LM}{k})$-time weakly explicit deterministic algorithm.
- An explicit deterministic algorithm optimal for most ranges of k and D.
- An explicit randomized algorithm optimal for $k = \Omega(\log \log L)$.

2   Group Testing

We draw from group testing theory to design efficient solutions in $\mathcal{BEEP}$ (see Sect. 5). The objective of *group testing* is to identify a subset of defective items in a set, by testing multiple items at a time instead of resorting to individual testing. One example is the christmas tree lighting problem: to search for a broken bulb among a group of six, one can arrange electrically in series three bulbs and apply a voltage. If they light up, then they are in good condition, and the broken bulb is one of the three others. Some classical applications of group testing are blood testing, DNA library screening, signal processing, streaming algorithms and wireless multiple-access communications [12].

Formal Definition. A formal definition of the (d, I)-*combinatorial group testing* (CGT) problem follows. Consider I items, represented by the integers in $\{1, \ldots, I\}$, and any arbitrary subset B of d items. The items in B are said to be *defective*. The only way to differentiate defective items from good (i.e., non-defective) items is through testing. For efficiency reasons, tests consider sets of items (*pools*) instead of individual items. When testing a pool, a positive result (output 1) indicates that at least one item in the pool is defective, whereas a negative result (output 0) indicates that no item in the pool is defective. Tests are considered to be error-free. A solution to the CGT problem is a *group testing strategy*, that is, a sequence of t tests (for some positive integer t) such that the set B can be computed from the results by using a *decoder*. One way of computing B is to use the *naive decoder*: a set B' is initialized to the set of all items (i.e., $\{1, \ldots, I\}$) after which for every negative test result (output 0), the items of the test's pool are removed from B'. It is important to note that the group testing strategy is tightly related to the decoder: more complex decoders could lead to fewer tests.

Explicitness in Group Testing. In group testing literature, testing strategies are devised to identify defective items from a pool, and efforts have been made to minimize the number of tests, and stages of adaptivity, required by the strategies. Again, however, it transpires that the best deterministic strategies rely on existentially-proven combinatorial objects, and so are not efficiently constructible or decodable, by the tester.

Consequently, computationally tractable solutions are sought, for practical reasons. In the group testing literature, an *explicit* strategy is one in which each test sequence can be constructed and the output decoded, in time polynomial in I and d. Also of interest is a weaker notion, which we refer to as *weak explicitness*, where construction and decoding time is polynomial in I and exponential in d. The terminology used here corresponds to that used for multi-broadcast. More precisely, when an explicit (respectively weakly explicit) testing strategy is used to obtain a solution to multi-broadcast, the result is an explicit (resp. weakly explicit) algorithm.

Related Work for Group Testing. In the most frequent setting in group testing, *non-adaptive* (i.e., offline) group testing, all tests are designed offline: a test's outcome does not influence the following tests. Non-adaptive group testing allows tests to be performed in parallel. However, it was proven in [14] that test strategies in non-adaptive group testing require $\Omega(d^2 \cdot \frac{\log I}{\log d})$ tests. An explicit construction with $O(d^2 \cdot \log I)$ tests for the non-adaptive setting is given in [21]. On the other hand, in a *fully adaptive setting* (i.e., online setting), where each test's pool depends on the results of all previous tests, the information theoretic lower bound implies that test strategies require $\Omega(d \log \frac{I}{d})$ tests, but all tests must be performed sequentially. An optimal fully-adaptive test strategy is given in [16]. Intermediately, *adaptive* group testing refers to multiple *stages* of tests: all tests of a stage are defined independently from the results of the stage, but can depend on the results of previous stages' tests. Thus tests in the same stage can be done in parallel but successive stages must be treated sequentially. Surprisingly enough when compared with non-adaptive group testing, it is possible to construct two-stage test strategies with $\Theta(d \log \frac{I}{d})$ tests [3,6]. In particular, a weakly explicit construction for such two-stage testing strategies (with $O(d \log \frac{I}{d})$ tests) is given in [6]. Additionally, explicit constructions are given in [4,17,20] with a nearly optimal number of tests. In particular, [20] gives an explicit construction for strategies with $O(d^{1+\epsilon} \log I)$ tests for any value $\epsilon > 0$.

3   Model and Definitions

3.1   Definitions

The *communication network* is represented by a simple static connected undirected graph $G = (V, E)$, where V is the node set and E the edge set. The *network size* $|V|$ is denoted by n and the *diameter* by D. Nodes have unique identifiers (IDs). This property is essential in order to break symmetry in deterministic algorithms. The *identifier* of a node $v \in V$, $id(v)$, is an integer from $\{1, \ldots, L\}$ where L is some upper bound on the identifiers unknown to nodes. Then, the *maximum length* over all identifiers in G is $\lceil \log L \rceil$ (also unknown).

We use the terminology of formal language theory. The empty word is denoted by ϵ. The operator $\|$ is for the *word concatenation*. For any positive integer i, 0^i denotes the concatenation of i symbols 0's (where $0^0 = \epsilon$). The *length* of a word x is denoted by $|x|$. For any word x and integer $j \in \{1, \ldots, |x|\}$, $x[j]$ denotes the jth bit of x. For any two words x and y of the same length, we define the (bitwise OR) *superposition* of x and y (and say that x and y are (OR) *superposed*) as the binary word w of length $|w| = |x|$ such that $\forall i \in \{1, \ldots, |w|\}$, $w[i] = 0 \Leftrightarrow x[i] = y[i] = 0$. We naturally extend the superposition to the case of several words of the same length. Additionally, for any two words x and y of the same length, x is said to be *included* in y if $\forall i \in \{1, \ldots, |x|\}$, $x[i] = 1 \Rightarrow y[i] = 1$.

Multi-broadcast. Let S be a subset of k nodes (for some $k > 1$) called *sources* and having (possibly identical) messages in $\{1, \ldots, M\}$, where M is unknown to all nodes. For any node v, $m(v)$ denotes its message. If v is not a source let $m(v) = \epsilon$. Equivalently, $m(v)$ refers to an integer in $\{1, \ldots, M\}$ or to its binary representation of length at most $\lceil \log M \rceil$.

In the *multi-broadcast* (with provenance) problem, all nodes must receive from each of the k sources its message with its ID. More precisely, they must compute the set $\{(m(v), id(v)) \mid v$ is a source $\}$.

Matrix Notations. For any $a \times b$ matrix M and any integers $i \in \{1, \ldots, a\}$ and $j \in \{1, \ldots, b\}$, the entry of M in row i and column j is denoted by $M[i, j]$. Additionally, the ith row of m is denoted by $M[i, :]$ and the the jth column of m is denoted by $M[:, j]$. For any integer d, let I_d be the $d \times d$ identity matrix, that is, the matrix with entry 1 on the diagonal and 0 otherwise.

3.2   Model Definitions

In the *beeping model* ($\mathcal{BEEP}$), an execution proceeds in synchronous rounds, i.e., there are synchronized local clocks and all nodes start at the same time in a *synchronous start*. In each round nodes synchronously execute the following steps. First, each node beeps or listens. Beeps are transmitted to all neighbors of the beeping node. Then, if a node beeped (in the previous step of the same round), it learns no information from its neighbors. Otherwise, it knows whether or not at least one of its neighbors beeped (during the previous step of the same round). Finally, each node performs local computations. The synchronous start assumption can be replaced by a slightly weaker variant called *wake-on-beep* [1], for an additive factor of $O(D)$ rounds.

4   A General Scheme for Multi-broadcast

A natural solution for multi-broadcast is as follows. First, a leader node (with the maximum ID) is elected, allowing the network to rely on broadcast and convergecast (respectively, sending a message from and to the leader). Once a leader has been elected, the ID range L is known to all nodes. Relying on communications via the leader, it is now possible to efficiently compute global bounds on the network's diameter D and the message range M. Then, the k sources are identified and ordered, as efficiently as possible, by all nodes. Henceforth, this is referred to as the *source identification component*. Finally, the sources convergecast their messages to the leader (pipelined so that the messages arrive to the leader contiguously in order), and the leader broadcasts the string of messages back through the network. Since all nodes agree on the sources' order, all nodes now have all the messages together with the corresponding IDs of the sources. We outline this scheme in Algorithm 1.

Algorithm 1. Multi-Broadcast Scheme

1: Perform Leader Election
2: Estimate Network Parameters
3: Perform Source Identification
4: Collect Source Messages
5: Broadcast Source Messages

All the steps of Algorithm 1, with the exception of Source Identification, can be performed efficiently, explicitly, and deterministically using known procedures from previous works on $\mathcal{BEEP}$:

- Leader election can be performed with $O(D + \log L)$ round complexity [13]. The algorithm requires unique identifiers and elects the node with the maximum identifier. The output is a boolean indicating whether the executing node is the leader or not.
- Estimating diameter D can be performed in $O(D)$ rounds [10]. The algorithm requires a leader, and outputs in all nodes an estimate $\tilde{D}$ with $D \leq \tilde{D} \leq 2D$. Henceforth, we assume that D is known because $\tilde{D}$ can be used instead of D with only a constant-factor overhead.
- Message range M can be similarly estimated in $O(D + \log M)$ time [10].
- Collecting source messages can be done using the COLLECTMESSAGES procedure from [10]. This procedure requires a leader and upper bounds of D and the maximum length, in bits, of the messages to be collected, denoted by p. It takes as input a set of messages held by nodes in the network. On completion, the leader receives the OR superposition of all the messages, and the running time is $O(D + p)$ rounds.
  We apply this procedure by collecting messages of $p = k\lceil \log M \rceil$ bits, one from each source, in which source numbered i in lexicographical order places its input message into the bit interval $[i\lceil \log M \rceil, (i + 1)\lceil \log M \rceil)$, with 0's in every other position (the values of k and the order i are computed during the previously performed source identification component). The superposition of these words is therefore simply the concatenation of all source messages in order. The running time is $O(D + p) = O(D + k \log M)$.
- Broadcasting source messages can be performed using the BEEP-WAVE procedure of [10]. This procedure allows a leader to broadcast a p-bit message to all nodes in $O(D + p)$ time. Applying the procedure to the concatenation of all k source messages in order yields an $O(D + k \log M)$ time.

All these auxiliary procedures terminate such that nodes start each subsequent procedure synchronously. Consequently, source identification is the only remaining step for which there is no efficient procedure, and it is here that the perspective of group testing allows us to make improvements. We denote the round complexity of a potential source identification algorithm by T_{SI}. Efficient source identification solutions are presented in Sect. 5 and their round complexities are given by Theorems 5 and 8. Moreover, the scheme for source identification when k is unknown is presented in Sect. 5.3.

Theorem 1. *Multi-broadcast can be solved in $O(D + \log L + k \log M + T_{SI})$ rounds in $\mathcal{BEEP}$.*

Proof. Applying the above procedures to the scheme in Algorithm 1, the total running time of steps 1 and 2 is $O(D + \log L + \log M)$. After these steps, a leader is elected and all nodes know common constant-factor upper bounds for D, L and M. The subsequent procedure for source identification takes T_{SI} rounds, and results in all nodes being aware of all source IDs. Finally, steps 4 and 5 are then correctly performed, completing multi-broadcast in a further $O(D + k \log M)$ rounds. The total running time is therefore $O(D + \log L + k \log M + T_{SI})$.

5 Source Identification and Group Testing

We now show how the problem of source identification can be reduced to that of combinatorial group testing (defined in Sect. 2). Recall that we have k source nodes with unique IDs from $[L]$, a specified leader node which is known to all nodes in the network, and universal knowledge of (linear upper bounds on) L and D. Upon completing source identification, we require that the leader node has knowledge of all the source IDs (i.e., of S).

Efficient and simple group testing strategies can be obtained by using *list disjunct matrices* (LDM). Such strategies, called LDM-strategies, are presented in Sect. 5.1 and are the building blocks of the source identification algorithm, described in two stages. First, a simplified scheme (when the number of sources k is known) is presented in Sect. 5.2. Then an extended scheme for unknown k is presented in Sect. 5.3. This extended scheme computes a CLDM-strategy (an extension of an LDM-strategy), and its time complexity (resp. computation cost) depends on the CLDM-strategy's parameters (resp., explicitness property). Weakly-explicit and explicit constructions of CLDM-strategies with optimal or nearly optimal parameters are proposed in Sect. 5.4, resulting in efficiently constructible source identification and multi-broadcast solutions.

5.1 Group Testing Strategies and LDM-strategies

Recall that the (d, I)-combinatorial group testing problem (CGT) consists of finding a subset B of d defective items within a set of I items. Good strategies for CGT use at least 2 stages (see Related work in Sect. 2). In a two-stage strategy, a first stage determines a subset B_1 of $\{1, \ldots, I\}$ with $B_1 \supset B$ and $|B_1| = \hat{I}$, and the second stage determines a subset B_2 of $\{1, \ldots, \hat{I}\}$ with $B_2 \supset f_1(B)$ and $|B_2| = d$ (where f_1 maps B_1 to $\{1, \ldots, \hat{I}\}$ in lexicographical order).

Definition 1. *Let B be some unknown subset of d defective items within a set of $\hat{I}$ items. A testing strategy using s stages and t tests over all s stages to determine a superset $B' \supset B$ of size at most $d + \ell - 1$ is called a $(d, \ell, \hat{I})$ s-stage t-test testing strategy.*

In group testing, it is common to build strategies using list disjunct matrices. A single list disjunct matrix defines a single stage testing strategy, and a sequence of s list disjunct matrices defines an s-stage testing strategy (for some integer s).

Definition 2. *A $(d, \ell, \hat{I}, t)$-list disjunct matrix is a $t \times \hat{I}$ binary matrix M such that for any disjoint subsets $T, R \subseteq \{1, \ldots, \hat{I}\}$ with $|T| = d$, $|R| = \ell$, there is a row i of the matrix with $\sum_{j \in T} M[i, j] = 0$ and $\sum_{j \in R} M[i, j] > 0$.*

Lemma 1. *A $(d, \ell, \hat{I}, t)$-list disjunct matrix defines a $(d, \ell, \hat{I})$ single stage t-test testing strategy: each row $M[i, :]$ defines the pool of the ith test (for $1 \leq i \leq t$).*

Definition 3. *A (d, I)-LDM-strategy using s stages and t tests is a sequence $M_1, \ldots, M_s$ of list disjunct matrices with parameters $(d, \ell_1, I_1, t_1), \ldots, (d, \ell_s, I_s, t_s)$ satisfying:*

- $I_1 = I$,
- $d + \ell_i - 1 = I_{i+1}$ *for all* $1 \leq i < s$,

- $\ell_s = 1$,
- $\sum_{i \leq s} t_i = t$.

Lemma 2. *A (d, I)-LDM-strategy using s stages and t tests is a $(d, 1, I)$ s-stage t-test testing strategy and thus solves (d, I)-CGT.*

If d is known then a (d, I)-LDM-strategy can be computed (see Sect. 5.4 for some constructions) and this LDM-strategy defines an s-stage t-test testing strategy solving (d, I)-CGT.

5.2 Source Identification for Known k

In this section, we give a simplified version (Algorithm 2) of the source identification solution, in which we know the number of sources k. This assumption is removed in the extended scheme presented in Sect. 5.3. Algorithm 2 relies on efficient constructions of LDM-strategies (for example, a 2-stage $O(k \log \frac{L}{k})$ weakly explicit LDM-strategy), which are presented later in Sect. 5.4.

Source Identification Scheme (Algorithm 2). The source identification algorithm first computes a (k, L)-LDM-strategy $\mathcal{F}$ using s stages and t tests (which requires knowing k and L), after which sources are identified in s phases. Let $\mathcal{F} = M_1, \ldots, M_s$ where M_u (for $1 \leq u \leq s$) has parameters (k, ℓ_u, L_u, t_u), $L_1 = L$ and $\ell_s = 1$. Details on constructions of good LDM-strategies are deferred to Sect. 5.4. Using a weakly explicit LDM-strategy results in a weakly explicit source identification solution, and an explicit LDM-strategy in an explicit source identification solution.

Nodes start with no knowledge about which nodes could be the sources, and in each phase they obtain more information by implementing a stage of the group testing strategy defined by $\mathcal{F}$ (see Lemma 2). Let f be initialized to the identity function on $\{1, \ldots, L\}$ in the first phase. The function f is updated so that in every phase u, it renames some of the identifiers in $\{1, \ldots, L\}$ to $\{1, \ldots, L_u\}$ (including all source IDs).

The algorithm executes s phases. In each phase u (for $1 \leq u \leq s$), a node v sets $c_u(v)$ to $M_u[:, f(id(v))]$ (i.e., the $f(id(v))^{th}$ column of M_u) if it is a source, and 0^{t_u} otherwise (see lines 5–6). The superposition w of the words c_u is collected by the leader and then broadcast to all network nodes through the use of the auxiliary functions described in Sect. 4 (see lines 7–8). Consequently, nodes compute $S_u = \{x \in \{1, \ldots, L_u\} \mid x$ is included in $w\}$ and update f (see lines 11–12). More precisely, f is updated to $f_u \circ f$, where f_u renames the elements of S_u to $\{1, \ldots, L_{u+1}\}$ according to their lexicographical order: the yth element of S_u is mapped to y. After all s phases are finished, nodes compute $S = f^{-1}(S_s)$.

Implementation of the Testing Strategy. Each phase u for $1 \leq u \leq s$ implements the stage u of the testing strategy. Nodes use the tests of stage u to determine some subset S_u of $\{1, \ldots, L_u\}$ which contains $f(S)$ (where $|f(S)| = |S|$ because no defective item is eliminated by the naive decoder, see Sect. 5.1). Indeed, the leader collects all messages c_u and broadcasts their superposition w to all nodes, which is the superposition of at most k columns of M_u. Each bit $w[i]$ (for $1 \leq i \leq t_u$) can be seen as the test result of test i of stage u in the testing strategy. In the last phase, S_s is a subset of $\{1, \ldots, L_s\}$ with $|S_s| = k + \ell_s - 1 = k$. Therefore, $S_s = f_{s-1} \circ \ldots \circ f_1(S)$.

Algorithm 2. Source Identification Scheme (with known k)

1: **Inputs**: k and upper bounds for L, M and D
2: Compute $M_1, \ldots, M_s$ and their parameters $(k, \ell_1, L_1, t_1), \ldots, (k, \ell_s, L_s, t_s)$
3: $f := id(v)$
4: **for** phase $u := 1 \,; u \leq s \,; u++$ **do**
5:     **if** v is a source node **then** $c_u := M_u[:, f]$
6:     **else** $c_u := 0^{t_u}$
7:     Collect all binary words c_u by OR superposition into w at the leader
8:     Broadcast the superposition w
9:     Get $S_u = \{x \in \{1, \ldots, L_u\} \mid x$ is included in $w\}$
10:     **if** $u < s$ **then**
11:        Let f_u be a function from S_u to $\{1, \ldots, L_{u+1}\}$ in lexicographical order.
12:        **if** v is a source node **then** $f = f_u(f)$
13: Return $S = f_1^{-1} \circ \ldots \circ f_{s-1}^{-1}(S_s)$ ▷ S is the set of source IDs

---

Theorem 2. *Assume Algorithm 2 computes a (k, L)-LDM-strategy $\mathcal{F}$ using s stages and t tests. Then it solves source identification in $O(Ds + t)$ rounds in $\mathcal{BEEP}$.*

Proof. Algorithm 2 solves source identification since the testing strategy defined by $\mathcal{F}$ correctly identifies all k source nodes. In phase u ($1 \leq u \leq s$), the leader gather binary words of t_u bits from the nodes in $O(D + t_u)$ rounds. Then the leader broadcasts the superposition in $O(D + t_u)$ rounds. Over all s phases, the round complexity is $O(\sum_{u \leq s}(D + t_u)) = O(Ds + t)$ rounds.

Therefore, a good source identification solution should use an LDM-strategy with both small s and small t. The related work in Sect. 2 describes such strategies. However, these either require high computation cost (i.e., weak explicitness) or non-optimal (but nearly optimal) s and t [20].

5.3   Extending the Source Identification Scheme to Unknown k

An extended scheme (of Algorithm 2), working when k is unknown, is presented below. The scheme computes an s-stage L-CLDM-strategy (see Definition 5) instead of a (k, L)-LDM-strategy, where the former object is a sequence of constructions that produces an $(\hat{k}, L)$-LDM-strategy for any number of defective items $\hat{k} \leq L$, and can thus be computed when k is unknown. Details on constructions of good CLDM-strategies are deferred to Sect. 5.4.

Definition 4. *A $(\hat{d}, \hat{I})$-list disjunct matrix construction is a function $\mathcal{C}$ with input $(\hat{d}, \hat{I})$ and output (M, ℓ, t) where M is a $(\hat{d}, \ell, \hat{I}, t)$-list disjunct matrix.*

Definition 5. *A I-CLDM-strategy is a sequence $\mathcal{C}_1, \ldots, \mathcal{C}_s$ of constructions of list disjunct matrices satisfying: $\forall \hat{d} \leq I$, let $\mathcal{C}_1(\hat{d}, I) = (M_1, \ell_1, t_1)$ and for $1 < i \leq s$, $\mathcal{C}_i(\hat{d}, I_i) = (M_i, \ell_i, t_i)$ for $I_i = \hat{d} + \ell_{i-1} - 1$, then $M_1, \ldots, M_s$ is a $(\hat{d}, I)$-LDM-strategy.*

Scheme for Source Identification with Unknown k. The extended scheme first computes an s-stage L-CLDM-strategy $\mathcal{F}_\mathcal{C} = \mathcal{C}_1, \ldots, \mathcal{C}_s$. Following which, sources are identified in s phases, and each phase consists of at most $\lceil \log k \rceil$ subphases. Similarly to Algorithm 2, nodes start with no knowledge about which nodes could be the sources, and in each phase u they obtain more information by implementing at most $\lceil \log k \rceil$ consecutive single stage testing strategies on $\{1, \ldots, L_u\}$. Notice that the set of items $\{1, \ldots, L_u\}$ tested upon does not change throughout the different single stage testing strategies (i.e., subphases) of the phase u. Let f be initialized to the identity function on $\{1, \ldots, L\}$ in the first phase. The function f is updated so that in every phase u, it renames some of the identifiers in $\{1, \ldots, L\}$ to $\{1, \ldots, L_u\}$ (including all source IDs).

Subphase Implementation. In sub-phase r of phase u, if $r = 1$ then node v computes $\hat{k}_u^1$, as the smallest power of 2 ($\hat{k}_u^1 = 2^{g_u}$ for some integer g_u) such that $\mathcal{C}_u(\hat{k}_u^1, L_u) = (M_u^1, \ell_u^1, t_u^1)$ satisfies $t_u^1 \geq D$. This prerequisite ensures that the round complexity of phase u in this extended scheme is the same as that in Algorithm 2. For any other subphase $r > 1$, node v computes $\hat{k}_u^r = 2^{r-1} \hat{k}_u^1$.

Following which, a node v first computes $\hat{k}_u^r$ and $\mathcal{C}_u(\hat{k}_u^r, L_u) = (M_u^r, \ell_u^r, t_u^r)$. Then, it sets c_u to $M_u^r[:, f(id(v))]$ (i.e., the $f(id(v))^{th}$ column of M_u^r) if it is a source, and 0^{t_u} otherwise. The superposition w of the words c_u is collected by the leader and then broadcast to all network nodes through the use of the auxiliary functions described in Sect. 4. Then, nodes compute $S_u^r = \{x \in \{1, \ldots, L_u\} \mid x$ is included in $w\}$. If $|B_u^r| \geq \hat{k}_u^r + \ell_u^r$, nodes execute subphase $r+1$ with $\hat{k}_u^{r+1} = 2\hat{k}_u^r$ and still on items $\{1, \ldots, L_u\}$. Otherwise, nodes finish the current phase and if

$u < s$ then nodes execute the following phase $u + 1$ with $L_{u+1} = \hat{k}_u^r + \ell_u^r - 1$ (on items $\{1, \ldots, L_{u+1}\}$) and the function f is updated to $f_u \circ f$, where f_u renames the elements of S_u^r to $\{1, \ldots, L_{u+1}\}$ according to their lexicographical order: the yth element of S_u is mapped to y.

The last subphase of a phase implements the only successful single stage testing strategy of the phase. Moreover, if $k_u^r > k$ then the single stage testing strategy defined by M_u^r is guaranteed to return a subset S_u^r of less than $\hat{k}_u^r + \ell_u^r - 1$ items. Consequently, each phase has at most $\lceil \log k \rceil$ subphases.

This method can be used to solve (k, L)-CGT with unknown k, at the cost of a multiplicative factor $\lceil \log k \rceil$ for both stages and tests in comparison to the corresponding (k, L)-LDM-strategy computed when k is known. Fortunately, when CLDM-strategies are used in our source identification solution, this multiplicative factor does not affect the round complexity (see Lemma 3 and Theorem 3, whose proofs are deferred to the full version of this paper).

Lemma 3. *Each phase u of the extended source identification scheme takes $R_u = O(\sum_{r \leq r'} t_u^r)$ rounds for $r' = \max\{1, \lceil \log k \rceil - g_u\}$. Let t_u be defined by $C_u(k, L_u)$. If C_u satisfies $t_u^1 = O(D)$ and if $r' > 1$, $\sum_{r \leq r'} t_u^r = O(t_u)$, then it follows that $R_u = O(D + t_u)$.*

The conditions of Lemma 3 are satisfied by all 3 CLDM-strategies proposed in Sect. 5.4. Consequently, the following theorem holds for each:

Theorem 3. *Assume that the s-stage L-CDM-strategy $\mathcal{F}_C$ used in the scheme satisfies Lemma 3 for each phase u $(1 \leq u \leq s)$. The extended scheme solves source identification with unknown k in $O(Ds + t)$ rounds, where t is defined by the (k, L)-LDM-strategy computed by $\mathcal{F}_C$ (with $\hat{k} = k$).*

5.4   Efficiently Constructible Source Identification Solutions

Various CLDM-strategies resulting in efficient deterministic source identification solutions are presented in this section. Theorem 2 from Sect. 5.2 emphasizes that both stages and tests should be as low as possible. However strategies with a single stage require a non-optimal $\Omega(d^2 \cdot \frac{\log I}{\log d})$ tests (see Related work in Sect. 2), thus the CLDM-strategies proposed here have at least 2 stages.

Several constructions of list disjunct matrices are presented, with a trade-off between computational cost and optimal parameters (optimal number of tests). First we give a weakly explicit construction with optimal parameters, resulting in a weakly-explicit (2-stage $O(k \log \frac{L}{k})$-tests) CLDM-strategy and thus a weakly explicit round-optimal source identification solution. Following which, we give two explicit constructions with nearly optimal parameters and use them to construct two different explicit CLDM-strategies. Their combination results in an explicit nearly optimal (optimal for most ranges of D and k) source identification solution.

Lemma 4. *For any integers $\hat{k}, \hat{L}$ with $\hat{L} > \hat{k}$, the identity matrix $I_{\hat{L}}$ is a $(\hat{k}, 1, \hat{L}, \hat{L})$-list disjunct matrix. Thus, there exists a construction function $C_{Ind}(\hat{k}, \hat{L}) = (I_{\hat{L}}, 1, \hat{L})$ with computation cost $poly(\hat{k}, \hat{L})$.*

The matrix construction $\mathcal{C}_{Ind}$ defines a testing strategy with individuals tests on all $\hat{L}$ items. Although this strategy is not efficient when $\hat{L} >> \hat{k}$, it is very efficient once $\hat{L} = O(\hat{k} \log \frac{\hat{L}}{k})$. The challenging part is therefore to reduce L items which could possibly be defective to a 'shortlist' of $\hat{L} = O(k \log \frac{L}{k})$ items.

Weakly Explicit Construction with Optimal Parameters. We use an optimal weakly-explicit group testing result from [6]:

Theorem 4 ([6]). *There exists an optimal construction function* $\mathcal{C}_W(\hat{k}, \hat{L}) = (M_W, \hat{k}, O(\hat{k} \log \frac{\hat{L}}{k}))$ *with computation cost* $O(\hat{k}^3 \hat{L}^{2\hat{k}+1} \log \hat{L})$.

The CLDM-strategy $\mathcal{F}_1 = \mathcal{C}_W, \mathcal{C}_{Ind}$ is a weakly explicit 2-stage $O(k \log \frac{L}{k})$-test CDLM-strategy. As a side note, $\mathcal{F}_1$ defines what is referred to as a *trivial two-stage testing strategy* in group testing (see Related work in Sect. 2): $\mathcal{C}_W$ determines most non-defective items, after which $\mathcal{C}_{Ind}$ can be used to determine the k defective items (among the remaining $O(k)$ items). When $\mathcal{F}_1$ is given to the source identification scheme in Sect. 5.3, the result is a weakly explicit algorithm with optimal round complexity for source identification.

Theorem 5. *The extended source identification scheme using a testing strategy defined by* $\mathcal{F}_1$ *is a weakly explicit algorithm solving source identification in optimal* $O(D + k \log \frac{L}{k})$ *rounds. Consequently, combining this result and the multi-broadcast scheme in Sect. 4, the result is a weakly explicit algorithm solving multi-broadcast in optimal* $O(D + k \log \frac{LM}{k})$ *rounds.*

Explicit Constructions with Near Optimal Parameters. Unfortunately, there are no known explicit constructions for group testing strategies using $O(k \log \frac{L}{k})$ tests and a constant number of stages. As a result, the best known results in group testing [20] do not give optimal multi-broadcast algorithms in $\mathcal{BEEP}$. However, by combining two explicit CLDM-strategies, we can design a multi-broadcast algorithm in $\mathcal{BEEP}$ optimal for most ranges of D and k. For $D >> k \log L$ we can use an existing explicit construction from [20]:

Theorem 6 ([20]). *For any constant* $\epsilon > 0$, *there exists a construction function* $\mathcal{C}_E(\hat{k}, \hat{L}) = (M_E, \hat{k}^{1+\epsilon}, \hat{k}^{1+\epsilon} \log \hat{L})$ *with computation cost* $poly(\hat{k}, \hat{L})$.

For $D << k \log L$ we present a new construction (proof deferred to the full version of this paper):

Theorem 7. *Given integers* $\hat{k}, \hat{L}$ *with* $\hat{L} \geq 2\hat{k}$, *let* q *denote* $\lfloor \log_{2\hat{k}} \hat{L} \rfloor$. *There exists a construction function* $\mathcal{C}_{DIG}(\hat{k}, \hat{L}) = (M_{DIG}, \hat{k}^q, 2\hat{k}q)$ *with computation cost* $poly(\hat{k}, \hat{L})$.

Two explicit CLDM-strategies are presented here:

- The first strategy $\mathcal{F}_2 = \mathcal{C}_E, \mathcal{C}_{Ind}$ is an explicit 2-stage $O(k^{1+\epsilon} \log L)$-test CLDM-strategy. It is, similarly to $\mathcal{F}_1$, a trivial two-stage testing strategy. When the source identification scheme in Sect. 5.3 uses a testing strategy defined by $\mathcal{F}_2$, the result is an explicit algorithm for source identification with optimal round complexity when $D = \Omega(k^{1+\epsilon} \log L)$.

– The second strategy $\mathcal{F}_3$ is a sequence of $O(\log k \log \frac{\log L}{\log k}) + 1$ constructions, where constructions $\mathcal{C}_i = \mathcal{C}_{DIG}$ for $1 \leq i \leq O(\log k \log \frac{\log L}{\log k})$
and the last construction is $\mathcal{C}_{Ind}$. $\mathcal{F}_3$ is an explicit CLDM-strategy using
$O(\log k \log \frac{\log L}{\log k}) + 1$ stages and $O(k \log \frac{L}{k})$ tests. When the source identification scheme in Sect. 5.3 uses a testing strategy defined by $\mathcal{F}_3$, the result is
an explicit algorithm for source identification with optimal round complexity
when $D = O(\frac{k \log \frac{L}{k}}{\log k \log \frac{\log L}{\log k}})$.

By executing these two source identification solutions (one defined by $\mathcal{F}_2$, the
other by $\mathcal{F}_3$) in parallel (i.e., one round of the first algorithm, then one of the
second, and so on), the following result can be obtained.

Theorem 8. *Source identification can be solved using an explicit algorithm with
optimal round complexity when either $D = O(\frac{k \log \frac{L}{k}}{\log k \log \frac{\log L}{\log k}})$ or $D = \Omega(k^{1+\epsilon} \log L)$
(for any constant $\epsilon > 0$). As a result, multi-broadcast can be solved using an
explicit algorithm with optimal round complexity for most ranges of k and D.*

6   Explicit Solutions Using Randomized Group Testing

While asymptotically optimal explicit 2-stage randomized group testing strate-
gies exist (e.g. constructing a $(\hat{d}, O(\hat{d}), \hat{I}, O(\hat{d} \log \frac{\hat{I}}{d}))$ list-disjunct matrix by setting each entry to 1 independently with probability $\Theta(1/\hat{d})$), these strategies are
not directly implementable in our $\mathcal{BEEP}$ framework. This is because they rely on
shared randomness, i.e. the tester must have access to the randomness used to
construct the matrix in order to decode it. However, one practical way to achieve
this in $\mathcal{BEEP}$ is to have the leader node generate the random bits to be used, and
broadcast them to the network. This will result in a time cost (in rounds) equivalent to the number of the generated random bits. To minimize this cost and
obtain an efficient randomized multi-broadcast algorithm in $\mathcal{BEEP}$, we present
a new group testing result demonstrating that an optimal testing strategy can
be generated using very few random bits:

Theorem 9. *Given $\hat{d}, \hat{I}$ with $\hat{I} \geq 2\hat{d}$, and $O(\log \hat{I}(1 + \frac{\log \log \hat{I}}{\log d}))$ independent uniformly random bits, one can construct an explicit 2-stage group testing strategy
$\mathcal{F}_P$ such that for any set T of $\hat{d}$ defective items, the strategy recovers T using
$O(\hat{d} \log \frac{\hat{I}}{d})$ tests and succeeding with high probability $(1 - 1/poly(\hat{I}))$.*

This strategy can be used in the same source identification framework as
those in Sect. 5, starting with an estimate $\hat{k}$ such that $\hat{k} \log \frac{L}{k} = \Theta(D)$, and successively doubling until the algorithm succeeds. The resulting algorithm solves
source identification in $O(D + k \log \frac{L}{k} + \log L \log \log L)$ rounds, with high probability (i.e., with probability $(1 - 1/poly(L))$). The proofs of Theorems 9 and 10
are deferred to the full version of this paper.

Theorem 10. *Source identification can be solved in $\mathcal{BEEP}$ with an explicit randomized algorithm in $O(D + k \log \frac{L}{k} + \log L \log \log L)$ rounds, succeeding with
high probability. This round complexity is optimal whenever $k = \Omega(\log \log L)$.*

References

1. Afek, Y., Alon, N., Bar-Joseph, Z., Cornejo, A., Haeupler, B., Kuhn, F.: Beeping a maximal independent set. Distrib. Comput. **26**(4), 195–208 (2013)
2. Bar-Yehuda, R., Israeli, A., Itai, A.: Multiple communication in multihop radio networks. SIAM J. Comput. **22**(4), 875–887 (1993)
3. Bonis, A., Gasieniec, L., Vaccaro, U.: Optimal two-stage algorithms for group testing problems. SIAM J. Comput. **34**(5), 1253–1270 (2005)
4. Cheraghchi, M.: Noise-resilient group testing: limitations and constructions. Discrete Appl. Math. **161**(1), 81–95 (2013)
5. Chlebus, B.S., Kowalski, D.R., Pelc, A., Rokicki, M.A.: Efficient distributed communication in ad-hoc radio networks. In: ICALP, pp. 613–624 (2011)
6. Cicalese, F., Vaccaro, U.: Superselectors: efficient constructions and applications. In: ESA, pp. 207–218 (2010)
7. Cornejo, A., Kuhn, F.: Deploying wireless networks with beeps. In: DISC, pp. 148–162 (2010)
8. Czumaj, A., Davies, P.: Communicating with beeps. In: OPODIS, pp. 1–16 (2016)
9. Czumaj, A., Davies, P.: Deterministic communication in radio networks. SIAM J. Comput. **47**(1), 218–240 (2018)
10. Czumaj, A., Davies, P.: Communicating with beeps. J. Parallel Distrib. Comput. (2019)
11. Dorfman, R.: The detection of defective members of large populations. Ann. Math. Statist. **14**(4), 436–440 (1943)
12. Du, D.Z., Hwang, F.K.: Combinatorial Group Testing and Its Applications. World Scientific, Singapore (1993)
13. Dufoulon, F., Burman, J., Beauquier, J.: Beeping a deterministic time-optimal leader election. In: DISC, pp. 20:1–20:17 (2018)
14. D'yachkov, A., Rykov, V., Rashad, A.: Superimposed distance codes. Prob. Control Inf. Theory **18**(4), 237–250 (1989)
15. Ghaffari, M., Haeupler, B.: Near optimal leader election in multi-hop radio networks. In: SODA, pp. 748–766 (2013)
16. Hwang, F.K.: A method for detecting all defective members in a population by group testing. J. Am. Stat. Assoc. **67**(339), 605–608 (1972)
17. Indyk, P., Ngo, H.Q., Rudra, A.: Efficiently decodable non-adaptive group testing. In: SODA, pp. 1126–1142 (2010)
18. Li, C.H.: A sequential method for screening experimental variables. J. Am. Stat. Assoc. **57**(298), 455–477 (1962)
19. Luo, J., Guo, D.: Neighbor discovery in wireless ad hoc networks based on group testing. In: 46th Annual Allerton Conference on Communication, Control, and Computing, pp. 791–797 (2008)
20. Ngo, H.Q., Porat, E., Rudra, A.: Efficiently decodable error-correcting list disjunct matrices and applications. In: ICALP, pp. 557–568 (2011)
21. Porat, E., Rothschild, A.: Explicit nonadaptive combinatorial group testing schemes. IEEE Trans. Inf. Theory **57**(12), 7982–7989 (2011)

Tracking Routes in Communication Networks

Davide Bilò[1][iD], Luciano Gualà[2][iD], Stefano Leucci[3(✉)][iD],
and Guido Proietti[4,5][iD]

[1] Department of Humanities and Social Sciences, University of Sassari, Sassari, Italy
davide.bilo@uniss.it
[2] Department of Enterprise Engineering,
University of Rome "Tor Vergata", Rome, Italy
guala@mat.uniroma2.it
[3] Department of Algorithms and Complexity, Max Planck Institut für Informatik,
Saarbrücken, Germany
stefano.leucci@mpi-inf.mpg.de
[4] Department of Information Engineering, Computer Science and Mathematics,
University of L'Aquila, L'Aquila, Italy
guido.proietti@univaq.it
[5] Institute for System Analysis and Computer Science "Antonio Ruberti"
(IASI CNR), Rome, Italy

Abstract. The minimum tracking set problem is an optimization problem that deals with monitoring communication paths that can be used for exchanging point-to-point messages using as few tracking devices as possible. More precisely, a tracking set of a given graph G and a set of source-destination pairs of vertices, is a subset T of vertices of G such that the vertices in T traversed by any source-destination shortest path P uniquely identify P. The minimum tracking set problem has been introduced in [Banik et al., CIAC 2017] for the case of a single source-destination pair. There, the authors show that the problem is APX-hard and that it can be 2-approximated for the class of planar graphs, even though no hardness result is known for this case. In this paper we focus on the case of multiple source-destination pairs and we present the first $\widetilde{O}(\sqrt{n})$-approximation algorithm for general graphs. Moreover, we prove that the problem remains NP-hard even for cubic planar graphs and all pairs $S \times D$, where S and D are the sets of sources and destinations, respectively. Finally, for the case of a single source-destination pair, we design an (exact) FPT algorithm w.r.t. the maximum number of vertices at the same distance from the source.

1 Introduction

In the context of network monitoring and surveillance, the problem of tracking moving objects is of primary interest. Think, for example, of a communication network in which messages are exchanged between pairs of hosts. One might wish to track the routing patterns of the messages, i.e., for each exchanged message,

© Springer Nature Switzerland AG 2019
K. Censor-Hillel and M. Flammini (Eds.): SIROCCO 2019, LNCS 11639, pp. 81–93, 2019.
https://doi.org/10.1007/978-3-030-24922-9_6

one would like to know its sender, its recipient, and the exact path it followed in the network. Quite naturally, we assume that messages follow a shortest path between the source and destination host in the network. One way to achieve this goal is that of equipping *a small number of* intermediate hosts with suitable detectors that activate whenever a message is routed though their hosts, so that the set of activated detectors uniquely identifies the message's path.

The above task can be formalized as the *minimum tracking set problem* [4], which is exactly the focus of this paper. Given an undirected graph $G = (V(G), E(G))$ representing the network, and a set $\mathcal{P} = \{(s_1, t_1), \ldots, (s_k, t_k)\}$ of k source-destination pairs of vertices of G, a *tracking set* (TS) of G w.r.t. $\mathcal{P}$ is a subset T of $V(G)$ such that the set of vertices in T traversed by any source-destination shortest path P in G, uniquely identifies P, that is, for any two distinct shortest paths P, Q between any two (possibly coinciding) source-destination pairs in $\mathcal{P}$, we have that $V(P) \cap T \neq V(Q) \cap T$, where $V(P)$ (resp. $V(Q)$) denotes the set of vertices traversed by P (resp. Q), endpoints included.[1] The goal is that of finding a tracking set of G w.r.t. $\mathcal{P}$ having minimum cardinality.

The above definition implicitly assumes that the order of activation of the detectors is unknown. A natural variant asks to uniquely identify P given the *ordered* sequence of vertices in T traversed by P. We will refer to such a set T as an *ordered tracking set* (OTS), and to the corresponding optimization problem as the *minimum ordered tracking set problem*. While a TS is necessarily also a OTS, the converse is not true in general.

The above problem have been considered in the special case in which $\mathcal{P}$ consists of a single source-destination pair (s, t) [4].[2] Notice that, in this case, every OTS is also a TS, hence the two variants of the problem coincide. Indeed, any shortest path P from s to t needs to traverse the vertices in $V(P) \cap T$ in increasing order of distance from s in G. The authors of [4] show, among other results, that the minimum tracking set problem is APX-hard and that it can be 2-approximated for the class of planar graphs, even though no hardness result is known for this case. Moreover, in [2] it is shown that checking whether an instance admits a TS of size at most h can be done in $O(2^{h^2} n^{O(1)})$ time, and thus this problem is fixed parameter tractable. Notice that no approximation algorithm is known for general instances, even for the case of single source-destination pair.

Our Results. In this paper we focus on the case of multiple source-destination pairs and we present the first $\widetilde{O}(\sqrt{n})$-approximation algorithm for general graphs. More precisely, in Sect. 3, we obtain a $O(\sqrt{n})$-approximation for the minimum TS problem which we are able to extend to a $O(\sqrt{n \log n})$-approximation

---

[1] Observe that a TS always exists unless $\mathcal{P}$ contains two pairs of the form (s, t) and (t, s). We then assume that our TS instances never contain such pairs.

[2] Observe that it is not possible to reduce the multi-source multi-destination case to the single-pair case by simply adding a super-source and a super-destination connected to all the sources and all the destinations, respectively, as erroneously claimed in [4].

for the minimum OTS problem. Moreover, in Sect. 4, we prove that both problems remain NP-hard even for cubic planar graphs when the set $\mathcal{P}$ contains all pairs between a given set of sources $S \subseteq V(G)$ and a set of destinations $D \subseteq V(G)$, i.e., $\mathcal{P} = S \times D$. Finally, for the case of a single source-destination pair, we design an (exact) fixed parameter (FPT) algorithm w.r.t. the maximum number h of vertices at the same distance from the source, having a running time of $O^*(2^{h^2})$, as discussed in Sect. 5.[3] Due to space limitations, proofs are omitted from this version of the paper and will appear in the full version.

Other Related Results. Besides the aforementioned ones, the authors of [4] also provide two other results. They show that, given a graph G, and a tracking set T w.r.t. a given pair (s, t), it is possible to efficiently pre-process G and T in order to build a data structure that is able to quickly answer queries of the form: given a subset of T, reconstruct the corresponding path P in time proportional to the number of edges of P. Finally, they provide an exact polynomial time algorithm for the *Catching the intruder* problem, in which we are given a graph G, a pair of vertices (s, t), and a subset of *forbidden* vertices, and we want to find a set T of vertices of minimum cardinality such that, a shortest path from s to t passes through a forbidden vertex if and only if it passes through a vertex of T.

A generalization of the minimum TS problem has been recently considered in [2], for which hardness results are provided. Moreover, the variant of the minimum OTS problem in which the objective is to track *all* paths between a given pair of nodes is studied in [3]. For this variant, the authors show that the problem of finding a solution of size at most h is NP-complete and fixed parameter tractable w.r.t. h.

A problem similar in spirit to the minimum TS problem is the *network verification* problem, informally the problem of establishing the accuracy of a high-level description of its physical topology, by making as few measurements as possible on its nodes [1,5]. More precisely, this task can be formalized as an optimization problem that, given a graph and a query model specifying the information returned by a query at a node, asks for finding a minimum-size subset of nodes to be queried so as to univocally identify the graph. It turns out that the verification problem with the *all-shortest-paths* query model is equivalent to the problem of placing landmarks on a graph [7]. In this problem, we want to place landmarks on a subset of the nodes in such a way that distinct nodes have different distance vectors to the landmarks, and the minimum number of landmarks to be placed is called the *metric dimension* of a graph [6].

2   Preliminaries

Let $G = (V(G), E(G))$ be a graph. If P is a simple path from u to v in G that traverses two vertices x and y, in this order, we will denote by $P[x : y]$ the subpath of P between x and y. We will use $P[: x]$ and $P[x :]$ as a shorthand

---

[3] The O^* notation suppresses polynomial multiplicative factors w.r.t. n.

for $P[u : x]$ and $P[x : v]$, respectively. If P is a path from u to v and Q is a path from v to w, we will use $P \circ Q$ to denote the path from u to w obtained by concatenating P with Q.

Let $P(x, y)$ denote the set of paths from x to y in G and $P^2(x, y) = P(x, y) \times P(x, y)$ be the set of all possible (ordered) pairs of (possibly coinciding) paths from x to y in G. We say that a pair $(P_1, P_2) \in P^2(x, y)$ is independent if P_1 and P_2 are vertex disjoint except for their endpoints x and y.[4]

Given two shortest paths P_1, P_2 from x to y, we say that the *unordered* pair $C = \{P_1, P_2\}$ is a *relevant cycle* if P_1 and P_2 have length at least 2 and (P_1, P_2) is independent. We say that x (resp. y) is the *upper endpoint* (resp. *lower endpoint*) of C. A vertex $v \in V(G)$ *covers* a relevant cycle C if v is an *internal vertex* of C, i.e., $v \in V(P_1) \cup V(P_2) \setminus \{x, y\}$. A set $T \subseteq V(G)$ covers C if it contains at least one vertex $v \in T$ that covers C. We let $C(v)$ be the number of relevant cycles covered by v.

Given a pair of vertices $s, t \in V(G)$, a *relevant cycle w.r.t. (s, t)* is a relevant cycle $\{P_1, P_2\}$ such that both P_1 and P_2 are subpaths of paths in $P(s, t)$.

The following lemmas will be useful in the sequel:

Lemma 1. *For any two distinct shortest paths P_1, P_2 between the same pair of vertices, there exists a relevant cycle $\{P_1', P_2'\}$ where P_1' is a subpath of P_1 and P_2' is a subpath of P_2.*

Lemma 2 ([4]). *Given a graph G, and $s, t \in V(G)$ with $s \neq t$, a set $T \subseteq V(G)$ is a TS w.r.t. (s, t) iff T covers all the relevant cycles w.r.t. (s, t).*

3   An Approximation Algorithm

This section is devoted to designing an approximation algorithm for the minimum TS and minimum OTS problems. For the sake simplicity, we will start by devising an approximation algorithm for the minimum TS problem when $\mathcal{P}$ contains a single pair (s, t). We will then extend our algorithm to handle arbitrary sets $\mathcal{P}$ while leaving the approximation ratio asymptotically unchanged. With some additional technical work, we are able to extend our multiple-pair algorithm for the minimum TS problem to the minimum OTS problem, while only losing a $O(\sqrt{\log n})$-factor in the approximation ratio. Due to space limitations, this latter extension is omitted and will appear in the full version of the paper.

Our algorithm for the minimum TS problem with a single source-destination pair (s, t) simulates the execution of the greedy algorithm for the set-cover problem for the instance $I = (\mathcal{I}, \mathcal{S})$ in which the set of items $\mathcal{I}$ contains all the relevant cycles of G w.r.t. (s, t), and the collection of sets $\mathcal{S}$ contains one set S_v per vertex $v \in V(G)$, where S_v is the set of relevant cycles w.r.t. (s, t) covered by v. The greedy algorithm maintains a partial set-cover X and iteratively adds to X the set $S \in \mathcal{S}$ that maximizes $| \cup_{S' \in X \cup \{S\}} S'|$, i.e., the set that covers

---

[4] Notice that, as a consequence of this definition, the pair (P_1, P_2) in which P_1 and P_2 coincide and consist of the single edge (u, v) is independent.

as many new elements as possible (with ties broken arbitrarily). The algorithm stops as soon as the current solution X covers all the elements in $\mathcal{I}$. Notice that, in general, the size of the above set-cover instance can be exponentially larger than the size of our tracking-path instance. However, we can simulate the greedy algorithm by only *implicitly* maintaining I and the partial solution X. Notice indeed that X is uniquely identified by the set $T = \{v : S_v \in X\}$ in which a set S_v is represented by the vertex v. In order to select the next set S, we only need to compute the number $C(v, T)$ of newly-covered relevant cycles w.r.t. (s, t) for each vertex $v \in V \setminus T$. We will now show how these values $C(v, T)$ can be efficiently computed. The pseudocode of our algorithm is shown in Algorithm 1.

3.1   Computing the Number of Independent Pairs

From now on, we will denote by $\ell(v)$ the *level* vertex $v \in V(G)$, i.e., the distance of v from s in G. In order to compute $C(v, T)$, we first consider the related problem of counting the number $N_G(x, y)$ of independent pairs of paths in G between two distinct vertices x and y such that $\ell(x) < \ell(y)$.[5] In the rest of this section we will assume that the input graph has been preprocessed so that G is actually the subgraph of the input graph obtained by the union of all the shortest paths between s and t.

We first handle some trivial cases: If $\ell(y) \leq \ell(x)$, then $N(x, y) = 0$, while if $\ell(y) = \ell(x) + 1$ then $N(x, y) = 1$ if $(x, y) \in E(G)$ and 0 otherwise. We will therefore assume that $\ell(y) \geq \ell(x) + 2$.

Consider now a pair $(P_1, P_2) \in P_G^2(x, y)$ that is not independent and let $z \neq x, y$ be the vertex of minimum level that is traversed by both P_1 and P_2. We say that (P_1, P_2) is a z-pair.

It immediately follows that, for $z \in V(G)$ with $\ell(x) < \ell(z) < \ell(y)$, (P_1, P_2) is a z-pair iff $z \in V(P_1) \cap V(P_2)$, the subpaths $P_1[x : z]$ and $P_2[x : z]$ are independent, and $(P_1[z : y], P_2[z : y]) \in P_G^2(z, y)$. We therefore have that the number of z-pairs is exactly $N(x, z) \cdot |P(z, y)|^2$.

Since the sets of z-pairs, for $\ell(x) < \ell(z) < \ell(y)$, partition the set of all non-independent pairs in $P_G^2(x, y)$, we can write:

$$N(x, y) = |P(x, y)|^2 - \sum_{\substack{z \in V \\ \ell(x) < \ell(z) < \ell(y)}} N(x, z) \cdot |P(z, y)|^2. \qquad (1)$$

All the values $|P(u, v)|$, $u, v \in V(G)$ can be precomputed in polynomial time using a simple dynamic programming algorithm. Indeed, by fixing v and by considering the possible choices of $u \in V(G)$, with $\ell(u) \leq \ell(v)$ in decreasing order w.r.t. $\ell(u)$ (where ties are broken arbitrarily), we have that $|P(u, v)| = 1$ if $u = v$, and $|P(u, v)| = \sum_{(u,x) \in E(G)} |P(x, v)|$ otherwise.

The above observations immediately result in a polynomial-time dynamic programming algorithm to compute $N(x, y)$. Indeed, it suffices to compute all

---

[5] When the graph G is clear from context, we may omit the subscript.

Algorithm 1. GreedyTracking(G, s, t)

$T \leftarrow \emptyset$;
// The values $C(v, T)$ are computed using Eq. (3.2).
while $\exists v \in V(G) \setminus T : C(v, T) > 0$ **do**
    $v \leftarrow \arg \max_{v \in V(G) \setminus T} C(v, T)$;
    $T \leftarrow T \cup \{v\}$;
return T;

---

the (at most $O(n^2)$) values $N(u, v)$ where $\ell(x) \leq \ell(u) < \ell(v) \leq \ell(y)$ in non-decreasing order of $\ell(v) - \ell(u)$. Each $N(u, v)$ is either in one of the trivial cases described at the beginning of this subsection or can be found in linear time using Eq. (1).

3.2 Computing the Number of Covered Relevant Cycles

We now tackle the problem of counting the number $C(v)$ of relevant cycles covered by v, we will then show how to extend this approach to compute $C(v, T)$. To this aim, we will focus on the number $C(x, u, v, y)$ of relevant cycles that have x and y as their upper and lower endpoints, respectively, and are covered by both vertex u and vertex v. Since, for each relevant cycle C covered by u, there exists *exactly one* other vertex $v \in C$ such that $\ell(u) = \ell(v)$, we can then compute $C(v)$ using the identity:

$$C(v) = \sum_{\substack{x, y \in V(G) \\ \ell(x) < \ell(v) < \ell(y)}} \sum_{\substack{u \in V(G) \setminus \{v\} \\ \ell(u) = \ell(v)}} C(x, u, v, y). \tag{2}$$

In order to compute $C(x, u, v, y)$ we observe that it is the product of the two quantities $C^+(x, u, v)$ and $C^-(u, v, y)$, where $C^+(x, u, v)$ and $C^-(u, v, y)$ denote the "upper" and "lower" parts of the cycle C. More formally, $C^+(x, u, v)$ is the number of pairs (P_1, P_2) such that: (i) P_1 is a shortest path from x to u in G, (ii) P_2 is a shortest path from x to v in G, and (iii) P_1 and P_2 are vertex-disjoint except for vertex x. Similarly, $C^-(u, v, y)$ is the number of pairs (P_1, P_2) that satisfy: (i) P_1 is a shortest path from u to y in G, (ii) P_2 is a shortest path from v to y in G, and (iii) P_1 and P_2 are vertex-disjoint except for vertex y.

Consider now the graph $G^+_{u,v}$ that is obtained by adding a new vertex t' and the edges (u, t') and (v, t') to the subgraph of G induced by all the vertices of level at most $\ell(v)$. Notice that $C^+(x, u, v)$ is exactly the number of *unordered* independent pairs of paths between x and t' in $G^+_{u,v}$, i.e., $C^+(x, u, v) = N_{G^+_{u,v}}(x, t')/2$, where the factor $1/2$ accounts for the fact that $N_{G^+_{u,v}}(s, t')$ is defined w.r.t. *ordered* pairs. A symmetrical argument allows us to write $C^-_{u,v}(u, v, y) = N_{G^-_{u,v}}(s', y)/2$, where $G^-_{u,v}$ is the subgraph of G induced by the vertices of a level at least $\ell(v)$, plus an additional vertex s' that is connected to both u and v. Combining the above observations with Eq. 2, we obtain:

$$C(v) = \frac{1}{4} \sum_{\substack{x,y \in V(G) \\ \ell(x)<\ell(v)<\ell(y)}} \sum_{\substack{u \in V(G) \setminus \{v\} \\ \ell(u)=\ell(v)}} N_{G^+_{u,v}}(x,t') \cdot N_{G^-_{u,v}}(s',y). \tag{3}$$

Clearly, $C(v)$ can be computed in polynomial time as Eq. (3) is essentially a sum of $O(n^3)$ terms, each of which can be found in polynomial time.

Handling Already Covered Cycles. We can easily extend the above algorithm to compute $C(v,T)$ once we observe that a relevant cycle C containing v is newly-covered iff T contains at most the lower and the upper endpoints of C.

Since the vertex u in Eq. (3) is on the same level as v, it cannot belong to T and we can update the range of the inner sum to $V(G) \setminus (T \cup \{v\})$. Moreover, the paths P_1 and P_2 in graph $G^+_{u,v}$ (resp. $G^-_{u,v}$) cannot contain any vertex in T, except possibly for x (resp. y), while all the remaining pairs of paths form a valid "upper" (resp. "lower") part of a newly-covered relevant cycle. To summarize, we can update Eq. (3) as follows:

$$C(v,T) = \frac{1}{4} \sum_{\substack{x,y \in V(G) \\ \ell(x)<\ell(v)<\ell(y)}} \sum_{\substack{u \in V(G) \setminus (T \cup \{v\}) \\ \ell(u)=\ell(v)}} N_{G^+_{u,v}-(T \setminus \{x\})}(x,t') \cdot N_{G^-_{u,v}-(T \setminus \{y\})}(s',y).$$

It is easy that the above equation coincides with (3) in the special case $T = \emptyset$ and that $C(v,T)$ can still be found in polynomial time.

3.3   Analysis of the Algorithm

We start by providing a general lower bound to the size OPT of an optimal TS for an instance $\langle G, \mathcal{P} \rangle$ of minimum tracking set, where $\mathcal{P} = \{(s_1,t_1),\ldots,(s_k,t_k)\}$.

Lemma 3. OPT $\geq \log \sum_{i=1}^{k} |P(s_i,t_i)|$.

From this, it turns out that the following holds:

Theorem 1. *GreedyTracking is a polynomial-time $\sqrt{3n}$-approximation algorithm for the minimum TS problem when $\mathcal{P} = \{(s,t)\}$.*

3.4   Extension to the Multiple Pairs Case

Let $\langle G, \{(s_1,t_1),\ldots,(s_k,t_k)\}\rangle$ be an instance of the multiple-pair tracking path problem and let G_i be the directed acyclic graph obtained as the union of all the (directed) shortest paths from s_i to t_i in G.

Observation 1. *A (multiple-pair) tracking set for the instance $\langle G, \{(s_1,t_1),\ldots, (s_k,t_k)\}\rangle$ is a (single-pair) tracking set for each $\langle G_i, s_i, t_i \rangle$, where $i = 1,\ldots,k$.*

Our algorithm is split into two phases. In the first phase we select a tracking set T_1 that covers all the relevant cycles of each single-pair instance $\langle G_i, s_i, t_i \rangle$ (w.r.t. (s_i, t_i)), while in the second phase we focus on selecting a set $T_2 \subseteq V(G)$ covering a collection of polynomially many objects (called *unrelated paths* and *chromosomes* in the sequel), that collectively encode all the pairs of paths that are not yet covered by T_1.

Phase 1. We (implicitly) construct a set-cover instance $I = (\mathcal{I}, \mathcal{S})$ where $\mathcal{I}$ is the disjoint union of all relevant cycles in $\langle G_i, s_i, t_i \rangle$, for $i = 1, \ldots, k$ and $\mathcal{S}$ consists of one set S_v per vertex $v \in V(G)$ containing all the cycles of $\mathcal{I}$ that are covered by v. Given a partial tracking set $X \subseteq V(G)$, we can compute the number of *newly-covered* items by S_v as $\sum_{i=1}^{k} C_{G_i}(v, T)$. We name the set computed by the above greedy strategy T_1.

Lemma 4. $|T_1| \leq \sqrt{3n} \cdot \text{OPT}$.

Phase 2. Before describing phase 2, we find it useful to give some preliminary definitions.

Definition 1. *An* unrelated pair *of G is a pair (i, j) with $1 \leq i < j \leq k$ such that there exist two vertex-disjoint shortest paths $P_i \in P(s_i, t_i)$, $P_j \in P(s_j, t_j)$.*

We say that unrelated pair (i, j) is covered by a set $T \subseteq V(G)$ if for all vertex-disjoint paths $P_i \in P(s_i, t_i)$, $P_j \in P(s_j, t_j)$, we have $(V(P_i) \cup V(P_j)) \cap T \neq \emptyset$.

Lemma 5. *Let $T \subseteq V(G)$ be a set that covers all the relevant cycles in $\langle G_1, s_i, t_i \rangle$, for $i = 1, \ldots, k$. Given a pair (i, j), we can decide in polynomial time whether (i, j) is an uncovered unrelated pair. If this is the case, then there is a unique pair of paths $P_i \in P(s_i, t_i)$, $P_j \in P(s_j, t_j)$ that are vertex-disjoint and do not contain any vertex in T. The pair P_i, P_j can be found in polynomial time.*

Definition 2. *A* forward chromosome *(resp.* backward chromosome*) of G is a quadruple $\langle i, j, x, y \rangle$ with $1 \leq i < j \leq k$ such that:*

- *there are two shortest paths $P_i \in P(s_i, t_i)$ and $P_j \in P(s_j, t_j)$ (resp. $P_j \in P(t_j, s_j)$) that both traverse x and y, in order (possibly $x = y$).*
- *The subpaths $P_i[: x]$ and $P_j[: x]$ are vertex-disjoint, except for x. Notice that this includes the case $s_i = s_j = x$ (resp. $s_i = t_j = x$).*
- *The subpaths $P_i[y :]$ and $P_j[y :]$ are vertex-disjoint, except for y. Notice that this includes the case $t_i = t_j = y$ (resp. $t_i = s_j = y$).*

We say that forward or backward chromosome $\chi = \langle i, j, x, y \rangle$ is covered by $T \subseteq V$ if, for all the pairs of paths P_i, P_j from the above definition, the set $V(P_i[: x]) \cup V(P_j[: x]) \cup V(P_i[y :]) \cup V(P_j[y :]) \setminus \{x, y\}$ contains at least one vertex from T. It turns out that, if T is chosen as in Phase 1, then the above set is uniquely determined by χ, as the following lemma claims (see also Fig. 1).

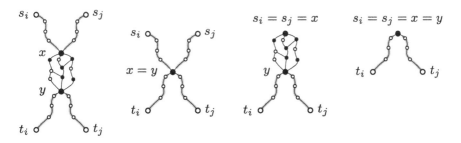

Fig. 1. Four different ways in which $P_i \in P(s_i, t_i)$ and $P_j \in P(s_j, t_j)$ can interact to form the forward chromosome $\chi = \langle i, j, x, y \rangle$. In this example, black vertices belong to the tracking T_1 computed in Phase 1, while the unique paths in $\Delta(\chi)$ are highlighted in gray, except for the vertices in $\{x, y\}$.

Lemma 6. *Let $T \subseteq V(G)$ be a set that covers all the relevant cycles in $\langle G_1, s_i, t_i \rangle$, for $i = 1, \ldots, k$. Given a quadruple $\chi = \langle i, j, x, y \rangle$, we can decide in polynomial time whether $\langle i, j, x, y \rangle$ is an uncovered (forward or backward) chromosome. If this is the case, then there exist a unique set $\Delta(\chi) = \{P_i[: x], P_j[: x], P_i[y :], P_j[y :]\}$ where P_i and P_j are as in Definition 2 and do not contain any vertex in $T \setminus \{x, y\}$. The set $\Delta(\chi)$ can be computed in polynomial time.*

Relevant cycles w.r.t. all (s_i, t_i), unrelated pairs, and (forward and backward) chromosomes provide a characterizations of the sets T that are also tracking sets, as shown in the following.

Lemma 7. *A set $T \subseteq V(G)$ is a tracking set for $\langle G, \{(s_1, t_1), \ldots, (s_k, t_k)\} \rangle$ if and only if all the following conditions are satisfied:*

(i) T covers all relevant cycles in $\langle G_i, s_i, t_i \rangle$, for all $i = 1, \ldots, k$; and
(ii) T covers all unrelated pairs of G; and
(iii) T covers all (forward and backward) chromosomes of G.

In Phase 2 of our approximation algorithm we consider the set $\mathcal{I}$ consisting of all the unrelated pairs and all (forward and backward) chromosomes of G that are uncovered by T_1. Since the number of unrelated pairs is at most k^2, and the number of chromosomes is at most $k^2 n^2$, the cardinality of $\mathcal{I}$ is $O(k^2 n^2)$. Moreover, thanks to Lemma 5 we can: (i) enumerate all the possible unrelated pairs (resp. chromosomes) of G a decide, in polynomial time, whether they are covered by T_1; and (ii) compute (in polynomial time) the set of vertices $v \in V(G)$ that would cause any such uncovered unrelated pair (resp. chromosome) to become covered. We define $\mathcal{S}$ as a collection containing one set S_v for each vertex $v \in V(G)$, where $S_v = \{x \in \mathcal{I} : v \text{ covers } x\}$.

Since, by Lemma 7, any multiple-pair tracking set must cover each element in $\mathcal{I}$ at least once, we have that the size of an optimal solution S^* to the set cover instance $(\mathcal{I}, \mathcal{S})$ is a lower bound to OPT. We then compute a $1 + \ln |\mathcal{I}| = O(\log n)$ approximation $\widetilde{S}$ of the optimal solution for $(\mathcal{I}, \mathcal{S})$ and select a partial tracking set $T_2 = \{v \in V : S_v \in \widetilde{S}\}$. Then, from Lemma 4, we have:

Theorem 2. $T_1 \cup T_2$ *is a* $O(\sqrt{n})$*-approximate multiple-pair tracking set.*

4   NP-hardness

In this section we prove that the multiple-pair minimum TS and minimum OTS problems are NP-hard even when G is a cubic planar graph and $\mathcal{P} = S \times D$, where $S, D \subseteq V(G)$ are the sets of *sources* and *destinations*, respectively.

Our reduction is from the vertex-cover problem on planar graphs with minimum degree 2 and maximum degree 3, which is known to be NP-hard [8]. An instance of (the decision version of) vertex-cover consists of a connected graph G' along with an integer k'. The goal is that of deciding whether there exists a subset $C \subseteq V(G')$ of at most k' vertices that covers all the edges, i.e., $\forall (u, v) \in E(G'), u \in C$ or $v \in C$.

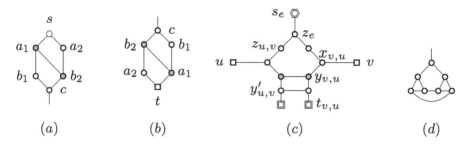

$$(a) \qquad\qquad (b) \qquad\qquad (c) \qquad\qquad (d)$$

Fig. 2. The gadgets used in the NP-hardness reduction: (a) and (b) respectively show the tracking source and tracking sink gadgets; (c) is the edge gadget corresponding to the edge $e = (u, v) \in E(G')$, where s_e is replaced by the tracking source gadget, while each of $t_{u,v}$ and $t_{v,u}$ is replaced by the tracking sink gadget; (d) is the dummy gadget that is used only to guarantee that the constructed graph G is cubic. Our construction guarantees the existence of an optimal tracking set that contains all gray vertices. Furthermore, every tracking set must contain at least one vertex in $\{u, x_{u,v}, z_{u,v}, v, x_{v,u}, z_{v,u}\}$ for every edge $(u, v) \in E(G')$.

Before describing how the graph G of our multiple-pair minimum TS/OTS instance is constructed, we first describe two useful gadgets, namely the *tracking source* and the *tracking sink* (see Fig. 2(a) and (b)). The tracking source gadget contains one source vertex (labeled s and depicted with an hexagonal vertex) and a vertex c that is used to connect the gadget to the rest of the graph. The tracking sink gadget is similar: it contains one destination vertex (labeled t and depicted with a square vertex) and a vertex c that is used to connect the gadget to the rest of the graph. In our reduction, we will use one tracking source gadget for each source in S and one tracking sink gadget for each of some destinations in D. Both gadgets are designed in such a way that every tracking set T must necessarily contain at least two of the vertices among a_1, a_2, b_2, and b_2, as otherwise there would exist two paths between s and c that traverse the same set of vertices in T (in the same order). Moreover, for the tracking source (resp., tracking sink) gadget, if T contains a_1 and b_2 then any path from s (resp., t) must traverse at least one tracked vertex of the gadget.

In the following we refer to a tracking source (resp. tracking sink) gadget as if it were a single vertex and will use two concentric hexagons (resp. two concentric squares) to distinguish it in the figures.

The graph G of our instance of minimum TS/OTS is obtained from G' by replacing each edge of $e = (u, v)$ with the *edge gadget* shown in Fig. 2(c). The set S consists of all the sources in the tracking source gadgets labeled s_e (with $e \in E(G')$), while the set D contains all the vertices in $V(G')$ plus all the destination in the tracking destination gadgets $t_{u,v}$ and $t_{v,u}$ (with $(u, v) \in E(G')$). Since G' is a planar cubic graph, the constructed graph G is also planar and the degree of each vertex is equal to either 2 or 3. We can append the *dummy gadget* (shown in Fig. 2(d)) to each vertex of degree 2 in order to obtain a new planar cubic graph G. It turns out that G' admits a vertex cover of size at most k' iff G admits a TS/OTS of size at most $k' + 8|E(G')|$, from which it follows:

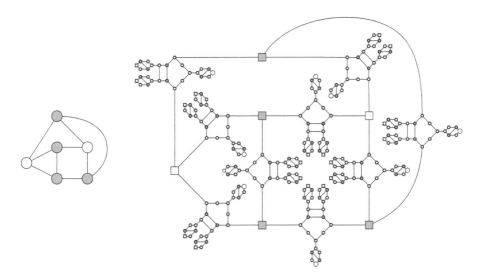

Fig. 3. An instance of vertex cover on a cubic planar graph G' (left) and its corresponding instance of minimum tracking set on graph G' (right). The set of source-destination pairs is $\mathcal{P} = S \times D$, where the vertices in S are represented as hexagons and the vertices in T are represented as squares. For the sake of simplicity, dummy gadgets are not shown. Gray vertices denote a vertex cover (left) and the corresponding tracking set (right).

Theorem 3. *The minimum TS/OTS problem is NP-hard for cubic planar graphs even when $\mathcal{P} = S \times D$.*

An example instance of vertex cover on a cubic planar graph G' and the corresponding planar graph G resulting from our reduction are shown in Fig. 3.

5   Exact Algorithm

In this section we design an exact algorithm that solves the problem for a single (s,t) pair. As before, we assume that G is the union of all the shortest paths from s to t in the input graph, and we call $\ell(v)$ the level of vertex v, i.e. the distance of v from s in G. Our algorithm requires time $O^*(2^{h^2})$, where h is an upper bound on the number of vertices of $V_i = \{v \in V(G) : \ell(v) = i\}$, for every $i = 0, \ldots, \ell(t)$.

The algorithm is based on dynamic programming and computes a tracking set for the pairs $V_i \times \{t\}$, for every i, in a bottom up fashion. However, as an additional input to the problem, we are given a *conflict graph* H that models the (unordered) pairs of distinct sources that are *indistinguishable*. Intuitively, two sources s_1 and s_2 of V_i are indistinguishable w.r.t. a tracking set for (s,t) if there are two shortest paths, one from s to s_1 and the other one from s to s_2, that both pass through the same subset of vertices of the tracking set. The idea of the algorithm is the following: if two sources s_1 and s_2 are indistinguishable, then we need to compute a tracking set for the pairs (s_1,t) and (s_2,t); if s_1 and s_2 are distinguishable, then it is enough to compute a set that is a tracking set for the pair (s_1,t) and a tracking set for the pair (s_2,t), but not necessarily a tracking set for both (s_1,t) and (s_2,t) at the same time![6]

We say that T is a tracking set w.r.t. i and H if the following conditions are satisfied: (i) T is a tracking set w.r.t. (s',t), for every $s' \in V_i$; and (ii) for every two distinct sources $s_1, s_2 \in V_i$ and every pair of paths $P_1 \in P(s_1,t)$ and $P_2 \in P(s_2,t)$, if $(s_1,s_2) \in E(H)$, then $V(P_1) \cap T \neq V(P_2) \cap T$.

The following lemma characterizes the solutions for the base cases.

Lemma 8. *Let H be a conflict graph defined on $V_{\ell(t)-1}$. Then T is a tracking set w.r.t. $\ell(t) - 1$ and H iff T is a vertex cover of H.*

Let $0 \leq i < \ell(t) - 1$ be fixed, let $S \subseteq V_i$ and let H be a conflict graph defined on V_i. We define the conflict graph $\text{CONFLICT}(S, H)$ for the level $i+1$ as follows: for any two distinct sources $s', s'' \in V_{i+1}$, $\text{CONFLICT}(S, H)$ has edge (s', s'') iff there are two edges $(s_1, s'), (s_2, s'') \in E(G)$, with $s_1, s_2 \in V_i$, such that either $s_1 = s_2$ or $((s_1, s_2) \in E(H)$ and $s_1, s_2 \notin S)$. Notice that, given a set of vertices T up to level i such that $T \cap V_i = S$ and H is conflict graph defined on V_i which contains the edge (s_1, s_2) iff s_1 and s_2 are indistinguishable w.r.t. T, we have that $\text{CONFLICT}(S, H)$ contains (s', s'') iff s' and s'' are indistinguishable w.r.t. T. The following key lemma shows the dependencies among T, H, and $\text{CONFLICT}(S, H)$.

Lemma 9. *Let $0 \leq i < \ell(t) - 1$, let H be a conflict graph defined on V_i, let T be a tracking set w.r.t. i and H, and let $S = V_i \cap T$. Then $T \setminus S$ is a tracking set w.r.t. $i+1$ and $\text{CONFLICT}(S, H)$. Furthermore, if T' is a tracking set w.r.t. $i+1$ and $\text{CONFLICT}(S, H)$, then $T' \cup S$ is also a tracking set w.r.t. i and H.*

---

[6] This means that there may be two paths, one in $P(s_1,t)$ and the other one in $P(s_2,t)$ that traverse the same subset of vertices of the tracking set.

The dynamic programming algorithm computes an optimum solution $\text{OPT}(i, H)$ for every i from $\ell(t) - 1$ downto 0 and for every possible conflict graph H defined on V_i. As proved in Lemma 8, $\text{OPT}(\ell(t) - 1, H)$ corresponds to a minimum vertex cover of H and can be computed in time $O(2^h h^2)$. For every other $i < \ell(t) - 2$, $\text{OPT}(i, H)$ is computed by guessing the vertices $S \subseteq V_i$ that are part of an optimal tracking set w.r.t. the pair (s, t) (all the 2^h possibilities are tried out), and by adding to S the vertices of $\text{OPT}(i + 1, \text{CONFLICT}(S, H))$. More precisely, thanks to Lemma 9, we have to select $\text{OPT}(i, H)$ so as

$$\big|\text{OPT}(i, H)\big| = \min_{S \subseteq V_i} \Big(|S| + \big|\text{OPT}(i + 1, \text{CONFLICT}(S, H))\big| \Big).$$

Since the number of possible conflict graphs per level is $2^{\binom{h}{2}} \leq 2^{h^2/2}$ and the time needed to find a solution $\text{OPT}(i, H)$ is $O^*(2^h)$, we have the following result.

Theorem 4. *Let $h = \max_{i=0,\dots,\ell(t)} \big|\{v \in V : \ell(v) = i\}\big|$. Then an optimum tracking set for (s, t) can be found in $O^*(2^{h^2})$ time.*

References

1. Bampas, E., Bilò, D., Drovandi, G., Gualà, L., Klasing, R., Proietti, G.: Network verification via routing table queries. J. Comput. Syst. Sci. **81**(1), 234–248 (2015)
2. Banik, A., Choudhary, P.: Fixed-parameter tractable algorithms for tracking set problems. In: Proceedings of the 4th International Conference on Algorithms and Discrete Applied Mathematics, CALDAM 2018, pp. 93–104 (2018)
3. Banik, A., Choudhary, P., Lokshtanov, D., Raman, V., Saurabh, S.: A polynomial sized kernel for tracking paths problem. In: Proceedings of the 13th Latin American Symposium on Theoretical Informatics, LATIN 2018, pp. 94–107 (2018)
4. Banik, A., Katz, M.J., Packer, E., Simakov, M.: Tracking paths. In: Proceedings of the 10th International Conference on Algorithms and Complexity, CIAC 2017, pp. 67–79 (2017)
5. Beerliova, Z., et al.: Network discovery and verification. IEEE J. Sel. Areas Commun. **24**(12), 2168–2181 (2006)
6. Harary, F., Melter, R.A.: On the metric dimension of a graph. Ars Combin. **2**(191–195), 1 (1976)
7. Khuller, S., Raghavachari, B., Rosenfeld, A.: Landmarks in graphs. Discrete Appl. Math. **70**(3), 217–229 (1996)
8. Lichtenstein, D.: Planar formulae and their uses. SIAM J. Comput. **11**(2), 329–343 (1982)

Positional Encoding by Robots
with Non-rigid Movements

Kaustav Bose⑩, Ranendu Adhikary$^{(\boxtimes)}$⑩, Manash Kumar Kundu⑩,
and Buddhadeb Sau

Department of Mathematics, Jadavpur University, Kolkata, India
{kaustavbose.rs,ranenduadhikary.rs,manashkrkundu.rs,
buddhadeb.sau}@jadavpuruniversity.in

Abstract. Consider a set of autonomous computational entities, called *robots*, operating inside a polygonal enclosure (possibly with holes), that have to perform some collaborative tasks. The boundary of the polygon obstructs both visibility and mobility of a robot. Since the polygon is initially unknown to the robots, the natural approach is to first explore and construct a map of the polygon. For this, the robots need an unlimited amount of persistent memory to store the snapshots taken from different points inside the polygon. However, it has been shown by Di Luna et al. [DISC 2017] that map construction can be done even by oblivious robots by employing a positional encoding strategy where a robot carefully positions itself inside the polygon to encode information in the binary representation of its distance from the closest polygon vertex. Of course, to execute this strategy, it is crucial for the robots to make accurate movements. In this paper, we address the question whether this technique can be implemented even when the movements of the robots are unpredictable in the sense that the robot can be stopped by the adversary during its movement before reaching its destination. However, there exists a constant $\delta > 0$, unknown to the robot, such that the robot can always reach its destination if it has to move by no more than δ amount. This model is known in literature as *non-rigid* movement. We give a partial answer to the question in the affirmative by presenting a map construction algorithm for robots with non-rigid movement, but having $O(1)$ bits of persistent memory and the ability to make circular moves.

Keywords: Autonomous robots · Map construction ·
Non-rigid movement · Polygon with holes ·
Look Compute-Move cycle · Distributed algorithm

1 Introduction

Distributed coordination of autonomous mobile robots has been extensively studied in literature in the last two decades. Fundamental problems like GATHERING [1,6,8,11], PATTERN FORMATION [3,5,12,13] etc., have been studied in the setting where the robots are deployed in the plane with infinite extent and without

© Springer Nature Switzerland AG 2019
K. Censor-Hillel and M. Flammini (Eds.): SIROCCO 2019, LNCS 11639, pp. 94–108, 2019.
https://doi.org/10.1007/978-3-030-24922-9_7

any obstacles. Recently in [10], MEETING, which is a simpler version of the GATHERING problem, has been investigated for robots inside a polygonal enclosure containing polygonal obstacles, where their boundaries limit both visibility and mobility of a robot. This setting models many real life scenarios like moping robots inside a room, robots employed in factories or an art gallery etc. To solve the various distributed problems in this model, the robots may have to first explore and construct a map of the environment. For this, the robots need an unlimited amount of persistent memory. However, in [10], it has been shown that map construction can be done even by oblivious robots with rigid movements, i.e., where a robot can accurately move by any distance. Their strategy is based on a positional encoding technique, where the robot carefully moves within the polygonal enclosure in such a way that their memory is implicitly encoded in its distance from the closest polygon vertex. In this paper, we show that this technique can be adapted to the non-rigid setting (where the movements of the robots can be interrupted by the adversary) as well, provided that the robot has a constant number of persistent bits and the ability to make circular moves.

2   Model and Definitions

Polygon. A *polygon* P is a non-empty, connected, and compact region in $\mathbb{R}^2$ whose boundary $\partial(P)$ is a set of finitely many disjoint simple closed polygonal chains. There is one connected component of $\partial(P)$, called the *external boundary*, which encloses all others (if any), which are called *holes*. Vertices and edges of a polygon can be defined in the standard way. $V(P)$ and $E(P)$ will respectively denote the set of vertices and edges of the polygon. For any two points $x, y \in P$, we say that x *and* y *are visible to each other* if the line segment joining them lies in P, i.e., $\overline{xy} \subset P$. We shall assume that there is some global coordinate system, with respect to which, the coordinates of the polygon vertices are algebraic numbers.

Robot. By a *robot*, we mean an anonymous mobile computational entity modeled as a dimensionless point inside P. A robot positioned at $x \in P$ can observe a point $y \in P$ if and only if x and y are visible to each other. The robot is endowed with $O(1)$ bits of persistent memory. This model is known in literature as FSTATE [9], where the internal state of the robot can assume a finite number of 'colors'. $\mathcal{S}$ will denote the set of all possible states of the robot. A robot, when active, operates according to the so-called *LOOK-COMPUTE-MOVE* cycle. In each cycle, a previously idle robot wakes up and executes the following steps. In the LOOK phase, the robot takes a snapshot of the region of P that it can currently see. The snapshot is expressed in the local coordinate system of the robot having the origin at its current position. In the COMPUTE phase, based on the snapshot and its internal state, the robot performs computations according to a deterministic algorithm to decide (1) a destination point $y \in P$, (2) a trajectory to y from its current location $x \in P$, which is either a straight line segment, or circular arc, (3) a state $s \in \mathcal{S}$. Then in the MOVE phase, the robot sets its internal state to s and moves towards the point y along the decided trajectory.

When a robot transitions from one LCM cycle to the next, all of its local memory (past computations and snapshots) are completely erased, and only its internal state is retained. Depending on whether or not the adversary can stop a robot before it reaches its computed destination, there are two movement models in literature, namely *rigid* and *non-rigid*, respectively. In the rigid model, a robot is always able to reach its desired destination without any interruption. In the case of non-rigid movements, there exists a constant $\delta > 0$, such that if the robot decides to move by an amount (path length) smaller than δ, then the robot will reach it; otherwise, it will move by at least δ amount. The value of δ is not known to the robot.

Geometric Definitions and Notations. Let v be any vertex of P, and u, w be its two adjacent vertices. We shall say that u *is the preceding vertex of v* and w *is the succeeding vertex of v* if one can reach from $\overline{vu}$ to $\overline{vw}$ by moving around v (staying inside P) in the counterclockwise direction (according to the sense of handedness of the robot). For any vertex $p_i \in V(P)$, unless mentioned otherwise, p_{i-1} and p_{i+1} will respectively denote the vertices preceding and succeeding p_i.

For a set $X = \{x_1, x_2, \ldots, x_n\}$ of distinct points in $\mathbb{R}^2$, $n \geq 2$, the *Voronoi region* of any $x_i \in X$, denoted by $Vor_X(x_i)$ or simply $Vor(x_i)$, is the set of all points in $\mathbb{R}^2$ which are closer to x_i than any other point in X, that is, $Vor_X(x_i) = \{y \in \mathbb{R}^2 \mid d(y, x_i) \leq d(y, x_j), \forall i \neq j\}$. Points shared by two Voronoi regions $Vor_X(x_i)$ and $Vor_X(x_j)$ constitute the *Voronoi edge* defined by x_i and x_j. Similarly, we can define Voronoi regions for a set $L = \{l_1, l_2, \ldots, l_n\}, n \geq 2$ of straight line segments (any two of which can intersect only at their endpoints). We will define

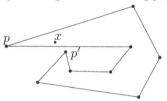

Fig. 1. The polygon vertex closest to x is p', but it is not visibile to x. Its closest visible vertex is p.

the Voronoi region of $l_i \in L$ as $LVor_L(l_i) = \{y \in \mathbb{R}^2 \mid d(y, l_i) \leq d(y, l_j), \forall i \neq j\}$ where $d(y, l_k) = Inf\{d(y, z) \mid z \in l_k\}$. In the context for our problem, there is a minor technical issue that needs to be addressed. For a polygon P, the polygon edge closest to a point $x \in P$ is of course visible to it. But the vertex closest to x may not be visible from x (See Fig. 1). In the remainder of the paper, unless mentioned otherwise, whenever we say 'closest vertex', it should be understood as 'closest visible vertex'. We will also define the *polygon Voronoi region* of a vertex p_i, denoted by $PVor_P(p_i)$, as the set of points $x \in P$ such that p_i is visible to x and p_i is closer to x than any other vertex visible from x. $Vor_{V(P)}(p_i)$ or $Vor_P(p_i)$ will denote the usual Voronoi region of p_i for the set $V(P)$.

For any point x, and any real number $r > 0$, $D(x, r)$ denotes the closed disc $\{y \in \mathbb{R}^2 \mid d(y, x) \leq r\}$. For any three points c, y, z such that $d(y, c) = d(z, c)$, we shall denote by $arc(y, z, c)$, the circular arc centered at c drawn from y to z in counterclockwise direction. Also, $arc(y, \theta, c)$ will denote the circular arc $arc(y, z, c)$ where $\angle ycz = \theta$. A point $x \in P$ is said to be *properly close* to $p_i \in V(P)$, if for any point $z \in arc(x, y, p_i)$, where $y \in \overline{p_i p_{i+1}}$ with $d(y, p_i) = d(x, p_i)$, the following holds: (1) $z \in PVor_{V(P)}(p_i)$ and (2) p_{i+1} is visible from z.

We can define a coordinate system by any ordered pair of distinct points in the polygon. The coordinate system defined by (u, v) will be the coordinate system with origin at u, $\overrightarrow{uv}$ as the positive X-axis, $d(u, v)$ as the unit distance and the positive Y-axis according to the chirality or handedness of the robot.

3   A Brief Overview of the Positional Encoding Technique

Computational Model. We assume that each robot internally runs a *Blum-Shub-Smale machine* [2] extended with a square-root primitive. A Blum-Shub-Smale machine is a random-access machine whose registers can store arbitrary real numbers and can operate directly on them. Its computational primitives are the four basic arithmetic operations on real numbers, and it can test whether a real number is positive. Depending on the application, it is also customary to extend the basic model with additional primitives, such as root extractions, trigonometric functions, etc. In our case, we only require the square-root primitive that will be needed in geometric computations.

Encoding Algebraic Reals. Consider an algebraic real number α. The minimal polynomial of α over $\mathbb{Q}$ is the unique monic polynomial in $\mathbb{Q}[x]$ of least degree which has α as a root. Let $\mathfrak{m}(x) = x^n + a_{n-1}x^{n-1} + \ldots + a_1 x + a_0 \in \mathbb{Q}[x]$ be the minimal polynomial of α over $\mathbb{Q}$. Now $\mathfrak{m}$ has n complex roots. However, the real roots can be arranged in ascending order. So, let α be the ith real root of $\mathfrak{m}$. Then α can be uniquely represented by $(n, i, a_{n-1}, \ldots, a_0)$. Now any rational number $(-1)^s \frac{p}{q}$, with $p, q > 0$, $s \in \{0, 1\}$, can represented as a 3-tuple of non-negative integers as $(s, p, q) \in \mathbb{Z}_{\geq 0}^3$. Thus α can be represented by an array of $3n+2$ non-negative integers. We can represent each non-negative integer m as the bit string $0^m 1$. Let us denote by $\beta(\alpha)$, the bit string obtained by concatenating the bit strings of the $3n+2$ non-negative integers. Now for any non-negative integer λ, let $r(\alpha, \lambda) < 1$ be the real number whose (usual) binary representation is $0.0^\lambda 1\beta(\alpha)$. We shall say that $r(\alpha, \lambda)$ *encodes* α.

Lemma 1. *If $0 < d < 1$ be a real number such that $d = r(\alpha, \lambda)$, for some algebraic real α and non-negative integer λ, then $\frac{d}{2} = r(\alpha, \lambda + 1)$. Therefore, $\frac{d}{2^k} = r(\alpha, \lambda + k)$, for any integer $k \geq 1$.*

Computing the Code. Suppose a basic Blum-Shub-Smale machine has an algebraic number α stored in its register and it has to construct its code $\beta(\alpha)$. The machine will generate all finite sequences of bits in lexicographic order. For each sequence, it will check if it is a well-formed code of an algebraic number; if it is, it will extract the coefficients of the polynomial $\mathfrak{q}$ from it. Then it computes $\mathfrak{q}(\alpha)$. Since α is algebraic, eventually a polynomial $\mathfrak{q}$ is found such that $\mathfrak{q}(\alpha) = 0$. Since $\mathfrak{q}$ must be a multiple of the minimal polynomial $\mathfrak{m}$ of α, we can determine it by finding its irreducible factor that has α as a root. Then Sturm's theorem [7] can be applied to find out how many real roots of the minimal polynomial are smaller than α. Thus we have obtained all that are required to encode α.

Computations on the Implicit Form. Once a number is encoded in this form, we cannot necessarily retrieve it in finite time. But we can approximate it arbitrarily well, for instance via Sturm's theorem. However, we can do Turing-computable bit manipulations on this implicit form to compute all kinds of common functions (e.g. basic arithmetic operations, root extractions of any degree etc.) on the algebraic number without decoding its explicit form.

Encoding Snapshots. A snapshot taken by a robot contains the visible portion of the polygon P, which is basically a union of line segments, each of which being a sub-segment of an edge of P. So, a snapshot can be represented as an array of real numbers, say $S = (x_1, y_1, x_1', y_1', x_2, y_2, x_2', y_2', \ldots)$, where (x_i, y_i) and (x_i', y_i') are the endpoints of the ith visible segment of $\partial(P)$. Note that none of these points is necessarily a vertex of P. We have discussed how to compute the code of a single algebraic number. Now we describe how we can encode a snapshot of P with algebraic vertices taken from a point $x \in P$. The vertices of P have algebraic coordinate with respect to some global coordinate system. Of course, the vertices may not have algebraic coordinates in the local coordinate system of the robot. Let Φ_x be the transformation from the global coordinate system to the local coordinate system of the robot. Note that x is not necessarily an algebraic point, and the parameters of Φ_x are not necessarily algebraic numbers either. Therefore, the coordinates and the distances between vertices of $\Phi_x(P)$ may not be algebraic. However, all the ratios of the distances are algebraic, as Φ_x, being a similarity transformation, preserves ratios between segment lengths. Then it follows that if the robot picks two visible vertices of $\Phi_x(P)$, say v and v', and transforms all the visible vertices of $\Phi_x(P)$ in the coordinate system (v, v'), then they will have algebraic coordinates. Then they can be encoded by a basic Blum-Shub-Smale machine as we discussed earlier. However, recall that a snapshot taken from x may not contain only vertices of $\Phi_x(P)$. We can identify the potentially non-vertex endpoints by a basic Blum-Shub-Smale machine, as a non-vertex point $(x_j, y_j) \in S$ is necessarily of the form $(x_j, y_j) = c(x_i, y_i), c > 1$ for some visible polygon vertex (x_i, y_i). These potentially non-vertex endpoints will be simply marked with an 'undefined' flag in the snapshot. The robot will pick two 'defined' points in the snapshot for the coordinate transformation. The coordinates of the 'defined' points of S will be transformed as discussed earlier, and each 'undefined' point will be simply replaced with a $(0, 0)$ or any algebraic point of our choice along with the 'undefined' flag. Then these coordinates can be encoded into a finite bit string, and then they can be concatenated into a single code for the entire snapshot. We can similarly encode multiple snapshots into a single bit string. Along with the snapshots, we can also pack as many other finitely described elements as we want.

Positional Encoding. Suppose that β is the code or bit string of the information that the robot wants to encode. Let d be a real number that encodes it, i.e., the binary representation of d is $0.0^\lambda 1\beta(\alpha)$ for some non-negative integer λ. The robot will encode the information by positioning itself in the polygon in such a way that its distance from the closest polygon vertex is d (according to its local coordinate system). From Lemma 1, it follows that the robot can encode the same information by placing itself at a distance $\frac{d}{2^k}$ from the vertex for any integer $k \geq 1$. This 'scalability' property allows the robot to get arbitrarily close to the vertex without losing information.

4   The Algorithm

In [10], the memory of a robot is encoded in the distance from its closest polygon vertex. Obviously, the robot needs rigid movements to accurately position itself at a point whose distance from the particular vertex correctly encodes the memory. In the non-rigid setting, we need some additional options where we can encode our memory. In particular, apart from the distance from some particular vertex, we shall also encode the memory in the tangent of the angle that the robot makes with an edge or a diagonal, at some vertex. In the remainder of the paper, whenever we say that the memory is encoded in some angle α, it is to be understood that the memory is encoded by the real number $tan(\alpha)$. Notice that since $tan(\alpha)$ monotonically tends to 0, as $\alpha < \frac{\pi}{2}$ tends to 0, we can use the scalability property of the encoding scheme to encode the memory in an angle as small as we want. The persistent bits or the internal states are used so that each time a robot wakes up, it knows 'where' its memory is encoded and which coordinate system the snapshots in the memory are expressed in. In each case, the robot also sets a particular polygon vertex, that is visible to it, as its *virtual vertex*. A summary of this is provided in Table 1.

Our map construction algorithm is similar to the one presented in [10]. The robot will keep exploring new vertices (but not touching it), and near each vertex, it will take a new snapshot and encode it, merging with the old snapshots. As it explores, it keeps track of the vertices that it has seen but not yet visited. Whenever it reaches a new connected component of the boundary, it explores it entirely in the counterclockwise direction (i.e., by moving from a vertex to its succeeding vertex). After exploring a connected component for the first time, it will take a second tour of it, in the same direction. After completely exploring a previously unexplored connected component, it will choose an unvisited vertex of a different component and move to it via a suitable path. The robot repeats this until there are no unvisited vertices recorded in its encoded memory. Implementation of this strategy in the non-rigid setting is based on four basic techniques. A brief overview of these techniques are presented in Sect. 4.1. From there follows the main result of the paper presented in Theorem 1. We refer the readers to the full version [4] of the paper for further details.

Theorem 1. *In* FSTATE, *a robot inside a polygon P with non-rigid movements can correctly construct and encode a map of the polygon in finite time.*

Table 1. The virtual vertex and encoded memory of the robot, corresponding to its internal state.

For any robot r at a point x inside the polygon P			
State	Virtual vertex	Memory	
		Encoded in	Coordinate system
s_1	p_i = the closest visible vertex	$d(x, p_i)$	(p_i, p_{i+1})
s_2	p_i = the closest visible vertex	$d(x, p_i)$	(p_{i-1}, p_i)
s_3	p_a = the nearer endpoint of the closest boundary segment, say $\overline{p_i p_{i+1}}$, $a \in \{i, i+1\}$	$tan(\angle xp_a p_b)$, where p_b is the other endpoint of $\overline{p_i p_{i+1}}$	(p_i, p_{i+1})
s_4	p_i = the closest visible vertex	$tan(\angle xp_i p_{i-1} - \frac{\pi}{2})$	(p_{i-1}, p_i)
s_5	p_i = the closest visible vertex	$tan(\pi - \angle xp_i p_{i+1})$	(p_i, p_{i+1})
s_6	p_i = the closest visible vertex	$tan(\angle xp_i O)$, where $p_i O$ is the angle bisector of $\angle p_{i-1} p_i p_{i+1}$	(p_{i-1}, p_i)
s_7	p_i = the closest visible vertex	$tan(\angle xp_i p_j)$, where either x lies on the interior of the Voronoi edge $PVor(p_i) \cap PVor(p_j)$ or $\overrightarrow{p_i x}$ intersects $PVor(p_i) \cap PVor(p_j)$ first	(p_i, p_j) or (p_j, p_i)

4.1 Four Basic Techniques

Moving from One Virtual Vertex to Another in the Same Connected Component of the Boundary

Suppose that p_i is the virtual vertex of the robot r with internal state s_1 (i.e., p_i is the vertex closest to r), and it has to approach the succeeding vertex p_{i+1}. If r had rigid movements, it could have simply moved to a point suitably close to p_{i+1} in one go, without any interruption. But since r has non-rigid movements, it can be stopped multiple times during its journey. Now consider the situation shown in Fig. 2a. To move towards p_{i+1} via any path, the robot has to pass through the Voronoi region of p_j. Hence, if r is stopped by the adversary while it is in the interior of $PVor_{V(P)}(p_j)$, it will set p_j as its virtual vertex. To resolve this, the robot will change its state to s_3 before moving. When its state is s_3, to set the virtual vertex, it considers the closest boundary segment, instead of the closest vertex. The endpoint of its closest boundary segment that is closer to it, is set as the virtual vertex. In case of a tie, any one of the endpoints can be chosen as the virtual vertex. The robot will move along a path as shown in Fig. 2b. Such a path can be defined by a tuple (p_i, p_{i+1}, α), where the path consists of two linear segments $\overline{p_i q}$ and $\overline{q p_{i+1}}$ of equal length with $\angle q p_i p_{i+1} = \angle q p_{i+1} p_i = \alpha$ and q lying on the perpendicular bisector of $\overline{p_i p_{i+1}}$. We shall denote the path as $\mathcal{P}(p_i, p_{i+1}, \alpha)$. The path should be chosen in such a way that any point on the path is closer to the boundary segment $\overline{p_i p_{i+1}}$ than any other point of $\partial(P)$. In other words, $\mathcal{P}(p_i, p_{i+1}, \alpha)$ should be inside $LVor_{E(P)}(\overline{p_i p_{i+1}})$.

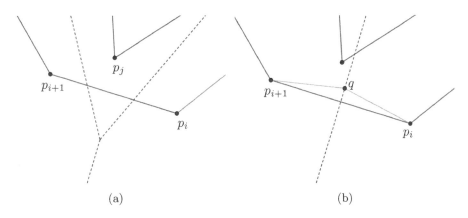

(a)   (b)

Fig. 2. (a) If a robot moves from p_i towards p_{i+1}, it has to pass through the Voronoi region of p_j. (b) The robot will move along the path $\mathcal{P}(p_i, p_{i+1}, \alpha)$ drawn in green. (Color figure online)

Now let us describe our strategy more formally. Suppose that a robot r is at a point x inside the polygon P, such that the following are true: (A1) $r.state = s_1$, (A2) x is properly close to the vertex p_i. Since $r.state = s_1$, p_i is the virtual vertex of r, and its memory is encoded in the distance $d(x, p_i)$ and expressed in the coordinate system defined by (p_i, p_{i+1}). Since r is properly close to p_i, if r moves around p_i along a circular arc in counterclockwise direction (i.e., keeping its distance from p_i fixed), p_i will remain its virtual vertex and also, all of $\overline{p_i p_{i+1}}$ will remain visible to it. So, r will move around p_i in counterclockwise direction to move to a point x' such that the following conditions are satisfied: (B1) the data encoded by $\alpha = \angle x' p_i p_{i+1}$ is same as the data encoded by $d(x', p_i) = d(x, p_i)$, both expressed in the coordinate system (p_i, p_{i+1}), (B2) the path $\mathcal{P}(p_i, p_{i+1}, \alpha)$ is inside $LVor_{E(P)}(\overline{p_i p_{i+1}})$. After reaching such a point x', r will change its state to s_3. It will then follow the path $\mathcal{P}(p_i, p_{i+1}, \alpha)$, where $\alpha = \angle x' p_i p_{i+1}$, to move towards p_{i+1}. However, we have not yet specified how close r should get to p_{i+1}. Our objective is to get close to p_{i+1}, take a new snapshot and encode the new snapshot (merged with the older ones) in its distance from p_{i+1}. We want these snapshots to be expressed in the coordinate system defined by (p_{i+1}, p_{i+2}), where p_{i+2} is the vertex succeeding p_{i+1}. But in order to do that, p_{i+2} should be visible to the robot. Notice that if some portion of $\overline{p_{i+1} p_{i+2}} \setminus \{p_{i+1}\}$ is visible to r, then it will be able to see all points of $\overline{p_{i+1} p_{i+2}}$ if it goes close enough to p_{i+1}.

However, if $\overline{p_{i+1}p_{i+2}} \setminus \{p_{i+1}\}$ is completely invisible to r, the segment $\overline{p_{i+1}p_{i+2}}$ will never be completely visible to it, no matter how close it gets to p_{i+1}. In this section, we will only discuss the first case. The later case is more complex and will be discussed in the next section.

So, consider the case where some portion of $\overline{p_{i+1}p_{i+2}} \setminus \{p_{i+1}\}$ is visible to r. In this case, r will move to a point x'' that is close enough to p_{i+1} so that the following conditions are satisfied: (C1) x'' is properly close to p_{i+1}, (C2) $d(x'', p_{i+1})$ is encoding the old snapshots merged with its current view (newly discovered vertices), all expressed in the coordinate system defined by (p_{i+1}, p_{i+2}). The robot will first move close enough to p_{i+1}, say at x''', so that the first condition is satisfied (See Fig. 3), i.e., x''' is properly close to p_{i+1}. Then the robot decides to further move towards p_{i+1} to a suitable point x'' in order to fulfill the last condition. There are two ways it may fail to achieve this. First, if $d(x'', x''') > \delta$, the adversary can stop it at some point x'''' in between. However, the old snapshots are still available as it is encoded in $\angle x'''' p_{i+1} p_i = \angle x''' p_{i+1} p_i$. So, r can identify that it has failed to reach its destination. Then it will recompute the destination and move towards it. Secondly, even if it reaches x'', a new vertex may be discovered which is not present in the data encoded in $d(p_{i+1}, x'')$. Therefore, r will again recompute a destination so that the newly discovered vertices are encoded (along with the old data). From the existence of $\delta > 0$ and the fact that the polygon has finitely many vertices, it follows that r can eventually reach a point x'' where it finds that $d(x'', p_{i+1})$ encodes precisely the data encoded by $\angle x'' p_{i+1} p_i$, merged with the new vertices of the polygon that are visible from x''. Observe that the visibility of both $\overline{p_{i+1}p_i}$ and $\overline{p_{i+1}p_{i+2}}$ are crucial at any point during this process. This is because the robot has to transform the data encoded in $\angle x'' p_{i+1} p_i$ from the coordinate system (p_i, p_{i+1}) to (p_{i+1}, p_{i+2}). When all three p_i, p_{i+1}, p_{i+2} are visible, the robot knows their exact positions and hence, it can perform this conversion, which is computable by a rational function, on (the implicit form of) the old snapshots. When the conditions C1, C2 are achieved, r will change its state to s_1. Clearly we are back to the situation where A1, A2 holds (p_i to be replaced with p_{i+1}), and hence r can now move to p_{i+1} in the same manner.

Discovering the Succeeding Vertex and Encoding a New Snapshot
Now consider the case where $\overline{p_{i+1}p_{i+2}} \setminus \{p_{i+1}\}$ is completely invisible to r (See Fig. 4). This is possible only if $\angle p_i p_{i+1} p_{i+2} > \pi$. Then no matter how close r gets to p_{i+1}, $\overline{p_{i+1}p_{i+2}} \setminus \{p_{i+1}\}$ will remain completely invisible to it. In this case, r will move to a point x'' that is close enough to p_{i+1}, such that the following conditions are satisfied. (D1) $d(x'', p_{i+1}) = d$ should encode all the old data encoded by $\angle x'' p_{i+1} p_i$ (both) expressed in the coordinate system (p_i, p_{i+1}). (D2) Let S be the semicircular disc of radius $2d$, centered at p_{i+1} and having diameter along the line $\overleftrightarrow{p_i p_{i+1}}$. Then S should not intersect with any portion of $\partial(P)$ except $\overline{p_i p_{i+1}}$. (D3) Every point on $\overline{p_i p_{i+1}}$ should be visible from every point on $arc(u, \pi, p_{i+1})$, where $u \in \overline{p_i p_{i+1}}$ with $d(u, p_{i+1}) = d$. When these conditions are satisfied, the robot will change its state to s_2. Clearly, p_{i+1} is its virtual vertex. Let y be a point on the line through p_{i+1} and perpendicular

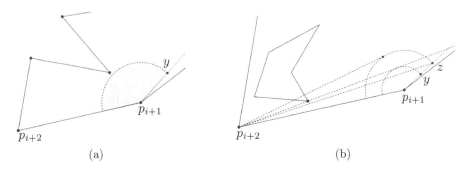

(a)   (b)

Fig. 3. (a) The shaded circular sector of radius $d = d(p_{i+1}, y)$ intersects no vertex other than p_{i+1}. Any point on $\overrightarrow{p_{i+1}y}$ less than $\frac{d}{2}$ distance away from p_{i+1} satisfies the first condition of proper closeness to p_{i+1}. (b) Any point on the interior of $\overline{p_{i+1}y}$ satisfies the second condition of proper closeness to p_{i+1}.

to $\overline{p_i p_{i+1}}$, with $d(y, p_{i+1}) = d(x'', p_{i+1}) = d$. The robot will then move to the point y along $arc(x'', y, p_{i+1})$. It implies from condition D2 that as r traverses along this arc (where it can be stopped several times by the adversary), p_{i+1} will remain its virtual vertex. Upon reaching the point y, $\overline{p_{i+1} p_{i+2}} \setminus \{p_{i+1}\}$ may still be completely invisible. In that case, r will have to move further along a circular arc and place itself on the extension of the segment $\overline{p_i p_{i+1}}$. But if r revolves with the same radius, its virtual vertex may change. Therefore it has to first reduce its distance from p_{i+1}. But recall that its distance from p_{i+1} is encoding its memory and hence, the data will be lost if this distance is changed. Therefore, before changing its distance from p_{i+1}, it will encode the data 'somewhere' else, such that it is preserved while it moves towards p_{i+1}. Notice that although moving around p_{i+1} with the same radius can change its virtual vertex, it can still move by a small enough angle without changing its virtual vertex. From its view from y, it can compute a point y'', such that the following conditions are satisfied: (E1) $d(p_{i+1}, y'') = d(p_{i+1}, y) = d$, (E2) $D(y'', d) \cap \partial(P) = \{p_{i+1}\}$, (E3) $\angle y'' p_{i+1} y < \frac{\pi}{2}$ encodes the same data encoded by d, both expressed in the coordinate system (p_i, p_{i+1}). Now r will first move to y'' along a circular arc and then change its state to s_4. Then it will reduce its distance from p_{i+1} to d', so that d' satisfies the following conditions. (F1) d' encodes the same data encoded by $\angle y'' p_{i+1} y$ both expressed in the coordinate system defined by $(p_i p_{i+1})$. (F2) Let z be the point on the extension of the segment $\overline{p_i p_{i+1}}$ with $d(z, p_{i+1}) = d'$. Then $D(z, d') \cap \partial(P) = \{p_{i+1}\}$. When these conditions are satisfied, it will change its state to s_2. Now r will move to z by moving around p_{i+1} in counterclockwise direction maintaining the distance d' from it. Upon reaching z, it can see at least some portion of $\overline{p_{i+1} p_{i+2}} \setminus \{p_{i+1}\}$. Suppose that it still can not see p_{i+2}. But since it can see some portion of $\overline{p_{i+1} p_{i+2}} \setminus \{p_{i+1}\}$, it can compute the point z' on the extension of the segment $\overline{p_{i+2} p_{i+1}}$ with $d(z', p_{i+1}) = d'$. Now r will move around p_{i+1} in clockwise direction towards z', but not touching it (say by choosing the middle of $arc(z', z, p_{i+1})$ as its destination and so on). Eventually, it will be able to see p_{i+2}. In fact, it can see both $\overline{p_{i+1}p_{i+2}}$ and $\overline{p_i p_{i+1}}$ entirely. Now r has to encode a new snapshot (merged with the old ones) in its distance from p_{i+1}. Before that it will encode its memory in the angle that it makes with the extension of $\overline{p_{i+2}p_{i+1}}$ at p_{i+1} by revolving further towards z', and then will change its state to s_5. Then it will move towards p_{i+1} so that conditions C1, C2 are satisfied. When they are achieved, r will change its state to s_1.

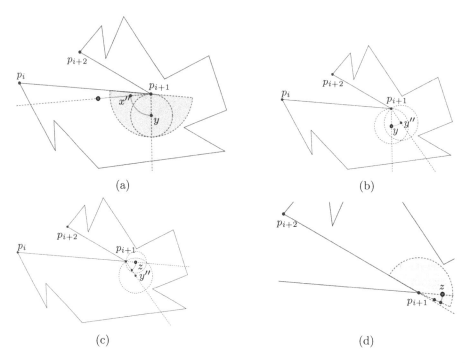

(a)

(b)

(c)

(d)

Fig. 4. The robot moving around p_{i+1} to discover the succeeding Vertex and encode a new snapshot. The trail of the robot is shown in blue. (Color figure online)

Taking a Second Tour of a Connected Component of the Boundary

From the two techniques discussed, it is clear how a robot can 'visit' all the vertices of a previously unexplored connected component C of $\partial(P)$. Also, whenever r encodes a new snapshot, it marks the position of its current virtual vertex with a 'visited once' flag. Upon completing its first tour of C, it will start a second tour of C in the same direction. In the second tour, the points from where the snapshots are taken, should constitute an 'approximation' of C, say $\overline{C}$, such that the closed polygonal curve $\overline{C}$ (1) does not self-intersect, (2) does not intersect $\partial(P)$, and (3) does not intersect any other previous approximations. This will ensure that eventually all polygon vertices are discovered (See [4]). Suppose that C is composed of m vertices $p_1, \ldots, p_m$. Assume that r has started exploring C from (close to) p_1. As described earlier, it will sequentially visit all the vertices and eventually arrive at a point close to p_m, from where p_1 is visible. It can clearly identify p_1 to be a previously visited vertex and will decide to start the second tour. Now clearly r has a full picture of C. So it can compute a distance d implicitly and include it in its memory, so that d has the following property. Let $\tilde{C}(d) = \{p'_1, \ldots, p'_m\}$ denote the approximation of $C = \{p_1, \ldots, p_m\}$ such that each $\overrightarrow{p'_i p'_{i+1}}$ is parallel to $\overline{p_i p_{i+1}}$ (p_{m+1} is to be understood as p_1) and the separation between them is d (See Fig. 5). Then d should be small enough, so that the approximation $\tilde{C}(d)$ satisfies all the three requirements. The points from where the robot will take snapshots during its second tour, will constitute an approximation $\overline{C} = \{p_1^1, p_1^2, \ldots p_m^1, p_m^2\}$ consisting of $2m$ points, with $\overline{C}$ lying in the region between C and $\tilde{C}(d)$.

We shall now discuss the procedure in detail. The robot will approach p_1 (with state s_3) in the same manner as described previously, but with an extra requirement that the path it follows should be lying in the region between C and $\tilde{C}(\frac{d}{2})$. Note that although d is computed in the implicit form, r can get an approximation of d in explicit form that is smaller than the actual value. First consider the case where $\angle p_m p_1 p_2$ is not reflex. Similar to the first tour, r goes to a point x so that the conditions C1 and C2 are satisfied (with $p_i = p_m, p_{i+1} = p_1, p_{i+2} = p_2$). We can refer to this in short by simply saying that 'r takes a snapshot at x'. The extra requirement in this case would be that $d(x, p_1) < \frac{d}{2}$. After this, r will change its state to s_1. Now r will move around p_1 to reach a point x' so that the condition B2 (with $p_i = p_1, p_{i+1} = p_2$) is satisfied, plus $\angle x' p_1 p_2$ should encode the view from x' merged with the older snapshots (encoded by $d(x', p_1)$) expressed in the coordinate system (p_1, p_2). Again using similar phrasing, we shall refer to this by saying 'r takes a snapshot at x''. Let us denote the points x and x' by p_1^1 and p_1^2. Note that our constructions ensure that the line segment $\overline{p_1^1 p_1^2}$ is inside the region C and $\tilde{C}(d)$. Now consider the case where $\angle p_m p_1 p_2$ is reflex. The robot will go to a point x that is close enough to p_1, such that the following conditions are satisfied: (G1) $d(x, p_1)$ encodes the old snapshots (encoded in $\angle x p_1 p_m$) expressed in the coordinate system defined by (p_m, p_1), (G2) $d(x, p_1) < \frac{d}{2}$. After reaching such a point x, r will change its state to s_2. Let $\overrightarrow{p_1 A}$ and $\overrightarrow{p_1 B}$ be the extensions of the segments $\overline{p_2 p_1}$ and $\overline{p_m p_1}$ respectively. Let $\overrightarrow{p_1 O}$ be the angular bisector of the angle $\angle A p_1 B$. Now r can move around p_1 to place itself at a point p_1^1 between the lines $\overrightarrow{p_1 A}$ and $\overrightarrow{p_1 O}$ such that the angle $\angle p_1^1 p_1 O$ encodes the view from p_1^1, merged with the older snapshots, all expressed in the coordinate system defined by (p_m, p_1). In other words, r takes a snapshot at p_1^1. Then r will change its state to s_6, move towards p_1 to encode the data in its distance from p_1, again change its state to s_2 and move around p_1 to take a snapshot at a point p_1^2 between the lines $\overrightarrow{p_1 O}$ and $\overrightarrow{p_1 B}$ encoding the snapshot (merged with the old ones) in the angle $\angle p_1^2 p_1 O$ expressed in the coordinate system defined by (p_m, p_1). Then r will again change its state to s_6 and move towards p_1 to encode the data in its distance from p_1, this time expressed in the coordinate system (p_1, p_2). After this, it will change its state to s_1. Continuing in this manner, the robot will revisit all the vertices of the component, and take snapshots at p_i^1 and p_i^2, near each vertex p_i. The polygonal chain $\overline{C}$ clearly satisfies all three desired properties.

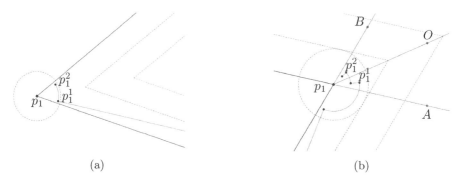

Fig. 5. The robot taking a second tour. The trail of the robot is shown in blue. The approximations $\tilde{C}(d)$ and $\tilde{C}(\frac{d}{2})$ are shown in pink and grey dotted lines respectively. (Color figure online)

Moving from One Connected Component to Another

A robot will move from a virtual vertex p_i to a vertex p_j belonging to a different connected component of $\partial(P)$ only if $\overline{p_i p_j} \subset PVor_P(p_i) \cup PVor_P(p_j)$. The robot r with state s_3 will approach p_i and encode its memory in its distance from p_i, expressed in the coordinate system defined by (p_{i-1}, p_i). The robot will then change its state to s_2. Note that p_j may not even be visible from its current position if $\angle p_{i-1} p_i p_{i+1}$ is reflex. If $\angle p_{i-1} p_i p_{i+1}$ is reflex and p_j lies in the open half-plane delimited by $\overleftrightarrow{p_{i-1} p_i}$ containing p_{i+1}, it will have to encode its memory in the coordinate system (p_i, p_{i+1}) by previously discussed techniques. It will then change its state to s_1. From its memory, it knows that the plan is to move to p_j. It will then move around p_i to move to a point x so that the following conditions are satisfied. (H1) The ray $\overrightarrow{p_i x}$ intersects the interior of the Voronoi edge $Vor_{S(x)}(p_i) \cap Vor_{S(x)}(p_j)$, where $S(x)$ denotes the polygon vertices visible from x. Suppose that the ray intersects the Voronoi edge $Vor_{S(x)}(p_i) \cap Vor_{S(x)}(p_j)$ at point A. (H2) The angle $\alpha = \angle x p_i p_j$ encodes its memory. All coordinates of the snapshots are expressed in the coordinate system defined by (p_i, p_j). The encoding will also contain a rational approximation of $\frac{1}{2}(p_i - p_j)$ expressed in the local coordinate system of r. The robot will then change its state to s_7, move along $\mathcal{P}(p_i, p_j, \alpha)$ towards p_j, i.e., it will first move to A and then to a point properly close to p_j. Consider the situation when r stops at a point z on the path $\mathcal{P}(p_i, p_j, \alpha)$. When r was at x, it verified H1 by checking from its snapshot that the disc $D(A, d)$ contains no polygon vertex other than p_i, p_j, where $d = d(A, p_i) = d(A, p_j)$. It implies from this that $\mathcal{P}(p_i, p_j, \alpha) \subset PVor_P(p_i) \cup PVor_P(p_j)$. Hence, at z, its closest visible vertex, and hence its virtual vertex, is either p_i or p_j. It computes intersections between the ray from its virtual vertex, passing through it, and the perpendicular bisectors of the lines joining its virtual vertex and other visible vertices; and then checks if the intersection point is on the corresponding Voronoi edge. It will find that the ray intersects the Voronoi edge defined by p_i and p_j first. However, r does not immediately know whether it is moving from p_i to p_j, or from p_j to p_i. However, it knows that its memory is encoded in the angle it makes with $\overline{p_i p_j}$ at its virtual vertex $\in \{p_i, p_j\}$. But r does not to know if it is encoded with the coordinate system (p_i, p_j) or (p_j, p_i). However, recall that the memory contains a rational approximation of $\frac{1}{2}(p_i - p_j)$, call it w, expressed in its local coordinate system. Now r computes $w + \frac{1}{2}(p_i + p_j)$, which gives an approximation of p_i. from which r determines that it is moving away from p_i, and also the fact that its encoded memory is expressed in (p_i, p_j). So, eventually it will move to a point properly close to p_j so that its distance from p_j encodes its memory expressed in either (p_j, p_{j+1}) or (p_{j-1}, p_j), and then change its state to s_1 or s_2 accordingly.

5   Conclusion

In this work, we have shown how a finite state robot with non-rigid movements can construct the map of a polygon by a positional encoding strategy. The techniques developed here, give a general movement strategy for finite state robots with non-rigid movements, to move about in the polygon, without losing its encoded memory. The map construction algorithm can be used as a subroutine to solve distributed algorithms for mobile robot systems under this model, where the knowledge of the polygon may be required. For instance, consider the GATH-ERING problem, where a set of autonomous, anonymous, asynchronous finite state mobile robots with no agreement in coordinate system and no communication capabilities, have to meet at some point in the polygon. Assume that the polygon is asymmetric. Then each robot will first construct and encode the map of the polygon. Since the polygon is asymmetric, the robots can deterministically pick a polygon vertex as their meeting point. Then using our techniques, the robots can move to that vertex. However, when the polygon is not asymmetric, GATHERING appears to be challenging even for robots with unlimited memory. For symmetric polygons, we can consider the relaxed version of GATHERING, called MEETING, where any two of the robots have to become mutually aware by seeing each other at their LOOK phases. Using our techniques, a patrolling strategy similar to [10] can be adapted to our setting to solve MEETING.

It would be very interesting to investigate whether map construction or MEETING can be solved by fully oblivious robots with non-rigid movements. Another direction would be to study the problems for oblivious robots with limited visibility. Also, our movement model allows the robots to make circular moves, as opposed to [10], where the robots can move only along a straight line. It would be interesting to see if the same result can be achieved without the ability to make circular moves.

Acknowledgements. The first three authors are supported by NBHM, DAE, Govt. of India, CSIR, Govt. of India and UGC, Govt. of India, respectively. We would like to thank the anonymous reviewers for their valuable comments which helped us improve the quality and presentation of the paper.

References

1. Agathangelou, C., Georgiou, C., Mavronicolas, M.: A distributed algorithm for gathering many fat mobile robots in the plane. In: ACM Symposium on Principles of Distributed Computing, PODC 2013, Montreal, QC, Canada, 22–24 July 2013, pp. 250–259 (2013). https://doi.org/10.1145/2484239.2484266
2. Blum, L., Cucker, F., Shub, M., Smale, S.: Complexity and Real Computation. Springer, Berlin (1998). https://doi.org/10.1007/978-1-4612-0701-6
3. Bose, K., Adhikary, R., Kundu, M.K., Sau, B.: Arbitrary pattern formation on infinite grid by asynchronous oblivious robots. In: Das, G.K., Mandal, P.S., Mukhopadhyaya, K., Nakano, S. (eds.) WALCOM 2019. LNCS, vol. 11355, pp. 354–366. Springer, Cham (2019). https://doi.org/10.1007/978-3-030-10564-8_28
4. Bose, K., Adhikary, R., Kundu, M.K., Sau, B.: Positional encoding by robots with non-rigid movements. CoRR abs/1905.09786 (2019). http://arxiv.org/abs/1905.09786
5. Cicerone, S., Di Stefano, G., Navarra, A.: Asynchronous arbitrary pattern formation: the effects of a rigorous approach. Distrib. Comput. 1–42 (2018). https://doi.org/10.1007/s00446-018-0325-7
6. Cieliebak, M., Flocchini, P., Prencipe, G., Santoro, N.: Distributed computing by mobile robots: gathering. SIAM J. Comput. $41(4)$, 829–879 (2012). https://doi.org/10.1137/100796534
7. Cohen, H.: A Course in Computational Algebraic Number Theory, vol. 138. Springer, Heidelberg (2013). https://doi.org/10.1007/978-3-662-02945-9
8. Flocchini, P., Prencipe, G., Santoro, N., Widmayer, P.: Arbitrary patternformation by asynchronous, anonymous, oblivious robots. Theor. Comput. Sci. $407(1–3)$, 412–447 (2008). https://doi.org/10.1016/j.tcs.2008.07.026
9. Flocchini, P., Santoro, N., Viglietta, G., Yamashita, M.: Rendezvous withconstant memory. Theor. Comput. Sci. 621, 57–72 (2016). https://doi.org/10.1016/j.tcs.2016.01.025
10. Luna, G.A.D., Flocchini, P., Santoro, N., Viglietta, G., Yamashita, M.: Meeting in a polygon by anonymous oblivious robots. In: 31st International Symposium on Distributed Computing, DISC 2017, pp. 14:1–14:15, Vienna, 16-20 October 2017. https://doi.org/10.4230/LIPIcs.DISC.2017.14
11. Pagli, L., Prencipe, G., Viglietta, G.: Getting close without touching :near-gathering for autonomous mobile robots. Distrib. Comput. $28(5)$, 333–349 (2015). https://doi.org/10.1007/s00446-015-0248-5
12. Suzuki, I., Yamashita, M.: Distributed anonymous mobile robots:formation of geometric patterns. SIAM J. Comput. $28(4)$, 1347–1363 (1999). https://doi.org/10.1137/S009753979628292X
13. Yamauchi, Y., Yamashita, M.: Randomized pattern formation algorithm for asynchronous oblivious mobile robots. In: Kuhn, F. (ed.) DISC 2014. LNCS, vol. 8784, pp. 137–151. Springer, Heidelberg (2014). https://doi.org/10.1007/978-3-662-45174-8_10

Arbitrary Pattern Formation
by Asynchronous Opaque Robots
with Lights

Kaustav Bose$^{(\boxtimes)}$ (ID), Manash Kumar Kundu (ID), Ranendu Adhikary (ID),
and Buddhadeb Sau

Department of Mathematics, Jadavpur University, Kolkata, India
{kaustavbose.rs,manashkrkundu.rs,ranenduadhikary.rs,
buddhadeb.sau}@jadavpuruniversity.in

Abstract. The ARBITRARY PATTERN FORMATION problem asks for a distributed algorithm that moves a set of autonomous mobile robots to form any arbitrary pattern given as input. The robots are assumed to be autonomous, anonymous and identical. They operate in Look-Compute-Move cycles under a fully asynchronous scheduler. The robots do not have access to any global coordinate system. The existing literature that investigates this problem, considers robots with unobstructed visibility. This work considers the problem in the more realistic *obstructed visibility* model, where the view of a robot can be obstructed by the presence of other robots. The robots are assumed to be punctiform and equipped with visible lights that can assume a constant number of predefined colors. We have studied the problem in two settings based on the level of consistency among the local coordinate systems of the robots: *two axis agreement* (they agree on the direction and orientation of both coordinate axes) and *one axis agreement* (they agree on the direction and orientation of only one coordinate axis). In both settings, we have provided a full characterization of initial configurations from where any arbitrary pattern can be formed.

Keywords: Distributed algorithm · Arbitrary pattern formation ·
Leader election · Autonomous robots · Opaque robots ·
Luminous robots · Obstructed visibility · Asynchronous scheduler ·
Look-compute-move cycle

1 Introduction

One of the recent trends of research in robotics is to use a swarm of simple and inexpensive robots to collaboratively execute complex tasks, as opposed to using one or few powerful and expensive robots. Robot swarms offer several advantages over single robot systems, such as scalability, robustness and versatility. Algorithmic aspects of decentralized coordination of robot swarms have been extensively

© Springer Nature Switzerland AG 2019
K. Censor-Hillel and M. Flammini (Eds.): SIROCCO 2019, LNCS 11639, pp. 109–123, 2019.
https://doi.org/10.1007/978-3-030-24922-9_8

studied in the literature over the last two decades. In theoretical studies, the traditional framework models the robot swarm as a set of autonomous, anonymous and identical computational entities freely moving in the plane. The robots do not have access to any global coordinate system. Each robot is equipped with sensor capabilities to perceive the positions of other robots. If the robots are equipped with camera sensors, then it is unrealistic to assume that each robot can observe all other robots in the swarm, as the line of sight of a robot can be obstructed by the presence of other robots. This setting is known as the *opaque robot* or *obstructed visibility* model, where it is assumed that a robot is able to see another robot if and only if no other robot lies in the line segment joining them.

ARBITRARY PATTERN FORMATION or $\mathcal{APF}$ is a fundamental coordination problem in swarm robotics where the robots are required to form any specific but arbitrary geometric pattern given as input. In this work, we study the problem in obstructed visibility model. The majority of the literature that studies this problem, considers robots with unobstructed visibility. Two recent works [14, 24] investigated two formation problems in the obstructed visibility model. In [14], UNIFORM CIRCLE FORMATION problem was studied, where the robots are required to form a circle by positioning themselves on the vertices of a regular polygon. Their approach is to first solve the MUTUAL VISIBILITY problem as a subroutine where the robots arrange themselves in a configuration in which each robot can see all other robots. Then they solved the original problem from a mutually visible configuration. The more general ARBITRARY PATTERN FORMATION problem was first studied in the obstructed visibility model by Vaidyanathan et al. in [24]. Their algorithm first solves MUTUAL VISIBILITY and then elects a leader by probabilistic method. In this work, our aim is to provide deterministic solutions. For robots with obstructed visibility and having only partial agreement in coordinate system, deterministic LEADER ELECTION is difficult and is of independent interest. Also, LEADER ELECTION and $\mathcal{APF}$ are deterministically unsolvable from some symmetric configurations. Therefore, trying to first solve MUTUAL VISIBILITY may create new symmetries, from where $\mathcal{APF}$ is deterministically unsolvable. Hence, if we want to first bring the robots to a mutually visible configuration, then we have to design an algorithm that does not create such symmetries. The existing algorithms in the literature for MUTUAL VISIBILITY do not have this feature. In both of these works, the robots are assumed to be *luminous*, i.e., they are equipped with persistent visible lights that can assume a constant number of predefined colors.

1.1   Our Contribution

In this paper, we study the ARBITRARY PATTERN FORMATION problem for a system of opaque and luminous robots in a fully asynchronous setting. We have shown that the problem can be solved from any initial configuration if the robots agree on the direction and orientation of both X and Y axes. If the robots agree on the direction and orientation of only X axis, $\mathcal{APF}$ is unsolvable when the initial configuration has a reflectional symmetry with respect to a line $\mathcal{K}$ which is parallel to the X axis and has no robots lying on $\mathcal{K}$. The same result holds even if the robots have unobstructed visibility. For all other initial configurations, $\mathcal{APF}$ is solvable.

1.2   Earlier Works

The study of ARBITRARY PATTERN FORMATION was initiated by Suzuki and Yamashita in [22,23]. In these papers, a complete characterization of the class of formable patterns was provided for autonomous and anonymous robots with an unbounded amount of memory. The problem was first studied in the weak setting of oblivious and asynchronous robots by Flocchini et al. in [15]. They showed that if the robots have no common agreement on coordinate system, then it is impossible to form an arbitrary pattern. If the robots have one axis agreement, then any odd number of robots can form an arbitrary pattern, but an even number of robots cannot, in the worst case. If the robots agree on both X and Y axes, then any pattern is formable from any configuration of robots. They also proved that it is possible to elect a leader for $n \geq 3$ robots if it is possible to form any arbitrary pattern. In [12,13], the authors studied the relationship between ARBITRARY PATTERN FORMATION and LEADER ELECTION. The later consists in distinguishing a unique robot as the leader. They proved that any arbitrary pattern can be formed from any initial configuration wherein the leader election is possible. More precisely, their algorithms work for four or more robots with chirality and for at least five robots without chirality. Combined with the result in [15], it follows that ARBITRARY PATTERN FORMATION and LEADER ELECTION are equivalent, i.e., it is possible to solve ARBITRARY PATTERN FORMATION for $n \geq 4$ with chirality (resp. $n \geq 5$ without chirality) if and only if LEADER ELECTION is solvable. In [8,16], the problem was studied allowing the pattern to have multiplicities. Recently, the case of $n = 4$ robots was fully characterized in [7] in the asynchronous setting, with and without chirality. They proposed a new geometric invariant that exists in any configuration with four robots, and using this invariant, they presented an algorithm that forms any target pattern from any solvable initial configuration. The problem of forming a sequence of patterns in a given order was studied in [10]. Randomized algorithms for pattern formation were studied in [26]. In [9,16], the so-called EMBEDDED PATTERN FORMATION problem was studied where the pattern to be formed is provided as a set of fixed and visible points in the plane. In [5], the problem was considered in a grid based terrain where the movements of the robots are restricted only along grid lines and only by a unit distance in each step. They showed that a set of fully asynchronous robots without agreement in coordinate system can form any arbitrary pattern, if their starting configuration is asymmetric.

All the aforementioned works considered robots with unlimited and unobstructed visibility. In the limited, but unobstructed, visibility setting, the problem was first studied in [25], and recently in [17]. In obstructed visibility, the GATHERING problem have been studied for fat robots in plane [2], and for point robots in three dimensional space [3]. A related problem in obstructed visibility model is the MUTUAL VISIBILITY problem. In this problem, starting from arbitrary configuration, the robots have to reposition themselves to a configuration in which every robot can see all other robots in the team. The problem has been extensively studied in the literature under various settings [1, 4, 11, 19–21]. ARBITRARY PATTERN FORMATION in the obstructed visibility model was first studied recently in [24], without any agreement in coordinate system, where the authors proved runtime bounds in terms of the time required to solve LEADER ELECTION. However, they did not provide any deterministic solution for LEADER ELECTION and is yet to be studied in the literature in the obstructed visibility model.

Organization.   The paper is organized as follows. In Sect. 2, a formal definition of the robotic model and the problem is given, along with some basic notations and terminology. In Sect. 3, we briefly discuss the algorithm for $\mathcal{APF}$ under one axis agreement. The main results of the paper are presented in Sect. 4. We refer the readers to the full version [6] of the paper for a more detailed description of the algorithms and formal proofs of correctness.

2   Preliminaries

2.1   Robot Model

A set of n mobile computational entities, called *robots*, are initially placed at distinct points in the Euclidean plane. The robots are assumed to be *anonymous* (they have no unique identifiers that they can use in a computation), *identical* (they are indistinguishable by their physical appearance), *autonomous* (there is no centralized control), *homogeneous* (they execute the same deterministic algorithm). The robots are modeled as points in the plane, i.e., they do not have any physical extent. The robots do not have access to any global coordinate system. Each robot is provided with its own local coordinate system centered at its current position, and its own notion of unit distance and handedness. However, the robots may have a priori agreement about the direction and orientation of the axes in their local coordinate systems. Based on this, we consider the two models: *two axis agreement* (they agree on the direction and orientation of both axes) and *one axis agreement* (they agree on the direction and orientation of only one axis).

This paper studies the pattern formation problem in the opaque and luminous robot model. The *opaque robot* model assumes that the robots have unlimited but obstructed visibility, i.e., a point p on the plane is visible to a robot r if and only if the line segment joining p and r does not contain any other robot. In the *luminous robot* model, introduced by Peleg [18], each robot is equipped with a visible light which can assume a constant number of predefined colors. The lights serve both as a weak explicit communication mechanism and a form of internal memory. We denote the set of colors available to the robots by $\mathcal{C}$.

The robots, when active, operate according to the so-called *LOOK-COMPUTE-MOVE* cycle. In each cycle, a previously idle or inactive robot wakes up and executes the following steps. In the LOOK phase, the robot takes a snapshot of the current configuration, i.e., it obtains the positions, expressed in its local coordinate system, of all robots visible to it, along with their respective colors. The robot also knows its own color. In the COMPUTE phase, based on the perceived configuration, the robot performs computations according to a deterministic algorithm to decide a destination point $x \in \mathbb{R}^2$ (expressed in its local coordinate system) and a color $c \in \mathcal{C}$. As mentioned earlier, the deterministic algorithm is same for all robots. In the MOVE phase, the robot then sets its light to c and moves towards the point x. After executing a LOOK-COMPUTE-MOVE cycle, a robot becomes inactive. Then after some finite time, it wakes up again to perform another LOOK-COMPUTE-MOVE cycle. Notice that after a robot sets it light to a particular color in the MOVE phase of a cycle, it maintains its color until the MOVE phase of the next LCM cycle. The robots are *oblivious* in the sense that when a robot transitions from one LCM cycle to the next, all of its local memory (past computations and snapshots) are erased, except for the color of the light. The robots are controlled by a fully asynchronous scheduler (ASYNC). In ASYNC, the robots are activated independently and each robot executes its cycles independently. The amount of time spent in LOOK, COMPUTE, MOVE and inactive states is finite but unbounded, unpredictable and not same for different robots. As a result, the robots do not have a common notion of time. Moreover, a robot can be seen while moving, and hence, computations can be made based on obsolete information about positions. Also, the configuration perceived by a robot during the LOOK phase may significantly change before it makes a move and therefore, may cause a collision. The scheduler that controls the activations and the durations of the operations can be thought of as an adversary, whose purpose is to disrupt the algorithm. In this paper, the robots are assumed to have RIGID movements, i.e., each robot is able to reach its desired destination without any interruption.

2.2 Definitions and Notations

In this section, we introduce the notations that will be used throughout the paper. We assume that a set $\mathcal{R}$ of n robots are placed at distinct positions in the Euclidean plane. For any time t, $\mathbb{C}(t)$ will denote the configuration of the robots at time t. For a robot r, its position at time t will be denoted by $r(t)$. When there is no ambiguity, r will represent both the robot and the point in the plane occupied by it. By 1_r, we shall denote the unit distance according to the local coordinate system of r. At any time t, $r(t).light$ or simply $r.light$ will denote the color of the light of r at t.

Suppose that a robot r, positioned at point p, takes a snapshot at time t_1. Based on this snapshot, suppose that the deterministic algorithm run in the COMPUTE phase instructs the robot to change its color (Case 1) or move to a different point (Case 2) or both (Case 3). In case 1, assume that it changes its color at time $t_2 > t_1$. In case 2, assume that it starts moving at time $t_3 > t_1$. Note that when we say that it starts moving at t_3, we mean that $r(t_3) = p$, but $r(t_3 + \epsilon) \neq p$ for sufficiently small $\epsilon > 0$. For case 3, assume that r changes it color at $t_2 > t_1$ and starts moving at $t_3 > t_2$. Then we say that r has a *pending move* at t if $t \in (t_1, t_2)$ in case 1 or $t \in (t_1, t_3]$ in case 2 and 3. A robot r is said to be *stable* at time t, if r is stationary and has no pending move at t. A configuration at time t is said to be a *stable configuration* if every robot is stable at t. A configuration at time t is said to be a *final configuration* if (1) every robot at t is stable, and (2) any robot taking a snapshot at t will not decide to move or change its color.

With respect to the local coordinate system of a robot, positive and negative directions of the X-axis will be referred to as *right* and *left*, while the positive and negative directions of the Y-axis will be referred to as *up* and *down*. For a robot r, $\mathcal{L}_V(r)$ and $\mathcal{L}_H(r)$ are respectively the vertical and horizontal line passing through r. We denote by $\mathcal{H}_U^O(r)$ (resp. $\mathcal{H}_U^C(r)$) and $\mathcal{H}_B^O(r)$ (resp. $\mathcal{H}_B^C(r)$) the upper and bottom open (resp. closed) half-plane delimited by $\mathcal{L}_H(r)$ respectively. Similarly, $\mathcal{H}_L^O(r)$ (resp. $\mathcal{H}_L^C(r)$) and $\mathcal{H}_R^O(r)$ (resp. $\mathcal{H}_R^C(r)$) are the left and right open (resp. closed) half-plane delimited by $\mathcal{L}_V(r)$ respectively. We define $\mathcal{H}^O(r_1, r_2)$ (resp. $\mathcal{H}^C(r_1, r_2)$) as the horizontal open (resp. closed) strip delimited by $\mathcal{L}_H(r_1)$ and $\mathcal{L}_H(r_2)$. For a robot r and a straight line $\mathcal{L}$ passing through it, r will be called *non-terminal on $\mathcal{L}$* if it lies between two other robots on $\mathcal{L}$, and otherwise it will be called *terminal on $\mathcal{L}$*.

2.3   Problem Definition

A swarm of n robots is arbitrarily deployed at distinct positions in the Euclidean plane. Initially, the lights of all the robots are set to a specific color called `off`. Each robot is given as input a pattern $\mathbb{P}$, which is a list of n distinct elements from $\mathbb{R}^2$. Notice that the robots, through the input, implicitly obtain n, the total number of robots. Without loss of generality, we assume that (1) the n elements in $\mathbb{P}$ are arranged in (ascending) dictionary order, i.e., $(x, y) < (x', y')$ iff either $x < x'$ or $x = x'$, $y < y'$, and (2) $(x, y) \in \mathbb{R}_{\geq 0}^2 = \{(a, b) \in \mathbb{R}^2 | a, b \geq 0\}$ for each element (x, y) in $\mathbb{P}$.

The goal of the ARBITRARY PATTERN FORMATION, or in short $\mathcal{APF}$, is to design a distributed algorithm so that there is a time t such that (1) $\mathbb{C}(t)$ is a final configuration, (2) the lights of all the robots at t are set to the same color, (3) $\mathbb{P}$ can be obtained from $\mathbb{C}(t)$ by a sequence of translation, reflection, rotation, uniform scaling operations, and (4) at any time $t' \in [0, t]$, no two robots occupy the same position in the plane, i.e., in other words, the movements are collision-free.

3   Arbitrary Pattern Formation Under One Axis Agreement

In this section, we shall discuss $\mathcal{APF}$ under the one axis agreement model in ASYNC. We assume that the robots agree on the direction and orientation of only X axis. The following theorem presents a basic impossibility result under this model.

Theorem 1. *$\mathcal{APF}$ is unsolvable in ASYNC if the initial configuration has a reflectional symmetry with respect to a line $\mathcal{K}$ which is parallel to the X axis with no robots lying on $\mathcal{K}$. The same result holds even if the robots have unobstructed visibility.*

Therefore, we shall assume that in the one axis agreement setting, the initial configuration does not have such a symmetry. We shall prove that with this assumption, $\mathcal{APF}$ is solvable in ASYNC. Our algorithm requires six colors, namely off, terminal, candidate, symmetry, leader, and done. As mentioned earlier, initially the lights of all the robots are set to off.

The goal of $\mathcal{APF}$ is that the robots have to arrange themselves to a configuration which is similar to $\mathbb{P}$ with respect to translation, rotation, reflection and uniform scaling. Since the robots do not have access to any global coordinate system, there is no agreement regarding where and how the pattern $\mathbb{P}$ is to be embedded in the plane. To resolve this ambiguity, we shall first elect a robot in the team as the leader. The relationship between leader election and arbitrary pattern formation is well established in the literature. Once a leader is elected, it is not difficult to reach an agreement on a suitable coordinate system. Then the robots have to reconfigure themselves to form the given pattern $\mathbb{P}$ with respect to this coordinate system. Thus, the algorithm is divided into two *stages*, namely *leader election* and *pattern formation from leader configuration*. The leader election stage is again logically divided into two *phases*, *phase 1* and *phase 2*. Since the robots are oblivious, in each LCM cycle, it has to infer in which stage or phase it currently is, from certain geometric conditions and the lights of the robots in the perceived configuration. These conditions are described in Algorithm 1. Notice that due to the obstructed visibility, two robots taking snapshots at the same time can have quite different views of the configuration. Therefore, it may happen that they decide to execute instructions corresponding to different stages or phases of the algorithm.

Algorithm 1. Arbitrary Pattern Formation

Input : The configuration of robots visible to me.
1 **Procedure** ARBITRARYPATTERNFORMATION()
2   **if** *there is a robot with light set to* leader **then**   // stage 2
3 | PATTERNFORMATIONFROMLEADERCONFIGURATION()
4   **else**   // stage 1
5   **if** *there are two robots with light set to* candidate *on the same vertical line or at least one robot with light set to* symmetry **then**
6 | PHASE2()
7   **else**
8 | PHASE1()

3.1 Leader Election

For a group of anonymous and identical robots, leader election is solved on the basis of the relative positions of the robots in the configuration. But this is only possible if the robots can see the entire configuration. Therefore, the naive approach would be to first bring the robots to a mutually visible configuration where each robot can see all other robots. But as mentioned earlier, this can create unwanted symmetries in the configuration from where arbitrary pattern formation may be unsolvable. Therefore, we shall employ a different strategy that does not require solving mutual visibility.

Formally, the aim of the leader election stage is to obtain a stable configuration where there is a unique robot r_l such that (1) $r_l.light = \text{leader}$, (2) $r.light = \text{off}$ for all $r \in \mathcal{R} \setminus \{r_l\}$, and (3) $r \in \mathcal{H}_R^O(r_l) \cap \mathcal{H}_U^O(r_l)$ for all $r \in \mathcal{R} \setminus \{r_l\}$. We shall call this a *leader configuration*, and call r_l the *leader*. As mentioned earlier, the leader election algorithm consists of two phases, namely phase 1 and phase 2 (See Algorithm 1). We describe the two phases in the following.

Algorithm 2. Phase 1 of Leader Election

Input : The configuration of robots visible to me.

1 **Procedure** PHASE1()
2     $r \leftarrow$ myself
3   **if** $r.light = off$ **then**
4   **if** *there are no robots in $\mathcal{H}_L^C(r)$ other than itself and all robots have their lights set to off* **then**
5 | BECOMELEADER()
6   **else if** LEFTMOSTTERMINAL() $= True$ **then**
7 | $(x^*, y^*) \leftarrow$ COMPUTEDESTINATION()
8 | $r.light \leftarrow$ terminal
9 | Move to (x^*, y^*)
10   **else if** $r.light = terminal$ **then**
11 | $r.light \leftarrow$ candidate
12   **else if** $r.light = candidate$ **then**
13   **if** *there is a robot with light candidate in $\mathcal{H}_R^O(r)$* **then**
14 | $r.light \leftarrow$ off
15   **else if** *there is a robot with light off in $\mathcal{H}_L^O(r)$* **then**
16 | $r.light \leftarrow$ off

17 **Procedure** BECOMELEADER()
18   **if** *there are no robots in $\mathcal{H}_B^C(r)$ other than itself* **then**
19 | $r.light \leftarrow$ leader
20   **else**
21       $r' \leftarrow$ the bottommost robot
22       $d \leftarrow |r'.y|$
23   Move $d + 1_r$ distance down vertically

24 **Function** LEFTMOSTTERMINAL()
25    $result \leftarrow False$
26   **if** *there are no robots in $\mathcal{H}_L^O(r)$* **then**
27   **if** *there is $\mathcal{H} \in \{\mathcal{H}_B^O(r), \mathcal{H}_U^O(r)\}$ such that $\mathcal{H} \cap \mathcal{L}_V(r)$ contains no robots* **then**
28 | $result \leftarrow True$
29   **else if** *there is exactly one robot r' in $\mathcal{H}_L^O(r)$ and $r'.light = candidate$* **then**
30   Let $\mathcal{H} \in \{\mathcal{H}_B^O(r), \mathcal{H}_U^O(r)\}$ be the open half-plane not containing r'
31   **if** *$\mathcal{H} \cap \mathcal{L}_V(r)$ contains no robots* **then**
32 | $result \leftarrow True$
33   return $result$
34 **Function** COMPUTEDESTINATION()
35   **if** *there are no robots in $\mathcal{H}_R^O(r)$* **then**
36 | return $(-1_r, 0)$
37   **else**
38       $\mathcal{L}' \leftarrow$ the leftmost vertical line containing a robot in $\mathcal{H}_R^O(r)$
39       $d \leftarrow$ the horizontal distance between r and $\mathcal{L}'$
40   return $(-d, 0)$

Algorithm 3. Phase 2 of Leader Election

Input : The configuration of robots visible to me.

1 **Procedure** PHASE2()
2    $r \leftarrow$ myself
3   **if** $r.light = candidate$ **then**
4       $r' \leftarrow$ another robot on $\mathcal{L} = \mathcal{L}_V(r)$ with light candidate or symmetry
5       $\mathcal{L}' \leftarrow$ the leftmost vertical line containing a robot in $\mathcal{H}_R^O(r)$
6       $\mathcal{K} \leftarrow$ the horizontal line passing through the mean of the positions of the robots lying on $\mathcal{L}'$
7       $d_r \leftarrow$ distance of r from $\mathcal{K}$
8       $d_{r'} \leftarrow$ distance of r' from $\mathcal{K}$
9       $d_{\mathcal{L}\mathcal{L}'} \leftarrow$ distance between $\mathcal{L}$ and $\mathcal{L}'$
10   set the positive direction of Y-axis of the local coordinate system towards r'
11   **if** $r'.light = symmetry$ **then**
12   **if** $d_r < d_{r'}$ **then**
13   Move $d_{r'} - d_r$ distance vertically in direction opposite to r'
14   **else if** $d_r = d_{r'}$ **then**
15          $r.light \leftarrow$ symmetry

16   **else if** $r'.light = candidate$ **then**
17   **if** r and r' are in the same closed half-plane delimited by $\mathcal{K}$ **then**
18   **if** $d_r > d_{r'}$ **then**
19   Move to COMPUTEDESTINATION2()
20   **else**
21   **if** $\lambda'(r) \prec \lambda'(r')$ **then**
22   Move to COMPUTEDESTINATION2()
23   **else if** $\lambda'(r) = \lambda'(r')$ **then**
24   **if** $d_r \geq d_{r'}$ **then**
25   **if** the number of robots is equal to n and there are no robots in $\mathcal{H}_B^O(r)$ **then**
26   **if** $\lambda(r) \prec \lambda(r')$ **then**
27   Move to COMPUTEDESTINATION2()
28   **else if** $\lambda(r) \succ \lambda(r')$ **then**
29   Move $\frac{d_{\mathcal{L}\mathcal{L}'}}{2}$ distance right horizontally
30   **else if** $\lambda(r) = \lambda(r')$ **then**
31   **if** $\mathcal{K}$ has no robots on it **then**
32   Move to COMPUTEDESTINATION2()
33   **else**
34                    $r.light \leftarrow$ symmetry
35   **else**
36   Move d_r distance vertically in direction opposite to r'

37   **else if** $r.light = symmetry$ **then**
38   **if** there is a robot in $\mathcal{H}_L^O(r)$ with light off **then**
39        $r.light \leftarrow$ off

40   **else if** $r.light = off$ **then**
41   **if** there are two robots r_1 and r_2 in $\mathcal{H}_L^O(r)$ with light symmetry on the same vertical line $\mathcal{L}$ **then**
42        $\mathcal{K} \leftarrow$ the horizontal line passing through the mid-point of the line segment joining r_1 and r_2.
43   **if** r is the leftmost robot lying on $\mathcal{K}$ **then**
44          $d \leftarrow$ distance of r from $\mathcal{L}$
45   Move $d + 1_r$ distance left

46 **Function** COMPUTEDESTINATION2()
47    $\mathcal{H} \leftarrow$ the open half-plane delimited by $\mathcal{L}_H(r')$ not containing r.
48   **if** $\mathcal{L}' \cap \mathcal{H}$ has no robots **then**
49   return $(-1_r, 0)$
50   **else**
51      $r'' \leftarrow$ the robot on $\mathcal{L}' \cap \mathcal{H}$ with maximum y-coordinate
52   return $(-\frac{1}{2}d_{\mathcal{L}\mathcal{L}'} \frac{r'.y}{r''.y - r'.y}, 0)$

Phase 1.   Since the robots agree on left and right, if there is a unique leftmost robot in the configuration, then it can identify this from its local view and elect itself as the leader. However, there might be more than one leftmost robots in the configuration. Assume that there are $k \geq 2$ leftmost robots in the configuration. Our aim in this phase is to reduce the number of leftmost robots to $k = 2$, or if possible, to $k = 1$. Suppose that the $k \geq 2$ leftmost robots are lying on the vertical line $\mathcal{L}$. We shall ask the two terminal robots on $\mathcal{L}$, say r_1 and r_2, to move horizontally towards left. If both robots move synchronously by the same distance, then the new configuration will have two leftmost robots. However, r_1 and r_2 can not distinguish between this configuration and the initial configuration, i.e., they can not ascertain from their local view if they are the only robots on the vertical line due to the obstructed visibility. To resolve this, we shall ask r_1 and r_2 to change their lights to `terminal` before moving. Now, consider the case where r_1 and r_2 move different distances with r_1 moving further. Suppose that r_2 reaches its destination first, and when it takes the next snapshot, it finds that r_1 (with light `terminal`) is on the same vertical line. So, r_2 incorrectly concludes that the two terminal robots have been brought on the same vertical line, while actually r_1 is still moving leftwards. To avert this situation, we shall use the color `candidate`. After moving, the robots will change their lights to `candidate` to indicate that they have completed their moves. So, if we have two robots with light `candidate` on the same vertical line, then we are done. On the other hand, if we end up with the two robots with light `candidate` not on the same vertical line, then the one on the left will become the leader. A pseudocode description of phase 1 is given in Algorithm 2.

Phase 2.   Phase 1 will end with either a leader configuration or a stable config-uration where there are two robots r_1 and r_2 such that (1) $r_1.light = r_2.light =$ `candidate`, (2) $r.light =$ `off` for all $r \in \mathcal{R} \setminus \{r_1, r_2\}$, (3) r_1 and r_2 are on the same vertical line, and (4) $r \in \mathcal{H}_R^O(r_1)$ for all $r \in \mathcal{R} \setminus \{r_1, r_2\}$. We shall refer to the later configuration as a *candidate configuration*. In the first case, leader election is done, and in the second case we enter phase 2. So assume that we have two leftmost robots, r_1 and r_2, lying on the vertical line $\mathcal{L}$ with their lights set to `candidate`.

Let $\mathcal{C}' = \mathcal{C} \setminus \{r_1, r_2\}$ be the configuration of the remaining robots. Let $\mathcal{L}'$ be the leftmost vertical line containing a robot in $\mathcal{C}'$, and $\mathcal{K}$ is the horizontal line passing through the mean of the positions of the robots lying on $\mathcal{L}'$. Let $\mathcal{H}_1$ and $\mathcal{H}_2$ be the two open half-planes delimited by $\mathcal{K}$. Let $\overline{\mathcal{H}_1}$ and $\overline{\mathcal{H}_2}$ be their closure. First assume that both r_1 and r_2 lie in the same closed half-plane, say $r_1, r_2 \in \overline{\mathcal{H}_1}$. Then the robot further from $\mathcal{K}$, say r_2, will move left. Then clearly we are back to phase 1, and eventually r_2 will become leader. Now assume that r_1 and r_2 are in different open half-planes. So, let $\mathcal{H}_1$ and $\mathcal{H}_2$ be the half-planes containing r_1 and r_2 respectively. For each $\mathcal{H}_i$, we define a coordinate system C_i in the following way. The point of intersection between $\mathcal{K}$ and $\mathcal{L}'$ is the origin, $\mathcal{L}' \cap \mathcal{H}_i$ is the positive Y-axis and the positive direction of X-axis is according to the global agreement. We express the positions of all the robots in $\mathcal{H}_i$ except

r_i with respect to the coordinate system C_i. Now arrange the positions in the dictionary order. Let $\lambda(r_i)$ denote the string thus obtained. Each term of the string is an element from $\mathbb{R}^2$. To make the length of the strings $\lambda(r_1)$ and $\lambda(r_2)$ equal, null elements Φ may be appended to the shorter string. For any non-null term (x, y) of a string, we set $(x, y) < \Phi$. We shall write $\lambda(r_1) \prec \lambda(r_2)$ iff $\lambda(r_2)$ is lexicographically larger than $\lambda(r_1)$. For each string $\lambda(r_i)$, let $\lambda'(r_i)$ be the string obtained from $\lambda(r_i)$ by deleting all terms with x-coordinate not equal to 0. That is, the terms of $\lambda'(r_i)$ corresponds to the robots on $\mathcal{L}' \cap \mathcal{H}_i$. Again, null elements Φ may be appended to make the length of the strings $\lambda'(r_1)$ and $\lambda'(r_2)$ equal. The plan is to choose a leader by comparing these strings. First, the robots will compare $\lambda'(r_1)$ and $\lambda'(r_2)$. Clearly, both of them can see all the robots on $\mathcal{L}'$, hence can compute $\lambda'(r_1)$ and $\lambda'(r_2)$. If $\lambda'(r_i) \prec \lambda'(r_j)$, r_i will move left. As before, r_i will become leader. If $\lambda'(r_1) = \lambda'(r_2)$, the robots have to compare the full strings $\lambda(r_1)$ and $\lambda(r_2)$. But in order to compute these strings, the complete view of the configuration is required. Let d_{r_1} and d_{r_2} be the distance of r_1 and r_2 from $\mathcal{K}$ respectively. If $d_{r_i} \geq d_{r_j}$, r_i will move vertically away from r_j by a distance d_{r_i} to get the full view of the configuration. It can be proved that after finitely many steps, it will be able to see all the robots in the configuration. When it can see all the robots in the configuration, it computes $\lambda(r_1)$ and $\lambda(r_2)$. If $\lambda(r_i) \prec \lambda(r_j)$, r_i will move left (r_i will become leader) and if $\lambda(r_i) \succ \lambda(r_j)$, r_i will move right (r_j will become leader). In the case where $\lambda(r_i) = \lambda(r_j)$ ($\mathbb{C}'$ is symmetric), r_i will move left if $\mathcal{K}$ has no robots on it, or otherwise will change its light to `symmetry`. In that case where $\mathbb{C}'$ is symmetric and $\mathcal{K}$ has at least one robot on it, the leftmost robot on $\mathcal{K}$ will move towards left to become the leftmost robot in $\mathbb{C}$ and will eventually become the leader. A pseudocode description of phase 2 is presented in Algorithm 3.

3.2 Pattern Formation from a Leader Configuration

A pseudocode description of this stage is given in Algorithm 4. At the start of this stage, we have a leader configuration. Let r_l be the leader having its light set to `leader`. Any robot that can see r_l starts executing Algorithm 4. In a leader configuration, all non-leader robots lie on one of the open half-planes delimited by the horizontal line passing through the leader r_l. This leads to an agreement on the direction of Y axis, as we can set the empty open half-plane to correspond to the negative direction of Y-axis or 'down'. Hence, we have an agreement on 'up', 'down', 'left' and 'right'. Order the robots in $\mathcal{R} \setminus \{r_l\}$ from bottom to up, and from left to right in case multiple robots on the same horizontal line. Let us denote these robots by $r_1, r_2, \ldots, r_{n-1}$ in this order. Notice that we still do not have a common notion of unit distance. So, we shall ask r_1 to move to the line $\mathcal{L}_V(r_l)$. The distance of this robot from the leader r_l will be set as the unit distance. Now it only remains to fix the origin. We shall set the origin at a point such that the coordinates of r_l are $(-1, -1)$. Now that we have a common fixed coordinate system, we embed the pattern $\mathbb{P}$ on the plane. Let t_i denote the point in the plane corresponding to $\mathbb{P}[i]$. Let us call these points $t_0, t_1, \ldots, t_{n-1}$ the *target points*. Since $\mathbb{P}$ is sorted in dictionary order, $t_0, t_1, \ldots, t_{n-1}$ are ordered from left to right, and from bottom to up in case there are multiple robots on the same vertical line. For each t_i, $i = 0, 1, \ldots, n - 1$, we define a point $p_i = (\Psi(i), -1)$ on $\mathcal{L}_H(r_l)$ in the following way. Let $t_i = (x_i, y_i)$. Let $\mathcal{L}_i$ be the vertical line passing through t_i, i.e., the line $X = x_i$. Let t_i be the kth target point on $\mathcal{L}_i$ from bottom, i.e., the kth item in $\mathbb{P}$ with the first coordinate equal to x_i. Then $\Psi(i) = x_i + \frac{(k-1)}{2(m_i-1)}\epsilon$, where m_i is equal to the total number of target points on $\mathcal{L}_i$, and ϵ is equal to the smallest horizontal distance between any two target points not on the same vertical line, or equal to 1 if all target points are on the same vertical line. The robots $r_2, \ldots, r_{n-1}$ will sequentially move to $p_2, \ldots, p_{n-1}$ respectively. Then $r_2, \ldots, r_{n-1}$ will sequentially leave $\mathcal{L}_H(r_l)$ to move to $t_2, \ldots, t_{n-1}$ respectively. Before moving, they will change their light to done. When r_1 finds that there is no other robot with light off in $\mathcal{H}_U^C(r_1)$ and there are no robots in $\mathcal{H}_B^O(r_1)$ except r_l, it decides to move to t_1. However, in this case, it will change its light to done after the move (See line 7 in Algorithm 4). When r_l sees no robots with light off, it decides to move to t_0. However, since there is no longer a robot on $\mathcal{L}_V(r_l)$, it can not ascertain the unit distance of the agreed coordinate system and therefore, can not determine the point t_0 in the plane. Hence, it has to locate the point t_0 by some other means. Let r be the leftmost (and bottommost in case of tie) robot that r_l can see. Clearly, r is at t_1. Hence, r_l knows the point in the plane with coordinates $\mathbb{P}[1]$, and also point with coordinates $(-1, -1)$ (its own position). From these two points, it can easily determine the point t_0, i.e., the point in the plane with coordinates $\mathbb{P}[0]$ in the agreed coordinate system. Therefore, it will change its light to done and move to t_0.

4   The Main Results

Theorem 2. *For a set of opaque and luminous robots with one axis agreement, $\mathcal{APF}$ is deterministically solvable if and only if the initial configuration is not symmetric with respect to a line $\mathcal{K}$ such that (1) $\mathcal{K}$ is parallel to the agreed axis and (2) $\mathcal{K}$ is not passing through any robot. Six colors are sufficient to solve the problem from any solvable initial configuration.*

If the robots agree on the direction and orientation of both axes, then leader election is easy. If there is a unique leftmost robot r, then as before it will become leader (by executing BECOMELEADER()). If there are multiple leftmost robots, then the bottommost one will move left. In the next snapshot, it will find itself eligible to become leader and start executing BECOMELEADER(). Therefore, leader election is solvable using only 2 colors, namely off and leader. Then stage 2 will be executed by Algorithm 4. Hence, we have the following result.

Theorem 3. *For a set of opaque and luminous robots with two axis agreement, $\mathcal{APF}$ is deterministically solvable from any initial configuration using three colors.*

Algorithm 4. Pattern Formation from Leader Configuration

Input   : The configuration of robots visible to me.

1 **Procedure** PATTERNFORMATIONFROMLEADERCONFIGURATION()
2    $r \leftarrow$ myself
3    $r_l \leftarrow$ the robot with light **leader**
4   **if** $r.light = off$ **then**
5   **if** $r_l \in \mathcal{H}_B^O(r) \cap \mathcal{H}_L^O(r)$ **then**
6   **if** *there is a robot with light* **done** **then**
7 | $r.light \leftarrow$ **done**
8   **else if** *there are no robots in* $\mathcal{H}_B^O(r) \cap \mathcal{H}_U^O(r_l) \cap \mathcal{H}_R^O(r_l)$ *and r is the leftmost robot on* $\mathcal{L}_H(r) \cap \mathcal{H}_R^O(r_l)$ **then**
9   **if** *there is no robot on* $\mathcal{L}_V(r_l)$ **then**
10 | $p \leftarrow$ the point of intersection of the lines $\mathcal{L}_H(r)$ and $\mathcal{L}_V(r_l)$
11 | Move to p
12   **else**
13   **if** *there are $k \geq 0$ robots on* $\mathcal{L}_H(r_l)$ *with light* **off** **then**
14 | Move to $(\bar{\Psi}(k+2), -1)$

15   **else if** $r_l \in \mathcal{L}_H(r)$, *there is a robot on* $\mathcal{L}_V(r_l)$, *and there are no robots with light* **off** *in* $\mathcal{H}_U^O(r) \cap \mathcal{H}_R^O(r_l)$ **then**
16   **if** $r.pos = (\Psi(i), -1)$ *and* PARTIALFORMATION(i) = *True* **then**
17 | $r.light \leftarrow$ **done**
18   Move to $\mathbb{P}[i]$

19   **else if** $r_l \in \mathcal{L}_V(r)$ **then**
20   **if** *there is no other robot with light* **off** *in* $\mathcal{H}_U^C(r)$ *and there are no robots in* $\mathcal{H}_B^O(r)$ *except* r_l **then**
21   Move to $\mathbb{P}[1]$

22   **else if** $r.light = leader$ **then**
23   **if** *there are no robots with light* **off** **then**
24 | $r.light \leftarrow$ **done**
25   Move to $\mathbb{P}[0]$

26 **Procedure** PARTIALFORMATION(i)
27   **if** $i = 2$ **then**
28 | return True
29   **else**
30   **if** *there is a robot with light* **done** *at* t_{i-1} **then**
31 | return True
32   **else**
33   return False

---

5   Concluding Remarks

In this work, we have provided a full characterization of initial configurations from where $\mathcal{APF}$ is deterministically solvable in ASYNC for a system of opaque and luminous robots with one and two axis agreement. The next step would be to study the problem with no agreement or agreement on only handedness. Another interesting question is whether the problem is solvable without light.

Acknowledgements. The first three authors are supported by NBHM, DAE, Govt. of India, UGC, Govt. of India and CSIR, Govt. of India, respectively. We would like to thank the anonymous reviewers for their valuable comments which helped us improve the quality and presentation of the paper.

References

1. Adhikary, R., Bose, K., Kundu, M.K., Sau, B.: Mutual visibility by asynchronous robots on infinite grid. In: Gilbert, S., Hughes, D., Krishnamachari, B. (eds.) ALGOSENSORS 2018. LNCS, vol. 11410, pp. 83–101. Springer, Cham (2019). https://doi.org/10.1007/978-3-030-14094-6_6
2. Agathangelou, C., Georgiou, C., Mavronicolas, M.: A distributed algorithm for gathering many fat mobile robots in the plane. In: Proceedings of the 2013 ACM Symposium on Principles of Distributed Computing, pp. 250–259. ACM (2013)
3. Bhagat, S., Chaudhuri, S.G., Mukhopadhyaya, K.: Gathering of opaque robots in 3D space. In: Proceedings of the 19th International Conference on Distributed Computing and Networking, ICDCN 2018, pp. 2:1–2:10, Varanasi, India, 4–7 January 2018. https://doi.org/10.1145/3154273.3154322
4. Bhagat, S., Mukhopadhyaya, K.: Optimum algorithm for mutual visibility among asynchronous robots with lights. In: Spirakis, P., Tsigas, P. (eds.) SSS 2017. LNCS, vol. 10616, pp. 341–355. Springer, Cham (2017). https://doi.org/10.1007/978-3-319-69084-1_24
5. Bose, K., Adhikary, R., Kundu, M.K., Sau, B.: Arbitrary pattern formation on infinite grid by asynchronous oblivious robots. In: Das, G.K., Mandal, P.S., Mukhopadhyaya, K., Nakano, S. (eds.) WALCOM 2019. LNCS, vol. 11355, pp. 354–366. Springer, Cham (2019). https://doi.org/10.1007/978-3-030-10564-8_28
6. Bose, K., Kundu, M.K., Adhikary, R., Sau, B.: Arbitrary pattern formation by asynchronous opaque robots with lights. CoRR abs/1902.04950 (2019). http://arxiv.org/abs/1902.04950
7. Bramas, Q., Tixeuil, S.: Arbitrary pattern formation with four robots. In: Izumi, T., Kuznetsov, P. (eds.) SSS 2018. LNCS, vol. 11201, pp. 333–348. Springer, Cham (2018). https://doi.org/10.1007/978-3-030-03232-6_22
8. Cicerone, S., Di Stefano, G., Navarra, A.: Asynchronous arbitrary pattern formation: the effects of a rigorous approach. Distrib. Comput. **32**(2), 1–42 (2018). https://doi.org/10.1007/s00446-018-0325-7
9. Cicerone, S., Di Stefano, G., Navarra, A.: Embedded pattern formation by asynchronous robots without chirality. Distrib. Comput. 1–25 (2018). https://doi.org/10.1007/s00446-018-0333-7
10. Das, S., Flocchini, P., Santoro, N., Yamashita, M.: Forming sequences of geometric patterns with oblivious mobile robots. Distrib. Comput. **28**(2), 131–145 (2015). https://doi.org/10.1007/s00446-014-0220-9
11. Di Luna, G.A., Flocchini, P., Chaudhuri, S.G., Poloni, F., Santoro, N., Viglietta, G.: Mutual visibility by luminous robots without collisions. Inf. Comput. **254**, 392–418 (2017)
12. Dieudonné, Y., Petit, F., Villain, V.: Leader election problem versus pattern formation problem. CoRR abs/0902.2851 (2009). http://arxiv.org/abs/0902.2851
13. Dieudonné, Y., Petit, F., Villain, V.: Leader election problem versus pattern formation problem. In: Lynch, N.A., Shvartsman, A.A. (eds.) DISC 2010. LNCS, vol. 6343, pp. 267–281. Springer, Heidelberg (2010). https://doi.org/10.1007/978-3-642-15763-9_26

14. Feletti, C., Mereghetti, C., Palano, B.: Uniform circle formation for swarms of opaque robots with lights. In: Izumi, T., Kuznetsov, P. (eds.) SSS 2018. LNCS, vol. 11201, pp. 317–332. Springer, Cham (2018). https://doi.org/10.1007/978-3-030-03232-6_21

15. Flocchini, P., Prencipe, G., Santoro, N., Widmayer, P.: Arbitrary pattern formation by asynchronous, anonymous, oblivious robots. Theor. Comput. Sci. **407**(1–3), 412–447 (2008). https://doi.org/10.1016/j.tcs.2008.07.026

16. Fujinaga, N., Yamauchi, Y., Ono, H., Kijima, S., Yamashita, M.: Pattern formation by oblivious asynchronous mobile robots. SIAM J. Comput. **44**(3), 740–785 (2015). https://doi.org/10.1137/140958682

17. Lukovszki, T., Meyer auf der Heide, F.: Fast collisionless pattern formation by anonymous, position-aware robots. In: Aguilera, M.K., Querzoni, L., Shapiro, M. (eds.) OPODIS 2014. LNCS, vol. 8878, pp. 248–262. Springer, Cham (2014). https://doi.org/10.1007/978-3-319-14472-6_17

18. Peleg, D.: Distributed coordination algorithms for mobile robot swarms: new directions and challenges. In: Pal, A., Kshemkalyani, A.D., Kumar, R., Gupta, A. (eds.) IWDC 2005. LNCS, vol. 3741, pp. 1–12. Springer, Heidelberg (2005). https://doi.org/10.1007/11603771_1

19. Sharma, G., Busch, C., Mukhopadhyay, S.: Mutual visibility with an optimal number of colors. In: Bose, P., Gąsieniec, L.A., Römer, K., Wattenhofer, R. (eds.) ALGOSENSORS 2015. LNCS, vol. 9536, pp. 196–210. Springer, Cham (2015). https://doi.org/10.1007/978-3-319-28472-9_15

20. Sharma, G., Vaidyanathan, R., Trahan, J.L., Busch, C., Rai, S.: Complete visibility for robots with lights in $O(1)$ time. In: Bonakdarpour, B., Petit, F. (eds.) SSS 2016. LNCS, vol. 10083, pp. 327–345. Springer, Cham (2016). https://doi.org/10.1007/978-3-319-49259-9_26

21. Sharma, G., Vaidyanathan, R., Trahan, J.L., Busch, C., Rai, S.: O(log n)-time complete visibility for asynchronous robots with lights. In: IEEE International 2017 Parallel and Distributed Processing Symposium (IPDPS), pp. 513–522. IEEE (2017)

22. Suzuki, I., Yamashita, M.: Distributed anonymous mobile robots. In: SIROCCO 1996, The 3rd International Colloquium on Structural Information & Communication Complexity, pp. 313–330, Siena, Italy, 6–8 June 1996

23. Suzuki, I., Yamashita, M.: Distributed anonymous mobile robots: formation of geometric patterns. SIAM J. Comput. **28**(4), 1347–1363 (1999). https://doi.org/10.1137/S009753979628292X

24. Vaidyanathan, R., Sharma, G., Trahan, J.L.: On fast pattern formation by autonomous robots. In: Izumi, T., Kuznetsov, P. (eds.) SSS 2018. LNCS, vol. 11201, pp. 203–220. Springer, Cham (2018). https://doi.org/10.1007/978-3-030-03232-6_14

25. Yamauchi, Y., Yamashita, M.: Pattern formation by mobile robots with limited visibility. In: Moscibroda, T., Rescigno, A.A. (eds.) SIROCCO 2013. LNCS, vol. 8179, pp. 201–212. Springer, Cham (2013). https://doi.org/10.1007/978-3-319-03578-9_17

26. Yamauchi, Y., Yamashita, M.: Randomized pattern formation algorithm for asynchronous oblivious mobile robots. In: Kuhn, F. (ed.) DISC 2014. LNCS, vol. 8784, pp. 137–151. Springer, Heidelberg (2014). https://doi.org/10.1007/978-3-662-45174-8_10

Breaking the Linear-Memory Barrier in **MPC**: Fast **MIS** on Trees with Strongly Sublinear Memory

Sebastian Brandt[1], Manuela Fischer[1(✉)], and Jara Uitto[1,2]

[1] ETH Zurich, Zurich, Switzerland
brandts@ethz.ch, {manuela.fischer,jara.uitto}@inf.ethz.ch
[2] University of Freiburg, Freiburg im Breisgau, Germany

Abstract. Recently, studying fundamental graph problems in the *Massively Parallel Computation (*MPC*)* framework, inspired by the *MapReduce* paradigm, has gained a lot of attention. An assumption common to a vast majority of approaches is to allow $\widetilde{\Omega}(n)$ memory per machine, where n is the number of nodes in the graph and $\widetilde{\Omega}$ hides polylogarithmic factors. However, as pointed out by Karloff et al. [SODA'10] and Czumaj et al. [STOC'18], it might be unrealistic for a single machine to have linear or only slightly sublinear memory.

In this paper, we thus study a more practical variant of the MPC model which only requires substantially sublinear or even subpolynomial memory per machine. In contrast to the linear-memory MPC model and also to streaming algorithms, in this low-memory MPC setting, a single machine will only see a small number of nodes in the graph. We introduce a new and strikingly simple technique to cope with this imposed locality. In particular, we show that the *Maximal Independent Set (*MIS*)* problem can be solved efficiently, that is, in $O(\log^3 \log n)$ rounds, when the input graph is a tree. This constitutes an almost exponential speed-up over the low-memory MPC algorithm in $\widetilde{O}(\sqrt{\log n})$-algorithm in a concurrent work by Ghaffari and Uitto [SODA'19] and substantially reduces the local memory from $\widetilde{\Omega}(n)$ required by the recent $O(\log \log n)$-round MIS algorithm of Ghaffari et al. [PODC'18] to n^α for any $\alpha > 0$, without incurring a significant loss in the round complexity. Moreover, it demonstrates how to make use of the all-to-all communication in the MPC model to almost exponentially improve on the corresponding bound in the LOCAL and PRAM models by Lenzen and Wattenhofer [PODC'11].

1 Introduction

Parallel Computation Paradigms for Massive Data: When confronted with huge data sets, purely sequential approaches become untenably inefficient. To address this issue, several parallel computation frameworks specially tailored for processing large scale data have been introduced. Inspired by the MapReduce paradigm [21], Karloff, Suri, and Vassilvitskii [27] proposed the *Massively*

© Springer Nature Switzerland AG 2019
K. Censor-Hillel and M. Flammini (Eds.): SIROCCO 2019, LNCS 11639, pp. 124–138, 2019.
https://doi.org/10.1007/978-3-030-24922-9_9

*Parallel Computation (*MPC*)* model, which was later refined in many works [3,9,10,20,26].

Massively Parallel Computation Model: In the MPC model, an input instance of size N is distributed across M machines with local memory of size S each. The computation proceeds in rounds, each round consisting of *local computation* at the machines interleaved with *global communication* (also called *shuffling*, adopting the MapReduce terminology) between the machines.

In the shuffling step, every machine is allowed to send as many messages to as many machines as it wants, as long as for every machine the total size of sent and received messages does not exceed its local memory capacity. The quantity of main interest is the round complexity: the number of rounds needed until the problem is solved, that is, until every machine outputs its part of the solution. This measure constitutes a good estimate for the actual running time, as local computation is presumed to be negligible compared to the cost-intensive shuffling, which requires a massive amount of data to be transferred between machines.

Sublinear Memory Constraint: Note that $S \geq N$ leads to a degenerate case that allows for a trivial solution. Indeed, as the data fits into the local memory of a single machine, the input can be loaded there, and a solution can be computed locally. Due to the targeted application of MPC in the presence of massive data sets, thus large N, it is often crucial that S is not only smaller than N but actually substantially sublinear in N. The total memory $M \cdot S$ in the system has to be at least N, so that the input actually fits onto the machines, but ideally not much larger. Summarized, one requires $S = \tilde{O}(N^\alpha)$ memory on each of the $M = \tilde{O}\left(N^{1-\alpha'}\right)$ machines, for $0 < \alpha' \leq \alpha < 1$.

Sublinear Memory for Graph Problems: Basically all known MPC techniques for graph problems need essentially linear in n—for instance, $\tilde{\Omega}(n)$ or mildly sublinear like $n^{1-o(1)}$—memory per machine, where n is the number of nodes in the input graph[1]. We refer to [12] for a brief discussion of this assumption. Note that for sparse graphs with $N = \tilde{O}(n)$ edges, this violates the sublinear memory constraint, getting close to the degenerate regime. This issue has been artificially circumvented by explicitly restricting the attention to dense graphs with $N = \tilde{\Omega}(n^{1+\alpha})$ edges, as to ensure sublinearity in N while still not having to relinquish the nice property that (essentially) all nodes fit into the memory of a single machine [27].

Besides being a stretch of the definition, this additionally imposed condition of denseness of the input graph does not seem to be realistic. In fact, as recently also pointed out by [20], most practical large graphs are sparse. For instance in the Internet, most of the nodes have a small degree. Even for dense graphs,

---

[1] In the context of graph problems, it is typical to assume that all incident edges of a node are stored on the same machine, resulting in two copies of an edge, one for each endpoint. We refer to [35, Sect. 1.1] for a thorough discussion. Also see the remark at the end of this section.

where in theory the sublinear memory constraint is met, practicability of the parameter range does not need to be ensured; linear or slightly sublinear in n might be prohibitively large.

Furthermore, it is a very natural question to ask whether there is a fundamental reason why the known techniques get stuck at the near-linear barrier. One important aspect of our work is, from the theory perspective, that it breaks this threshold and thereby opens up a whole new unexplored domain of research questions.

Low-Memory MPC **Model:** We study a more realistic regime of the parameters for problems on large graphs, captured by the following *low-memory* MPC *model*.

Low-Memory MPC Model for Graph Problems:
The input is a graph $G = (V, E)$ with n nodes and m edges of size $N = \widetilde{O}(n + m)$. Given $M = \widetilde{O}\left(\frac{N^{1+\alpha'}}{S}\right)$ machines with local memory $S = \widetilde{O}(n^\alpha)$ each, for arbitrary constants $\alpha > 0$ and $\alpha' \geq 0$, we raise the question of what problems on G can be solved efficiently—that is, in poly log log n rounds.

Note that for sparse graphs, this condition exactly matches the sublinear memory constraint, and hence does not allow a trivial solution, as opposed to the setting with essentially linear memory assumed by all traditional MPC algorithms. We point out that low memory variants of the MPC model have been studied before [16,36], resulting in $O(\log n)$-round algorithms for a variety of problems. For many of the fundamental graph problems, however, $O(\log n)$ is often particularly easy to achieve, for instance by directly adopting LOCAL algorithms. We thus restrict our attention to "efficient" algorithms, which we define to be a poly log log n function, given that the state-of-the-art algorithms in the MPC model tend to end in this regime of round complexities. Note that no general super-constant lower bounds are known [37].

Concurrent Related Work: Until very recently, MPC research had focused on linear-memory algorithms. After (a preliminary version of) this work, the low-memory setting gained a lot of attention. This led to a variety of new results for graph problems in this model. In the following, we briefly outline recent developments that have taken place after this work.

In follow-up works, the authors of this paper [15] as well as independently Behnezhad et al. [13] devise MIS and matching algorithms in uniformly sparse graphs in $O(\log^2 \log n)$ rounds. In independent concurrent works, Ghaffari and Uitto [25] and Onak [33] provide algorithms for the problems of maximal independent set and matching in general graphs in $\widetilde{O}(\sqrt{\log n})$ rounds. In [17], Chang et al. develop an $O(\sqrt{\log \log n})$-round low-memory MPC algorithm for $(\Delta+1)$-list coloring.

Remark 1. If a node cannot be stored on a single machine, as its degree is larger than S, one has to introduce some sort of a workaround, e.g., have several smaller-degree copies of the same node on several separate machines. In the end of Sect. 2, we argue how to get rid of this issue, in our problem setting, by a clean-up phase in the very beginning. To make the statements and arguments more readable, we throughout think of this clean-up as having taken place already. Instead, one could also work with the simplifying assumption that every machine has $S = \widetilde{O}(n^\alpha + \Delta)$ memory, so that this issue does not arise in the first place.

1.1 Limitations of Linear-Memory MPC Techniques

In the following, we briefly overview recent techniques from the world of Massive Parallel Computation algorithms, and give some indications as to why they are likely to fail in the low-memory setting. The restriction to substantially sublinear memory, to the best of our knowledge, indeed rules out all the known MPC techniques, which seem to hit a boundary at roughly $S = \widetilde{\Omega}(n)$: moving from essentially linear to significantly sublinear memory incurs an exponential overhead in their round complexity, regardless of the density of the graph. This blow-up in the running time gives rise to the question of to what extent this near-linear memory is necessary for efficient algorithms.

(Direct) PRAM/LOCAL Simulation: One easy way of devising MPC algorithms is by shoehorning parallel or distributed algorithms into the MPC setting. For not too resource-heavy PRAM algorithms, there is also a standard simulation technique [26,27] that automatically transforms them into MPC algorithms. This approach, however, suffers from several shortcomings. First and foremost, the reduction leads to an $\Omega(\log n)$ round complexity, which is exponentially above our efficiency threshold.

Round Compression: Another similar technique, called *round compression*, introduced by Assadi and Khanna [5,7], provides a generic way of compressing several rounds of a distributed algorithm into fewer MPC rounds, resulting in an (almost) exponential speed-up. However, this method heavily relies on storing intermediate values, leading to a blow-up of the memory. In particular, when requiring the algorithm to run in poly log log n rounds, superlinear memory per machine seems inevitable.

Filtering: The idea of the *filtering* technique [28,29] is to reduce the size of the input by carefully removing a large amount of edges from the input graph that do not contribute to the (optimal) solution of the problem. This reduction is done by either randomly sampling the edges, or by deterministically choosing sets of relevant edges, so that the resulting (partial) problem instance fits on a single machine, and hence can be solved there. This requires significantly superlinear memory, or logarithmically many rounds if memory is getting close to the linear regime. Moreover, the approach seems to get stuck fundamentally at $S = \widetilde{\Omega}(n)$, since it relies on one machine eventually seeing the whole (filtered) graph.

Coresets: One very recent and promising direction for MPC graph algorithms is the one of (randomized composable) coresets [6,7], in some sense building on the filtering approach. The idea is that not all the information of the graph is needed to (approximately) solve the problem. One thus can get rid of unimportant parts of the information. Solving the problem on this core, one then can derive a perfect solution or a good approximation to it, at much lower cost. This solution, however, is found by loading the coreset (or parts of it) on one machine, and then locally computing a solution, which again seems to be stuck at $S = \widetilde{\Omega}(n)$, for similar reasons as the filtering approach.

1.2   Local Techniques for Low-Memory MPC

In this section, we propose a direction that seems to be promising to pursue in order to devise efficient MPC algorithms in the substantially sublinear memory regime.

Inherent Locality and Local Algorithms: The low-memory MPC model, as compared to the traditional MPC graph model and the streaming setting, suffers from inherent locality: Since the memory of a single machine is too small to fit all the nodes simultaneously, it will never be able to have a global view of all the nodes in the graph. When devising techniques, we thus need to deal with this intrinsic local view of the machines. It seems natural to borrow ideas from local distributed graph algorithms, which are designed exactly to cope with this locality restriction. A direct simulation, however, in most cases only results in $\Omega(\text{poly} \log n)$-round algorithms. The problem is that these algorithms do not make use of the additional power of the MPC model, the global all-to-all communication, as the communication in those message-passing-based models is restricted to neighboring nodes.

Local Meets Global: We propose a strikingly simple technique to enhance local-inspired approaches with global communication, in order to arrive at efficient algorithms in the world of low-memory MPC which are exponentially faster than their local counterparts and whose memory requirements are polynomially smaller per machine than their traditional MPC counterparts. We describe this technique in the context of the MIS problem on trees, even though it is more general.

1.3   Our Results

In this paper, we focus on the *Maximal Independent Set (*MIS*)* problem, one of the most fundamental local graph problems. We propose efficient and surprisingly simple algorithms for the case of trees, which is particularly interesting for the following reason. While trees admit a trivial solution in the linear-memory MPC model, this cheat will not work in our low-memory setting. In some sense, it thus is the easiest non-trivial case, which makes it the most natural starting point for further studies. In fact, we strongly believe that our techniques can be extended to more general graph families and problems[2].

---

[2] Indeed, there is a follow-up work generalizing our approach from trees to uniformly sparse graphs and from MIS only to MIS and maximal matching [13,15].

We provide two different efficient algorithms for MIS on trees. Our first algorithm in Theorem 1 is strikingly simple and intuitive, but comes with a small overhead in the total memory of the system, meaning that $M \cdot S$ is superlinear in the input size n.

Theorem 1. *There is an $O(\log^2 \log n)$-round MPC algorithm that w.h.p.[3] computes an MIS on n-node trees in the low-memory setting, that is, with $S = \widetilde{O}(n^\alpha)$ local memory on each of $M = \widetilde{O}(n^{1-\alpha/3})$ machines, for any $0 < \alpha < 1$.*

Our second algorithm in Theorem 2 gets rid of this overhead at the cost of a factor of $\log \log n$ in the running time.

Theorem 2. *There is an $O(\log^3 \log n)$-round MPC algorithm that w.h.p. computes an MIS on n-node trees in the low-memory setting, that is, with $S = \widetilde{O}(n^\alpha)$ local memory on each of $M = \widetilde{O}(n^{1-\alpha})$ machines, for any $0 < \alpha < 1$.*

The algorithms in Theorems 1 and 2 almost match the conditional lower bound of $\Omega(\log \log n)$ for MIS (on general graphs) due to Ghaffari, Kuhn, and Uitto [24], which holds unless there is an $o(\log n)$-round low-memory MPC algorithm for connected components. This, in turn, is believed to be impossible under a popular conjecture [38].

Our algorithms improve almost exponentially on the $\widetilde{O}(\sqrt{\log n})$-round low-memory MPC algorithms in concurrent works—for bounded-arboricity by Onak [33] and for general graphs by Ghaffari and Uitto [25]—as well as on the algorithms directly adopted from the PRAM/LOCAL model: an $O(\log n)$-round algorithm for general graphs due to Luby [32] and independently Alon, Babai and Itai [1], and the $O(\sqrt{\log n} \cdot \log \log n)$-round algorithm for trees by Lenzen and Wattenhofer [31]. Note that for rooted trees, the PRAM/LOCAL algorithm by Cole and Vishkin [19] directly gives rise to an $O(\log^* n)$-round low-memory MPC algorithm.

Moreover, our result shows that the local memory can be reduced substantially from $\widetilde{\Omega}(n)$ to n^α or even $n^{1/\text{poly} \log \log n}$ (see Corollary 1) while not incurring a significant loss in the round complexity, compared to the recent $O(\log \log n)$-round MIS algorithm by Ghaffari et al. [23].

Throughout the paper, when we mention the low-memory MPC setting, we refer to the parameter range for S as given in Theorems 1 and 2, that is, $S = \widetilde{O}(n^\alpha)$, where $\alpha > 0$ is an arbitrary constant. However, α does not need to be a constant. Indeed, we can even go to subpolynomial memory $S = n^{o(1)}$.

Corollary 1. *For any $\alpha = \Omega(1/\text{poly} \log \log n)$, an MIS on an n-node tree can be computed on $M = \widetilde{O}(n^{1-\alpha/3})$ machines with $S = \widetilde{O}(n^\alpha)$ local memory each in $O\left(\frac{1}{\alpha} \cdot \log^2 \log n\right)$ MPC rounds.*

---

[3] As usual, w.h.p. stands for *with high probability*, and means with probability at least $1 - n^{-c}$, for any $c \geq 1$.

1.4 Our Approach in a Nutshell

In the following, we give a short (and slightly imprecise) sketch of the steps of our algorithm. Our approach is based on the *shattering* technique which recently has gained a lot of attention in the LOCAL model of distributed computing [8] and goes back to the early nineties [11]. The idea of *shattering* is to randomly break the graph into several significantly smaller components by computing a partial solution. The problem on the remaining components then is solved by a *post-shattering* algorithm.

Shattering

The goal of our *shattering* technique is to compute an independent set such that after the removal of these independent set nodes and all their neighbors, the remaining graph, w.h.p., consists of components of size at most $\text{poly} \log n$. This is done in two steps: first, the maximum degree, w.h.p., is reduced to $\text{poly} \log n$ using the *iterated subsample-and-conquer* method, and then a local shattering algorithm is applied to this low-degree graph.

(I) Degree Reduction via Iterated Subsample-and-Conquer

Our subsample-and-conquer method will w.h.p. reduce the maximum degree of a graph polynomially, from Δ to roughly $\Delta^{1/(1+\alpha)}$, as long as $\Delta = \Omega(\text{poly} \log n)$. After $O(\log_{1+\alpha} \log \Delta)$ iterations, the degree of our graph drops to $\text{poly} \log n$.

Subsample: We sample the nodes independently with probability roughly $\Delta^{-\frac{1}{1+\alpha}}$, where Δ is an upper bound on the current maximum degree[4] of the graph. This subsampling step guarantees, roughly speaking, the following three very desirable properties of the graph G' induced by the sampled nodes.

(i) The diameter of each connected component of G' is bounded by $O(\log_\Delta n)$.
(ii) The number of nodes in each connected component of G' is at most $n^{\alpha/3}$.
(iii) Every node with degree $\Delta^{1/(1+\alpha)}$ or higher in G has many neighbors in G'.

Conquer: We find a random MIS in all the connected components of G' in parallel. This can be done by gathering the connected components[5], locally picking one of the two 2-colorings of this tree uniformly at random, and adding the black, say, nodes to the MIS. We will see that properties (i) and (ii) are crucial to ensure that the gathering can be done efficiently. In particular, storing the components on a single machine is possible due to the small size of the components, and the gathering is fast due to the small diameter. Because of property (iii), the randomness in the choice of the MIS in every connected component, as well as the tree structure, all high-degree nodes in the original graph (sampled or not), w.h.p., will have an adjacent independent set node and thus, are removed from the graph for the next iteration.

---

[4] Note that in the MPC model it is easy to keep track of the maximum degree.
[5] Gathering the connected components means loading all the nodes of a connected component onto the same machine.

(II) Low-Degree Local Shattering
Once the degree has dropped to $\Delta' = \operatorname{poly}\log n$, we apply the shattering part of the LOCAL MIS algorithm of Ghaffari [22], which runs in $O(\log \Delta') = O(\log\log n)$ rounds and w.h.p. leads to connected components of size $\operatorname{poly}\Delta' \cdot \log n = \operatorname{poly}\log n$ in the remainder graph. Observe that the simulation of this algorithm in the MPC model is straightforward.

Post-shattering
We gather the connected components of size $\operatorname{poly}\log n$ and solve the remaining problem locally.

2   Algorithm Overview and Roadmap

In this section, we give the formal statements we need to prove our main result, and provide an overview of the structure of the remainder of the paper. We start with a result that is repeatedly used to gather all nodes of a connected component onto one machine, provided that they fit there. It will come in two variants, which naturally give rise to Theorems 1 and 2, respectively. The proof is deferred to Sect. 4 (part (a)) and the full version [14] (part (b)).

Lemma 1 (Gathering). *Let G be an n-node graph and G' any n'-node subgraph of G consisting of connected components of size at most $k = O\left(n^{\alpha/3}\right)$ and diameter at most d. Then there are*

(a) an $O(\log d)$-round low-memory MPC algorithm with $M = \widetilde{O}\left(n^{1-\alpha/3}\right)$ machines and

(b) an $O(\log d \cdot \log\log n)$-round low-memory MPC algorithm with $M = \widetilde{O}\left(n^{1-\alpha}\right)$ machines, if $n' \cdot d^3 = O(n)$,

that compute an assignment of nodes to machines so that all the nodes of a connected component of G' are on the same machine.

Next, we will provide the results corresponding to the two main parts of our algorithm, the *shattering* and the *post-shattering*.

Lemma 2 (Shattering). *There are*

(a) an $O(\log\log n \cdot \log\log \Delta)$-round low-memory MPC algorithm that uses $M = \widetilde{O}(n^{1-\alpha/3})$ machines and

(b) an $O(\log^2\log n \cdot \log\log \Delta)$-round low-memory MPC algorithm with $M = \widetilde{O}(n^{1-\alpha})$ machines

that compute an independent set on an n-node tree with maximum degree Δ so that the remainder graph, after removal of the independent set nodes and their neighbors, w.h.p. has only components of size at most $\operatorname{poly}\log n$.

The proof of this *Shattering Lemma* can be found in Sect. 3. The following *Post-Shattering Lemma* is a direct consequence of the *Gathering Lemma*.

Lemma 3 (Post-Shattering). *There are*

(a) *an* $O(\log k)$*-round low-memory* MPC *algorithm with* $M = \tilde{O}\left(n^{1-\alpha/3}\right)$ *machines and*
(b) *an* $O(\log k \cdot \log \log n)$*-round low-memory* MPC *algorithm with* $M = \tilde{O}\left(n^{1-\alpha}\right)$ *machines*

that find an MIS *in an* n*-node graph consisting of connected components of size* $k = O\left(n^{\alpha/3}\right)$.

Proof. By Lemma 1, we can gather the connected components in $O(\log k)$ rounds. Then, an MIS of each connected component can be computed locally. Note that Theorem 1.1 by Ghaffari [22] certifies that the number of nodes remaining after our shattering process can be made small enough to satisfy the conditions required by Lemma 1.

Note that the naive simulation of the corresponding LOCAL post-shattering algorithm [22,34] would lead to a round complexity of $2^{O(\sqrt{\log \log n})}$.

We now put together the results to prove Theorems 1 and 2.

Proof (Proof of Theorems 1 and 2). We apply the *shattering* algorithm from Lemma 2 to get an independent set, with connected components of size $k = \text{poly} \log n$ in the remainder graph. Then we run the *post-shattering* algorithm from Lemma 3 to find an MIS in all these components. The combination of the initial independent set found by the *shattering* and all the MIS found by the *post-shattering* results in an MIS in the original tree.

Memory per Machine below Δ: If the degree of a node is larger than the local memory, one needs to store several lower-degree copies of this node on different machines. Here, we give a short argument for why one can assume without loss of generality that all incident edges of a node are stored on the same machine. Notice that in a tree with n nodes, there can be at most $n^{1-\alpha/2}$ nodes with degree at least $n^{\alpha/2}$. If we now just ignore all these high-degree nodes and find an MIS among the remaining nodes, the resulting graph, after removal of all MIS nodes and their neighbors, has at most $n^{1-\alpha/2}$ nodes. Repeating this argument roughly $2/\alpha$ times gives an MIS in the whole input graph.

3   Shattering

Lemma 4 (Iterated Subsample-and-Conquer). *There are*

(a) *an* $O(\log_{1+\alpha} \log \Delta)$*-round low-memory* MPC *algorithm with* $M = \tilde{O}(n^{1-\alpha/3})$ *machines and*
(b) *an* $O(\log_{1+\alpha} \log \Delta \cdot \log \log n)$*-round low-memory* MPC *algorithm with* $M = \tilde{O}(n^{1-\alpha})$ *machines*

that compute an independent set on an n*-node tree with maximum degree* Δ *such that the remainder graph, after removal of the independent set nodes and their neighbors, w.h.p. has maximum degree* $\text{poly} \log n$.

The proof of this lemma can be found in Sect. 3.1.

Lemma 5 (Low-Degree Local Shattering [22]). *There is an $O(\log \Delta)$-round LOCAL algorithm that computes an independent set on an n-node graph with maximum degree Δ so that the remainder graph, after removal of all nodes in the independent set and their neighbors, w.h.p. has connected components of size $poly\Delta \cdot \log n$.*

We now combine these two results to prove Lemma 2.

Proof (Proof of Lemma 2). We apply the algorithm of Lemma 4, w.h.p. yielding an independent set with a remainder graph that has maximum degree $\Delta' = \text{poly} \log n$. On this low-degree graph, we simulate the LOCAL algorithm of Lemma 5 in a straight-forward manner, which takes $O(\log \Delta') = O(\log \log n)$ rounds and w.h.p. leaves us with connected components of size $poly\Delta' \cdot \log n = \text{poly} \log n$.

3.1 Degree Reduction via Iterated Subsampling

We prove the following result, and then show how it can be used to prove Lemma 4. For the purposes of the proof of Lemma 6 we assume that Δ is a large enough $\text{poly} \log n$ in order to be able to apply Lemma 1. Notice that from the perspective of the final runtime, the exponent of the logarithm turns into a constant factor hidden in the O-notation.

Lemma 6. *There are*

(a) *an $O(\log \log n)$-round low-memory MPC algorithm with $M = \widetilde{O}(n^{1-\alpha/3})$ machines and*

(b) *an $O(\log^2 \log n)$-round low-memory MPC algorithm with $M = \widetilde{O}(n^{1-\alpha})$ machines*

that compute an independent set on an n-node tree G with maximum degree $\Delta = \Omega(\text{poly} \log n)$ such that the remainder graph, after removal of the independent set nodes and their neighbors, w.h.p. has maximum degree at most $\Delta^{(1+\delta')\delta}$, for some $\delta = \Theta(1/(1+\alpha))$ and any $\delta' > 0$.

Proof. We first outline the algorithm and then slowly go through the steps of the algorithm again while proving its key properties.

Algorithm: Every node is sampled independently with probability $\Delta^{-\delta}$ into a set V'. The connected components of $G' = G[V']$ are gathered by Lemma 1, and one of the two 2-colorings is picked uniformly at random, independently for every connected component. This can be done locally. All the black nodes, say, are added to the MIS, and are removed from the graph along with their neighbors.

Subsampling: We first prove that the random subsampling leads to nice properties of the graph induced by subsampled nodes.

Lemma 7. *After the subsampling, w.h.p., the following holds.*

(i) Every connected component of G' has diameter $O\left(\frac{1}{\delta} \cdot \log_\Delta n\right)$.
(ii) Every connected component of G' consists of $n^{O((1-\delta)/\delta)}$ nodes.
(iii) Every node with degree $\Omega\left(\Delta^{(1+\delta')\delta}\right)$ in G has degree $\Omega(poly\log n)$ in G'.

Proof. Consider an arbitrary path of length $\ell = \Omega\left(\frac{1}{\delta} \cdot \log_\Delta n\right)$ in G. This path is in G' only if all its nodes are subsampled into V', which happens with probability at most $\Delta^{-\delta \cdot \ell} = \frac{1}{poly\, n}$. A union bound over all—at most n^2 many—paths in the tree T shows that, w.h.p., the length of every path, and hence in particular also the diameter of every connected component, in G' is bounded by $O\left(\frac{1}{\delta} \cdot \log_\Delta n\right)$. Since the degree among the subsampled nodes is bounded by $O\left(\Delta^{1-\delta}\right)$, w.h.p., which is a simple application of Chernoff and union bound, it follows that every connected component consists of at most $O\left(\Delta^{(1-\delta)\cdot\ell}\right) = n^{O((1-\delta)/\delta)}$ nodes. Finally, another simple Chernoff and union bound argument shows that every node with degree $\Omega\left(\Delta^{(1+\delta')\delta}\right)$ in the graph G has at least $\Omega\left(\Delta^{\delta'\cdot\delta}\right) = \Omega(poly\log n)$ neighbors in G', which concludes the proof of Lemma 7 $\square$

Gathering: Since G' consists of components that have a low diameter by Lemma 7 (i) and that are small enough to fit on a single machine by Lemma 7 (ii)—provided that $\delta = \Theta\left(1/(1+\alpha)\right)$ is chosen such that the components have size $O\left(n^{\alpha/3}\right)$—we can gather them efficiently by Lemma 1, in either $O(\log\log n)$ or $O(\log^2 \log n)$ rounds. The random MIS can then be easily computed locally.

Random MIS: It remains to show that every high-degree node in G, w.h.p., has at least one adjacent node that joins the random MIS, which leads to the removal of this high-degree node from the graph. Note that this is trivially true for all subsampled nodes, by maximality of an MIS.

Now consider an arbitrary non-subsampled node v with degree $\Omega\left(\Delta^{(1+\delta')\delta}\right)$ and its $\Omega(poly\log n)$ subsampled neighbors, by Lemma 7 (iii). Observe that, since we are in a tree and thus in particular in a triangle-free graph, there cannot be edges between these neighbors. Therefore no two neighbors of a non-subsampled node belong to the same connected component in G', which means that all the neighbors in V' of v are colored independently, and hence are added to an MIS independently with probability $1/2$. By the Chernoff inequality, w.h.p. at least one of v's neighbors must have been added to an MIS, and a union bound over all nodes concludes the proof of the degree reduction, and hence of Lemma 6.

Proof (Proof of Lemma 4). This follows from $\log_{\frac{1}{(1+\delta')\delta}} \log \Delta = \log_{1+\alpha} \log \Delta$ many applications of Lemma 6. $\square$

4   Gathering Connected Components

In this section, we provide a proof of the *Gathering Lemma*. Our approach is essentially a tuned version of the Hash-to-Min algorithm by Chitnis et al. [18]

and the graph exponentiation idea by Lenzen and Wattenhofer [30]. Notice that, however, Chitnis et al. only show an $O(\log n)$ bound for the round complexity; it is not possible to just use their method as a black box. The section is divided into two subsections, where we first give a simple and fast but memory-inefficient algorithm and then present a slightly slower algorithm that only needs a constant space overhead.

In very recent works, independent of this paper, Andoni et al. [2] and Assadi et al. [4] studied, among other problems, finding connected components in the low-memory setting of MPC. In particular, Andoni et al. give algorithms to find connected components and to root a forest with constant success probability, with $O(m)$ total memory in time $O(\log d \cdot \log \log n)$. While their results are more general, ours have the advantages of being (arguably) much simpler and deterministic. Furthermore, to turn their algorithm to work with high probability, the straightforward approach requires a logarithmic overhead in the total memory.

In this section, we present the naive gathering algorithm in part (a) of Lemma 1. The proof of part (b) is deferred to the full version [14].

Proof (Proof of Lemma 1, part (a)). We first present the algorithm. The underlying idea of the algorithm is to find a minimum-ID[6] node within every component and to create a virtual graph that connects all the nodes of that component to this minimum-ID node, the *leader*.

Gathering Algorithm: In every round, every node u completes its 1-hop neighborhood to a clique. Once a round is reached in which there are no more edges to be added, u stops and selects its minimum-ID neighbor as its leader.

Observe that once there is a round in which u does not add any edges, the component of u forms a clique, and thus all nodes in this component have the same leader, namely the minimum-ID node in this clique. Next, we prove that this algorithm terminates quickly.

Lemma 8. *The gathering algorithm takes $O(\log d)$ rounds on a graph with diameter d.*

Proof. Consider any shortest path $u_1, \ldots, u_\ell$ of length $2 \leq \ell \leq d$. After the first round, every u_i gets connected to u_{i-2} and u_{i+2} for $2 < i < \ell - 1$. Thus, the diameter of the new graph is at most $\lceil 2d/3 \rceil$. After $O(\log d)$ iterations, the diameter within each component has reduced to 1, and the algorithm halts.

It remains to show that not too many edges are added, so that the virtual graph of any component still fits into the memory of a single machine.

Lemma 9. *The number of edges in the virtual graph created by the gathering algorithm in a component of size k is $O(k^3)$.*

---

[6] We assume without loss of generality that every node has a unique identifier. If not, every node can draw an $O(\log n)$-bit identifier at random, which w.h.p. will be unique.

Proof. During the execution of the algorithm, each node in a component may create an edge between any other two nodes in the corresponding component, thus at most k^3.

Since we require the components to be of size at most $O(n^{\alpha/3})$, the previous claim guarantees that the virtual graph of any connected component indeed fits into the memory. So as to not overload any machine with too many components, we assume that the shuffling distributes the components to the machines in an arbitrary feasible way, e.g., greedily[7].

Remark 2. A weakness of the gathering algorithm is that we need $O(k^3)$ memory to store a connected component of size k, even if this component originally just consisted of as few as $k-1$ edges. This is because a single edge can exist on up to k machines. In the worst case, the required memory is blown up by a power 3. This leads to a super-linear overall memory requirement, that is, we need roughly $N^{1+2\alpha/3}$ total memory in the system. Notice that this can be implemented either by adding more machines or by adding more memory to the machines, since we do not care on which machines the resulting components lie, as long as they fit the memory.

References

1. Alon, N., Babai, L., Itai, A.: A fast and simple randomized parallel algorithm for the maximal independent set problem. J. Algorithms **7**(4), 567–583 (1986)
2. Andoni, A., Stein, C., Song, Z., Wang, Z., Zhong, P.: Parallel graph connectivity in log diameter rounds. In: The Proceedings of the Symposium on Foundations of Computer Science (FOCS), pp. 674–685 (2018)
3. Andoni, A., Nikolov, A., Onak, K., Yaroslavtsev, G.: Parallel algorithms for geometric graph problems. In: Proceedings of the Symposium on Theory of Computing (STOC), pp. 574–583 (2014)
4. Assadi, S., Sun, X., Weinstein, O.: Massively parallel algorithms for finding well-connected components in sparse graphs. arXiv e-prints (2018)
5. Assadi, S.: Simple round compression for parallel vertex cover. arXiv preprint: 1709.04599 (2017). http://arxiv.org/abs/1709.04599
6. Assadi, S., Bateni, M., Bernstein, A., Mirrokni, V.S., Stein, C.: Coresets meet EDCS: algorithms for matching and vertex cover on massive graphs. In: Proceedings of the Thirtieth Annual ACM-SIAM Symposium on Discrete Algorithms, SODA 2019, San Diego, California, USA, 6–9 January 2019, pp. 1616–1635 (2019)
7. Assadi, S., Khanna, S.: Randomized composable coresets for matching and vertex cover. In: The Proceedings of the Symposium on Parallel Algorithms and Architectures (SPAA), pp. 3–12 (2017)
8. Barenboim, L., Elkin, M., Pettie, S., Schneider, J.: The locality of distributed symmetry breaking. J. ACM (JACM) **63**(3), 20 (2016)

---

[7] An alternative and simple way to prevent overloading is to add an $O(\log n)$ factor of memory per machine and consider a random assignment of components to machines as a balls-into-bins process.

9. Beame, P., Koutris, P., Suciu, D.: Skew in parallel query processing. In: Proceedings of the 33rd ACM SIGMOD-SIGACT-SIGART Symposium on Principles of Database Systems, pp. 212–223. ACM (2014)
10. Beame, P., Koutris, P., Suciu, D.: Communication steps for parallel query processing. J. ACM (JACM) **64**(6), 40 (2017)
11. Beck, J.: An algorithmic approach to the Lovász local lemma. Random Struct. Algorithms **2**(4), 343–365 (1991)
12. Behnezhad, S., Derakhshan, M., Hajiaghayi, M.: Brief announcement: semi-MapReduce meets congested clique. arXiv preprint arXiv:1802.10297 (2018)
13. Behnezhad, S., Derakhshan, M., Hajiaghayi, M., Karp, R.M.: Massively parallel symmetry breaking on sparse graphs: MIS and maximal matching. CoRR abs/1807.06701 (2018). http://arxiv.org/abs/1807.06701
14. Brandt, S., Fischer, M., Uitto, J.: Breaking the linear-memory barrier in MPC: Fast MIS on trees with strongly sublinear memory. arXiv:1802.06748 (2018)
15. Brandt, S., Fischer, M., Uitto, J.: Matching and MIS for uniformly sparse graphs in the low-memory MPC model. CoRR abs/1807.05374 (2018). http://arxiv.org/abs/1807.05374
16. Ceccarello, M., Pietracaprina, A., Pucci, G., Upfal, E.: Space and time efficient parallel graph decomposition, clustering, and diameter approximation. In: The Proceedings of the Symposium on Parallel Algorithms and Architectures (SPAA), pp. 182–191 (2015)
17. Chang, Y.J., Fischer, M., Ghaffari, M., Uitto, J., Zheng, Z.: The complexity of $(\delta + 1)$ coloring in congested clique, massively parallel computation, and centralized local computation. arXiv preprint arXiv:1808.08419 abs/1808.08419 (2018). http://arxiv.org/abs/1808.08419
18. Chitnis, L., Das Sarma, A., Machanavajjhala, A., Rastogi, V.: Finding connected components in map-reduce in logarithmic rounds. In: ICDE 2013, pp. 50–61. IEEE Computer Society (2013)
19. Cole, R., Vishkin, U.: Deterministic coin tossing and accelerating cascades: micro and macro techniques for designing parallel algorithms. In: Symposium on Theory of Computing, pp. 206–219 (1986)
20. Czumaj, A., Łacki, J., Madry, A., Mitrović, S., Onak, K., Sankowski, P.: Round compression for parallel matching algorithms. arXiv preprint: 1707.03478 (2017)
21. Dean, J., Ghemawat, S.: MapReduce: simplified data processing on large clusters. Commun. ACM **51**(1), 107–113 (2008)
22. Ghaffari, M.: An improved distributed algorithm for maximal independent set. In: the Proceedings of ACM-SIAM Symposium on Discrete Algorithms (SODA) (2016)
23. Ghaffari, M., Gouleakis, T., Konrad, C., Mitrovic, S., Rubinfeld, R.: Improved massively parallel computation algorithms for MIS, matching, and vertex cover. In: Proceedings of the International Symposium on Principles of Distributed Computing (PODC) (2018, to appear)
24. Ghaffari, M., Kuhn, F., Uitto, J.: Personal communication (2019)
25. Ghaffari, M., Uitto, J.: Sparsifying distributed algorithms with ramifications in massively parallel computation and centralized local computation. In: the Proceedings of ACM-SIAM Symposium on Discrete Algorithms (SODA), pp. 1636–1653 (2019)
26. Goodrich, M.T., Sitchinava, N., Zhang, Q.: Sorting, searching, and simulation in the MapReduce framework. In: Asano, T., Nakano, S., Okamoto, Y., Watanabe, O. (eds.) ISAAC 2011. LNCS, vol. 7074, pp. 374–383. Springer, Heidelberg (2011). https://doi.org/10.1007/978-3-642-25591-5_39

27. Karloff, H., Suri, S., Vassilvitskii, S.: A model of computation for MapReduce. In: the Proceedings of ACM-SIAM Symposium on Discrete Algorithms (SODA), pp. 938–948 (2010)

28. Kumar, R., Moseley, B., Vassilvitskii, S., Vattani, A.: Fast greedy algorithms in MapReduce and streaming. ACM Trans. Parallel Comput. (TOPC) **2**(3), 14 (2015)

29. Lattanzi, S., Moseley, B., Suri, S., Vassilvitskii, S.: Filtering: a method for solving graph problems in MapReduce. In: The Proceedings of the Symposium on Parallel Algorithms and Architectures (SPAA), pp. 85–94 (2011)

30. Lenzen, C., Wattenhofer, R.: Brief announcement: exponential speed-up of local algorithms using non-local communication. In: 29th Symposium on Principles of Distributed Computing (PODC), Zurich, Switzerland, July 2010

31. Lenzen, C., Wattenhofer, R.: MIS on trees. In: Proceedings of the International Symposium on Principles of Distributed Computing (PODC), pp. 41–48 (2011)

32. Luby, M.: A simple parallel algorithm for the maximal independent set problem. SIAM J. Comput. **15**(4), 1036–1053 (1986)

33. Onak, K.: Round compression for parallel graph algorithms in strongly sublinear space. CoRR abs/1807.08745 (2018). http://arxiv.org/abs/1807.08745

34. Panconesi, A., Srinivasan, A.: Improved distributed algorithms for coloring and network decomposition problems. In: Proceedings of the Symposium on Theory of Computing (STOC), pp. 581–592. ACM (1992)

35. Pandurangan, G., Robinson, P., Scquizzato, M.: Fast distributed algorithms for connectivity and MST in large graphs. In: Proceedings of the 28th ACM Symposium on Parallelism in Algorithms and Architectures, pp. 429–438. ACM (2016)

36. Pietracaprina, A., Pucci, G., Riondato, M., Silvestri, F., Upfal, E.: Space-round tradeoffs for MapReduce computations. In: Proceedings of the International Conference on Supercomputing, pp. 235–244. ACM (2012)

37. Roughgarden, T., Vassilvitskii, S., Wang, J.R.: Shuffles and circuits: (on lower bounds for modern parallel computation). In: Proceedings of the 28th ACM Symposium on Parallelism in Algorithms and Architectures, pp. 1–12. ACM (2016)

38. Yaroslavtsev, G., Vadapalli, A.: Massively parallel algorithms and hardness for single-linkage clustering under l_p distances. In: Proceedings of the 35th International Conference on Machine Learning, ICML 2018, Stockholmsmässan, Stockholm, Sweden, 10–15 July 2018, pp. 5596–5605 (2018)

Collaborative Delivery on a Fixed Path with Homogeneous Energy-Constrained Agents

Jérémie Chalopin[1], Shantanu Das[1], Yann Disser[2], Arnaud Labourel[1(✉)], and Matúš Mihalák[3]

[1] Aix Marseille Univ, Université de Toulon, CNRS, LIS, Marseille, France
{jeremie.chalopin,shantanu.das,arnaud.labourel}@lis-lab.fr
[2] Department of Mathematics, TU Darmstadt, Darmstadt, Germany
disser@mathematik.tu-darmstadt.de
[3] Department of Data Science and Knowledge Engineering, Maastricht University, Maastricht, Netherlands
matus.mihalak@maastrichtuniversity.nl

Abstract. We consider the problem of collectively delivering a package from a specified source to a designated target location in a graph, using multiple mobile agents. Each agent starts from a distinct vertex of the graph, and can move along the edges of the graph carrying the package. However, each agent has limited energy budget allowing it to traverse a path of bounded length B; thus, multiple agents need to collaborate to move the package to its destination. Given the positions of the agents in the graph and their energy budgets, the problem of finding a feasible movement schedule is called the *Collaborative Delivery* problem and has been studied before.

One of the open questions from previous results is what happens when the delivery must follow a fixed path given in advance. Although this special constraint reduces the search space for feasible solutions, the problem of finding a feasible schedule remains NP hard (as the original problem). We consider the optimization version of the problem that asks for the optimal energy budget B per agent which allows for a feasible delivery schedule, given the initial positions of the agents. We show the existence of better approximations for the fixed-path version of the problem (at least for the restricted case of single pickup per agent), compared to the known results for the general version of the problem, thus answering the open question from the previous paper.

We provide polynomial time approximation algorithms for both directed and undirected graphs, and establish hardness of approximation for the directed case. Note that the fixed path version of collaborative delivery requires completely different techniques since a single agent may be used multiple times, unlike the general version of collaborative delivery studied before. We show that restricting each agent to a single pickup allows better approximations for fixed path collaborative delivery

This work was partially supported by the ANR project ANCOR (anr-14-CE36-0002-01).

K. Censor-Hillel and M. Flammini (Eds.): SIROCCO 2019, LNCS 11639, pp. 139–153, 2019.
https://doi.org/10.1007/978-3-030-24922-9_10

compared to the original problem. Finally, we provide a polynomial time algorithm for determining a feasible delivery strategy, if any exists, for a given budget B when the number of available agents is bounded by a constant.

1   Introduction

We consider a team of mobile agents which need to collaboratively deliver a package from a source location to a destination. The difficulty of collaboration can be due to several limitations of the agents, such as limited communication, restricted vision or the lack of persistent memory, and this has been the subject of extensive research (see [21] for a recent survey). When considering agents that move physically (such as mobile robots or automated vehicles), a major limitation of the agents are their energy resources, which restricts the distance that the robot can travel. This is particularly true for small battery operated robots or drones, for which the energy limitation is the real bottleneck. We consider a set of mobile agents where each agent i has a budget B_i on the distance it can move, as in [1,5,11,12,14,19]. We model the environment as a directed or undirected edge-weighted graph G, with each agent starting on some vertex of G and traveling along edges of G, until it runs out of energy and stops forever. In this model, the agents are obliged to collaborate as no single agent can usually perform the required task on its own.

Given a graph G with designated source and target vertices, and k agents with given starting locations and energy budgets, the decision problem of whether the agents can collectively deliver a single package from the source to the target node in G is called COLLABORATIVEDELIVERY. Chalopin et al. [11,12] showed that COLLABORATIVEDELIVERY is weakly NP-hard on paths and strongly NP-hard on general graphs. When the agents are homogenous, each agent has the same uniform budget initially. The optimization version of this problem asks for the minimum energy budget B per agent, that allows a feasible schedule for delivering the package. Throughout this paper we consider agents with uniform budgets. There exist constant factor approximations [5,11] for the optimal budget needed for solving COLLABORATIVEDELIVERY.

Unlike previous papers, this paper considers a version of the problem where the package must be transported through a designated path that is provided as input to the algorithm. This is a natural assumption, e.g. for delivery of valuable packages which must go on a "safe" route, allowing them to be tracked. We call this variant FIXEDPATH COLLABORATIVEDELIVERY. Even with this additional constraint, the problem remains NP-hard for general graphs due to the result in [11]. Note that on trees, the two problems are equivalent and both problems are known to be weakly NP-hard. However, for arbitrary graphs, the two problems are quite different. In particular, in the FIXEDPATH COLLABORATIVEDELIVERY, each agent may be used multiple times, while in the original version each agent participates at most once in any optimal delivery schedule (see [11]). In this paper, we attempt to find the difference between the two problems in terms of approximability.

Our Contributions. We show that the best possible approximation of the optimal budget B for FIXEDPATH COLLABORATIVEDELIVERY is between 2 and 3 for directed graphs and at most 2.5 for undirected graphs. In contrast, the best known approximation ratio for the general version of COLLABORATIVEDE-LIVERY is 2 for undirected graphs [11], and there is no known lower bound on approximability.

In the fixed path version of the problem agents may be used multiple times in a feasible delivery schedule, i.e., the same agent may move the package along several disjoint segments of the path. Thus, it is not surprising that our solution for FIXEDPATH COLLABORATIVEDELIVERY has a higher approximation ratio than the general version of the problem where each agent is used at most once.

For better comparison, we can make the FIXEDPATH COLLABORATIVEDE-LIVERY problem easier by restricting each agent to a single pickup of the package. This easier version of the problem was considered recently in [25] which provided a 3-approximation algorithm. In this paper we improve upon this and provide a 2-approximation algorithm for directed graphs and a $(2 - 1/2^k)$-approximation algorithm for undirected graphs. We also show that there exists no polynomial-time approximation algorithm with better approximation ratio than $\frac{3}{2}$ for directed graphs.

Finally, for the case where the number of agents k is a constant, we show that the decision version of FIXEDPATH COLLABORATIVEDELIVERY can be solved in pseudo-polynomial time. For this setting, we also provide a fully polynomial-time approximation scheme (FPTAS) giving a $(1 + \epsilon)$-approximation to the optimal budget, for any $\epsilon > 0$.

Our Model. We consider finite, connected (or strongly connected), edge-weighted graphs $G = (V, E)$ with $n = |V|$ vertices. For undirected graphs, the weight $w(e)$ of an edge $e \in E$ defines the energy required to cross the edge in either direction. For directed graphs, there may be up to two directed arcs between any pair of vertices and the weight of each arc is the energy required to traverse the arc from its tail to its head. We have k mobile agents which are initially placed on arbitrary nodes $p_1, \ldots, p_k$ of G, called the starting positions. Each agent has an initially assigned energy budget $B > 0$ which allows each agent to move along the edges of the graph for a total distance of at most B (if an agent travels only on a part of an edge, its travelled distance is downscaled proportionally to the part travelled). We say that agents have *uniform budget B*.

The agents are required to move a package from a given source node s to a target node t. An agent can pick up the package when it is at the same location as the package; we say that the agent is carrying the package. An agent carrying the package can drop it at any location that it visits, i.e., either at a node or even at a point inside an edge/arc. The agents do not need to return to their starting locations, after completing their task. We assume that the graph and the starting locations are initially known and the objective is to compute a strategy for movements of the agents which allows the delivery of the package from s to t (along a given $s - t$ path P). We denote by $d(x, y) = d_G(x, y)$ the distance between two nodes x, y in G (i.e. the sum of the weights on the shortest path from x to y). The length of path P is the sum of the weights on the path, denoted by $w(P) = d_P(s, t)$.

Definitions. Given a graph G with edge-weights w, vertices $s \neq t \in V(G)$, starting nodes $p_1, \ldots, p_k$ for the k agents, and an energy budget B, we define COLLABORATIVEDELIVERY as the decision problem of whether the agents can collectively deliver the package.

A solution to COLLABORATIVEDELIVERY is given in the form of a *delivery schedule* which prescribes for each agent whether it moves and if so, the locations in which it has to pick up and drop off the package. A delivery schedule is *feasible* if the package can be delivered from s to t and each agent moves at most distance B.

Given (G, w, s, t) and the locations $p_1, \ldots, p_k$ of the agents in G, the optimization version of COLLABORATIVEDELIVERY is to compute the minimum value of B for which there exists a feasible delivery schedule. The problem of FIXED-PATH COLLABORATIVEDELIVERY provides an additional parameter: an $(s - t)$ path P in G, and the feasible delivery schedules are restricted to those where the package travels on the given path P.

Related Work. The model of energy-constrained robot was introduced by Betke et al. [9] for single agent exploration of grid graphs. Later Awerbuch et al. [2] studied the same problem for general graphs. In both these papers, the agent is allowed to return to its starting node to refuel, and between two visits to the starting node the agent can traverse at most B edges. Duncan et al. [18] studied a similar model where the agent is tied with a rope of length B to the starting location and they optimized the exploration time, giving an $\mathcal{O}(m)$ time algorithm. A more recent paper [15] provides a constant competitive algorithm for the same exploration problem when the value of energy budget B is not much more than the distance to farthest node.

For energy-constrained agents without the option of refuelling, multiple agents may be needed to explore even graphs of restricted diameter. Given a graph G and k agents starting from the same location, each having an energy constraint of B, deciding whether G can be explored by the agents is NP-hard, even if graph G is a tree [22]. Dynia et al. studied the online version of the problem [19,20]. They presented algorithms for exploration of trees by k agents when the energy of each agent is augmented by a constant factor over the minimum energy B required per agent in the offline solution. Das et al. [14] presented online algorithms that optimize the number of agents used for tree exploration when each agent has a fixed energy bound B. On the other hand, Dereniowski et al. [17] gave an optimal time algorithm for exploring general graphs using a large number of agents. When both k and B are fixed, Bampas et al. [3] studied the problem of maximizing the number of nodes explored by the agents, called the *maximal exploration* problem.

When multiple agents start from arbitrary locations in a graph, optimizing the total energy consumption of the agents is computationally hard for several formation problems which require the agents to place themselves in desired configurations (e.g. connected or independent configurations) in a graph [10,16]. Anaya et al. [1] studied centralized and distributed algorithms for the information exchange by energy-constrained agents, in particular the problem of transferring information from one agent to all others (*Broadcast*) and from all agents to one agent (*Convergecast*). For both problems, they provided hardness results for trees and approximation algorithms for arbitrary graphs. Czyzowicz et al. [13] recently showed that the problems of collaborative delivery, broadcast and convergecast remain NP-hard for general graphs even if the agents are allowed to exchange energy when they meet. Further results on collective delivery with energy exchange showed that the problem remains hard even when B is a small constant [4].

As mentioned before, the collaborative delivery problem was first studied by Chalopin et al. [11] in arbitrary undirected graphs for both uniform or non-uniform budgets. When the agents have non-uniform budgets, they provided the so-called *resource-augmented algorithms* where the budgets of the agents are augmented by a small constant factor to allow polynomial time solutions for all feasible instances of the original problem. The surprising result that collaborative delivery non-uniform budgets is weakly NP-hard even for a line was proved in [12] where a quasi-pseudo-polynomial time algorithm was provided.

Bärtschi et al. [5] considered the returning version of the problem, where each agent needs to return to its starting location. They showed that, in this case, the problem can be solved in polynomial time for trees, but the problem is still NP-hard for arbitrary planar graphs. They provided 2-resource-augmented algorithm for general graphs in this setting and showed that it is the best possible solution that can be computed in polynomial time. Other variants of collaborative delivery that have been considered are when agents have distinct rate of energy consumption [6] or when the agents have distinct speeds [7]. In these cases the optimization criteria is to minimize the total energy consumption and/or the total time taken for delivery. Another related work [8] studied the collective delivery problem for selfish agents that try to optimize their personal gain.

2   Lower Bounds on Optimal Budget

In this section we prove some lower bounds on the approximation factor for any polynomial time algorithm that solves collaborative delivery with uniform budgets on a fixed path.

We give a reduction from an NP-hard variant of SAT [24]. Note the difference to the polynomially solvable $(3,3)$-SAT, where each variable appears in exactly three clauses [26].

$(\leq 3,3)$-SAT
Input: A formula with a set of clauses C of size three over a set of variables X, where each variable appears in at most three clauses.
Problem: Is there a truth assignment of X satisfying C?

Observe that we may assume that each variable appears at most twice in positive literals and at most once in a negative literal, otherwise we can either eliminate or negate the variable.

Theorem 1. *The minimum uniform budget required to solve* FIXEDPATH COL-LABORATIVEDELIVERY *on directed graphs cannot be approximated to within a factor better than* 2 *in polynomial time, unless* P = NP.

Proof. We reduce from $(\le 3,3)$-SAT by constructing, for every sufficiently small $\varepsilon > 0$ and every instance of $(\le 3,3)$-SAT, an instance of FIXEDPATH COLLABORATIVEDELIVERY that has a solution with budget $B \le 2 - \varepsilon$ if and only if the $(\le 3,3)$-SAT instance has a satisfying assignment. In this case, our instance always admits a solution with budget $B = 1$. Since $(\le 3,3)$-SAT is NP-hard, this then implies that no $(2 - \varepsilon)$-approximation algorithm can exist, unless P = NP.

In the following, fix $0 < \varepsilon < 1$ and consider an instance of $(\le 3,3)$-SAT with variables $x_1, \ldots, x_t$ and clauses $C_1, \ldots, C_m$. We construct a (directed) instance of FIXEDPATH COLLABORATIVEDELIVERY with $k = (3+q)t$ agents, where $q := \lceil 3/\varepsilon \rceil$, starting at vertices $p_1, \ldots, p_k$. The agents $p_{3i-2}, p_{3i-1}, p_{3i}$ for $i \in \{1, \ldots, t\}$ are associated with the (at most) two positive literals and the single negative literal of variable x_i, in this order, that appear in the clauses. In case variable x_i only appears in a single positive literal, the agent p_{3i-1} does not represent any literal. The other agents are so-called *blockers*. We incrementally construct the fixed s-t-path $P = (v_0, v_1, \ldots, v_{m+2(q+1)t})$ that the package has to be transported along.

The first m arcs of P correspond to the clauses $C_1, \ldots, C_m$. Each arc $e = (v_{j-1}, v_j)$ with $j \in \{1, \ldots, m\}$ has weight $w(e) = 1$ and is associated with clause C_j. For every literal of a variable x_i that appears in C_j, we let p_{ij} denote the starting position of the (unique) agent associated with this literal, and we introduce an arc $e_{ij} = (p_{ij}, v_{j-1})$ of weight $w(e_{ij}) = 0$.

Now we add the variable gadgets to the path P. Let $q_i := m + 2(q+1)(i-1)$. The gadget associated with each variable x_i (cf. Fig. 1) is the subpath $P_i = (v_{q_i}, \ldots, v_{q_{i+1}})$ of P consisting of $2q + 2$ edges. The first q arcs have weight $\varepsilon/3$ each, the central two arcs $e_i = (v_{q_i+q}, v_{q_i+q+1})$ and $e'_i = (v_{q_i+q+1}, v_{q_i+q+2})$ have weights $w(e_i) = \varepsilon/3$ and $w(e'_i) = 1 - \varepsilon/3$, and the final q arcs have weight $1 - \varepsilon/3$ each. For $\ell \in \{1, \ldots, q\}$, we connect the starting position of the $((i-1)q + \ell)$-th blocker to $v_{q_i+\ell-1}$ with an arc of weight 0, and we add a shortcut arc (that cannot be taken by the package) $(v_{q_i+\ell}, v_{q_{i+1}-\ell})$ of weight 0. Finally, we connect the three agents associated with variable x_i as follows: We add an arc (p_{3i-2}, v_{q_i+q}) of weight $1 - \varepsilon/3$, an arc (p_{3i-1}, v_{q_i+q+1}) of weight $\varepsilon/3$, and an arc (p_{3i}, v_{q_i+q}) of weight 0.

We first claim that in every solution with $B \le 2 - \varepsilon$ we can assume that, without loss of generality, for every $i \in \{1, \ldots, t\}$ and every $\ell \in \{1, \ldots, q\}$, the $((i-1)q+\ell)$-th blocker transports the package across the arc $(v_{q_i+\ell-1}, v_{q_i+\ell})$, then takes the shortcut arc $(v_{q_i+\ell}, v_{q_{i+1}-\ell})$, and finally transports the package across the arc $(v_{q_{i+1}-\ell}, v_{q_{i+1}-\ell+1})$. To see this, consider the last arc $(v_{q_{i+1}-1}, v_{q_{i+1}})$ of P'_i. Since the arcs preceding the vertices v_{q_i} and $v_{q_{i+1}-1}$ along P both have length at least $1 - \varepsilon/3$, no agent other than the two blockers connected to v_{q_i} and v_{q_i+1} can reach $v_{q_{i+1}-1}$ with more than $B - (1 - \varepsilon/3) \leq 1 - 2\varepsilon/3$ budget remaining, which is insufficient to cross the last arc of P_i'. Since there is no disadvantage in using the $((i-1)q+1)$-st blocker rather than the $((i-1)q+2)$-nd, we may assume that the $((i-1)q+1)$-st blocker transports the package as claimed. By repeating this argument (slightly adapted for the iq-th blocker), we can fix all subsequent blockers, too. Note that each blocker requires only $B = 1$.

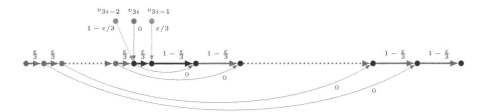

Fig. 1. Illustration of the variable gadget. Thick, horizontal arcs are part of the fixed path of the package. Colors indicate responsibilities: blue is for blockers and green/red is for agents associated with positive/negative literals. (Color figure online)

After fixing all blockers, we can observe that every agent with budget $B \leq 2 - \varepsilon$ can only transport the package along an arc inside a single clause or variable gadget: This is because transporting the package inside a clause gadget requires one unit of budget, and entering/leaving a variable gadget before or after transporting the package across one of its two central arcs also takes at least one unit of budget (all other arcs of a variable gadget are handled by blockers).

Finally, and crucially, observe that, in order to transport the package across the two central edges of the variable gadget for x_i, either the two agents p_{3i-2} and p_{3i-1} associated with the positive literals of x_i, or the agent p_{3i} associated with the negative literal are needed, since blockers cannot help (see above). We interpret the former situation as x_i being set to *false*, and the latter situation as x_i being set to *true*. Note that either assignment can be accomplished with $B = 1$.

If a variable is set to *true*, the two agents corresponding to positive literals are free to transport the package across the single (!) clause gadget each of them can reach. Otherwise, the agent corresponding to the negative literal is free to do this. In both cases, we interpret this as the clause being satisfied by the corresponding variable. Note that satisfying a clause again requires only $B = 1$.

Clearly, we can turn a satisfying assignment for $(\leq 3, 3)$-SAT into a feasible solution of FIXEDPATH COLLABORATIVEDELIVERY with $B = 1$. Conversely, every feasible solution of FIXEDPATH COLLABORATIVEDELIVERY with $B \leq 2 - \varepsilon$ corresponds to a satisfying assignment for $(\leq 3, 3)$-SAT. Note that q is constant for fixed ε, hence our construction can be done in polynomial time.

3 Approximation Algorithms for Fixed Path Delivery

In this section, we give approximation algorithms solving FIXEDPATH COLLAB-ORATIVEDELIVERY for both directed and undirected graphs. In the following, we assume that we are given the optimal value of B for a given instance of the problem and we provide a polynomial time algorithm to compute a delivery strategy that uses an energy budget of at most $\alpha \cdot B$ for some constant $\alpha > 1$. When B is not known, we can guess the optimal value of B by using a binary search in the interval $[D/k, D]$ where D is the length of the given fixed path plus the distance from node s to the nearest agent. The binary search terminates when we find the smallest B for which our algorithm provides a valid strategy for a budget of $\alpha \cdot B$. Clearly this provides an α-approximation algorithm for the optimization problem.

3.1 Directed Graphs: 3-Approximation

Theorem 2. *There is a 3-approximation algorithm for* FIXEDPATH COLLABO-RATIVEDELIVERY *on directed graphs.*

Proof. Consider an instance $(G, w, P, \{p_i \mid 1 \leq i \leq k\})$ of FIXEDPATH COLLAB-ORATIVEDELIVERY on directed graphs and let S be an optimal solution of this instance with uniform budget B. For i from 0 to $\ell = \left\lfloor \frac{d_P(s,t)}{B} \right\rfloor$, we define m_i as the point on P at distance iB from s. Observe that $\ell = O(\min(n, k))$ since the path P is of length less or equal than kB, P has at most $n - 1$ arcs and each arc in P has a weight at most B. For i from 0 to $\ell - 1$, let I_i be the interval $[m_i, m_{i+1}]$ on path P. In the solution S, there is an agent a_j starting at position p_j that moves the package from s to some point in I_0. Observe that since the length of each interval is B, for any set $\mathcal{I}$ of l intervals at least l different agents must carry the package inside $\cup_{I \in \mathcal{I}} I$, i.e., the trajectory of these agents intersects interval $\cup_{I \in \mathcal{I}} I$ in S. If the number of such agents for a set $\mathcal{I}$ is exactly l, it means each agent covers exactly an interval of size B and there is no other agent picking the package at the end of the last interval. This can only happen if $\mathcal{I} = \cup_{i=0}^{\ell-1} I_i$, $m_\ell = t$ and for all $0 \leq i \leq \ell - 1$, there is an agent at m_i. This case is easy to check and if it happens, one can construct an easy optimal solution. Hence, we can assume, w.l.o.g., that any set $\mathcal{I}$ of l intervals at least $l + 1$ different agents must carry the package inside $\cup_{I \in \mathcal{I}} I$. Hence, there exists a bijection f between a set $R \subseteq [1, k] \setminus \{j\}$ and $[0, \ell - 1]$ such that for each $i \in R$, agent a_i carries the package inside interval $I_{f(i)}$ in S. Observe that $d_G(p_j, m_0) \leq B$ since agent a_j can reach $s = m_0$ with budget B in the solution S. For all $i \in R$, we have $d_G(p_i, m_{f(i)+1}) \leq 2B$ since agent a_i can reach some point in $I_{f(i)}$ with budget B in solution S and then reach $m_{f(i)+1}$ by moving inside P for a distance at most B. We can deduce that there is a bijection g between the set $R' = R \cup \{j\}$ and $[0, \ell]$ such that $d_G(p_i, m_{g(i)}) \leq 2B$. One can find such a bijection g using the following algorithm :

1. Construct a weighted bipartite graph $H = (A \cup M, E, w_H)$ with $A = [0, k]$, $M = [0, \ell]$, $E = M \times A$ and for all $i \in M, j \in A$, $w_H(ij) = d_G(m_i, p_j)$.

This can be done in $O(k(m + n \log n))$ using a Dijkstra's algorithm [23] starting from each starting position of an agent. Observe that graph H has $O(\min(n, k) + k) = O(n + k)$ vertices and $O(min(n, k).k)$ edges.

2. Compute a maximal matching in H that minimizes the maximum weight. For each weight ω, one can compute in time $O((n + k)^2 \log(n + k) + k(n + k) \min(n, k))$ a maximal matching [23] in the graph H without edges of weight greater than ω. Hence, one can decide if there is a maximal matching in H with maximum weight ω and by using binary search, one can compute a maximal matching in H which minimizes the maximum weight, in time $O(\log k((n + k)^2 \log(n + k) + k(n + k) \min(n, k)))$.

3. For each edge ij in the matching, we fix $g(i) = j$. This gives us a bijection g between some set R' of size $\ell + 1$ and M. This bijection minimizes the maximal distance $d_G(p_i, m_{g(i)})$ and this value must be less than $2B$ since there is at least one such bijection.

From such a bijection g, we can deduce a 3-approximated solution of our instance: for each $i \in [0, \ell]$, agent $a_{g^{-1}(i)}$ moves to point m_i (cost less than $2B$) and then carries the package to point m_{i+1} if $i < \ell$ or t otherwise (cost less than B).

3.2   Undirected Graphs: 2.5-Approximation

Theorem 3. *There is a 2.5-approximation algorithm for* FIXEDPATH COLLABORATIVEDELIVERY *on undirected graphs.*

Proof. The proof is similar to that of Theorem 2, the intervals are slightly different in order to use the possibility for an agent to move on the path P in both directions.

Consider an instance $(G, w, P, \{p_i \mid 1 \leq i \leq k\})$ of FIXEDPATH COLLABORATIVEDELIVERY on undirected graphs and let S be an optimal solution with budget B. For i from 0 to $\ell = \left\lfloor \frac{d_P(s,t)}{B} \right\rfloor$, we define m_i as the point on P at distance iB from s (same definition as in proof of Theorem 2). For i from 1 to ℓ, we define m'_i as the point on P at distance $iB - B/2$ from s. We set $m'_0 = s$. Let $\ell' = \ell + 1$ and $m'_{\ell'}$ be the point of P at distance $\ell B + B/2$ from s, if $d_P(m_\ell, t) > \frac{B}{2}$, and let $\ell' = \ell$ and $m'_{\ell'} = t$ otherwise. For i from 0 to $\ell' - 1$, let I_i be the interval $[m'_i, m'_{i+1}]$ on path P. Observe that $|I_0| = B/2$, and for each $i \in [1, \ell' - 1]$, $|I_i| = B$. Hence the union of l intervals have a length strictly greater than $(l - 1)B$. With a similar argument as proof of Theorem 2, there exists a bijection f between a set $R \subseteq [1, k]$ and $[1, \ell' - 1]$ such that for each $i \in R$, agent a_i carries the package inside interval $I_{f(i)}$ in S. The starting position p_i of agent a_i is at distance at most B from some point $s_{f(i)}$ in $I_{f(i)}$. Observe that for all $i \in [0, \ell]$, we have $d_P(m_i, m'_i) \leq B/2$ and $d_P(m_i, m'_{i+1}) \leq B/2$. It follows that every point in I_i and so s_i is at distance at most $B/2$ of m_i. By the triangular inequality, we have that for all $i \in [0, \ell]$ $d_P(p_i, m_{f(i)}) \leq \frac{3}{2}B$. One can find a bijection f having this property with the same algorithm as in the proof of Theorem 2.

From a bijection f, we can deduce a 2.5-approximated solution of our instance: for each $i \in [0, \ell]$, agent $a_{f^{-1}(i)}$ moves to point m_i (cost less or equal than $\frac{3}{2}B$) and then carries the package to point m_{i+1} if $i < \ell$ or t otherwise (cost less or equal than B).

4   Special Case: Single Pickup Per Agent

In this section, we consider a slightly easier version of the problem when each agent can pickup the package at most once. We first present a lower bound of $\frac{3}{2}$ on the approximation ratio of optimizing FIXEDPATH COLLABORATIVEDELIVERY.

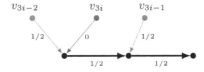

Fig. 2. Illustration of the clause gadget for the case where agents cannot pickup the package more than once.

4.1   Lower Bound

Theorem 4. *The minimum uniform budget required to solve* FIXEDPATH COLLABORATIVEDELIVERY *on directed graphs cannot be approximated to within a factor better than* 1.5 *in polynomial time, unless* $P = NP$, *even when each agent may pickup the package at most once.*

Proof. We use the same construction as in the proof of Theorem 1, but we set $\varepsilon = 3/2$ and $q = 0$ (cf. Fig. 2). All claims in the proof of Theorem 1 remain valid for any $B < 3/2$. Note that, since we eliminated all blockers, no agent has to pickup the package more than once in the optimum solution.

4.2   Approximation Algorithm for Single Pickup Per Agent

Lemma 1. *Given any instance of the decision problem for* FIXEDPATH COLLABORATIVEDELIVERY *that admits a solution where each agent can pickup the package at most once; then we can compute in polynomial time a 2-approximate delivery strategy. When the graph is undirected, we can compute a* $(2 - 1/2^k)$-*approximation in polynomial time.*

Proof. Suppose there exists a feasible solution S for the problem using uniform budget B and single pickup per agent. Consider the fixed $(s - t)$ path P and partition it into segments using the points $X = (m_1, m_2 \ldots m_l = t)$ on P, such that $l = \lceil w(P)/B \rceil$, the length of segment (m_i, m_{i+1}) is B, $\forall 1 \le i < l$, and the length of the first segment $(s, m_1) \leq B$. We have the following observations for strategy S: (1) Any agent that moves the package over point m_i in strategy S must have enough energy to reach point m_i, and (2) Any single agent can not transport the package over two distinct points in X since the distance between two points is at least B.

Case (i): In strategy S, the agent that picks up the package at s is not the same agent that moves the package over m_1. In this case, there exists a matching between the agents and the points $X^+ = (s = m_0, m_1, m_2, \ldots m_l = t)$ such that each agent can reach the point that it is mapped to. We call any such matching a type M_0 matching. Case (ii): In strategy S, a single agent delivers the package from s to m_1 with its original energy budget B. In this case, there exists a matching between the agents and the points in X (w.l.o.g. agent a_i is mapped to point p_i), such that, agent a_1 has enough energy to move the package from s to m_1 and $\forall i > 1$, agent a_i can reach m_i, using budget B. We call any such matching a type M_1 matching. Note that if S is a feasible solution to the problem using single pickup per agent, then there exists a matching of type M_0 or M_1. If we can find such a matching, then, using budget B per agent, we can move the package to point p_1 and move each agent a_i to the respective point m_i in path P. If the budget of each agent is augmented by factor 2, then using the additional budget B, the agent a_i that is mapped to point m_i can actually deliver the package to the next point m_{i+1}. This gives a 2-approximate solution to the problem (for directed and undirected graphs).

For undirected graphs, we will now construct a delivery strategy where each agent has a budget $2B - B/2^l$. As per previous discussion, using the original budget B each agent a_i can reach point m_i and the package can be moved to point m_1. Each agent a_i now has available energy budget of at least $B - B/2^l$ after arriving at the designated point m_i.

Consider the points $m'_i = m_i + B - (2^i - 1)B/2^l$, $1 \leq i \leq l-1$. The agent a_1 delivers the package from point m_1 to m'_1. For $1 < i < l$, each agent a_i located at point m_i returns to m'_{i-1} to pick up the package and then moves the package to point m'_i. This requires an additional budget of $B - (2^i - 1)B/2^l + 2 \times 2^i B/2^l = B(1 - 1/2^l)$, for each of these agents. Finally, note that the distance between point m'_{l-1} and the target $t = m_l$ is at most $B/2 - B/2^l$, and so the agent a_l can move from m_l to m'_{l-1} to pick up the package and deliver it to the target, using $2 \times (B/2 - B/2^l) < B(1 - 1/2^l)$ additional energy.

Since $k \geq l$, this provides a $(2 - 1/2^k)$-approximate solution strategy for any instance which has a feasible solution using k agents and a single pickup per agent.

The computation of the schedule requires constructing a bipartite graph between k agents and at most k points, and then solving maximum matching in this bipartite graph. The former task requires $O(n^3)$ time using an all-pair shortest path algorithm to compute distances in the original graph. The second task of computing the matching requires at most $O(k^2)$ time.

As in the previous section, we use a binary search to find the smallest B for which there exists a matching of type M_0 or M_1 from the above lemma. This gives us a $(2 - 1/2^k)$-approximate (respectively 2-approximate) solution to the optimization problem for undirected (resp. directed) graphs. Hence we have the following results:

Theorem 5. *The minimum uniform budget required to solve* FIXEDPATH COL-LABORATIVEDELIVERY *with single pickup per agent on undirected graphs can be approximated to a factor* $(2 - 1/2^k)$, *in polynomial time.*

Theorem 6. *The minimum uniform budget required to solve* FIXEDPATH COL-LABORATIVEDELIVERY *with single pickup per agent on directed graphs can be approximated to a factor* 2, *in polynomial time.*

5   Delivery with Few Agents

In this section we consider the special case when agents are allowed to exchange the package at vertices only. Using the dynamic programming technique, we design an algorithm that for a given B, computes whether there exists a feasible schedule with uniform budget B, and has a running time that is exponential in k and pseudo-polynomial in n (the run-time will depend on B). To find a minimum B such that there exists a feasible schedule, we can use binary search on B, which adds multiplicative $\log B$ increase to the run-time.

We keep a boolean table $T_v[j|p_1^v, \ldots, p_k^v|B_1^v, \ldots, B_k^v]$ denoting whether there exists a feasible schedule that delivers the package from s to vertex v on the path P such that

1. the last agent that delivers the package to vertex v is agent a_j,
2. the positions of the k agents, when the package arrives at v, are $p_1^v, \ldots, p_k^v$, and
3. the remaining budgets of the agents are $B_1^v, \ldots, B_k^v$.

We initialize $T_s[0|p_1, \ldots, p_k|B, \ldots, B] = $ TRUE and initialize $T_s[\ldots] = $ FALSE for all other values of j and p_i^s and B_i^s, $i = 1, \ldots, k$. Here, $j = 0$ denotes that no agent has been used yet. We also abuse the notation and use p_0 to denote s. Clearly, $T_v[j|p_1^v, \ldots, p_k^v|B_1^v, \ldots, B_k^v] = $ TRUE if and only if $p_j^v = v$, and there exists a vertex u on the path P before vertex v and an agent's index $j' \neq j$ such that there is a feasible schedule where agent a_j walks from position p_j^u to pick-up the package at vertex u from agent $a_{j'}$ and carries it from vertex u to vertex v. I.e., we have $T_v[j|p_1^v, \ldots, p_k^v|B_1^v, \ldots, B_k^v] = $ TRUE if and only if there exists u and j' and an entry in the table T such that $T_u[j'|p_1^u, \ldots, p_k^u|B_1^u, \ldots, B_k^u] = $ TRUE and $p_j^v = v$, $p_{j'}^v = p_{j'}^u = u$, $p_i^v = p_i^u$ for every $i \neq j, j'$, $B_j^v = B_j^u - d(p_j^u, u) - d_P(u, v)$, and $B_i^v = B_i^u$ for every $i \neq j$. Recall that $d_P(u, v)$ denotes the distance from u to v on the path P.

At the end, when the whole table is computed, we check whether there is an entry at target vertex t such that $T_t[\ldots] = $ TRUE, in which case there is a feasible schedule for the uniform budget B, and there is no feasible schedule otherwise. To compute the feasible schedule, standard bookkeeping techniques can be applied. There are $n \cdot n^k \cdot B^k$ entries in T that need to be computed. To compute one entry $T_v[j|p_1^v, \ldots, p_k^v|B_1^v, \ldots, B_k^v]$, we need to check the existence of j' and u with the above mentioned properties, which can be done in time $O(k \cdot n)$. Hence, the total run-time of the algorithm is $O(k \cdot n^{k+2} \cdot B^k)$. Thus, we have shown the following:

Theorem 7. *There is an algorithm that decides whether a feasible schedule for uniform budget B exists and runs in $O(k \cdot n^{k+2} \cdot B^k)$ time.*

By using the data rounding technique, we turn the developed algorithm into a fully polynomial-time approximation scheme (FPTAS). Let $\epsilon > 0$ be a (small) error margin for which we want to design a $(1+\epsilon)$-approximation algorithm (for computing a minimum feasible uniform budget B).

We define an alternative weight unit $\mu := \epsilon \frac{w(P)/k + X}{m^2}$, where $w(P)$ is the weight of the fixed path P, X is the minimum distance of any agent to the path P, and m is the number of edges of the graph G. We measure the weights $w(e)$ in the integer multiples of μ, rounded-up, i.e., we define $\bar{w}(e) := \lceil w(e)/\mu \rceil$.

We solve the problem in the new edge weights $\bar{w}(e)$ using the dynamic programming approach, where we also measure budget in multiples of μ. Let $\bar{B}$ be the computed optimum uniform budget for the modified edge-weights. Our algorithm returns $B^A = \bar{B} \cdot \mu$ as the solution for the original edge-weights. Let $\bar{P}_1, \ldots, \bar{P}_k$ be the walks that the k agents walk in the optimum solution for the modified edge-weights. Hence, $\bar{B} = \max_i\{\bar{w}(\bar{P}_i)\}$, and thus $\bar{B} \cdot \mu = \max_i\{\bar{w}(\bar{P}_i) \cdot \mu\}$. Observe also that B^A is a feasible budget, since every path $\bar{P}_i$ can be walked with budget B^A, since the original length of $\bar{P}_i$ is $w(\bar{P}_i) \leq \mu \cdot \bar{w}(\bar{P}_i) \leq \mu \bar{B}$.

Let B^* be the optimum budget for the original edge-weights, and let $P_1^*, \ldots, P_k^*$ be the walks of the k agents in some optimum solution. Hence, $B^* = \max_i\{w(P_i)\}$. We now argue that B^A is not much larger than B^*. We have $B^A = \mu \cdot \bar{B} = \mu \cdot \max_i\{\bar{w}(\bar{P}_i)\} \overset{(1)}{\leq} \mu \cdot \max_i\{\bar{w}(P_i^*)\} = \max_i\{\mu \cdot \bar{w}(P_i^*)\} \overset{(2)}{\leq} \max_i\{w(P_i^*) + m^2\mu\} = m^2\mu + \max_i\{w(P_i^*)\} = m^2\mu + B^* = m^2\left(\epsilon \frac{w(P)/k+X}{m^2}\right) + B^* \overset{(3)}{\leq} \epsilon \cdot B^* + B^* = (1+\epsilon)B^*$. Here, inequality (1) is because $\max_i \bar{P}_i$ is the optimum feasible solution in weights $\bar{w}$; inequality (2) follows because any walk appears at most m times on the path P, and between any two appearances, the walk contains at most m edges (this part of the walk is a simple path), inequality (3) follows because B^* needs to be at least $w(P)/k + X$ (the average traveled distance per agent on P plus the distance to get from the initial position to the path P).

We now analyze the run-time of the algorithm. Observe first that $B^* \leq \min_i d(p_i, s) + w(P) \leq (X + w(P)) + w(P) \leq 2(X + w(P))$. Therefore, measured in the units μ, we search for $\bar{B}$ in the range between 1 and $2(X + w(P))/\mu \leq \frac{2m^2 k}{\epsilon}$. Hence, one run of the dynamic programming on the modified weights takes time $O(k \cdot n^{k+2} \cdot (\frac{2m^2 k}{\epsilon})^k)$. Using the binary search to find optimum $\bar{B}$ adds a multiplicative logarithmic factor of $\log\left(\frac{2m^2 k}{\epsilon}\right)$. Thus, we have shown the following.

Theorem 8. *For any $\epsilon > 0$, there is an algorithm that computes a feasible uniform budget B that is at most $(1 + \epsilon)B^*$, where B^* is the optimum uniform budget, and runs in $O\left(k \cdot n^{k+2} \cdot (\frac{2m^2k}{\epsilon})^k \log\left(\frac{2m^2k}{\epsilon}\right)\right)$ time.*

Corollary 1. *There exists an FPTAS for the variant where the number of agents is constant.*

6   Conclusions

The problem of collectively delivering a package by mobile agents is a difficult problem even when the path for moving the package is given in advance. However, for the case of single pickup per agent, we were able to find better approximation algorithms for the fixed path version of collaborative delivery. These results leave many open questions: how to reduce the gap between the upper and lower bounds for the various versions of the problem? How to extend the results to agents with *non-uniform* budgets and find resource-augmented algorithms for fixed path delivery? Finally, what is the effect of restricting package handovers to nodes only and not anywhere inside the edges.

References

1. Anaya, J., Chalopin, J., Czyzowicz, J., Labourel, A., Pelc, A., Vaxès, Y.: Convergecast and broadcast by power-aware mobile agents. Algorithmica **74**(1), 117–155 (2016)
2. Awerbuch, B., Betke, M., Rivest, R.L., Singh, M.: Piecemeal graph exploration by a mobile robot. Inf. Comput. **152**(2), 155–172 (1999)
3. Bampas, E., Chalopin, J., Das, S., Hackfeld, J., Karousatou, C.: Maximal exploration of trees with energy-constrained agents. CoRR, abs/1802.06636 (2018)
4. Bampas, E., Das, S., Dereniowski, D., Karousatou, C.: Collaborative delivery by energy-sharing low-power mobile robots. In: Fernández Anta, A., Jurdzinski, T., Mosteiro, M.A., Zhang, Y. (eds.) ALGOSENSORS 2017. LNCS, vol. 10718, pp. 1–12. Springer, Cham (2017). https://doi.org/10.1007/978-3-319-72751-6_1
5. Bärtschi, A., et al.: Collaborative delivery with energy-constrained mobile robots. Theor. Comput. Sci. (2017)
6. Bärtschi, A., et al.: Energy-efficient delivery by heterogeneous mobile agents. In: 34th Symposium on Theoretical Aspects of Computer Science (STACS), pp. 10:1–10:14 (2017)
7. Bärtschi, A., Graf, D., Mihalák, M.: Collective fast delivery by energy-efficient agents. In: 43rd International Symposium on Mathematical Foundations of Computer Science (MFCS 2018). Leibniz International Proceedings in Informatics (LIPIcs), vol. 117, pp. 56:1–56:16 (2018)
8. Bärtschi, A., Graf, D., Penna, P.: Truthful mechanisms for delivery with agents. In: 17th Workshop on Algorithmic Approaches for Transportation Modelling, Optimization, and Systems (ATMOS 2017). OpenAccess Series in Informatics (OASIcs), vol. 59, pp. 2:1–2:17 (2017)
9. Betke, M., Rivest, R.L., Singh, M.: Piecemeal learning of an unknown environment. Mach. Learn. **18**(2), 231–254 (1995)

10. Bilò, D., Disser, Y., Gualà, L., Mihalák, M., Proietti, G., Widmayer, P.: Polygon-constrained motion planning problems. In: Flocchini, P., Gao, J., Kranakis, E., Meyer auf der Heide, F. (eds.) ALGOSENSORS 2013. LNCS, vol. 8243, pp. 67–82. Springer, Heidelberg (2014). https://doi.org/10.1007/978-3-642-45346-5_6

11. Chalopin, J., Das, S., Mihalák, M., Penna, P., Widmayer, P.: Data delivery by energy-constrained mobile agents. In: Flocchini, P., Gao, J., Kranakis, E., Meyer auf der Heide, F. (eds.) ALGOSENSORS 2013. LNCS, vol. 8243, pp. 111–122. Springer, Heidelberg (2014). https://doi.org/10.1007/978-3-642-45346-5_9

12. Chalopin, J., Jacob, R., Mihalák, M., Widmayer, P.: Data delivery by energy-constrained mobile agents on a line. In: Esparza, J., Fraigniaud, P., Husfeldt, T., Koutsoupias, E. (eds.) ICALP 2014. LNCS, vol. 8573, pp. 423–434. Springer, Heidelberg (2014). https://doi.org/10.1007/978-3-662-43951-7_36

13. Czyzowicz, J., Diks, K., Moussi, J., Rytter, W.: Communication problems for mobile agents exchanging energy. In: Suomela, J. (ed.) SIROCCO 2016. LNCS, vol. 9988, pp. 275–288. Springer, Cham (2016). https://doi.org/10.1007/978-3-319-48314-6_18

14. Das, S., Dereniowski, D., Karousatou, C.: Collaborative exploration by energy-constrained mobile robots. In: Scheideler, C. (ed.) Structural Information and Communication Complexity. LNCS, vol. 9439, pp. 357–369. Springer, Cham (2015). https://doi.org/10.1007/978-3-319-25258-2_25

15. Das, S., Dereniowski, D., Uznanski, P.: Energy constrained depth first search. CoRR, abs/1709.10146 (2017)

16. Demaine, E.D., Hajiaghayi, M., Mahini, H., Sayedi-Roshkhar, A.S., Oveisgharan, S., Zadimoghaddam, M.: Minimizing movement. ACM Trans. Algorithms **5**(3), 1–30 (2009)

17. Dereniowski, D., Disser, Y., Kosowski, A., Pająk, D., Uznański, P.: Fast collaborative graph exploration. Inf. Comput. **243**, 37–49 (2015)

18. Duncan, C.A., Kobourov, S.G., Anil Kumar, V.S.: Optimal constrained graph exploration. In: 12th ACM Symposium on Discrete Algorithms, SODA 2001, pp. 807–814 (2001)

19. Dynia, M., Korzeniowski, M., Schindelhauer, C.: Power-aware collective tree exploration. In: Grass, W., Sick, B., Waldschmidt, K. (eds.) ARCS 2006. LNCS, vol. 3894, pp. 341–351. Springer, Heidelberg (2006). https://doi.org/10.1007/11682127_24

20. Dynia, M., Łopuszański, J., Schindelhauer, C.: Why robots need maps. In: Prencipe, G., Zaks, S. (eds.) SIROCCO 2007. LNCS, vol. 4474, pp. 41–50. Springer, Heidelberg (2007). https://doi.org/10.1007/978-3-540-72951-8_5

21. Flocchini, P., Prencipe, G., Santoro, N.: Distributed Computing by Oblivious Mobile Robots. Morgan & Claypool, San Rafael (2012)

22. Fraigniaud, P., Gąsieniec, L., Kowalski, D.R., Pelc, A.: Collective tree exploration. Networks **48**(3), 166–177 (2006)

23. Fredman, M.L., Tarjan, R.E.: Fibonacci heaps and their uses in improved network optimization algorithms. J. ACM **34**(3), 596–615 (1987)

24. Garey, M.R., Johnson, D.S.: Computers and Intractability: A Guide to the Theory of NP-Completeness. W. H. Freeman & Co., New York (1979)

25. Giannakos, A., Hifi, M., Karagiorgos, G.: Data delivery by mobile agents with energy constraints over a fixed path. CoRR, abs/1703.05496 (2017)

26. Tovey, C.A.: A simplified NP-complete satisfiability problem. Discrete Appl. Math. **8**(1), 85–89 (1984)

Asynchronous Rendezvous with Different Maps

Serafino Cicerone[1]([⊠]), Gabriele Di Stefano[1], Leszek Gąsieniec[2], and Alfredo Navarra[3]

[1] Department of Information Engineering, Computer Science and Mathematics, University of L'Aquila, Via Vetoio, 67100 L'Aquila, Italy
{serafino.cicerone,gabriele.distefano}@univaq.it
[2] Department of Computer Science, University of Liverpool, Ashton Street, Liverpool L69 3BX, UK
l.a.gasieniec@liverpool.ac.uk
[3] Department of Mathematics and Computer Science, University of Perugia, Via Vanvitelli 1, 06123 Perugia, Italy
alfredo.navarra@unipg.it

Abstract. This paper provides a study on the rendezvous problem in which two anonymous mobile entities referred to as *robots* r_A and r_B are asked to meet at an arbitrary node of a graph $G = (V, E)$. As opposed to more standard assumptions robots may not be able to visit the entire graph G. Namely, each robot has its own *map* which is a connected subgraph of G. Such mobility restrictions may be dictated by the topological properties combined with the intrinsic characteristics of robots preventing them from visiting certain edges in E.

We consider four different variants of the rendezvous problem introduced in [Farrugia et al. *SOFSEM'15*] which reflect on restricted maneuverability and navigation ability of r_A and r_B in G. In the latter, the focus is on models in which robots' actions are synchronised. The authors prove that one of the maps must be a subgraph of the other. I.e., without this assumption (or some extra knowledge) the rendezvous problem does not have a feasible solution. In this paper, while we keep the containment assumption, we focus on asynchronous robots and the relevant bounds in the four considered variants. We provide some impossibility results and almost tight lower and upper bounds when the solutions are possible.

1 Introduction

The *Rendezvous* problem comprises the task of meeting two anonymous mobile robots which start at different nodes of a graph or different locations in the Euclidean space. Many variants (with different assumptions) of rendezvous have

Work supported in part by the European project "Geospatial based Environment for Optimisation Systems Addressing Fire Emergencies" (GEO-SAFE), contract no. H2020-691161, and by the Italian National Group for Scientific Computation (GNCS-INdAM).

K. Censor-Hillel and M. Flammini (Eds.): SIROCCO 2019, LNCS 11639, pp. 154–169, 2019.
https://doi.org/10.1007/978-3-030-24922-9_11

been studied in the past. An exhaustive survey on the problem can be found in [15], and some further advances in [1,3–6,9,11,13,16]. In this paper, we are interested in the design of deterministic algorithms for asynchronous robots moving across edges in the underlying graph of network connections. The deterministic and asynchronous variant of rendezvous in graphs has been first introduced in [8]. Later in [7], the problem has been fully characterised and the adopted model utilised the minimal setting under which the rendezvous can be accomplished. The authors of [7] give also the answer to the question posed in [8] whether there exists a deterministic algorithm for rendezvous of two asynchronous robots in any finite connected graph without knowing any upper bound on its size. The minimal assumptions to enable rendezvous include:

- The input anonymous graph has no labels on points. Instead, at each node of degree d, the relevant end points of incident edges are sorted and labelled by port numbers $1, \ldots, d$. The local labelling of ports at each node is fixed, i.e., every robot sees the same local labelling. However, no coherence between local labellings is assumed. I.e., one edge can have two different port numbers at its opposite ends. When a robot leaves a node, it is aware of the port number by which it leaves and when it enters a node, it is aware of the entry port number. It can also verify, at each node, whether a given positive integer is a port number at this node.
- Each robot has a unique ID, but no knowledge on the ID of the other robot.
- Robots can meet on nodes or along edges, i.e., forcing robots to meet on nodes may prevent them from rendezvous.

In the model described above robots do not know G nor the initial distance between them in G. They cannot mark neither the nodes nor the edges. Rendezvous has to be accomplished for any local labelling of ports. The robots terminate their walks at the time of meeting one another. The rendezvous algorithm works also for infinite graphs. In fact, in finite graph the resolution of the rendezvous is often trivial or it can be reduced to the graph *exploration* problem. For example, utilising search methods proposed in [10] one can force to meet the two robots in finite tree. Namely, the rendezvous can be easily reached once both robots discover the centre(s) of the tree.

In this paper, we are interested in a different model in which robots have no IDs and most importantly they may not be allowed to access the whole graph. The roaming space of each robot is limited to a specific subset of nodes and edges. The reasons to adopt such restriction may vary, however, the restriction itself is natural and was used earlier, e.g., in the evacuation problem [2] where an entity may represent a disabled person not able to adopt steep stairs or an escalator.

The rendezvous problem with heterogeneous (different accessibility restrictions) entities was formally introduced in [12] under the name of *rendezvous with different maps*. In the most general variant two asynchronous and anonymous robots r_A and r_B provided with two different maps G_A and G_B, both isomorphic to (possibly different) subgraphs G'_A and G'_B of the finite input graph G. The meeting can happen only on nodes but it is assumed that traversal of edges is mutually exclusive. This assumption is equivalent to the one used in [7], where robots can also meet on edges.

The main difference between the standard rendezvous problem studied, e.g., in [7] and the rendezvous with different maps studied here is in the way robots build their trajectories. In particular, in the latter the robots do not have to construct their maps (discover reachable nodes and edges). The maps are provided to them beforehand and the relevant trajectories can be precomputed prior to the actual search stage. This is in contrast to the model utilised in [7] where the trajectory of a robot is computed "on-the-go" on the basis of the current local information about port numbers, node degrees and the ID. Thus the main difficulty in the model adopted here refers to inconsistency of the maps provided to the robots in which one robot may not be able to access certain nodes or edges reachable for the other. Similar challenges occur also in blind rendezvous [14].

According to [12] without some extra information (e.g., node IDs) rendezvous with maps cannot be accomplished if $G'_A \not\subseteq G'_B$, where r_A is the robot with the smaller map. Thus here we also assume $G'_A \subseteq G'_B$. In contrast to [12], we focus solely on *asynchronous* robots. We study four natural variants of the rendezvous with different maps, combining two natural assumptions/properties considered in [12]: (1) availability of relative (with no explicit labels) ordering of nodes, and (2) presence of robot weights vs edge weight tolerance. The four variants are determined by the presence (or absence) of those two properties that we are going to formally define later. We also discuss two hierarchies (that share the bottom and the top levels) formed by the four studied variants of the problem. At the top level of these hierarchies we assume presence of both properties. In the middle we have two incomparable levels where only one property is present. Finally at the bottom level we consider the absence of both properties.

We provide both the lower and the upper bounds with respect to the considered variants. We show that at the bottom level of the hierarchies very little can be done w.r.t. the rendezvous problem. In particular, the absence of the two properties makes the problem unsolvable in G with an arbitrary topology, and is tractable only in the case of simple topologies including paths and stars. We also show that in the two intermediate (and incomparable) variants rendezvous can be efficiently concluded in cycles and trees (the robots cannot rendezvous in cycles at the bottom level). Finally, we propose efficient (in terms of moves made) algorithm for the upper level requiring only $O(N \log N)$ steps. This result is almost tight in view of the natural lower bound of $\Omega(N)$, where N denotes the cumulative number of vertices of the two maps G'_A and G'_B.

2   Model

We start with a summary and further extension of the computation model introduced in [12]. We consider rendezvous of anonymous (and indistinguishable with respect to the control mechanism) robots in networks modelled by finite undirected graphs. The network $G = (V, E)$ is a simple connected graph, where $|V| = n$ and $|E| = m$. The two robots r_A and r_B initiate search at different *starting nodes* $s_A \neq s_B$ in G. Each robot $r_X \in \{r_A, r_B\}$ has its own *map*

$G_X = (V_X, E_X)$ which is isomorphic to a specific subgraph $G'_X = (V'_X, E'_X)$ of G induced by the sets of nodes V'_X and edges E'_X *reachable* from s_X by robot r_X. In particular, the matching between the map of r_X and G'_X is deterministic and known to r_X. We emphasise that each robot r_X only knows its own map G_X and the starting node s_X. In other words r_A has no knowledge of G_B and s_B, and vice versa. Moreover, during search r_A cannot adopt edges outside of its map G_A and its trajectory is oblivious w.r.t. to the knowledge possessed by r_B. Note that, once r_X has computed its trajectory on its map, by the above assumptions, it can move on G'_X consistently, without ambiguities. Let $n_X = |V_X|$ be the number of nodes of map G_X, $m_X = |E_X|$ be the number of edges of G_X, while by N and M we denote $n_A + n_B$ and $m_A + m_B$, respectively. Finally, given a node $v \in V$, the set of its neighbours is denoted by $N_G(v) = \{v' \mid (v, v') \in E\}$.

We assume that the robots act in asynchronous fashion. Each robot computes its trajectory, the sequence of visited nodes and edges, independently and prior to the actual search. We assume that the use of edges is exclusive, i.e., two robots cannot be located (move in either directions) on the same edge at any time. When the robot is ready to move along a chosen edge it awaits the relevant "green light" signal (meaning the edge is now available) from the system. In consequence, rendezvous is possible only on nodes when one robot is immobilised indefinitely or awaits access of an edge through which the other robot is approaching. The time required to move across an edge is assumed to be finite but unknown. In turn, as the complexity of the solution we adopt the sum of the lengths of the robots' trajectories before rendezvous, i.e., the number of edges traversed in total.

In what follows we formalise four different variants of rendezvous with different maps. Each variant is determined by the availability of extra knowledge O and W (for the definition see below) w.r.t. the maps. For all considered variants, we assume that G'_A is a subgraph of G'_B. Otherwise, as already indicated, the rendezvous problem with different maps may not have a solution [12].

- Property O: the nodes of G are *totally ordered*. In particular, if $V = \{v_1, v_2, \ldots, v_n\}$ then $v_i < v_{i+1}$, for all $i = 1, 2, \ldots, n-1$. We say that Property O holds if this order is consistent with the order of nodes observed by robot r_X in G_X. That is, if $V_X = \{v_1^X, v_2^X, \ldots, v_{n_X}^X\}$, $v_p^X = v_i$, and $v_q^X = v_j$, where $v_i, v_j \in V$ and $i < j$, we also get $v_p^X < v_q^X$.
- Property W: each robot $r_X \in \{r_A, r_B\}$ has an associated weight $w_X \in \mathbb{R}^+$, and each edge $e \in E$ can tolerate weights up to the limit $w(e) \in \mathbb{R}^+$. In this setting let H_X denote the (possibly disconnected) subgraph of G induced by edges $e \in E$ such that $w(e) \geq w_X$. Then G'_X is the connected component of H_X which contains s_X. We assume that the maps $G_X = (V_X, E_X)$ contains information about the weights tolerated by the relevant edges (where $w(e^X) = w(e)$, for each $e^X \in E_X$ represents the edge $e \in E$).

We consider four variants based on properties O and W:

- WO **variant**, where both properties O and W hold,
- $\overline{\text{W}}$O **variant**, where only O holds,

- $\overline{\text{WO}}$ **variant**, where only W holds,
- $\overline{\text{WO}}$ **variant**, where neither O nor W holds.

By slightly abusing our notation the codes WO, $\overline{\text{WO}}$, $\overline{\text{WO}}$, and $\overline{\text{WO}}$ will be used not only to define the variants of the problem, but also the set of instances of the relevant variants. For example, WO will refer to all instances of rendezvous with different maps where each robot r_X knows: (1) its *weighted* map G_X, (2) the starting point s_X, and (3) it is aware that its node ordering is consistent with the node ordering of the other robot. Using this notation one can define a relationship $\sqsubseteq$ between the elements of $\mathbb{V} = \{\text{WO}, \overline{\text{WO}}, \overline{\text{WO}}, \overline{\text{WO}}\}$. For example, $\overline{\text{WO}} \sqsubseteq \overline{\text{WO}}$ means that for each instance $i \in \overline{\text{WO}}$ it is possible to identify a set $I \subseteq \overline{\text{WO}}$ of *instances induced by i* as follows: if $i = (G_A, s_A, G_B, s_B)$, then each instance in I is obtained from i by maintaining (G_A, s_A, G_B, s_B) and by adding any possible consistent ordering on nodes of G_A and G_B. One can observe that such relationship defines two hierarchies: $\overline{\text{WO}} \sqsubseteq \overline{\text{WO}} \sqsubseteq \text{WO}$ and $\overline{\text{WO}} \sqsubseteq \overline{\text{WO}} \sqsubseteq \text{WO}$. The following holds.

Remark 1. Let $\mathcal{V}_1, \mathcal{V}_2 \in \mathbb{V}$ such that $\mathcal{V}_1 \sqsubseteq \mathcal{V}_2$. If $i \in \mathcal{V}_1$ and $I \subseteq \mathcal{V}_2$ is the set of instances induced by i, then: (1) if i is solvable in $\mathcal{V}_1$, then each induced instance in I is solvable in $\mathcal{V}_2$; (2) if all the instances in I are unsolvable in $\mathcal{V}_2$, then i is unsolvable in $\mathcal{V}_1$.

In the remaining part of the paper we propose and analyse algorithmic solutions for the rendezvous problem with different maps. Our algorithms assume each robot r_X has the input map G_X and the initial position s_X. The output of an algorithm refers the rendezvous trajectory computed by each r_X on G_X. The complexity of the solution is defined as the sum of the lengths of trajectories adopted by both robots until rendezvous takes place. For the sake of simplicity, knowing that G_X and G'_X are isomorphic and that r_X is aware of the isomorphism, in the following we always write G_X rather than G'_X even when we refer to the moves along edges in G'_X and the properties of G'_X.

3   Preliminary Results

In this section we provide a general lower bound holding for all variants, and a more restrictive one which does not hold for WO. Then, we present a sufficient condition for solving the rendezvous problem that we exploit successively in our resolution algorithms for variants WO and $\overline{\text{WO}}$. Finally, we provide optimal algorithms and infeasibility results for maps with specific topologies in the weaker variants $\overline{\text{WO}}$ and $\overline{\text{WO}}$. Some of the proofs are omitted due to space constraints.

3.1   Lower Bounds

The following lemma provides a lower bound on the length of the trajectory performed by robots in any solving algorithm with respect to the WO variant.

Lemma 1. *In variant* WO, *rendezvous requires use of trajectories of length* $\Omega(N)$. *In particular, each* r_X *must visit all the* n_X *nodes of its map.*

Proof. Consider an instance of the problem where $n_A = 1$. Then, any rendezvous algorithm is stuck with r_A immobilised in the starting node s_A. Since r_B has no knowledge of the position of r_A, in the worst case it has to move throughout all the nodes of its map.   □

Thus by Remark 1 and Lemma 1 the lower bound $\Omega(N)$ applies also in any variant in $\mathbb{V}$.

Lemma 2. *In variants $\overline{\mathsf{WO}}$ and $\mathsf{W\overline{O}}$ rendezvous requires use of trajectories of length $\Omega(M)$. In particular, each r_X must traverse all the m_X edges of its map.*

Proof. First note that neither in variant $\overline{\mathsf{WO}}$ nor in $\mathsf{W\overline{O}}$ robots have enough information to meet in (agreed in advance on) a common target node for their rendezvous.

In variant $\overline{\mathsf{WO}}$ consider the case in which the map of r_B is formed of $m_B = \Omega(n_B^2)$ edges and the map of r_A is a single edge $e = \{v_1, v_2\}$. According to Lemma 1, any rendezvous algorithm $\mathcal{A}$ must force robots to visit all nodes in their map. Thus also r_A has to visit both nodes v_1 and v_2 by traversing the only edge at most once. If r_B traverses only $o(m_B)$ edges and stops, the adversary picks e among edges not traversed by r_B with the endpoints different to the node where r_B rests eventually. This is possible if the map of r_B is dense enough. During rendezvous, r_A is allowed first to access e and is kept there until r_B stops. Since the final node on r_B's trajectory is different to v_1 and v_2 rendezvous is not reached. In the complementary case, i.e., when r_B visits its whole map we assume that e is the last edge visited by r_B. Here also the adversary allows r_A to enter this edge first where r_A waits until r_B comes to visit this edge. This will force r_B to visit all its m_B edges, that is $\Omega(M)$ edges.

In variant $\mathsf{W\overline{O}}$ consider a 3-layer graph $G = (V, E)$, where the set of nodes V is partitioned into three subsets V_1, V_2 and V_3 of the same size $\frac{n}{3}$. Also the set of edges is partitioned into E_1 and E_2, such that graphs $(V_1 \cup V_2, E_1)$ and $(V_2 \cup V_3, E_2)$ are complete bipartite graphs. We also assume that edge tolerance within each set E_i, for $i = 1, 2$, is uniform, however, edges in E_2 tolerate w_A but those in E_1 don't. In contrast, all edges in E tolerate w_B. Assume also that $s_A \in V_2$ and $s_B \in V_1$. By Lemma 1, r_X cannot stay in s_X. Let e be the edge r_A traverses first. The adversary temporarily entraps r_A on e. If the trajectory computed by r_B is of length at least $\Omega(n^2)$, due to the uniform weight tolerance on edges in E_2 the adversary can pick e, s.t., occurs on r_B's trajectory only after $\Omega(n^2)$ steps. In the complementary case, when the trajectory of r_B is of length $o(n^2)$, the adversary picks e outside of the trajectory of r_B. In this case the adversary instructs r_B to move first entrapping it in the last edge e' of its trajectory. If the protocol for r_A is perpetual or of length $\Omega(n^2)$ due to uniformity of edges the adversary can force this protocol to walk $\Omega(n^2)$ edges before entering e', and the rendezvous takes place only if $e' \in E_2$. If the protocol for r_A is of length $o(n^2)$ the adversary keeps r_B away from e' and stops at its destination node v. Finally, r_A is released to finish walk at v'. As the robots cannot agree in advance to meet on the same target node, i.e., $v \neq v'$, there is no rendezvous in this case. By the generality of edges e and e' mentioned in the above arguments, the claim follows.   □

It follows from Lemma 2 that in variants $\overline{\text{WO}}$ and $\text{W}\overline{\text{O}}$ (and by Remark 1 also in $\overline{\text{WO}}$) any algorithm has to move robots through all edges of their respective maps. Whereas, in variant WO this is not true as robots could exploit knowledge about nodes' ordering and edges' weight tolerance.

3.2 A Sufficient Condition for Solving Rendezvous

In this section we provide a sufficient condition for solving rendezvous with different maps. To this respect it worth to mention [1], where a characterisation of pairs of walks that enforce rendezvous against an asynchronous adversary is given. We first formalise concepts of walks and sub-walks in a graph. A *walk* in a graph G is an ordered sequence of edges of G, $W = ((v_{i_1}, v_{i_2}), (v_{i_2}, v_{i_3}), \ldots, (v_{i_{k-1}}, v_{i_k}))$, where the second node of an edge is the first node of the subsequent edge; in W, v_{i_1} is the starting node and v_{i_k} is the final node. By $|W|$, we denote the number of edges forming W. Given two walks W' and W'' in G, we write $W' \subseteq W''$ when W' is a *sub-walk* of W'', i.e., W' is a (not necessarily contiguous) sub-sequence of edges in W''. If a walk W contains all edges of G then it is called a *complete walk* of G.

Lemma 3. *Let W_X be a complete walk of map G_X computed by r_X starting in s_X, according to an algorithm $\mathcal{A}$. If $W_A \subseteq W_B$, then r_A and r_B will meet eventually, even in variant $\text{W}\overline{\text{O}}$.*

Proof. According to $\mathcal{A}$, robot r_X moves along walk W_X starting at s_X and finishing in the final node of W_X, unless the rendezvous is accomplished earlier. Since $W_A \subseteq W_B$, robot r_B has to visit all edges in W_A in the same order as robot r_A does. Thus no adversary can force r_B to overpass r_A on W_A despite actions of robots being asynchronous.   □

3.3 On the Complexity of Rendezvous in Variants $\overline{\text{WO}}$ and $\text{W}\overline{\text{O}}$

We start the discussion of rendezvous with different maps in variant $\overline{\text{WO}}$.

As already discussed, one can find in [7] full characterisation of the standard asynchronous rendezvous problem, including the minimal assumptions under which the rendezvous can be accomplished. These include (1) port numbering consistent with the degree of the nodes for the two maps, (2) unique IDs of robots, and (3) meeting allowed at nodes and edges. Consider now variant $\overline{\text{WO}}$ variant with an instance in which $G_A = G_B$. In such case, one can claim that "rendezvous with different maps" is equivalent to "standard rendezvous problem" when neither port numbering nor node IDs are provided.

Thus using the argument above and [7] we get the following theorem.

Theorem 1. *In variant* $\overline{\text{WO}}$ *rendezvous is not feasible.*

Note that rendezvous can be obtained in more specific topologies. We discuss some cases below. It is worth to mention that the rendezvous algorithm for trees sketched in the introduction does not work when different maps are in use, as the centres computed for different maps may not coincide.

Lemma 4. *In variant* $\overline{\text{WO}}$, *if network G is a path or a star graph then rendezvous can be solved optimally.*

The next result affirms that in case the input map is a cycle the rendezvous problem cannot be solved in the $\overline{\text{WO}}$ variant. In fact, cycles will play the central role in discussion on how the rendezvous complexity changes in the relevant variants.

Lemma 5. *In variant* $\overline{\text{WO}}$, *if G is a cycle rendezvous cannot be resolved.*

Proof. Consider the case with $G_A = G_B$ both being a cycle with an even number of nodes. Assume an instance where the two robots lie at some antipodal nodes of the cycle. The adversary can force a symmetric behaviour of the two robots. That is, whatever one robot does according to the provided algorithm, the other makes exactly the same symmetric move. As robots are always located at some antipodal positions the meeting will never take place.   □

Consider now the subset $I \subseteq \overline{\text{WO}}$ containing all instances with $G_A = G_B$. If $I' \subset \text{WO}$ contains all instances with $w_A = w_B$ drawn from I, we conclude using Theorem 1 that also in variant WO rendezvous is not always feasible.

Theorem 2. *In variant* WO *rendezvous with maps is not always feasible.*

The following lemma provides a feasibility results for variant WO when $w_A < w_B$ and the topology of G is restricted to cycles.

Lemma 6. *In variant* WO, *if G is a cycle and $w_A < w_B$ then there exists an algorithm that allows robots to meet along walks of length $O(N \cdot |b_A|)$, where b_X is the binary representation of weight w_X.*

Proof. Since G is a cycle and $G_A \subseteq G_B \subseteq G$ then G_X is either a path or a cycle.

If G_X is a path then r_X applies the strategy provided in the proof of Lemma 4: *from s_X, r_X goes to an arbitrary endpoint of the path and then walk along all the edges to reach the other endpoint. If G_X is a cycle, the algorithm works as follows. Consider the binary representation b_X of w_X. Initially, robot r_X traverses the whole cycle (returning to s_X) in any direction; then, for each bit of b_X and starting from the least significant bit: if the current bit is 1, the robot performs a complete visit of the cycle in one direction, if the bit is 0, then the robot does the same in the opposite direction.*

If G_A is a path, the two robots meet within the first two visits of the cycle made by r_B, hence with a trajectory of length at most $2N$. If G_A is a cycle and $w_A < w_B$, the two trajectories differ as either (1) b_A and b_B have different sizes or (2) they agree on the same direction but differ for on least one bit or (3) they do not agree on the same direction but the most significant bit equals 1 for both. In the first case, r_B traverses G more times than r_A if they do not meet before, so they must meet eventually. In the second and the third cases, they traverse the cycle in the opposite directions at least once, and this is enough to force their meeting. The complexity of this algorithm is $O(N \cdot |b_A|)$, as $|b_A| \leq |b_B|$. $\square$

4   A $O(N \log N)$ Algorithm for Variant WO

In [12], the authors define an algorithm for the case of synchronous robots that solves the rendezvous in variant WO utilising trajectories of length $O(N)$. A similar technique for asynchronous robots leads to trajectories of length $O(N^2)$. In what follows we propose a novel algorithm for asynchronous robots with the complexity $O(N \log N)$. The new algorithm is based on new techniques an it requires better understanding of the considered problem.

We start by observing that in variant WO one can define the total order $\prec^{WO}$ on edges in E, where $G = (V, E)$ is the input network. This ordering is defined as follows: edges are first ordered according to their (increasing) weights, and in case of ties edges with smaller endpoints are earlier in the order. Formally, given two edges $e' = (v_i, v_j)$ and $e'' = (v_{i'}, v_{j'})$ then $e' \prec^{WO} e''$ if and only if (1) $w(e') < w(e'')$, or (2) $w(e') = w(e'')$ and $\min(i, j) < \min(i', j')$, or (3) $w(e') = w(e'')$, $\min(i, j) = \min(i', j')$ and $i + j < i' + j'$.

Let $E = \{e_1, e_2, \ldots, e_m\}$ where $e_i \prec^{WO} e_{i+1}$, for each $i = 1, 2, \ldots, m-1$ (i.e., indices are consistent with the order $\prec^{WO}$). Hence, if $G(i)$ is the subgraph of G induced by edges $e_i, e_{i+1}, \ldots, e_m$, the following properties hold: (1) $G(i)$ may be disconnected, and (2) $G(i+1)$ is a subgraph of $G(i)$.

Notice that the same notation adopted for elements of E is used to refer to edges in a map G_X, that is, if $E_X = \{e_1^X, e_2^X, \ldots, e_{m_X}^X\}$, then $e_i^X \prec^{WO} e_{i+1}^X$ for each $i = 1, 2, \ldots, m_X - 1$.

We introduce a rendezvous method called *two-steps approach*. In the first step, *a rendezvous algorithm $\mathcal{A}$ reduces the search space by computing a convenient submap $H_X \subseteq G_X$. In the second, $\mathcal{A}$ instructs each robot to meet inside H_X.*

The intuition behind this approach is *the smaller/simpler the search space, rendezvous becomes more efficient.* According to Lemma 1, H_X must contain all n_X nodes of G_X, thus the search space reduction can only affect edges from G_X. Also, since H_X must be connected, it contains at least $n_X - 1$ edges in the form of a spanning tree of G_X.

The search space reduction in variant WO is given below. Please note, this method cannot be used in the other three variants since it relies on order $\prec^{WO}$ allowing to create the spanning tree T_X.

Definition 1. *Consider variant WO with maps G_X. Denote by T_X the maximal spanning tree of G_X obtained by Kruskal's algorithm, where edges are drawn in the reverse order to $\prec^{WO}$.*

The following lemma determines a relationship between the maximal spanning trees T_A and T_B.

Procedure: MAKEWALK

Input : Tree T_X, initial robot's position s_X

Output : A walk W starting at s_X and passing through all nodes in T_X.

1 Let $N_{T_X}(s_X) = \{v_{i_1}, v_{i_2}, \ldots, v_{i_k}\}$, such that $i_1 < i_2 < \ldots < i_k$;

2 $W = \text{list}()$; // W is initialised as an empty list

3 **for** $1 \leq j \leq k$ **do**

4 | Let $e = (s_X, v_{i_j})$;

5 | W.append(e).concat(MAKESUBWALK(T_X, v_{i_j}, s_X));

6 **return** W;

Procedure: MAKESUBWALK

Input : Tree T_X, current node s, previous node $f \in N_{T_X}(s)$

Output : A closed walk W starting and finishing in s, passing through all nodes
    in $T_X \setminus T_f$, where T_f is the maximal subtree of T_X rooted at f.

1 Let $N_{T_X}(s) = \{v_{i_1}, v_{i_2}, \ldots, v_{i_k}\}$, such that $i_1 < i_2 < \ldots < i_k$;

2 Let $succ(v_{i_j}) = \begin{cases} v_{i_{j+1}} & \text{if } j < k \\ v_{i_1} & \text{otherwise} \end{cases}$

3 Let $next = succ(f)$;

4 Let $e' = (s, next)$ and $e'' = (s, f)$;

5 $W = \text{list}(e')$; // W is initialised as a list containing just e'

6 **for** $k - 1$ *times* **do**

7 | W.concat(MAKESUBWALK($T_X, next, s$)).append(e'');

8 | $next = succ(next)$;

9 **return** W;

Fig. 1. Procedure MAKEWALK executed by robot $r_X \in \{r_A, r_B\}$ starting on node s_X of T_X. The requested walk W_X starting from and ending at the initial robot's position s_X, and containing all edges in T_X is obtained by exploiting the recursive Procedure MAKESUBWALK.

Lemma 7. T_A *is a subtree of* T_B.

Proof. According to the imposed order to the edges, assume by contradiction there exists an edge $e \in T_B$ whose endpoints are both in V_A but $e \notin T_A$. It follows that for any $e' \in T_A$ we have $e' \prec^{\text{WO}} e$. In fact edge e appears in T_B because either Property W imposes $w(e') < w(e)$ or $w(e') = w(e)$ and $e' \prec^{\text{WO}} e$ is due to Property O. Thus by applying the Kruskal's algorithm according to the reverse order to $\prec^{\text{WO}}$, since r_B selects e then $w(e) \geq w(e')$. It follows that $w(e) = w(e')$ and hence $e' \prec^{\text{WO}} e$ because of Property O. Since r_A and r_B share the same node ordering, then e should be selected by r_A before any other edge, and this contradicts the hypothesis. □

In Fig. 1 we present a pseudo-code of procedure MAKEWALK adopting complete walk along edges of tree T_X. In particular, given T_X and a starting node s_X, by calling MAKEWALK(T_X, s_X) we obtain a walk W_X that starts at s_X, passes through all the edges of the tree (in each direction in the form of well defined Euler tour), and finishes at s_X. This property is crucial for any rendezvous algorithm based on traversing T_X several times. The following lemma provides a useful relationship between W_A and W_B.

Lemma 8. *Let* $W_A = \text{MAKEWALK}(T_A, s_A)$, *and* $W_B = \text{MAKEWALK}(T_B, s_B)$. *Then,* $W_A \subseteq 2 \cdot W_B$, *where* $2 \cdot W_B$ *is the concatenation of two occurrences of* W_B.

Proof. Assume first $s_A = s_B$. From Lemma 7 we have that $T_A \subseteq T_B$. Procedure MAKEWALK ensures that also the ordered tree T_A is a subtree of the ordered tree T_B, that is, nodes in the walks maintain their relative ordering. It follows that $W_A \subseteq W_B$. Since in general $s_A \neq s_B$ and the length of W_X is $2n_X - 2$, then W_B may contain a suffix $(v_j^A, v_{j+1}^A), (v_{j+1}^A, v_{j+2}^A), \dots, (v_{2n_A-4}^A, v_{2n_A-3}^A)$ of W_A before its prefix $(v_1^A, v_2^A), (v_2^A, v_3^A), \dots, (v_{j-2}^A, v_{j-1}^A)$, for some $1 < j < 2n_A - 2$. However, by concatenating two occurrences of W_B, the suffix and the prefix of W_A will appear in the right order at least once. In other words, by traversing W_B twice, we can guarantee rendezvous by visiting W_A in the right order at least once. ☐

Algorithm WO-ASYNCH (cf Fig. 2) exploits Procedure MAKEWALK to build complete walks that fulfill condition of Lemma 8. The following theorem states that rendezvous in variant WO can be solved with complexity $O(N \log N)$, for the input network G with an arbitrary topology.

Theorem 3. *In variant* WO, *for any network* G, *Algorithm* WO-ASYNCH *guarantees rendezvous along a trajectory of length* $O(N \log N)$.

Proof. Algorithm WO-ASYNCH can be divided into four parts: (1) in the first two lines the spanning tree T_X is computed along with its integer logarithmic size (i.e., $k_X = \lceil \log |T_X| \rceil$), (2) Line 4, where the walk $W_X = \text{MAKEWALK}(T_X, s_X)$ is computed, (3) the block of Lines 5–14, where a target t_X is computed, and (4) block of Lines 15–18, where the complete walk W_X^+ is computed and performed. Such a walk W_X^+ consists of $2k_X$ concatenations of W_X plus a sub-sequence of W_X (i.e., the *final step*) needed to reach the target t_X. We now analyze two cases, according to the sizes k_A and k_B.

- Case $k_B > k_A$. We show that W_A^+ is a sub-walk of W_B^+, and hence from Lemma 3 the claim holds. In W_A^+ the sequence W_A is repeated $2k_A$ times plus a subsequence of W_A (due to the final step). In W_B^+, the sequence W_B is repeated $2k_B \geq 2(k_A + 1)$ times, which is at least $2k_A + 2$ times. From Lemma 8, the first two repetitions of W_B assure that W_A is contained in $2W_B$, that is a subsequence of W_B^+. It follows that $2k_A + 2$ sequences of W_B include $2k_A + 1$ sequences of W_A. Since W_A^+ is a subsequence of $2k_A + 1$ repetitions of W_A, then W_A^+ is contained in W_B^+.
- Case $k_B = k_A$. We show that robots r_A and r_B can select a common node t_X to rendezvous. When r_B executes Part (3) of the algorithm, it computes a tree denoted as $T(j)$, which corresponds to the smallest subtree of T_B having size k_B.

Algorithm: WO-ASYNCH
Input: Map $G_X = (V_X, E_X)$, starting node s_X, robot's weight w_X

```
/* Part (1): compute Tx and its integer logarithmic size      */
```
1 compute the maximal spanning tree T_X (by using the ordering $\prec^{WO}$);
2 **if** $|T_X| == 1$ **then exit** ;
3 let $k_X = \lceil \log |T_X| \rceil$;
```
/* Part (2): compute walk Wx                                   */
```
4 let $W_X = \text{MAKEWALK}(T_X, s_X)$;
```
/* Part (3): compute target tx                                 */
```
5 let $i_X = \text{argmin}_i \{w(e_i^X) \mid e_i^X \in T_X\}$;
6 let $j = nil$;
7 **foreach** i **in** $(i_X + 1, i_X + 2, \ldots, m_X)$, in order **do**
8 $\quad$ let $T(i)$ be a largest subtree of T_X induced by nodes in $G(i)$;
9 $\quad$ **if** $k_X > \lceil \log |T(i)| \rceil$ **then**
10 $\quad\quad$ $j = i - 1$;
11 $\quad\quad$ **break**

12 **if** $j == nil$ **then** $j = m_X$;
13 let e be any edge in $T(j)$ having maximum order ;
14 let t_X be the endpoint of e with largest index ;
```
/* Part (4): compute walk Wx+ by using Wx and tx              */
```
15 **while** not (W_X has been fully traversed $2 \cdot k_X$ times $\vee$ rendezvous is accomplished) **do**
16 $\quad$ traverse the next edge in W_X

17 **while** not (r_X is on t_X $\vee$ rendezvous is accomplished) **do**
18 $\quad$ traverse the next edge in W_X

Fig. 2. Algorithm WO-ASYNCH executed by robot $r_X \in \{r_A, r_B\}$.

Since $k_B = k_A$ and since Lemma 7 holds, $T(j)$ can be assumed as a sub-tree of T_B isomorphic to T_A. We now prove that $|T_B| < 2|T_A|$ holds, which means that there cannot exists two distinct sub-trees of T_B isomorphic to T_A. According to the current hypothesis, we get that the following relationships hold: (1) $|T_B| \geq |T_A|$, and (2) $\lceil \log(|T_B|) \rceil = \lceil \log(|T_A|) \rceil$. Denoting by n' the integer value $\lceil \log(|T_B|) \rceil = \lceil \log(|T_A|) \rceil$ we can represent $\log(|T_B|)$ and $\log(|T_A|)$ as follows:

$$\log(|T_B|) = (n' - 1) + b, \ \log(|T_A|) = (n' - 1) + a, \text{ with } 0 < a < b \leq 1.$$

One can observe that $\log(|T_B|) - \log(|T_A|) = b - a < 1$, which in turn implies $\log(|T_B|/|T_A|) < 1$, $|T_B|/|T_A| < 2$, and finally the required relationship $|T_B| < 2|T_A|$. It follows that if r_B selects at Lines 13 and 14 the largest in order edge e belonging to $T(j)$, and node t_X as the endpoint of e with the largest index, then the same target node will be selected by both r_A and r_B (i.e., $t_A = t_B$).

Summarising, if $k_B > k_A$, the algorithm forces robots to meet during the $2k_B$ repetitions of walk W_B. And if $k_B = k_A$, the algorithm forces robots to

---

Algorithm: Tree-$\overline{\text{WO}}$-Asynch
Input : Map defined by the tree $T_X = (V_X, E_X)$, starting node s_X

1 let $k_X = |T_X|$;
2 let $W_X = \text{MakeWalk}(T_X, s_X)$;
3 let t_X be the node in V_X with maximum order ;
4 **while** *not (W_X has been fully traversed $2 \cdot k_X$ times $\vee$ rendezvous is accomplished)* **do**
5   traverse the next edge in W_X

6 **while** *not (r_X is on t_X $\vee$ rendezvous is accomplished)* **do**
7   traverse the next edge in W_X

Fig. 3. Algorithm Tree-$\overline{\text{WO}}$-Asynch executed by robot $r_X \in \{r_A, r_B\}$.

meet at $t_A = t_B$. The trajectory of each robot is of length at most $2 \cdot k_B \cdot |T_B| = 2 \cdot \lceil \log |T_B| \rceil \cdot |T_B|$. Thus the total complexity of rendezvous is $O(N \log N)$. □

Theorem 3 indicates that Algorithm WO-Asynch is almost optimal as it solves rendezvous with trajectories of length $O(N \log N)$, and according to Lemma 1 the relevant lower bound is $\Omega(N)$. In terms of further improvements one could proceed along two different directions. One could try to find a more efficient algorithm for an arbitrary topology, or focus on some restricted classes of graphs. With respect to the latter, as the currently best rendezvous algorithm relies on spanning trees, the restricted cases would likely have to refer to sub-classes of trees. And indeed observe that the results provided in Sect. 3 for path graphs and star graphs also hold in variant WO.

5   Algorithms for Variant $\overline{\text{WO}}$

For variant $\overline{\text{WO}}$, in [12] one can find a rendezvous algorithm with double exponential (in N) complexity. We improve this result in specific classes of graphs.

First observe that in this variant it is possible to define a total ordering $\prec^{\overline{\text{WO}}}$ on edges in E, where $G = (V, E)$ is the input network. This ordering is defined as follows: edges are ordered by utilising the total ordering of nodes. Formally, given two edges $e' = (v_i, v_j)$ and $e'' = (v_{i'}, v_{j'})$ such that $v_i < v_j$ and $v_{i'} < v_{j'}$ then $e' \prec^{\overline{\text{WO}}} e''$ if and only if (1) $v_i < v_{i'}$, or (2) $v_i = v_{i'}$ and $v_j < v_{j'}$.

Observe that even if we have the total order $\prec^{\overline{\text{WO}}}$, in variant $\overline{\text{WO}}$ we cannot use the two-steps approach proposed in Sect. 4. In fact, if we compute again the maximal spanning tree (say T_X) of G_X by using the Kruskal's according to the reverse ordering of $\prec^{\overline{\text{WO}}}$, the required property $T_A \subseteq T_B$ is not present any longer in general graphs. This follows from different properties of maps in variants $\overline{\text{WO}}$ and WO; in particular, here all edges in $G_B \setminus G_A$ incident to G_A have a lower weight than any edge in G_A. Nevertheless, one can adopt the two-steps approach in special classes of maps, including trees. Algorithm Tree-$\overline{\text{WO}}$-Asynch shown in Fig. 3 can be used to solve rendezvous using trajectories of polynomial length.

Theorem 4. *In variant* $\overline{\text{WO}}$, *when* G *is a tree Algorithm* TREE-$\overline{\text{WO}}$-ASYNCH *allows robots to meet along a trajectory of length* $O(N^2)$.

In the remaining, we propose efficient rendezvous algorithms for some other restricted topologies.

Lemma 9. *In variant* $\overline{\text{WO}}$, *when* G *is a cycle one can design an optimal rendezvous algorithm.*

Proof. Since G is a cycle and $G_A \subseteq G_B \subseteq G$ then each map G_X is either a path or the whole cycle. The rendezvous algorithm adopts the following strategy.

If G_X is a cycle, robot r_X starts at s_X, and makes a complete walk in arbitrary direction visiting all nodes before returning to s_X. Then, r_X walks to the largest in provided order node t_X. If G_X is a path, r_X applies the strategy utilised in Lemma 4, i.e., robot r_X visits first an arbitrary endpoint of the path, then walks to the opposite endpoint on this path.

It is easy to see that robots do meet eventually, either on the final target node t_X or because r_B overpasses r_A. In both cases, the complexity is bounded by $O(N)$. $\qquad\square$

Lemma 10. *In variant* $\overline{\text{WO}}$, *if both* G_A *and* G_B *are complete graphs (or complete bipartite graphs), there exists rendezvous algorithm with the complexity* $O(N^3)$.

Proof. Assume that both G_A and G_B are complete graphs. Each robot r_X computes its walk (rendezvous trajectory) W_X as follows:

1. Assume robot r_X is initially located at $s_X = v_i$, which becomes a *base node*.
2. From the current base node v_i, r_X visits back and forth all its neighbours starting from v_{n_X} down to v_{i+2}, and then it moves to the next base node v_{i+1} (in the periodic order);
3. Robot r_X repeats the same strategy until all nodes on its map served as base nodes.

On the basis of W_X, robot r_X computes the complete walk W_X^+ consisting of k_X concatenations of W_X plus a sub-sequence of W_X needed to reach the target node t_X which is the largest in the provided order. We now prove that if the two robots visit their own maps adopting W_X^+, the rendezvous is accomplished. We consider two cases based on the sizes of k_A and k_B. If $|k_B| > |k_A|$, by construction of W_X^+ we get $W_A \subseteq W_B$, and due to Lemma 3 rendezvous must be accomplished. If $|k_B| = |k_A|$, the thesis trivially follows since the target t_X are the same, $t_A = t_B$. Since the trajectory of each robot is at most $(k_X + 1) \cdot |W_X|$, and $|W_X| = O(N^2)$, the total complexity is $O(N^3)$.

One can observe that the above algorithm can be easily adapted when both G_A and G_B are complete bipartite graphs. $\qquad\square$

6   Conclusion

We studied deterministic rendezvous of two asynchronous robots in the network modelled by graphs with restrictions imposed on edges. The restrictions prevent robots from visiting certain parts of the network. We considered four variants based on all possible combinations of presence/absence of two properties: (1) coherent ordering of nodes and (2) weighted robots/edges. We provided some impossibility results, lower bounds, and efficient algorithmic solutions. Two important problems remain open. The first is to establish whether our algorithm in variant WO is optimal. The second is to decide whether there exists a rendezvous algorithm in variant $\overline{\text{WO}}$ with the polynomial (in N) complexity, or the exponential approach provided in [12] cannot be improved.

References

1. Bampas, E., et al.: On asynchronous rendezvous in general graphs. Theor. Comput. Sci. **753**, 80–90 (2019)
2. Borowiecki, P., Das, S., Dereniowski, D., Kuszner, Ł.: Distributed evacuation in graphs with multiple exits. In: Suomela, J. (ed.) SIROCCO 2016. LNCS, vol. 9988, pp. 228–241. Springer, Cham (2016). https://doi.org/10.1007/978-3-319-48314-6_15
3. Chalopin, J., Dieudonné, Y., Labourel, A., Pelc, A.: Fault-tolerant rendezvous in networks. In: Esparza, J., Fraigniaud, P., Husfeldt, T., Koutsoupias, E. (eds.) ICALP 2014. LNCS, vol. 8573, pp. 411–422. Springer, Heidelberg (2014). https://doi.org/10.1007/978-3-662-43951-7_35
4. Chalopin, J., Dieudonné, Y., Labourel, A., Pelc, A.: Rendezvous in networks in spite of delay faults. Distrib. Comput. **29**(3), 187–205 (2016)
5. Czyzowicz, J., Kosowski, A., Pelc, A.: How to meet when you forget: log-space rendezvous in arbitrary graphs. Distrib. Comput. **25**(2), 165–178 (2012)
6. Czyzowicz, J., Kosowski, A., Pelc, A.: Time versus space trade-offs for rendezvous in trees. Distrib. Comput. **27**(2), 95–109 (2014)
7. Czyzowicz, J., Labourel, A., Pelc, A.: How to meet asynchronously (almost) everywhere. ACM Trans. Algorithms **8**(4), 37:1–37:14 (2012)
8. De Marco, G., Gargano, L., Kranakis, E., Krizanc, D., Pelc, A., Vaccaro, U.: Asynchronous deterministic rendezvous in graphs. Theor. Comput. Sci. **355**(3), 315–326 (2006)
9. Dereniowski, D., Klasing, R., Kosowski, A., Kusznera, L.: Rendezvous of heterogeneous mobile agents in edge-weighted networks. Theor. Comput. Sci. **608**(3), 219–230 (2015)
10. Dessmark, A., Fraigniaud, P., Pelc, A.: Deterministic rendezvous in graphs. In: Di Battista, G., Zwick, U. (eds.) ESA 2003. LNCS, vol. 2832, pp. 184–195. Springer, Heidelberg (2003). https://doi.org/10.1007/978-3-540-39658-1_19
11. Dieudonné, Y., Pelc, A., Villain, V.: How to meet asynchronously at polynomial cost. SIAM J. Comput. **44**(3), 844–867 (2015)
12. Farrugia, A., Gąsieniec, L., Kuszner, L., Pacheco, E.: Deterministic rendezvous in restricted graphs. In: Italiano, G.F., Margaria-Steffen, T., Pokorný, J., Quisquater, J.-J., Wattenhofer, R. (eds.) SOFSEM 2015. LNCS, vol. 8939, pp. 189–200. Springer, Heidelberg (2015). https://doi.org/10.1007/978-3-662-46078-8_16

13. Flocchini, P., Santoro, N., Viglietta, G., Yamashita, M.: Rendezvous with constant memory. Theor. Comput. Sci. **621**, 57–72 (2016)
14. Gu, Z., Wang, Y., Hua, Q.S., Lau, F.C.M.: Blind rendezvous problem. In: Gu, Z., Wang, Y., Hua, Q.S., Lau, F.C.M. (eds.) Rendezvous in Distributed Systems. Springer, Singapore (2017). https://doi.org/10.1007/978-981-10-3680-4_5
15. Pelc, A.: Deterministic rendezvous in networks: a comprehensive survey. Networks **59**(3), 331–347 (2012)
16. Viglietta, G.: Rendezvous of two robots with visible bits. In: Flocchini, P., Gao, J., Kranakis, E., Meyer auf der Heide, F. (eds.) ALGOSENSORS 2013. LNCS, vol. 8243, pp. 291–306. Springer, Heidelberg (2014). https://doi.org/10.1007/978-3-642-45346-5_21

Gathering Synchronous Robots in Graphs: From General Properties to Dense and Symmetric Topologies

Serafino Cicerone[1], Gabriele Di Stefano[1], and Alfredo Navarra[2(✉)]

[1] Dipartimento di Ingegneria e Scienze dell'Informazione e Matematica, Università degli Studi dell'Aquila, 67100 L'Aquila, Italy
{serafino.cicerone,gabriele.distefano}@univaq.it
[2] Dipartimento di Matematica e Informatica, Università degli Studi di Perugia, 06123 Perugia, Italy
alfredo.navarra@unipg.it

Abstract. The *Gathering* task by means of a swarm of robots disposed on the vertices of a graph requires robots to move toward a common vertex from where they do not move anymore.

When dealing with very weak robots in terms of capabilities, considering synchronous or asynchronous settings may heavily affect the feasibility of the problem. In fact, even though dealing with asynchronous robots in general requires more sophisticated strategies with respect to the synchronous counterpart, sometimes it comes out that asynchronous robots simply cannot solve the problem whereas synchronous robots can. We study general properties of graphs that can be exploited in order to accomplish the gathering task in the *synchronous* setting, obtaining an interesting sufficient condition for the feasibility, applicable to any topology. We then consider dense and symmetric graphs like complete and complete bipartite graphs where asynchronous robots cannot solve much. In such topologies we *fully characterize* the solvability of the gathering task in the synchronous setting by suitably combining some strategies arising by the general approach with specific techniques dictated by the considered topologies.

1 Introduction

In this paper we consider the *Gathering* task by means of a swarm of very weak robots initially disposed on different vertices of a graph. The task requires robots to move toward a common vertex from where they do not move anymore.

Robots are assumed to be: *Anonymous:* no unique identifiers; *Autonomous:* no centralized control; *Oblivious:* no memory of past events; *Homogeneous:* they

The work has been supported in part by the European project "Geospatial based Environment for Optimisation Systems Addressing Fire Emergencies" (GEO-SAFE), contract no. H2020-691161, and by the Italian National Group for Scientific Computation (GNCS-INdAM).

© Springer Nature Switzerland AG 2019
K. Censor-Hillel and M. Flammini (Eds.): SIROCCO 2019, LNCS 11639, pp. 170–184, 2019.
https://doi.org/10.1007/978-3-030-24922-9_12

all execute the same deterministic algorithm; *Silent:* no means of communication; *Disoriented:* no common orientation. Robots operate in standard *Look-Compute-Move* (LCM) cycles. In one cycle a robot takes a snapshot of the current global configuration (Look) in terms of robots' locations. Successively, in the Compute phase it decides whether to move toward a neighboring vertex or not, and in the positive case it moves (Move).

Cycles might be performed synchronously or asynchronously. Standard models are:

- *Fully-Synchronous* (FSYNC): The *activation* phase (i.e. the execution of a LCM-cycle) of all robots can be logically divided into global rounds. In each round all the robots are activated, obtain the same snapshot of the environment, compute and perform their move.
- *Semi-Synchronous* (SSYNC): It coincides with the FSYNC model, with the only difference that not all robots are necessarily activated in each round.
- *Asynchronous* (ASYNC): The robots are activated independently, and the duration of each phase is finite but unpredictable.

The amount of time to complete a full LCM-cycle is assumed to be finite but unpredictable. In particular, in both the SSYNC and ASYNC cases it is usually assumed the existence of an *adversary* which determines the computational cycles timing. Such timing is assumed to be *fair*, that is, each robot performs its LCM-cycle within finite time and infinitely often. Without such an assumption the gathering would be unsolvable as the adversary could prevent some robots to ever move.

It is very common (as dictated by impossibility results) that in combination with the LCM-model, robots are endowed with the so-called *multiplicity detection* capability (see e.g. [8,21]). Basically, when more than one robot resides on the same vertex x, then x is said to be occupied by a *multiplicity*. A robot is said to have the (global strong) multiplicity detection ability when it can detect the exact number of robots composing a multiplicity at any given vertex. Other weaker forms of multiplicity detection could be defined but they would lead to wider impossibility results.

While the gathering problem has been deeply investigated and fully characterized for robots moving on the Euclidean plane [8] (also with respect to given meeting points [4,5,7]), not much is known for the graph environment, apart from a few of specific topologies. Concerning feasibility aspects, the aim has been usually that of finding the minimal set of assumptions under which robots are able to solve the problem. One of the main observations for gathering in the Euclidean plane has been to show that ASYNC robots are as much powerful as FSYNC ones [8], except for the only case of exactly two robots. This case is unsolvable in the ASYNC or SSYNC contexts whereas it is solvable by two FSYNC robots.

In graphs, a full characterization for the gathering task is missing, even for FSYNC robots. Actually, the main results concern (i) ASYNC robots and (ii) specific topologies.

The considered topologies so far are trees [9,10,17], rings [11–15,17,19–21], regular bipartite graphs [18], finite [9] or infinite [16] grids and hypercubes [3], also from an optimization perspective. Most of the considered topologies are very symmetric when dealing with anonymous graphs, that is all vertices look equivalent. This choice has been done so that robots cannot exploit much topological properties. For instance if a tree or a finite grid admits only one center, then all robots can detect it and move there, even asynchronously. Contrary, in rings, infinite grids or hypercubes, all vertices are equivalent and the synchronicity may heavily impact on feasibility.

A main observation coming out from the literature about gathering in graphs is that feasibility is very constrained with respect to synchronicity. On the one hand, dealing with ASYNC robots is much harder than considering SSYNC or FSYNC ones. On the other hand, it may happen that ASYNC robots simply cannot solve some instances, hence sensibly reducing the scope of research for resolution algorithms. In other words, the graph context seems requiring deep investigation on the different results one may achieve when switching from the ASYNC to the SSYNC or FSYNC cases.

Our Results. First, we study general properties of graphs that can be exploited in order to accomplish the gathering task in the SSYNC (and hence also in the FSYNC) setting. The investigation leads to obtain an interesting sufficient condition for the feasibility, applicable to any topology. We then consider dense and symmetric graphs like complete and complete bipartite graphs where ASYNC or SSYNC robots cannot solve much. In such topologies we fully characterize the solvability of the gathering task in the FSYNC setting by suitably combining some strategies arising by the general approach with specific techniques dictated by the considered topologies. Also, we evaluate the number of LCM-cycles required by our algorithms to accomplish the gathering task. Due to space limitations, some proofs are omitted or just sketched.

2   Problem Definition and General Impossibility Results

The topology where robots are placed on is represented by a simple and connected graph $G = (V, E)$, with vertex set V and edge set E. The cardinality of V is represented as $|V|$ or $|G|$. A function $\lambda : V \to \mathbb{N}$ represents the number of robots on each vertex of G, and we call $C = (G, \lambda)$ a *configuration* whenever $\sum_{v \in V} \lambda(v)$ is bounded and greater than zero. A vertex $v \in V$ such that $\lambda(v) > 0$ is said *occupied*, *unoccupied* otherwise. A subset $V' \subseteq V$ is said *occupied* if at least one of its elements is occupied, *unoccupied* otherwise. A configuration is *initial* if each robot lies on a different vertex (i.e., $\lambda(v) \leq 1$ for each $v \in V$). A configuration is *final* if all the robots are on a single vertex (i.e., $\exists u \in V : \lambda(u) > 0$ and $\lambda(v) = 0, \forall v \in V \setminus \{u\}$). The Gathering problem can be formally defined as the problem of transforming an initial configuration into a final one. Throughout the paper we assume that each initial configuration is composed of at least two robots (otherwise the problem is trivially solved). A *gathering algorithm* for this problem is a deterministic distributed algorithm that brings the robots in the system to a final configuration in a finite number of LCM-cycles from any given initial configuration, regardless of the adversary. Formally, an algorithm $\mathcal{A}$ solves the Gathering problem for an initial configuration C if, for any execution $\mathbb{E} : C = C(0), C(1), \dots$ of $\mathcal{A}$, there exists a time instant $i > 0$ such that $C(i)$ is final and no robots move after i, i.e., $C(t) = C(i)$ holds for all $t \geq i$. Given an initial configuration $C = (G, \lambda)$, if there exists a gathering algorithm for C we say that C is *gatherable*, otherwise we say that C is *ungatherable*. For FSYNC robots, the *time complexity* of a gathering algorithm $\mathcal{A}$ is the maximum amount of time units (that is the number of LCM-cycles) required by $\mathcal{A}$ for processing any gatherable initial configuration. For the gathering problem, a natural lower bound for the time complexity of any algorithm is $\Omega(D_G)$, where D_G is the diameter of the graph underlying the configuration.

During an execution, $\Lambda(t)$ denotes the number of occupied vertices at time t; formally, $\Lambda(t) = |\{u \in V : \lambda(v) > 0\}|$.

We now recall from [17] the notions of configuration automorphisms and symmetries to be applied to general graphs, and accordingly we also recall general impossibility results.

Configuration Automorphisms and Symmetries. Two undirected graphs $G = (V_G, E_G)$ and $H = (V_H, E_H)$ are *isomorphic* if there is a bijection φ from V_G to V_H such that $\{u, v\} \in E_G$ if and only if $\{\varphi(u), \varphi(v)\} \in E_H$. An *automorphism* on a graph G is an isomorphism from G to itself, that is a permutation of the vertices of G that maps edges to edges and non-edges to non-edges. The set of all automorphisms of G forms a group called *automorphism group* of G and denoted by $\mathrm{Aut}(G)$. Two vertices $u, v \in V$ are *equivalent* if there exists an automorphism $\varphi \in \mathrm{Aut}(G)$ such that $\varphi(u) = v$.

The concept of isomorphism can be extended to configurations in a natural way: two configurations (G, λ) and (G', λ') are isomorphic if G and G' are isomorphic via a bijection φ and for each vertex v in G, $\lambda(v) = \lambda'(\varphi(v))$. An *automorphism* on a configuration (G, λ) is an isomorphism from (G, λ) to itself and the set of all automorphisms of (G, λ) forms a group that we call *automorphism group* of (G, λ), denoted by $\mathrm{Aut}((G, \lambda))$.

If $|\mathrm{Aut}(G)| = 1$, that is G admits only the identity automorphism, then G is said *asymmetric*, otherwise it is said *symmetric*. Analogously, if $|\mathrm{Aut}((G, \lambda))| = 1$, we say that the configuration (G, λ) is *asymmetric*, otherwise it is *symmetric*. Two distinct robots r and r' in a configuration (G, λ) are *equivalent* if there exists $\varphi \in \mathrm{Aut}((G, \lambda))$ that makes equivalent the vertices in which they reside. Note that $\lambda(u) = \lambda(v)$ whenever u and v are equivalent. Moreover, if u and v are equivalent, a robot r cannot distinguish its position at vertex u from robot r' located at vertex $v = \varphi(u)$. As a consequence, no algorithm can distinguish between two equivalent robots.

Given $\varphi \in \mathrm{Aut}((G, \lambda))$ different from the identity, the *cyclic subgroup* of order p generated by φ is given by $H = \{\varphi^0, \varphi^1 = \varphi \circ \varphi^0, \varphi^2 = \varphi \circ \varphi^1, \dots, \varphi^{p-1} = \varphi \circ \varphi^{p-2}\}$, where φ^0 is the identity automorphism, $\varphi^i \neq \varphi^0$ for each $0 < i < p$, and $\varphi^p = \varphi^0$. In (G, λ), the *orbit* of a vertex v of G is $Hv = \{\gamma(v) \mid \gamma \in H\}$.

Note that (1) the orbits Hv, for each $v \in V$, form a partition of V, and (2) vertices/robots belonging to any orbit are pairwise equivalent.

The next theorem provides a sufficient condition for a configuration to be ungatherable, but we first need the following definition:

Definition 1. [17] *Given a graph* $G = (+6V, E)$, *let* $C = (G, \lambda)$ *be a configuration. An automorphism* $\varphi \in Aut(C)$ *is said* partitive *on* $V' \subseteq V$ *if the cyclic subgroup* $H = \{\varphi^0, \varphi^1 = \varphi \circ \varphi^0, \varphi^2 = \varphi \circ \varphi^1, \ldots, \varphi^{p-1} = \varphi \circ \varphi^{p-2}\}$ *generated by* φ *has order* $p > 1$ *and is s.t.* $|Hu| = p$ *for each* $u \in V'$.

The next two claims use this definition to provide general impossibility results for the gathering problem on graphs. Since they refer to FSYNC robots, they also hold for both SSYNC and ASYNC robots.

Theorem 1. [17] *Let* $G = (V, E)$ *be any graph and let* $C = (G, \lambda)$ *be any non-final configuration. If there exists* $\varphi \in Aut(C)$ *partitive on* V *then* C *is ungatherable.*

It is worth to remark that the above theorem requires the existence of an automorphism φ, which in turn is based on the function λ defining the exact number of robots on each vertex. Hence, Theorem 1 holds when the robots are endowed with the global-strong multiplicity detection. Stating a negative result, it follows that such a theorem holds even when considering weaker robots (i.e., without global-strong multiplicity detection).

The following corollary obtained from [17] shows that some configurations can be gathered only at some predetermined vertices.

Corollary 1. *Let* $G = (V, E)$ *be any graph,* $C = (G, \lambda)$ *be any configuration, and* $V' \subset V$ *unoccupied. If there exists an automorphism* $\varphi \in Aut(C)$ *that is partitive on* $V \setminus V'$, *then each gathering algorithm for* C *(if any) must move robots toward* V'.

3 A Sufficient Condition for Gathering in Arbitrary Graphs

In this section we provide a sufficient condition for gathering FSYNC robots in arbitrary graphs. This result exploits new concepts we define in the following, like *recognizable subgraphs* and *d-primality*.

Recognizable Subgraphs. Informally, a subgraph H of a graph G is said *recognizable* if any automorphism of G maps H on itself, that is, H cannot be confused with other subgraphs. Formally:

Definition 2. *A subgraph* $H = (V_H, E_H)$ *of a graph* $G = (V, E)$ *is recognizable if* $V_H = \{\varphi(v) \mid v \in V_H\}$ *for each automorphism* $\varphi \in Aut(G)$. *A recognizable subgraph* H *is minimal if there not exists a proper subgraph* H' *of* H *such that* H' *is recognizable.*

In the following, we provide interesting properties of recognizable subgraphs. Let $G = (V, E)$ be a graph and let $G[V']$ denotes the subgraph induced by $V' \subseteq V$. By definition: (1) the empty subgraph is recognizable; (2) G is recognizable; (3) if V_1 and V_2 are the vertex sets of recognizable subgraphs of G, then $G[V_1 \cup V_2]$ is recognizable; (4) if G is asymmetric then any subgraph with a single vertex is recognizable.

Lemma 1. *Let $G = (V, E)$ be a graph. The following properties hold: (1) if $H = (V_H, E_H)$ is a recognizable subgraph of G, then $G[V \setminus V_H]$ is recognizable; (2) if $G[V_1]$ and $G[V_2]$ are two distinct minimal recognizable subgraphs of G, with $V_1 \subseteq V$ and $V_2 \subseteq V$, then $V_1 \cap V_2$ is empty.*

Proof. Omitted. □

Theorem 2. *The set containing all the minimal recognizable subgraphs of $G = (V, E)$ forms a unique partition of V.*

Proof. The partition can be found with the following procedure. If G is minimal recognizable, we are done. If G is not minimal recognizable, let $G[V_1]$ a minimal recognizable subgraph of G and let $\{V_1, V \setminus V_1\}$ a partition of V. By Lemma 1, $G[V \setminus V_1]$ is recognizable. If it is also minimal recognizable we are done, otherwise recursively apply the procedure to $G[V \setminus V_1]$. Regarding the uniqueness, by contradiction let us suppose that there exist $\{V_1, V_2, \ldots, V_p\}$ and $\{V_1', V_2', \ldots, V_{p'}'\}$ distinct partitions of V such that $G[V_i]$, $G[V_j']$ are minimal recognizable graphs, for each $i = 1, 2, \ldots, p$ and $j = 1, 2, \ldots, p'$. Then, there exists at least a vertex v such that $v \in V_{i*}$ and $v \in V_{j*}'$ and the two sets V_{i*} and V_{j*}' are distinct. Then $V_{i*} \cap V_{j*}' \neq \emptyset$, a contradiction to Lemma 1. □

The following corollary shows that each pair of vertices inside a minimal recognizable subgraph of G are equivalent.

Corollary 2. *Let $G = (V, E)$ be a graph and $\{V_1, V_2, \ldots, V_p\}$ be the unique partition of V induced by the minimal recognizable subgraphs. For each pair of vertices $u, v \in V_i$, $i = 1, 2, \ldots, p$, u and v are equivalent.*

Proof. Omitted. □

Let $G = (V, E)$ be a graph and H be a minimal recognizable subgraph of G. If H is disconnected, all the connected components are pairwise isomorphic. We denote by $c(H)$ and $s(H)$ the number of connected components of H and the size of each connected component of H, respectively.

*d-**Primality and Batches.*** In order to make use of a recognizable subgraph H for gathering purposes, we need to relate the number k of robots of a given configuration and the topology of H.

Definition 3. *Let k and d be two positive integers. We say that k is d–prime if $lpf(k) > d$, where $lpf(k)$ denotes the least prime factor of k.*

Notice that when k is d–prime the following properties hold: (1) $k > d > 0$; (2) if $d = 1$ then every $k > 1$ is d–prime; (3) for each integer $2 \leq d' \leq d$, d' does not divide k.

Definition 4. *Given a graph $G = (V, E)$, let $C = (G, \lambda)$ be a configuration with k robots, $\mathcal{A}$ be a gathering algorithm for C, and $\mathbb{E} : C = C(0), C(1), \ldots$ be any execution of $\mathcal{A}$ starting from C. For an integer i, $0 < i \leq k$, the batch $B_i(t)$, $t \geq 0$, is the subset of vertices $\{u \in V \mid \lambda(u) = i\}$ in $C(t)$.*

We simply use B_i when we are not interested to any specific time instant. The size of B_i is $|B_i|$, i.e., the number of vertices in B_i. Given a batch B_i, we use the following additional notions:

- the *order* of B_i is i, i.e., there are i robots in each vertex in B_i.
- $B_{\min}$ and $B_{\max}$ denote the non-empty batches with minimum and maximum order, respectively.

By the above definitions, in a configuration C with k robots: (1) $|B_1| = k$ iff C is initial, and (2) $B_k \neq \emptyset$ (i.e., $B_{\max}$ has order k) iff C is final.

Lemma 2. *Let $G = (V, E)$ be a graph with n vertices, and let H be a minimal recognizable subgraph of G. Let $C = (G, \lambda)$ be a non-final configuration composed of k robots, $2 \leq k \leq n$, all residing on vertices of H, such that k is $\max\{c(H), s(H)\}$–prime. If the robots are on different connected components of H, then (1) two connected components of H contain a different number of robots, otherwise (2) in H there are batches with different orders.*

Proof. Since k is $\max\{c(H), s(H)\}$–prime, then $lpf(k) > \max\{c(H), s(H)\}$. This implies that k is both $c(H)$–prime and $s(H)$–prime. These relationships imply $k \geq lpf(k) > s(H)$ and $k \geq lpf(k) > c(H)$; in other words, k is greater than both $c(H)$ and $s(H)$ and both $c(H)$ and $s(H)$ do not divide k.

If the robots are on different components of H, $c(H) > 1$. Since k is $c(H)$-prime, $c(H)$ does not divide k and this implies case (1). If the robots are on the same component, since k is $s(H)$-prime, then the number $1 < s' \leq s(H)$ of occupied vertices does not divide k. Then, case (2) occurs.   □

The Sufficient Condition. We now use the notions of recognizable graphs, d-primality, and *graph canonization* to provide a sufficient condition to the solvability of the gathering problem by means of FSYNC robots. This condition is applicable to any graph topology.

In graph theory the *graph canonization* is the problem of finding a canonical form of a given graph G. A canonical form is function that associate to G a labeled graph $Canon(G)$ such that $Canon(G)$ is isomorphic to G and every graph G' that is isomorphic to G is such that $Canon(G') = Canon(G)$ [2]. The labels are from a linearly ordered set (e.g., $\{1, 2, \ldots, n\}$). The graph canonization problem is at least as computationally hard as the graph isomorphism problem, which is not known to be solvable in polynomial time nor to be NP-complete.

However in [1] a linear time algorithm is reported that with probability at least $1 - \exp(-O(n \log n / \log \log n))$ produce a canonical labeling. This justify the good behavior in practice of the deterministic algorithm proposed by McKay [22] whose complexity is studied in [23].

Theorem 3. *Let $G = (V, E)$ be a graph with n vertices, and let $C = (G, \lambda)$ be an initial configuration composed of k SSYNC robots, $2 \le k \le n$. If there exists a minimal recognizable subgraph H of G such that k is $\max\{c(H), s(H)\}$–prime, then C is gatherable.*

Sketch of the Proof. Consider a canonical labeling $Canon(G)$ and let $\bar{H} = (\bar{V}, \bar{E})$ be the minimal recognizable subgraph such that k is $\max\{c(\bar{H}), s(\bar{H})\}$–prime and having the vertex with the minimum label. All the robots can agree on $\bar{H}$.

We show there exists an algorithm $\mathcal{A}_g$ able to gather all robots in C on a vertex of $\bar{V}$. If $\mathbb{E} : C = C(0), C(1), \ldots$ denotes any execution of $\mathcal{A}_g$, then $\mathcal{A}_g$ will apply the same move m to each active robot in any obtained configuration $C(t)$. Such a move m is defined as follows:

- *(target of m)* if there are no robots in $\bar{V}$, the target $T(t)$ for m coincides with $\bar{V}$. If there are robots on different components of $\bar{H}$, $T(t)$ corresponds to vertices of the connected components that contains the largest number of robots. If all the robots are on a single component of $\bar{H}$ then the target $T(t)$ is the set containing each vertex $v \in \bar{V}$ such that $\lambda(v)$ is maximum.
- *(robots moving according to m)* robots allowed to move are all those not on vertices of $T(t)$;
- *(trajectory)* if a moving robot r is not on a vertex in $T(t)$, it moves toward an adjacent vertex along a shortest path to a vertex in $T(t)$ having a minimum distance from r. If all the robots are on a single connected component of $\bar{H}$, the shortest path is taken among the paths having all the vertices in that component.

Note that it might happen that during a LCM-cycle, only robots lying on the vertices of $T(t)$ are activated by the adversary, that is no one moves during that cycle. However, by the fairness assumption, all other robots will be activated within finite time.

According to $\mathcal{A}_g$, at starting time $t_0 = 0$, as long as there are no robots in $\bar{H}$, m allows all robots to move toward $\bar{V}$ along shortest paths. This implies that there exists a first time $t_1 \ge t_0$ when at least one robot is inside $\bar{H}$. For each time $t \ge t_1$, subgraph $\bar{H}$ is not empty because, from time t_1 onward, $T(t) \subseteq \bar{V}$ and robots on $T(t)$ do not move. This property simple follows from the definition of $T(t)$ that requires each vertex in $T(t)$ to be occupied when there are robots in $\bar{H}$.

From t_1 on, according to m, all the robots move toward the set of vertices of the components of $\bar{H}$ with the largest number of robots. When all the robots are inside $\bar{H}$, by Lemma 2, they will convergence at a time $t_2 \ge t_1$ toward a single component of $\bar{H}$ with the largest number of robots.

From t_2 on, according to m, the robots will move inside the same component toward the batch with maximum order. In particular, Lemma 2 guarantees that there are batches with different orders, and this property makes move m well defined: robots within batches with non-maximum order move toward the batch with maximum order. During this phase, robots remain within the same component and the process continues until only batch B_k is formed. This means that the gathering is eventually accomplished.   □

Theorem 3 provides a powerful means for gathering purposes as long as the input graph admits some topological properties like a limited number of centers, medians, bounded degree nodes and so forth. As we are going to see, still some of the outcoming techniques can be exploited for FSYNC robots even in very symmetric graphs like complete and complete bipartite graphs.

4   Gathering in Complete and Complete Bipartite Graphs

In this section, we provide a full characterization of gathering in two dense and symmetric topologies that are complete graphs and complete bipartite graphs.

Complete Graphs. From [6] it is known that ASYNC robots cannot accomplish the gathering task if the underlying graph is a complete graph. The result can be easily extended to SSYNC robots (proof omitted). Assuming FSYNC robots, we now show there are instances that instead are gatherable.

Theorem 4. *Let G be a complete graph with n vertices, and let $C = (G, \lambda)$ be an initial configuration with $k \geq 2$ FSYNC robots. C is gatherable if and only if k is $(n - k)$–prime.*

Sketch of the Proof. ($\Leftarrow$) There exists a simple algorithm $\mathcal{A}_{clique}$ able to gather all robots in C when k is $(n - k)$–prime. It uses the following move: *if the configuration is initial (i.e., no multiplicity occurs), then each robot moves toward an arbitrary unoccupied vertex, otherwise each robot not in $B_{\max}$ moves toward an arbitrary vertex in $B_{\max}$.* After the first move, since k is $(n - k)$–prime, then $C(1)$ contains batches with different orders. Moreover, the number of occupied vertices in $C(1)$ is at most $n - k$. Thanks to this property and the assumption that k is $(n - k)$–prime, the move can be applied again and $B_{\max}$ is reduced at each step thereafter until it becomes the set of a single vertex. Hence, the gathering is accomplished.

($\Rightarrow$) By Corollary 1, any gathering algorithm must move robots toward the unoccupied vertices. If k is not $(n - k)$–prime then $k = ik'$ for some integer i such that $2 \leq i \leq n - k$. Then the adversary can move k' robots on each of some i vertices in $C(1)$. Hence in $C(1)$ there exists only one non-empty batch, namely $B_{k'}$, and this batch has size $i \geq 2$. After that, the adversary can always keep the symmetry of the configuration by keeping this batch during the rest of the execution. This prevent the gathering.   □

Corollary 3. *In any gatherable configuration defined on a n-vertex complete graph, the gathering problem can be solved by FSYNC robots in $O(\log n)$ time.*

Proof. Let $C = (G, \lambda)$ be any gatherable initial configuration with k robots and defined on a complete graph G with n vertices. Consider the algorithm $\mathcal{A}_{clique}$ proposed in the proof of Theorem 4 to process C. After the first move of $\mathcal{A}_{clique}$, each robot not in $B_{\max}$ moves toward an arbitrary vertex in $B_{\max}$. Being C gatherable, Theorem 4 implies that k is $(n - k)$–prime. Since k is $(n - k)$–prime, this move always reduces the size of $B_{\max}$. Hence, the maximum amount of time is required by any execution of $\mathcal{A}_{clique}$ when $B_{\max}$ is as large as possible. According to the move, the following relationship holds for each time instant $i \geq 1$: $|B_{\max}(i + 1)| \leq \min\{|B_{\max}(i)|, \sum_{j \neq \max} |B_j(i)|\}$. By this property, it follows that the size of $B_{\max}$ at time instant $i + 1$ is maximal when at time i the size of $B_{\max}$ is comparable to the number of occupied vertices containing moving robots. Hence, to produce the longest execution for $\mathcal{A}_{clique}$, half of the robots should move each time toward $B_{\max}$. This produces an execution that requires $O(\log k)$ time units. Since k is $(n - k)$–prime, we get $k \geq lpf(k) > n - k$ and hence $k > n/2$.   □

Complete Bipartite Graphs. We now consider the gathering problem of FSYNC robots on complete bipartite graphs. In the remainder we use the following notation: if $G = (V_1 \cup V_2, E)$ is a complete bipartite graph and $C = (G, \lambda)$ is any initial configuration, then n_1 and n_2 denote the number of vertices of V_1 and V_2, respectively; k_1 and k_2 denote the number of robots on V_1 and V_2, respectively; if $k_1 > 0$ and $k_2 > 0$, then $lcpf(k_1, k_2)$ denotes the *least common prime factor* of k_1 and k_2; and we say that partition V_i, $i = 1, 2$, is *unoccupied* (*occupied*, resp.) if $k_i = 0$ ($k_i > 0$, resp.).

The following definition extends to bipartite graphs some concepts introduced in Definition 4.

Definition 5. *Let $G = (V_1 \cup V_2, E)$ be a complete bipartite graph, $C = (G, \lambda)$ be a configuration with $k_1 + k_2$ robots, $\mathcal{A}$ be a gathering algorithm for C, and $\mathbb{E} : C = C(0), C(1), \dots$ be an execution of $\mathcal{A}$ that starts from C. Then:*

- *$B_i^1(t) = B_i(t) \cap V_1$ and $B_i^2(t) = B_i(t) \cap V_2$, that is, $B_i^1(t)$ and $B_i^2(t)$ are the projections of $B_i(t)$ into V_1 and V_2, respectively;*
- *$B_{\min}^1(t) = B_{\min}(t) \cap V_1$ and $B_{\min}^2(t) = B_{\min}(t) \cap V_2$;*
- *$B_{\max}^1(t) = B_{\max}(t) \cap V_1$ and $B_{\max}^2(t) = B_{\max}(t) \cap V_2$;*
- *$\Lambda^1(t)$ and $\Lambda^2(t)$ denote the number of occupied vertices at time t in V_1 and V_2, respectively;*

The next two lemmas concern different cases for gathering on complete bipartite graphs.

Lemma 3. *Let $G = (V_1 \cup V_2, E)$ be a complete bipartite graph with $n_1 + n_2$ vertices, and let $C = (G, \lambda)$ be an initial configuration composed of $k_1 + k_2$ FSYNC robots, with $k_1 > 0$ and $k_2 > 0$. If k_1 and k_2 are coprime, then C is gatherable.*

Sketch of the Proof. Consider the algorithm $\mathcal{A}_{cop}$ defined by the following moves:

- m_1: if $\Lambda^1(t) = 1$ and $\Lambda^2(t) = 1$, robots in $B_{\min}(t)$ move toward $B_{\max}(t)$;
- m_{2a}: if $\Lambda^1(t) = 1$ and $\Lambda^2(t) > 1$, robots move toward the unique vertex in $B^1_{\max}(t)$;
- m_{2b}: if $\Lambda^1(t) > 1$ and $\Lambda^2(t) = 1$, robots move toward the unique vertex in $B^2_{\max}(t)$;
- m_3: if $\Lambda^1(t) > 1$ and $\Lambda^2(t) > 1$, robots in V_1 (V_2, resp.) moves toward $B^2_{\max}(t)$ ($B^1_{\max}(t)$, resp.).

Since $k_1 > 0$ and $k_2 > 0$ by hypothesis, move m_3 is applied in $C(0)$. This leads to a configuration $C(1)$ where all robots in V_1 moved to V_2 and vice versa. As long as the condition that generates the move m_3 is verified, robots will continue to swap the partition they reside. As m_3 implies that the vertices occupied at time $t + 1$ form a subset of those occupied at time t, then it is clear that $\Lambda^i(t + 1) \leq \Lambda^i(t)$ holds. Assume now $\Lambda^i(t + 1) = \Lambda^i(t)$. According to move m_3, this last relationship implies that in V_i there was only the batch $B^i_{\max}(t)$ (i.e., all the occupied vertices in V_i had the same multiplicity). But being k_1 and k_2 coprime, when robots move to V_i necessarily they form batches with different orders, and this implies $|B^i_{\max}(t + 1)| < |B^i_{\max}(t)|$. This guarantees that there exists a time t' such that $C(t')$ has at least one partition containing exactly one occupied vertex. Then, at $C(t')$ one among moves m_1, m_{2a}, or m_{2b} will accomplish the gathering. □

Lemma 4. *Let $G = (V_1 \cup V_2, E)$ be a complete bipartite graph with $n_1 + n_2$ vertices, and let $C = (G, \lambda)$ be an initial configuration composed of $k_1 + k_2$ FSYNC robots, with $k_1 > 0$ and $k_2 > 0$. If k_1 and k_2 are not coprime, then C is gatherable if and only if both the following conditions hold:*

1. $n_1 - k_1 < lcpf(k_1, k_2)$ *or* $n_2 - k_2 < lcpf(k_1, k_2)$,
2. $k_1 \neq k_2$ *or* $n_1 \neq n_2$.

Sketch of the Proof. ($\Rightarrow$) If condition (2) does not hold, then $k_1 = k_2$ and $n_1 = n_2$. This means that C is partitive and hence ungatherable by Theorem 1. From now on, assume that condition (1) does not hold. This implies that k_1 and k_2 are not coprime, $n_1 - k_1 \geq d$ and $n_2 - k_2 \geq d$, where $d = lcpf(k_1, k_2)$. In $C(0)$, there exists only the batch $B^1_1(0)$ in V_1 and, symmetrically, only the batch $B^2_1(0)$ in V_2. Since k_1 and k_2 are not coprime, the size of each of such batches is a multiple of d.

Assume that robots are moved from V_1 to V_2. All the robots in one batch are moved and hence a multiple of d robots are moved. If the target is any occupied vertex, then all the robots in the same batch are equally distributed by the adversary in the elements of another batch; if the target is any unoccupied vertex, then all the robots in the same batch are equally distributed by the adversary in d vertices (this is possible since $n_2 - k_2 \geq d$).

This analysis still holds when robots are moved from V_2 to V_1 (by symmetry) and also when robots are swapped between V_1 to V_2 (in this case the analysis of the move toward occupied vertices is the same while it is more evident that there are always at least d unoccupied vertices in each side). In any case, a batch with d elements eventually remains.

($\Leftarrow$) Consider the algorithm $\mathcal{A}_{\neg cop}$ defined by the following moves:

- move $m_{\overline{1}}$. Let $\Lambda^1(t) = 1$ and $\Lambda^2(t) = 1$: if $k_1 \neq k_2$ then each robot in $B_{\min}(t)$ moves toward the unique vertex in $B_{\max}(t)$ else each robot in the larger partition between V_1 and V_2 moves toward the unique occupied vertex in the other partition;
- move $m_{\overline{2a}}$. Let $\Lambda^1(t) = 1$ and $\Lambda^2(t) > 1$: each robot moves toward the unique vertex in $B^1_{\max}(t)$;
- move $m_{\overline{2b}}$. Let $\Lambda^1(t) > 1$ and $\Lambda^2(t) = 1$: each robot moves toward the unique vertex in $B^2_{\max}(t)$;
- move $m_{\overline{3}}$. Let C be not initial, $\Lambda^1(t) > 1$ and $\Lambda^2(t) > 1$: each robot in V_1 moves toward $B^2_{\max}(t)$ and each robot in V_2 moves toward $B^1_{\max}(t)$;
- move $m_{\overline{4}}$. Let C be initial, $\Lambda^1(t) > 1$ and $\Lambda^2(t) > 1$: each robot in V_1 (V_2, resp.) moves toward an arbitrary unoccupied vertex in V_2 (V_1, resp.).

Move $m_{\overline{4}}$ is applied in $C(0)$. This leads to a configuration $C(1)$ where all robots in V_1 moved to the unoccupied vertices in V_2 and vice versa. According to condition 1, in $C(1)$ there are batches with order greater than one. This implies that in $C(1)$ move $m_{\overline{3}}$ is applied, thus obtaining a configuration $C(2)$ where again all robots in V_1 moved to V_2 and vice versa.

As long as move $m_{\overline{3}}$ is applied, robots will continue to swap the partition they reside until, at a time $t' > 0$, there will be at least one partition containing just one occupied vertex. Then, at $C(t')$ one among moves $m_{\overline{1}}$, $m_{\overline{2a}}$, or $m_{\overline{2b}}$ will accomplish the gathering. $\square$

Theorem 5. *Let $G = (V_1 \cup V_2, E)$ be a complete bipartite graph with $n_1 + n_2$ vertices, and let $C = (G, \lambda)$ be an initial configuration composed of $k_1 + k_2$ FSYNC robots. C is gatherable if and only if one of the following conditions hold:*

1. *($k_2 = 0$, k_1 is n_2–prime) or ($k_1 = 0$, k_2 is n_1–prime);*
2. *$k_1 > 0$, $k_2 > 0$, k_1 and k_2 are coprime;*
3. *$k_1 > 0$, $k_2 > 0$, k_1 and k_2 are not coprime, $n_1 - k_1 < lcpf(k_1, k_2)$ or $n_2 - k_2 < lcpf(k_1, k_2)$, $k_1 \neq k_2$ or $n_1 \neq n_2$.*

Proof. ($\Leftarrow$) We show that Algorithm $\mathcal{A}_{bip}$ described in Fig. 1 is able to gather all robots in C when C fulfills one of the conditions (that are mutually exclusive) expressed in the statement.

Assume that condition 1 holds. In particular, w.l.o.g. assume $k_2 = 0$ and k_1 is n_2–prime. In such a case it is interesting to observe that $n_1 \neq n_2$ holds. In fact, since k_1 is n_2–prime then $n_2 < lpf(k_1) \leq k_1 \leq n_1$. In particular, k_1 is n_2–prime implies $n_1 > n_2$, and hence $\mathcal{A}_{bip}$ calls $\mathcal{A}_g$ at Line 2 for moving robots in $C(0)$. Each time that $\mathcal{A}_{bip}$ restarts for handling configurations $C(1), C(2), \ldots$, always $\mathcal{A}_g$ is executed. Since $n_1 > n_2$, then V_2 is a minimal recognizable subgraph of G and k_1 is n_2–prime by hypothesis. Hence, by Theorem 3 the gathering is eventually accomplished on a vertex of V_2.

If condition 2 holds, then $\mathcal{A}_{bip}$ calls $\mathcal{A}_{cop}$ at Line 9. As remarked in the proof of Lemma 3, $\mathcal{A}_{cop}$ always swaps robots between V_1 and V_2 until the last move

---

Algorithm: $\mathcal{A}_{bip}$
Input: Configuration $C = (G, \lambda)$, where G is a complete bipartite graph, fulfilling conditions of Theorem 5.

1 **if** $n_1 > n_2$ *and* $k_1 + k_2$ *is* n_2-*prime* **then**
2 $\quad$ call $\mathcal{A}_g$ // gathering accomplished on a vertex in V_2
3 **end**
4 **else if** $n_2 > n_1$ *and* $k_1 + k_2$ *is* n_1-*prime* **then**
5 $\quad$ call $\mathcal{A}_g$ // gathering accomplished on a vertex in V_1
6 **end**
7 **else**
8 $\quad$ **if** $k_1 > 0$, $k_2 > 0$, k_1 *and* k_2 *are coprime* **then**
9 $\quad\quad$ call $\mathcal{A}_{cop}$
10 $\quad$ **end**
11 $\quad$ **if** $k_1 > 0$, $k_2 > 0$, k_1 *and* k_2 *are not coprime*, $n_1 - k_1 < lcpf(k_1, k_2)$ *or* $n_2 - k_2 < lcpf(k_1, k_2)$, $k_1 \neq k_2$ *or* $n_1 \neq n_2$ **then**
12 $\quad\quad$ call $\mathcal{A}_{\neg cop}$
13 $\quad$ **end**
14 **end**

Fig. 1. Algorithm $\mathcal{A}_{bip}$ for gathering FSYNC robots in a complete bipartite graph G. It uses algorithms $\mathcal{A}_g$ (from proof of Theorem 3), $\mathcal{A}_{cop}$ (from proof of Lemma 3), and $\mathcal{A}_{\neg cop}$ (from proof of Lemma 4).

completes the gathering. So, if $\mathcal{A}_{bip}$ calls $\mathcal{A}_{cop}$ in $C(0)$, then $\mathcal{A}_{cop}$ will be always called until the gathering is accomplished.

If condition 3 holds, then $\mathcal{A}_{bip}$ calls $\mathcal{A}_{\neg cop}$ at Line 12. Similarly to the previous case, $\mathcal{A}_{\neg cop}$ always swaps robots between V_1 and V_2 until the last move completes the gathering. So, if $\mathcal{A}_{bip}$ calls $\mathcal{A}_{\neg cop}$ in $C(0)$, then $\mathcal{A}_{\neg cop}$ will be always called until the gathering is accomplished as proved by Lemma 4.

($\Rightarrow$) Assume that none of the conditions expressed in the statement applies. We show that C is ungatherable by analyzing two cases, according whether V_1 and V_2 are both occupied or not.

If one between V_1 or V_2 is unoccupied, then neither k_1 is n_2-prime nor k_2 is n_1-prime (otherwise condition 1 holds). Let us analyze the case in which k_1 is n_2-prime, being the other one symmetric. In this case the proof is similar to that of Theorem 4. In fact, since G is a complete bipartite graph, there exists an automorphism in $\mathrm{Aut}(C)$ that makes all robots pairwise equivalent. In other words, if $\mathcal{A}$ is any gathering algorithm for C, then any move planned by $\mathcal{A}$ is performed by each robot. In particular, each move applied at $C = C(0)$ must move each robot in V_1 toward an arbitrary vertex in V_2. Since k_1 is not n_2-prime, then we get $n_2 \geq lpf(k_1)$, that is there exist at least $lpf(k_1) \geq 2$ unoccupied vertices in the partition V_2 in $C(0)$. Hence, the adversary may select $lpf(k_1)$ unoccupied vertices in V_2 as targets to create a configuration $C(1)$ consisting of $lpf(k_1)$ occupied vertices with $\frac{k_1}{lpf(k_1)}$ robots per vertex. It is easy to observe that in $C(1)$ there exists an automorphism that makes the $lpf(k)$ multiplicities pairwise equivalent. Then, from $C(1)$ any possible move will create a configuration $C(t)$, $t > 1$, isomorphic to $C(1)$, thus preventing the resolution of the gathering.

If both V_1 and V_2 are occupied, then k_1 and k_2 are not coprime otherwise condition 2 holds. Since neither condition 3 can apply, then at least one of the following properties holds: (1) $n_1 - k_1 \geq lcpf(k_1, k_2)$ and $n_2 - k_2 \geq lcpf(k_1, k_2)$, and (2) $k_1 = k_2$ and $n_1 = n_2$. This means that the 'only-if' case of Lemma 4 holds, and hence C turns out to be ungatherable.   □

As for complete graphs, by similar arguments, the next corollary can be sated.

Corollary 4. *In any gatherable configuration defined on a n-vertex complete bipartite graph, the gathering problem can be solved by* FSYNC *robots in $O(\log n)$ time.*

5   Concluding Remarks

We have considered the gathering problem of synchronous weak robots moving in graphs. First we have studied general properties that allow to solve the problem regardless the underlying topology. Then we have focused on dense and symmetric graphs like complete and complete bipartite graphs, where we fully characterize when the gathering can be accomplished by means of FSYNC robots. While in complete graphs SSYNC robots are not able to solve the gathering, it remains open what can they do in complete bipartite graphs. Moreover, the proposed algorithms require $O(\log n)$ time whereas the natural lower bound for the considered topologies is $\Omega(D_G)$, with the diameter D_G being 1 or 2. Is it possible to improve the algorithms to this respect or more suitable lower bounds can be obtained? Our investigation highlights how the graph environment is very sensible to synchronization issues. This opens a wide area of research since FSYNC or SSYNC robots have not been much considered in graphs.

References

1. Babai, L., Kucera, L.: Canonical labelling of graphs in linear average time. In: 20th Annual Symposium on Foundations of Computer Science, San Juan, Puerto Rico, 29–31 October 1979, pp. 39–46. IEEE Computer Society (1979)
2. Babai, L., Luks, E.M.: Canonical labeling of graphs. In: Proceedings of the 15th Annual ACM Symposium on Theory of Computing, Boston, Massachusetts, USA, 25–27 April 1983, pp. 171–183. ACM (1983)
3. Bose, K., Kundu, M.K., Adhikary, R., Sau, B.: Optimal gathering by asynchronous oblivious robots in hypercubes. In: Gilbert, S., Hughes, D., Krishnamachari, B. (eds.) ALGOSENSORS 2018. LNCS, vol. 11410, pp. 102–117. Springer, Cham (2019). https://doi.org/10.1007/978-3-030-14094-6_7
4. Cicerone, S., Di Stefano, G., Navarra, A.: Minimum-traveled-distance gathering of oblivious robots over given meeting points. In: Gao, J., Efrat, A., Fekete, S.P., Zhang, Y. (eds.) ALGOSENSORS 2014. LNCS, vol. 8847, pp. 57–72. Springer, Heidelberg (2015). https://doi.org/10.1007/978-3-662-46018-4_4
5. Cicerone, S., Di Stefano, G., Navarra, A.: MinMax-distance gathering on given meeting points. In: Paschos, V.T., Widmayer, P. (eds.) CIAC 2015. LNCS, vol. 9079, pp. 127–139. Springer, Cham (2015). https://doi.org/10.1007/978-3-319-18173-8_9

6. Cicerone, S., Di Stefano, G., Navarra, A.: Asynchronous robots on graphs: gathering. In: Flocchini, P., Prencipe, G., Santoro, N. (eds.) Distributed Computing by Mobile Entities: Current Research in Moving and Computing. LNCS, vol. 11340, pp. 184–217. Springer, Cham (2019). https://doi.org/10.1007/978-3-030-11072-7_8

7. Cicerone, S., Stefano, G.D., Navarra, A.: Gathering of robots on meeting-points: feasibility and optimal resolution algorithms. Distrib. Comput. **31**(1), 1–50 (2018)

8. Cieliebak, M., Flocchini, P., Prencipe, G., Santoro, N.: Distributed computing by mobile robots: gathering. SIAM J. Comput. **41**(4), 829–879 (2012)

9. D'Angelo, G., Di Stefano, G., Klasing, R., Navarra, A.: Gathering of robots on anonymous grids and trees without multiplicity detection. Theor. Comput. Sci. **610**, 158–168 (2016)

10. D'Angelo, G., Di Stefano, G., Navarra, A.: Gathering asynchronous and oblivious robots on basic graph topologies under the look-compute-move model. In: Alpern, S., Fokkink, R., Gąsieniec, L., Lindelauf, R., Subrahmanian, V. (eds.) Search Theory: A Game Theoretic Perspective, pp. 197–222. Springer, New York (2013). https://doi.org/10.1007/978-1-4614-6825-7_13

11. D'Angelo, G., Di Stefano, G., Navarra, A.: Gathering on rings under the look-compute-move model. Distrib. Comput. **27**(4), 255–285 (2014)

12. D'Angelo, G., Di Stefano, G., Navarra, A., Nisse, N., Suchan, K.: Computing on rings by oblivious robots: a unified approach for different tasks. Algorithmica **72**(4), 1055–1096 (2015)

13. D'Angelo, G., Navarra, A., Nisse, N.: A unified approach for gathering and exclusive searching on rings under weak assumptions. Distrib. Comput. **30**(1), 17–48 (2017)

14. D'Angelo, G., Stefano, G.D., Navarra, A.: Gathering six oblivious robots on anonymous symmetric rings. J. Discrete Algorithms **26**, 16–27 (2014)

15. D'Emidio, M., Di Stefano, G., Frigioni, D., Navarra, A.: Characterizing the computational power of mobile robots on graphs and implications for the Euclidean plane. Inf. Comput. **263**, 57–74 (2018)

16. Di Stefano, G., Navarra, A.: Gathering of oblivious robots on infinite grids with minimum traveled distance. Inf. Comput. **254**, 377–391 (2017)

17. Di Stefano, G., Navarra, A.: Optimal gathering of oblivious robots in anonymous graphs and its application on trees and rings. Distrib. Comput. **30**(2), 75–86 (2017)

18. Guilbault, S., Pelc, A.: Gathering asynchronous oblivious agents with local vision in regular bipartite graphs. Theor. Comput. Sci. **509**, 86–96 (2013)

19. Izumi, T., Izumi, T., Kamei, S., Ooshita, F.: Time-optimal gathering algorithm of mobile robots with local weak multiplicity detection in rings. IEICE Trans. **96–A**(6), 1072–1080 (2013)

20. Klasing, R., Kosowski, A., Navarra, A.: Taking advantage of symmetries: gathering of many asynchronous oblivious robots on a ring. Theor. Comput. Sci. **411**, 3235–3246 (2010)

21. Klasing, R., Markou, E., Pelc, A.: Gathering asynchronous oblivious mobile robots in a ring. Theor. Comput. Sci. **390**, 27–39 (2008)

22. McKay, B.D.: Practical graph isomorphism. Congressus Numerantium **30**, 45–87 (1981)

23. Miyazaki, T.: The complexity of McKay's canonical labeling algorithm. In: Groups and Computation, Proceedings of a DIMACS Workshop, New Brunswick, New Jersey, USA, 7–10 June 1995, pp. 239–256 (1995)

Time-Energy Tradeoffs for Evacuation by Two Robots in the Wireless Model

Jurek Czyzowicz[1], Konstantinos Georgiou[2], Ryan Killick[3],
Evangelos Kranakis[3(✉)], Danny Krizanc[4], Manuel Lafond[5], Lata Narayanan[6],
Jaroslav Opatrny[6], and Sunil Shende[7]

[1] Départemant d'informatique, Université du Québec en Outaouais, Gatineau,
Canada
[2] Department of Mathematics, Ryerson University, Toronto, Canada
[3] School of Computer Science, Carleton University, Ottawa, ON, Canada
kranakis@scs.carleton.ca
[4] Department of Mathematics and Computer Science, Wesleyan University,
Middletown, CT, USA
[5] Department of Computer Science, Université de Sherbrooke,
Sherbrooke, QC, Canada
[6] Department of Computer Science and Software Engineering, Concordia University,
Montreal, QC, Canada
[7] Department of Computer Science, Rutgers University, Camden, USA

Abstract. Two robots stand at the origin of the infinite line and are
tasked with searching collaboratively for an exit at an unknown location
on the line. They can travel at maximum speed b and can change speed
or direction at any time. The two robots can communicate with each
other at any distance and at any time. The task is completed when
the last robot arrives at the exit and evacuates. We study time-energy
tradeoffs for the above evacuation problem. The evacuation time is the
time it takes the last robot to reach the exit. The energy it takes for a
robot to travel a distance x at speed s is measured as xs^2. The total and
makespan evacuation energies are respectively the sum and maximum of
the energy consumption of the two robots while executing the evacuation
algorithm.

Assuming that the maximum speed is b, and the evacuation time is at
most cd, where d is the distance of the exit from the origin, we study the
problem of minimizing the total energy consumption of the robots. We
prove that the problem is solvable only for $bc \geq 3$. For the case $bc = 3$,
we give an optimal algorithm, and give upper bounds on the energy for
the case $bc > 3$.

We also consider the problem of minimizing the evacuation time when
the available energy is bounded by Δ. Surprisingly, when Δ is a constant,
independent of the distance d of the exit from the origin, we prove that

A full version of this work is available on the Computing Research Repository [12].
J. Czyzowicz, K. Georgiou, E. Kranakis, M. Lafond, L. Narayanan and J. Opatrny—
Research supported in part by NSERC Discovery grant.
R. Killick—Research supported by the Ontario Graduate Scholarship.

K. Censor-Hillel and M. Flammini (Eds.): SIROCCO 2019, LNCS 11639, pp. 185–199, 2019.
https://doi.org/10.1007/978-3-030-24922-9_13

evacuation is possible in time $O(d^{3/2} \log d)$, and this is optimal up to a logarithmic factor. When Δ is linear in d, we give upper bounds on the evacuation time.

Keywords: Energy · Evacuation · Linear · Robot · Speed · Time · Trade-offs · Wireless communication

1 Introduction

Linear search is an online problem in which a robot is tasked with finding an exit placed at an unknown location on an infinite line. It has long been known that the classic doubling strategy, which guarantees a search time of $9d$ for an exit at distance d from the initial location is optimal for a robot travelling at speed at most 1 (see any of the books [1, 2, 25] for additional variants, details and information). If even one more robot is allotted to the search then clearly an exit at distance d can always be found in time d by one of the robots. Therefore the problem of group search by multiple robots on the line is concerned with minimizing the time the *last* robot arrives at the exit; the problem is also called *evacuation*. It was first introduced as part of a study on cycle-search [10] and further elaborated on an infinite line for multiple communicating robots with crash [18] and Byzantine faults [16].

The time taken for group search on the line clearly depends on the communication capabilities of the robots. In the wireless communication model, the robots can communicate at any time and over any distance. In the face-to-face communication model, the robots can only communicate when they are in the same place at the same time. A straightforward algorithm achieves evacuation time $3d$ in the wireless model, and can be seen to be optimal, while it has been shown that in the face-to-face model, two robots cannot achieve better evacuation time than one robot [8].

In this paper, we consider the *energy* required for group search on the line. We use the energy model proposed in [11] in which the energy consumption of a robot travelling a distance x at speed s is proportional to xs^2. This model is motivated by the concept of viscous drag in fluid dynamics; see Sect. 1.1 for more details. The authors of [11], studied the question of the minimum energy required for group search on the line by two robots travelling at speed at most b while guaranteeing that both robots reach the exit within time cd, where d is the distance of the exit from the starting position of the robots. For the special case $b = 1, c = 9$, they proved the surprising result that two robots can evacuate with less energy than one robot, while taking the same evacuation time.

Our main approach throughout the paper is to investigate time-energy trade-offs for group search by two robots in the wireless communication model. Assuming that the maximum speed is b, and the evacuation time is at most cd, where d is the distance of the exit from the origin, we study the problem of minimizing the total energy consumption of the robots. We also consider the problem of minimizing the evacuation time when the available energy is bounded by Δ.

1.1 Model and Problem Definitions

Two robots are placed at the origin of an infinite line. An exit is located at unknown distance d from the origin and can be found if and only if a robot walks over it. A robot can change its direction or speed at any time, e.g., as a function of its distance from the origin, or the distance walked so far. Robots operate under the wireless model of communication in which messages can be transmitted between robots instantaneously at any distance. Feasible solutions are robots' trajectories in which, eventually, both robots evacuate, i.e. they both reach the exit. Given a location of the exit, the time by which the second robot reaches the exit is referred to as the *evacuation time*. We distinguish between constant-memory robots that can only travel at a constant number of hard-wired speeds, and unbounded-memory robots that can dynamically compute speeds and distances, and travel at any possible speed.

The energy model being used throughout the paper is motivated from the concept of viscous drag in fluid dynamics [4]. In particular, an object moving with constant speed s will experience a *drag* force F_D proportional[1] to s^2. In order to maintain the speed s over a distance x the object must do work equal to the product of F_D and x resulting in a continuous energy loss proportional to the product of the object's squared speed and travel distance. For simplicity we take the proportionality constant to be one, and define the energy consumption moving at constant speed s over a segment of length x to be xs^2. We extend the definition of energy for a robot moving in the same direction from point a to point b on the line, using speed $s(x) \in \mathbb{R}, x \in [a, b]$, as $\int_a^b s^2(x)\mathrm{d}x$. The total energy of a specific robot traversing more intervals, possibly in different directions, is defined as the sum of the energies used in each interval.

Given a collection of robots, the *total evacuation energy* is defined as the sum of the robots' energies used till both robots evacuate. Similarly, we define the *makespan evacuation energy* as the maximum energy used by any of the two robots.

For each $d > 0$ there are two possible locations for the exit to be at distance d from the origin: we will refer to either of these as input instances d for the group search problem. More specifically, we are interested in the following three optimization problems:

Definition 1. *Problem* $\mathrm{EE}_d^b(c)$: *Minimize the total evacuation energy, given that the evacuation time is no more than cd (for all instances d) and using speeds no more than b.*

Definition 2. *Problem* $\mathrm{TE}_d^b(\Delta)$: *Minimize the evacuation time, given that the total evacuation energy is no more than Δ (for all instances d), and using speeds at most b.*

---

[1] The constant of proportionality has (SI) units kg/m and depends, among other things, on the shape of the object and the density of the fluid through which it moves.

Definition 3. *Problem* $ME_d^b(\Delta)$*: Minimize the evacuation time, given that the makespan evacuation energy is no more than Δ (for all instances d), and using speeds at most b.*

For the last two problems, we consider two cases when the evacuation energy Δ is a constant and when it is linear in d.

1.2   Our Results

Consider the following intuitive and simple algorithm for wireless evacuation, which is a parametrized version of a well-known algorithm for the case of unit speed robots that achieve evacuation time $3d$.

Definition 4 (Algorithm Simple Wireless Search $\mathcal{N}_{s,r}$). *Robots move at opposite directions with speed s until the exit is found. The finder announces "exit found" and halts. The other robot changes direction and moves at speed r until the exit is reached.*

We analyze the behaviour of this algorithm for all three proposed problems, and determine the speeds that achieve the minimum evacuation energy (or time) among all algorithms of this class, while respecting the given bound on evacuation time (resp. energy). In some cases, the algorithms derived are shown to be optimal. In particular, our main results are the following:

1. We show that the problem $EE_d^b(c)$ admits a solution if and only if $cb \geq 3$. Furthermore, for every $c, b > 0$ with $cb = 3$, we show that the optimal total evacuation energy is $4b^2d$, and this is achieved by $\mathcal{N}_{s,r}$ with $s = r = b$ (Theorem 1).
2. For every $c, b > 0$ with $cb \geq 3$, we derive the optimal values of s and r for the algorithm $\mathcal{N}_{s,r}$ that minimize the total evacuation energy (Theorem 2).
3. We observe that if total or makespan energy Δ is a constant, problems $TE_d^b(\Delta)$ and $ME_d^b(\Delta)$ cannot be solved by robots that can only use a finite number of speeds. We prove that if Δ is bounded by a constant, the optimal evacuation time is $\Omega(d^{3/2})$ (see Theorem 4). Somewhat surprisingly, we give an algorithm with total evacuation time $O(d^{3/2} \log d)$ (see Theorem 5); thus the algorithm is optimal up to a logarithmic factor. Our algorithm requires the robots to continuously change their speed at every distance x from the origin. This is the only part that requires robots to have unbounded memory.
4. For the problems $TE_d^b(\Delta)$ and $ME_d^b(\Delta)$ with total or makespan energy $\Delta = O(d)$ and $b = 1$, we give upper bounds on the total evacuation time (see Theorems 5 and 7 respectively).

Due to space limitations, some proofs are omitted from this extended abstract. The interested reader may see [12] for a full version of the paper.

1.3 Related Work

In group search, a set of communicating robots interact and co-operate by exchanging information in order to complete the task which usually involves finding an exit placed at an unknown location within a given search domain. Some of the pioneering results related to our work are concerned with search on an infinite domain, like a straight line [3,5,6,24], while others with search on the perimeter of a closed domain like unit disk [10] or equilateral triangle or square [20]. The communication model being used may be either wireless [10] or F2F [7,17,20]. Search and evacuation problems with a combinatorial flavour have been recently considered in [13,14] and search-and-fetch problems in [22,23], while [9] studied average-case/worst-case trade-offs for a specific evacuation problem on the disk. The interested reader may also wish to consult a recent survey paper [15] on selected search and evacuation topics.

Traditional approaches to evaluating the performance of search have been mostly concerned with time. This is apparent in the book [2] and the research described in the seminal works on deterministic [3], stochastic [5,6] and randomized [24] search and continued up to the most recent research papers on linear search for robots with terrain dependent speeds [19] and robots with Byzantine [16] and crash fault behaviour [18] (see also the survey paper [15]). Aside from the research by [21], in which the authors are looking at the turn cost when robots change direction during the search, little or no research has been conducted on other measures of performance.

The first paper on search and evacuation to change this focus from optimizing the time to the energy consumption required to find the exit as well as to time/energy tradeoffs is due to [11]. The authors determine optimal (and in some cases nearly optimal) linear search algorithms inducing the lowest possible energy consumption and also propose a linear search algorithm that simultaneously achieves search time $9d$ and consumes energy $8.42588d$, for an exit located at distance d unknown to the robots. However, the previously mentioned paper [11] differs from our present work in that the authors focus exclusively on the face-to-face communication model while here we focus on the wireless model. In the present paper, we extend the results of [11] to the realm of the wireless communication model and study time/energy trade-offs for evacuating two robots on the infinite line. Despite their apparent similarities, the face-to-face and wireless communication models lead to completely different approaches for the design of efficient linear search algorithms.

2 Minimizing Energy Given Bounds on Evacuation Time and Speed

This section is devoted to the problem $\mathrm{EE}_d^b(c)$ of minimizing the total evacuation energy, given that the robots can travel at speed at most b and are required to complete the evacuation within time cd for every instance d where d is the distance of the exit from the origin. We start with establishing a necessary condition on the product bc.

Lemma 1. *No online (wireless) algorithm can solve* $\mathrm{EE}_d^b(c)$ *if* $bc < 3$.

Proof (Lemma 1). Fix $0 < \epsilon \leq 3$ and let $bc = 3 - \epsilon$. We show that no algorithm can solve problem $\mathrm{EE}_d^b(c)$. For the sake of contradiction, consider a wireless algorithm solving $\mathrm{EE}_d^{(3-\epsilon)/b}(b)$, and having evacuation time no more than $(3 - \epsilon)d/b$, if the exit is placed d away from the origin. For a large enough $d > 0$, we let the algorithm run till the first point among $\pm d$ is reached by a robot (and maybe they are reached simultaneously). Without loss of generality, assume that $+d$ is reached, say by robot R, no later than the other point. Note that for this point to be reached, at least time d/b has passed. Now, we place the exit at point $-d$. The additional time that R needs to reach the exit is $2d/b$, for a total time of $3d/b$, a contradiction to the stipulated evacuation time of $(3 - \epsilon)d/b$.

Next we show that algorithm $\mathcal{N}_{b,b}$ is an optimal solution to the problem $\mathrm{EE}_d^b(c)$ when $bc = 3$. We start with the following lemma:

Lemma 2. *Let* $b, c > 0$ *with* $bc = 3$ *and consider an evacuation algorithm such that robots use maximum speed* b *and evacuate by time* cd *for an exit at distance* d *from the origin. Then for every* $d > 0$, *the points* $d, -d$, *must be visited at time* d/b.

Proof (Lemma 2). Suppose not. Notice that the points $\pm d$ cannot be visited *before* time d/b using speed at most b. We look at two cases.

Case 1: There exists $d > 0$ such that neither d nor $-d$ is visited at time d/b. Consider the first time $t > d/b$ when either of them is visited, wlog let the point $+d$ be visited at time $t > d/b$ by robot R_1. We put the exit at $-d$. Then R_1 has to travel an additional distance of $2d$, and can use speed at most b, so needs time at least $2d/b$ to get to the exit. The total time taken by R_1 to evacuate is at least $t + 2d/b > 3d/b = cd$.

Case 2: There exists $d > 0$ such that d is visited at time d/b but $-d$ is not visited at this time (or vice versa). Wlog suppose R_1 is at point d at time d/b. Let $-d + 2\epsilon$ be the closest point to $-d$ that *has* been visited at time d/b where $\epsilon > 0$ since by assumption $-d$ is not visited at this time. We put the exit at $-d + \epsilon$. The time limit to evacuate is $c(d - \epsilon)$. At time d/b, R_1 is at distance $2d - \epsilon$ from the exit, so the total time for R_1 to reach the exit is at least

$$d/b + (2d - \epsilon)/b = 3d/b - \epsilon/b = cd - \frac{c\epsilon}{3} > cd - c\epsilon$$

In both cases, we showed that the robots cannot evacuate in the required time bound. This completes the proof by contradiction.

Theorem 1. *For every* $b, c > 0$ *with* $bc = 3$, *the algorithm* $\mathcal{N}_{b,b}$ *is the only feasible solution to* $\mathrm{EE}_d^b(c)$, *and is therefore optimal, and has total energy consumption* $4b^2d$.

Proof (Theorem 1). Lemma 2 implies that in order to achieve an evacuation time cd, both robots must use the maximum speed b and explore in different directions. If the exit is found at distance d by one of the robots, the time is d/b, and therefore, the other robot must travel at the maximum speed b in order to arrive at the exit in time cd. Thus, the only algorithm that can evacuate within time cd while using speed at most b is $\mathcal{N}_{b,b}$. A total distance of $4d$ is travelled by the two robots, all at speed b, therefore the total energy consumed is $4b^2d$.

Next we consider the case of $c, b > 3$ and determine the optimal choices of speeds s, r for $\mathcal{N}_{s,r}$, as well as the induced total evacuation energy and competitive ratio for problem $\mathrm{EE}_d^b(c)$.

Theorem 2. *Let $\delta = 2 + \sqrt[3]{2} \approx 3.25992$. For every $c, b > 0$, problem $\mathrm{EE}_d^b(c)$ admits a solution by algorithm $\mathcal{N}_{s,r}$ if and only if $cb \geq 3$. For the spectrum of c, b for which a solution exists, the following choices of speeds s, r are feasible and optimal for $\mathcal{N}_{s,r}$*

	$3 \leq cb \leq \delta$	$cb > \delta$
s	$\frac{b}{bc-2}$	$\frac{\delta}{\sqrt[3]{2}c}$
r	b	$\frac{\delta}{c}$

The induced total evacuation energy is $f(cb)\frac{2d}{c^2}$, where

$$f(x) := \begin{cases} \frac{x^2}{(x-2)^2} + x^2 & , 3 \leq x \leq \delta \\ \frac{1}{2}\left(2 + \sqrt[3]{2}\right)^3 & , x > \delta \end{cases}$$

It was observed in [11] that the optimal offline solution, given that d is known, equals $\frac{2d}{c^2}$. The competitive ratio is given by $\sup_d \frac{c^2}{2d} e(c, b, d) = f(cb)$ for algorithms inducing total evacuation energy $e(c, b, d)$. The competitive ratio of $\mathcal{N}_{s,r}$ for the choices of Theorem 2 is summarized in Fig. 1. Note that in particular, Theorem 2 claims that the competitive ratio only depends on the product cb, and when $cb = 3$, the competitive ratio is 18 and is decreasing in cb (strictly only when $cb < \delta$). The optimal speed choices for the unbounded problem $\mathrm{EE}_d^c(\infty)$ are exactly those that appear under case $cb > \delta$. The remaining of the section is devoted to proving Theorem 2.

First we derive closed formulas for the performance of $\mathcal{N}_{s,r}$. From the definition of energy used, and given that the robots move at speed 1, we deduce what the evacuation time and energy are when the exit is placed at distance d from the origin. The following two functions will be invoked throughout our argument below.

$$\mathcal{T}(s, r) := \frac{1}{s} + \frac{2}{r} \tag{1}$$

$$\mathcal{E}(s, r) := s^2 + r^2 \tag{2}$$

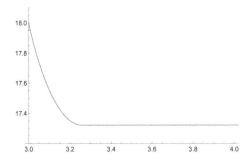

Fig. 1. The competitive ratio of $\mathcal{N}_{s,r}$ for the choices of Theorem 2.

Lemma 3. *Let b, c be such that there exist s, r for which $\mathcal{N}_{s,r}$ is feasible. Then, for instance d of $\mathrm{EE}_d^b(c)$, the induced evacuation time of $\mathcal{N}_{s,r}$ is $d \cdot \mathcal{T}(s,r)$ and the induced total evacuation energy is $2d \cdot \mathcal{E}(s,r)$.*

Next we show the spectrum of c, b for which $\mathcal{N}_{s,r}$ is applicable.

Lemma 4. *Algorithm $\mathcal{N}_{s,r}$ gives rise to a feasible solution to problem $\mathrm{EE}_d^b(c)$ if and only if $bc \geq 3$. For every such $b, c > 0$, the optimal choices of $\mathcal{N}_{s,r}^f$ can be obtained by solving Convex Program:*

$$\min_{s,r \in \mathbb{R}} \mathcal{E}(s,r) \qquad\qquad (\mathrm{NLP}_c^b)$$

$$s.t. \quad \mathcal{T}(s,r) \leq c$$

$$0 \leq s, r \leq b.$$

Moreover, if s_0, r_0 are the optimizers to NLP_c^b, then the competitive ratio of $\mathcal{N}_{s_0,r_0}$ equals $c^2 \cdot \mathcal{E}(s_0, r_0)$.

A corollary of Lemma 4 is that any candidate optimizer to NLP_c^b satisfying 1st order necessary optimality conditions is also a global optimizer. As a result, the proof of Theorem 2 follows by showing the proposed solution is feasible and satisfies 1st order necessary optimality conditions. This is done in Lemmata 5 and 6.

Towards proving that 1st order optimality conditions are satisfied, we argue first that for all $c, b > 0$ with $cb \geq 3$, the optimizers of NLP_c^b satisfy the time constraint tightly. Indeed, if not, then one could reduce any of the values among s, r to make the constraint tight, improving the induced energy. Hence, in the optimal solutions to NLP_c^b, any of $s, r \leq b$ could be additionally tight or not. In what follows, δ represents $2 + \sqrt[3]{2}$, as in the statement of Theorem 2.

Lemma 5. *For each $c, b > 0$ for which $3 \leq cb \leq \delta$, the optimal solution to NLP_c^b is given by $s = \frac{b}{bc-2}, r = b$.*

Lemma 6. *For each $c, b > 0$ for which $cb > \delta$, the optimal solution to NLP_c^b is given by $s = \frac{\delta}{\sqrt[3]{2}c}, r = \frac{\delta}{c}$.*

Proof (Theorem 2). By Lemmas 4 and 5, the optimal induced energy when $3 \leq cb \leq \delta$ is

$$2d\mathcal{E}\left(\frac{b}{bc-2}, b\right) = 2d\left(\frac{b^2}{(bc-2)^2} + b^2\right)$$

and the induced competitive ratio is

$$(cb)^2\left(1 + \frac{1}{(cb-2)^2}\right).$$

Finally, by Lemmas 4 and 6, the optimal induced energy when $cb > \delta$ is

$$2d\mathcal{E}\left(\frac{1 + 2^{2/3}}{c}, \frac{2 + \sqrt[3]{2}}{c}\right) = d\frac{\left(2 + \sqrt[3]{2}\right)^3}{c^2}.$$

Hence the competitive ratio is constant and equals

$$\frac{1}{2}\left(2 + \sqrt[3]{2}\right)^3 \approx 17.3217,$$

completing the proof of Theorem 2. $\qquad\blacksquare$

3 Minimizing Evacuation Time, Given Constant Evacuation Energy

In this section we consider the problem of minimizing evacuation time, given *constant* total (or makespan) evacuation energy. First we observe that if the robots can use only a finite number of speeds, there is no feasible solution to the problems $\mathrm{ME}_d^b(\Delta)$ or $\mathrm{TE}_d^b(\Delta)$.

Theorem 3. *If Δ is a constant, and the robots have access to only a finite number of speeds, there is no feasible solution to the problems $\mathrm{ME}_d^b(\Delta)$ or $\mathrm{TE}_d^b(\Delta)$*

Proof (Theorem 3). Suppose the robots can only use speeds in a finite set. Wlog let s be the minimum speed in the set. Define $d' = \Delta/s^2$, and place the exit at $d' + \epsilon$ for any $\epsilon > 0$. Travelling at any speed at or above s, it is impossible for even one of the robots to reach the exit with energy $\leq \Delta$. $\qquad\blacksquare$

Next we prove a lower bound on the evacuation time in this setting.

Theorem 4. *For every constant $e \in \mathbb{R}_+$, the optimal evacuation time for problem $\mathrm{ME}_d^b(e)$ is $\Omega(d^{3/2})$, asymptotically in d.*

Proof (Theorem 4). For any arbitrarily large value of d, we place the exit at distance d from the origin. For any robot to reach the exit before running out of battery, a robot can travel at speed at most $e/\sqrt{d}$. Therefore the time for even the first robot to reach the exit is at least $\frac{d}{e/\sqrt{d}} = d^{3/2}/e$.

Note that the above lower bound also holds for problem $\text{TE}_d^b(e)$ (if the total evacuation energy is no more than e, then also the makespan evacuation energy is no more than e). Next we prove that this naive lower bound is nearly tight (up to a $\log d$ factor). First we consider the case that $e \leq 1$. Then, we show how to modify our solution to also solve the problem when $e > 1$.

The key idea is to allow functional speed $s = s(x)$ to depend on the distance x of the robot from the origin. We will make sure that the choice of s is such that, for every large enough d, once the exit is located at distance d, there is "enough" leftover energy for the other robot to evacuate too. For that, we will choose the maximum possible speed r (which can now depend on d, and which will be constant) so as to evacuate without exceeding the maximum energy bounds. Notably, even though our algorithmic solution is described as a solution to $\text{TE}_d^b(e)$, it will be transparent in the proof that it is also feasible to $\text{ME}_d^b(e)$.

Theorem 5. *For every constant $e \leq 1$, problem $\text{TE}_d^b(e)$ admits a solution by $\mathcal{N}_{s,r}$, where (functional) speed s is chosen as*

$$s(x) = \frac{1}{\sqrt{2 + 2x}\,(1/e + \log(1+x))}.$$

When the exit is found (hence its distance d from the origin becomes known), speed r is chosen as

$$r = \sqrt{\frac{e}{2d\,(e\log(d+1)+1)}},$$

inducing evacuation time $O\left(d^{3/2}\log d\right)$, where in particular the constant in the asymptotic (in d) is independent of e.

Proof (Theorem 5). First we observe that since $e \leq 1$, $s(x) \leq 1$ for all $x \geq 0$. Given that d is at least, say, 1, it is also immediate that $r \leq 1$, hence the speed choices comply with the speed bound.

The exit placed at distance d from the origin is located by the finder in time

$$\int_0^d \frac{1}{s(x)}\mathrm{d}x = \frac{2\sqrt{2}\left((d+1)^{3/2}(3e\log(d+1)-2e+3)+2e-3\right)}{9e} \leq d^{3/2}\log d,$$

where the inequality holds for every $e \leq 1$, and for big enough d.

When the exit is located by a robot, the other robot is at distance $2d$ from the exit. Moreover, each of the robots have used energy

$$\int_0^d s^2(x)\mathrm{d}x = \frac{e}{2} - \frac{e}{2e\log(d+1)+2},$$

hence the leftover energy for the non-finder (i.e., the robot that did not find the exit) to evacuate is at least

$$e - 2\left(\frac{e}{2} - \frac{e}{2e\log(d+1)+2}\right) = \frac{e}{e\log(d+1)+1}.$$

The non-finder is informed of d, and hence can choose constant speed r so as to use exactly all of the leftover energy, i.e. by choosing r satisfying

$$\int_0^{2d} r^2 \mathrm{d}x = \frac{e}{e\log(d+1)+1}.$$

Note that our choice of r is also feasible to problem $\mathrm{ME}_d^b(e)$. Solving for r gives the value declared at the statement of the theorem. Finally, choosing this specific value of r, the non-finder needs additional $2d/r$ time to evacuate, which is at most

$$(2d)^{3/2}\sqrt{\frac{(e\log(d+1)+1)}{e}} \le (2d)^{3/2}\sqrt{\frac{\log(d+1)}{e}} \le d^{3/2}\log d,$$

where the last inequality holds for big enough d, since e is constant. So the overall evacuation time is no more than $2d^{3/2}\log d$, for big enough d, as promised.

It remains to address the case $e > 1$. For this, we recall that we solve $\mathrm{TE}_d^b(e)$ for large enough values of d, and we modify our solution so as to choose functional speed

$$\bar{s}(x) := \min\{s(x), 1\},$$

effectively using even less energy than before. The distance that is traversed at speed 1 depends only on constant e, and hence the additional evacuation time is $O(1)$ with respect to d.

4 Minimizing Evacuation Time with Bounded Linear Total Evacuation Energy

In this section we study the problem $\mathrm{TE}_d^1(\Delta)$ of minimizing the total evacuation time, where $\Delta = ed$ for some constant e. We show how to choose optimal speed values s, r for algorithm $\mathcal{N}_{s,r}$. Note that even though d is unknown to the algorithm, speeds s, r may depend on the known constant e, and the maximum speed $b = 1$.

In this section we prove the following theorem:

Theorem 6. *Let $\delta = 2 + \sqrt[3]{2} \approx 3.25992$. For every constant $e \in \mathbb{R}_+$, problem $\mathrm{TE}_d^1(ed)$ admits a solution by $\mathcal{N}_{s,r}$, where speeds s, r are chosen as follows*

	$e < \delta$	$e \in [\delta, 4)$	$e \ge 4$
s	$\sqrt{\dfrac{e}{2(1+2^{2/3})}}$	$\sqrt{\dfrac{e-2}{2}}$	1
r	$\sqrt{\dfrac{e}{(2+2^{1/3})}}$	1	1

The induced total evacuation time is given by $g(e)d$ where $g(e)$ is given by:

$$g(e) := \begin{cases} \sqrt{\dfrac{\left(2+2^{1/3}\right)^3}{e}} & , e < \delta \\ 2 + \sqrt{\dfrac{2}{e-2}} & , e \in [\delta, 4) \\ 3 & , e \ge 4 \end{cases}$$

First we observe that, given the values of $s = s(e), r = r(e)$, it is a matter of straightforward calculations to verify, assuming they are feasible and optimal, that the induced evacuation time is indeed equal to $g(e)d$ as promised. Given Lemma 3, we know that the optimal speed choices for algorithm $\mathcal{N}_{s,r}$, for problem $\text{TE}_d^1 (ed)$ are obtained as the solution to the following NLP.

$$\min_{s,r \in \mathbb{R}} \frac{1}{s} + \frac{2}{r} \qquad (\text{NLP}_e')$$

$$s.t. \quad 2(s^2 + r^2) \le e$$

$$0 \le s, r \le 1$$

The optimal solutions to NLP_e' can be obtained by solving complicated algebraic systems and by invoking KKT conditions, for the various values of e, as we also did for NLP_c^b. However, the advantage is that one can map the optimal solutions to NLP_c^1, see Theorem 2 and use $b = 1$, to feasible solutions to NLP_e'. Then, we just need to *verify* 1st order optimality conditions for the candidate optimizers. Since the NLP is convex, these should also be unique global optimizers.

Indeed, one of the critical structural properties pertaining to the optimizers of NLP_c^1 is that the time constraint $\frac{1}{s} + \frac{2}{r} \le cd$ is satisfied *tightly*. At the same time, the optimal speed values, as described in Theorem 2, as a function of c, achieve evacuation energy equal to $f(c)d\frac{2}{c^2}$. Attempting to find the correspondence between parameters c, e (and problems NLP_c^1, NLP_e'), we consider the transformation $f(c)\frac{2}{c^2} = e$. For the various cases of the piece-wise function f, the transformation gives rise to the piece-wise function g and optimal speeds s, r (as a function of e) of Theorem 6.

Overall, the previous approach provides just a mapping between the provable optimizers $s(c), r(c)$ to NLP_c^1, and candidate solutions $s(e), r(e)$ to NLP_e', and more importantly, it saves us from solving complicated algebraic systems induced by KKT conditions. What we verify next (which is much easier), is that feasibility and KKT conditions are indeed satisfied for the obtained candidate solutions $s(e), r(e)$. Since the NLP is convex, that also shows that $s(e), r(e)$, as stated in Theorem 6 are actually global optimizers to NLP_e'.

Lemma 7. *For every $e \in \mathbb{R}_+$, speeds $s(e), r(e)$, as they are defined in Theorem 6, are feasible to NLP_e'.*

Lemma 8. *For every $e \in \mathbb{R}_+$, speeds $s(e), r(e)$, as stated in Theorem 6, are the optimal solutions to NLP_e'.*

5   Minimizing Evacuation Time with Bounded Linear Makespan Evacuation Energy

In this section we study the problem $\text{ME}_d^1 (\Delta)$ of minimizing the makespan evacuation time, given that the makespan evacuation energy $\Delta = ed$ for some constant e. We show how to choose optimal speed values s, r for algorithm $\mathcal{N}_{s,r}$.

Note that even though d is unknown to the algorithm, speeds s, r may depend on the known value e, and the maximum speed $b = 1$.

Theorem 7. *For every constant $e \in \mathbb{R}_+$, problem $\mathrm{ME}_d^1(ed)$ admits a solution by $\mathcal{N}_{s,r}$, where speeds s, r are chosen as follows*

	$e < 3$	$e \geq 3$
s	$\sqrt{\frac{e}{3}}$	1
r	$\sqrt{\frac{e}{3}}$	1

The induced evacuation time is given by $g(e)d$ where

$$g(e) := \begin{cases} 3\sqrt{\frac{3}{e}} , e < 3 \\ 1 \qquad , e \geq 3 \end{cases}$$

Proof (Theorem 7). What distinguishes the performance, and feasibility, of $\mathcal{N}_{s,r}$ between $\mathrm{TE}_d^1(ed)$ and $\mathrm{ME}_d^1(ed)$, is that in the former, the total evacuation energy (equal to $d(2s^2 + 2r^2)$) is bounded by e, while in the latter the makespan evacuation energy (equal to $d(s^2 + 2r^2)$) is bounded by e. Hence, similar to the analysis for $\mathrm{TE}_d^1(ed)$, the optimal speed choices for $\mathcal{N}_{s,r}$ to $\mathrm{ME}_d^1(ed)$ are the optimal solutions to the following NLP.

$$\min_{s,r \in \mathbb{R}} \frac{1}{s} + \frac{2}{r} \qquad (\mathrm{NLP}_e'')$$

$$\text{s.t.} \quad s^2 + 2r^2 \leq e$$
$$0 \leq s, r \leq 1$$

Note that NLP_e'' is convex, hence any choice of feasible speeds satisfying 1st order optimality (KKT) conditions is also the unique global minimizer. Moreover, the choices of s, r of the statement of the theorem are clearly feasible to NLP_e''. Hence, it suffices to show that the choices of s, r do indeed satisfy KKT conditions.

When $e < 3$ we note that the energy constraint is tight, while both speed constraints are not tight. Hence, s, r are the unique optimizers if there exists $\lambda \geq 0$ satisfying

$$-\nabla \left(\frac{1}{s} + \frac{2}{r} \right) = \lambda \nabla (s^2 + 2r^2) \Leftrightarrow \begin{pmatrix} 1/s^2 \\ 2/r^2 \end{pmatrix} = \lambda \begin{pmatrix} 2s \\ 4r \end{pmatrix}$$

from which we conclude that $\lambda = 1/(2s^3) = 1/(2r^3) > 0$ as wanted (for $s = r = \sqrt{e/3}$).

When $e \geq 3$ we note that the speed constraints are both tight, while the energy constraint is tight only when $e = 3$. In that case, it suffices to show that there exist nonnegative λ_1, λ_2 satisfying

$$-\nabla \left(\frac{1}{s} + \frac{2}{r} \right) = \lambda_1 \begin{pmatrix} 1 \\ 0 \end{pmatrix} + \lambda_2 \begin{pmatrix} 0 \\ 1 \end{pmatrix}$$

Clearly, $\lambda_1 = 1/s^2 = 1 > 0$ and $\lambda_2 = 2/r^2 = 2 > 0$, which concludes the proof.

6   Conclusion

We investigated how the wireless communication model affects time/energy trade-offs for completion of the evacuation task by two robots. Our study raises several interesting problems worth investigating. In addition to improving the trade-offs, it would be interesting to consider search with multiple agents some of which may be faulty in linear [16,18] as well as cyclical [10] search domains.

References

1. Ahlswede, R., Wegener, I.: Search Problems. Wiley-Interscience, Chichester (1987)
2. Alpern, S., Gal, S.: The Theory of Search Games and Rendezvous. Springer, Boston (2003). https://doi.org/10.1007/b100809
3. Baeza Yates, R., Culberson, J., Rawlins, G.: Searching in the plane. Inf. Comput. **106**(2), 234–252 (1993)
4. Batchelor, G.K.: An Introduction to Fluid Dynamics. Cambridge Mathematical Library. Cambridge University Press, Cambridge (2000)
5. Beck, A.: On the linear search problem. Isr. J. Math. **2**(4), 221–228 (1964)
6. Bellman, R.: An optimal search. SIAM Rev. **5**(3), 274 (1963)
7. Brandt, S., Laufenberg, F., Lv, Y., Stolz, D., Wattenhofer, R.: Collaboration without communication: evacuating two robots from a disk. In: Fotakis, D., Pagourtzis, A., Paschos, V.T. (eds.) CIAC 2017. LNCS, vol. 10236, pp. 104–115. Springer, Cham (2017). https://doi.org/10.1007/978-3-319-57586-5_10
8. Chrobak, M., Gąsieniec, L., Gorry, T., Martin, R.: Group search on the line. In: Italiano, G.F., Margaria-Steffen, T., Pokorný, J., Quisquater, J.-J., Wattenhofer, R. (eds.) SOFSEM 2015. LNCS, vol. 8939, pp. 164–176. Springer, Heidelberg (2015). https://doi.org/10.1007/978-3-662-46078-8_14
9. Chuangpishit, H., Georgiou, K., Sharma, P.: Average case - worst case tradeoffs for evacuating 2 robots from the disk in the face-to-face model. In: Gilbert, S., Hughes, D., Krishnamachari, B. (eds.) ALGOSENSORS 2018. LNCS, vol. 11410, pp. 62–82. Springer, Cham (2019). https://doi.org/10.1007/978-3-030-14094-6_5
10. Czyzowicz, J., Gąsieniec, L., Gorry, T., Kranakis, E., Martin, R., Pajak, D.: Evacuating robots via unknown exit in a disk. In: Kuhn, F. (ed.) DISC 2014. LNCS, vol. 8784, pp. 122–136. Springer, Heidelberg (2014). https://doi.org/10.1007/978-3-662-45174-8_9
11. Czyzowicz, J., et al.: Energy/time trade-offs for linear-search. In: The 46th International Colloquium on Automata, Languages and Programming (ICALP 2019) (2019, to appear)
12. Czyzowicz, J.: Time-energy tradeoffs for evacuation by two robots in the wireless model. CoRR, abs/1905.06783 (2019)
13. Czyzowicz, J., et al.: God save the queen. In: 9th International Conference on Fun with Algorithms (FUN 2018). LIPIcs, vol. 100, pp. 16:1–16:20 (2018)
14. Czyzowicz, J., et al.: Priority evacuation from a disk using mobile robots. In: Lotker, Z., Patt-Shamir, B. (eds.) SIROCCO 2018. LNCS, vol. 11085, pp. 392–407. Springer, Cham (2018). https://doi.org/10.1007/978-3-030-01325-7_32
15. Czyzowicz, J., Georgiou, K., Kranakis, E.: Group search and evacuation. In: Flocchini, P., Prencipe, G., Santoro, N. (eds.) Distributed Computing by Mobile Entities: Current Research in Moving and Computing, Chap. 14. LNCS, vol. 11340, pp. 335–370. Springer, Cham (2019). https://doi.org/10.1007/978-3-030-11072-7_14

16. Czyzowicz, J., et al.: Search on a line by byzantine robots. In: Proceedings of 27th ISAAC, pp. 27:1–27:12 (2016)

17. Czyzowicz, J., Georgiou, K., Kranakis, E., Narayanan, L., Opatrny, J., Vogtenhuber, B.: Evacuating robots from a disk using face-to-face communication (extended abstract). In: Paschos, V.T., Widmayer, P. (eds.) CIAC 2015. LNCS, vol. 9079, pp. 140–152. Springer, Cham (2015). https://doi.org/10.1007/978-3-319-18173-8_10

18. Czyzowicz, J., Kranakis, E., Krizanc, D., Narayanan, L., Opatrny, J.: Search on a line with faulty robots. In: Proceeding of PODC, pp. 405–413. ACM (2016)

19. Czyzowicz, J., Kranakis, E., Krizanc, D., Narayanan, L., Opatrny, J., Shende, S.: Linear search with terrain-dependent speeds. In: Fotakis, D., Pagourtzis, A., Paschos, V.T. (eds.) CIAC 2017. LNCS, vol. 10236, pp. 430–441. Springer, Cham (2017). https://doi.org/10.1007/978-3-319-57586-5_36

20. Czyzowicz, J., Kranakis, E., Krizanc, D., Narayanan, L., Opatrny, J., Shende, S.: Wireless autonomous robot evacuation from equilateral triangles and squares. In: Papavassiliou, S., Ruehrup, S. (eds.) ADHOC-NOW 2015. LNCS, vol. 9143, pp. 181–194. Springer, Cham (2015). https://doi.org/10.1007/978-3-319-19662-6_13

21. Demaine, E.D., Fekete, S.P., Gal, S.: Online searching with turn cost. Theor. Comput. Sci. **361**(2), 342–355 (2006)

22. Georgiou, K., Karakostas, G., Kranakis, E.: Search-and-fetch with one robot on a disk - (track: wireless and geometry). In: Proceedings of 12th ALGOSENSORS 2016, pp. 80–94 (2016)

23. Georgiou, K., Karakostas, G., Kranakis, E.: Search-and-fetch with 2 robots on a disk - wireless and face-to-face communication models. In: Liberatore, F., Parlier, G.H., Demange, M. (eds.) Proceedings of the 6th International Conference on Operations Research and Enterprise Systems, ICORES 2017, Porto, Portugal, 23–25 February 2017, pp. 15–26. SciTePress (2017)

24. Kao, M.-Y., Reif, J.H., Tate, S.R.: Searching in an unknown environment: an optimal randomized algorithm for the cow-path problem. Inf. Comput. **131**(1), 63–79 (1996)

25. Stone, L.: Theory of Optimal Search. Academic Press, New York (1975)

Evacuating Two Robots from a Disk: A Second Cut

Yann Disser$^{(\boxtimes)}$ and Sören Schmitt$^{(\boxtimes)}$

Department of Mathematics, TU Darmstadt, Darmstadt, Germany
disser@mathematik.tu-darmstadt.de, soeren.schmitt@stud.tu-darmstadt.de

Abstract. We present an improved algorithm for the problem of evacuating two robots from the unit disk via an unknown exit on the boundary. Robots start at the center of the disk, move at unit speed, and can only communicate locally. Our algorithm improves previous results by Brandt et al. [CIAC'17] by introducing a second detour through the interior of the disk. This allows for an improved evacuation time of 5.6234. The best known lower bound of 5.255 was shown by Czyzowicz et al. [CIAC'15].

1 Introduction

We consider the problem of evacuating two robots from the unit disk via an unknown exit on the boundary. The robots start at the center of the disk and move at unit speed (with infinite acceleration). They have unlimited computing resources and we neglect the time taken to perform arbitrary calculations. However, robots are point-shaped and only perceive the information available at their respective locations. In particular, they can only exchange information (in no time) while colocated at the same point on the disk. Both robots have full knowledge of the algorithms executed by either robot, and they share the same coordinate system. The objective is to minimize the evacuation time, i.e., the time needed until both robots have reached the exit, in the worst case over all possible positions of the exit. Note that the evacuation time for an algorithm is equal to its competitive ratio, since the shortest path to any potential exit location has length one.

This evacuation problem was first introduced by Czyzowicz et al. [12], who showed that the basic algorithm that moves both robots along the boundary in opposing directions achieves an evacuation time of 5.74 and gave a lower bound of 5.199. In a follow-up paper, Czyzowicz et al. [9] presented two improved algorithms with evacuation times of 5.644 and 5.628. Both these algorithms introduce detours through the interior of the disk and may lead to a forced meeting before the exit is found. Additionally, Czyzowicz et al. [9] improved the lower bound to 5.255. Brandt et al. [5] introduced a general necessary condition for worst-case exit positions and gave a slightly improved algorithm, without forced meeting, that achieves an evacuation time of 5.625. Figure 1 shows the trajectories of both robots in each of these algorithms.

Yann Disser is supported by the 'Excellence Initiative' of the German Federal and State Governments and the Graduate School CE at TU Darmstadt.

K. Censor-Hillel and M. Flammini (Eds.): SIROCCO 2019, LNCS 11639, pp. 200–214, 2019.
https://doi.org/10.1007/978-3-030-24922-9_14

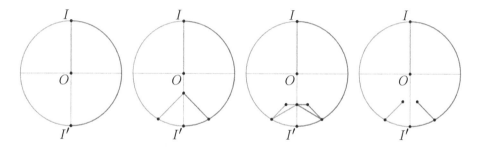

Fig. 1. From left to right: The evolution of the algorithms presented in [12], [9] and [5]. Green indicates the trajectories of robot R_1, red indicates the trajectories of robot R_2 and orange indicates that both robots move together. (Color figure online)

1.1 Our Results

The idea behind introducing a detour through the interior of the disk is to protect the algorithm against the worst-possible exit position: Since the robot that finds the exit needs to intercept the other robot, it makes sense to move towards the other robot before reaching the worst-case interception point, which worsens the evacuation time for some exit positions, but improves it for the worst case. Of course, we can apply this idea iteratively to improve the new worst case by introducing a second detour etc. Brandt et al. [5] discuss this idea and state: "However, the improvement in the evacuation time achieved by the collection of these very small cuts is negligibly small, even compared to the improvement given by our algorithm." We refute this statement by showing that introducing even only a single additional cut reduces the evacuation time by the same order of magnitude as the improvement by Brandt et al. [5] relative to the result of Czyzowicz et al. [9]. Specifically, we improve the evacuation time to 5.6234. This indicates that there might still be room for improvement in the upper bound when considering a large family of additional cuts. It is worth noting that our algorithm does not use a forced meeting of the agents on either detour to the interior of the disk (see Fig. 5).

1.2 Related Work

Robot evacuation has been studied for various settings, differing in number of robots and/or exits, robot capabilities, objective, shape of the region, initial knowledge etc. Most results were obtained for evacuation from the disk with wireless communication, i.e., for robots that can exchange information at all times. Czyzowicz et al. [20] and Pattanayak et al. [11] consider evacuation with multiple exits and known positions of the exists relative to each other. Lamprou et al. [18] consider two robots of different speeds. Regarding evacuation with more than two robots, Czyzowicz et al. [13] study the setting with three robots, one of which may be faulty. Czyzowicz et al. focus on evacuating a single robot, the "queen", that is supported by up to three [14] or more [8] "servants". Regarding other environments, Czyzowicz et al. [10] study evacuation from equilateral triangles and squares, and Borowiecki et al. [4] study the evacuation problem in graphs.

A problem closely related to evacuation is the search problem. Especially the problem of finding a specific point on the line has received considerable attention, e.g., [2,3,16], and other works have focused on searching the plane, e.g., [15]. Another related problem is the rendezvous or gathering problem, where robots initially located at different points need to find each other [1,6,7,17,21]. Finally, the problem where one robot is trying to catch the other is called the lion and man problem and was first studied for the unit disk [19].

2   Preliminaries

In this section we define the general notation for the following work. We use the following notation for line segments and arcs between two points A, B.

$\overline{AB}$ denotes the straight line segment between A and B.
$|\overline{AB}|$ denotes the length of the segment $\overline{AB}$.
$\overarc{AB}$ denotes the shorter arc from A to B along the boundary of the disk for A and B on the boundary.
$|\overarc{AB}|$ denotes the length of the arc $\overarc{AB}$ for A and B on the boundary.

A *cut* is the movement of a robot from the boundary of the disk into the interior and back to the point where the robot left the boundary. In general a cut can have any shape, but our algorithm only uses line segments. The *depth* of a cut is defined as half the distance traveled when moving along a cut. The *evacuation time* is the time until both robots have reached the exit.

Our task is to define trajectories for the robots that minimize the evacuation time. To obtain the evacuation time for a given exit we need to know where the robots exchange the information about the location of the found exit. This is done by the *meeting protocol*, a term coined in [9]. For an illustration, refer to Fig. 2.

Definition 1 (Meeting Protocol). *If at any time t_0 one of the robots finds the exit at point E, it computes the shortest additional time t so that the other robot, after traveling distance $t_0 + t$, is located at point M satisfying $|\overline{EM}| = t$. This ensures that the robot that found the exit can move along the segment $\overline{EM}$ to pick the other robot up at point M at time $t_0 + t$. After both robots meet they evacuate along the segment $\overline{ME}$ via the exit at E, resulting in an evacuation time of $t_0 + 2t$.*

With the meeting protocol we are able to calculate the evacuation time for a given exit. Note that, because the robots move at unit speed, we can use time and traveled distance interchangeably. From the point of view of the robot that finds the exit, the evacuation time is the sum of the time it takes the robot to find the exit, the time it takes the robot to pick up the other robot at their

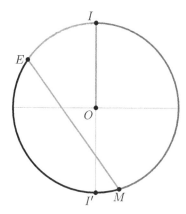

Fig. 2. Illustration of the meeting protocol while the robots perform the algorithm presented in [12]. Robot R_1 finds the exit at point E, uses the meeting protocol to calculate the meeting point and picks up robot R_2 at point M.

meeting point, and the time it takes to get back to the exit. From the point of view of the robot that gets picked up, the evacuation time is the sum of the time it takes the robot to get to the meeting point and the time it takes the robot to travel from the meeting point to the exit. Note that equations of the form $x + 2\sin((x + y)/2) = y$ have to be solved as part of the meeting protocol. In general there are no closed forms known for these equations and therefore we have to rely on numeric solutions.

To prove our stated evacuation time we will refer to a criterion established in [5]. We briefly recap the relevant definitions. To properly define certain relevant angles we first distinguish two cases regarding the direction of the movement of the two robots at the exit and the corresponding meeting point.

We say that the movement of the two robots at the exit and the corresponding meeting point is *conform* if the two robots would move to the same side of the infinite line through the exit and the corresponding meeting point if they did not find the exit, respectively were not picked up at the corresponding meeting point. If the two robots would move to different sides of the infinite line we say that their movement (at the exit and the corresponding meeting point) is *converse*. For an illustration, refer to Fig. 3. The authors of [5] note that the cases where one or both robots would move *on* the infinite line can be arbitrarily considered to belong to one of the two cases.

For the cases of conform and converse movement we now (under the assumption of local differentiability of the movement) define two angles regarding the movement of the two robots at the exit E and at the corresponding meeting point M, respectively, and the straight line segment s between E and M.

We assume that, locally, robot R_1 arrives at the exit E via the local linearization of its trajectory g and would continue on g if it did not find the exit at E. Analogously, we assume that, locally, robot R_2 arrives at the corresponding meeting point M via the local linearization of its trajectory h and would continue on h if it was not picked up at M.

In the case of conform movement, the angle between g and s is denoted by β and the angle between s and h by γ. In the case of converse movement the angle between g and s is denoted by β and the angle between h and s by γ. Note that only the definition of γ differs in the two cases. For an illustration, refer again to Fig. 3.

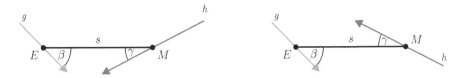

Fig. 3. Illustration of conform movement (left) and converse movement (right) of the robots.

Theorem 1 ([5, Corollary 2.5, Theorems 2.8, 2.9 and 2.10]). *If the trajectories of the two robots are differentiable around points E and M and $2\cos(\beta) + \cos(\gamma) \neq 1$ holds, then there is an exit position that yields a larger evacuation time than placing the exit at E.*

This means that, to consider an exit E as the worst-case candidate, it is necessary that either the trajectory of at least one robot is not differentiable (around the exit or the corresponding meeting point) or $2\cos(\beta) + \cos(\gamma) = 1$. Like the algorithms illustrated in Fig. 1, our algorithm will follow the idea of initially moving the robots to an arbitrary point I on the boundary (we denote the antipodal point by I') and then moving one robot counter-clockwise and the other robot clockwise along the boundary to find the exit. The search on the boundary will only be interrupted if they get picked up or perform a cut. We make a statement about algorithms that follow this general idea, which is helpful to calculate the angles β and γ.

Proposition 1. *If the robots start their search for the exit together at an arbitrary point I on the boundary, one robot moves counter-clockwise and the other robot moves clockwise, their movement is conform, they move along the boundary towards the point I' and their search along the boundary is only interrupted if they get picked up or perform cuts, then the following statement holds: If the corresponding meeting point M of an exit E lies on the boundary, then $\beta = \gamma = \pi - \frac{x+y}{2}$, where $x := |\widehat{IE}|$ and $y := |\widehat{IM}|$.*

Proof. We distinguish between the two cases: $x + y < \pi$ and $x + y \geq \pi$.

For the first case $x + y < \pi$ see the left side of Fig. 4. Because $|\overline{OE}| = 1$ and $|\overline{OM}| = 1$, the triangle $\triangle EOM$ is isosceles. Therefore the base angles η and η' are equal. With the statement, that the interior angles of a triangle add up to π, we can express $\eta = \eta'$ as $\frac{\pi - (x+y)}{2}$. Our next observation is that the direction vector of the tangent at E and the position vector of E are perpendicular (the same holds for the tangent at M and the position vector of M). This immediately yields $\beta = \frac{\pi}{2} + \eta$. Using our expression for η we get $\beta = \pi - \frac{x+y}{2}$. The same follows for γ.

For the second case $x + y \geq \pi$ see the right side of Fig. 4. We again have an isosceles triangle. Therefore the base angles η and η' are equal. Because of $x + y \geq \pi$ we have to calculate the interior angle of the triangle $\triangle EOM$ at O. We easily obtain the interior angle by $2\pi - (x + y)$. Now analogous to the first case, with the equality of the base angles and the statement about the interior angles of a triangle, we can express η as $(\pi - (2\pi - (x + y)))/2$. In this case the angle β can be obtained by $\frac{\pi}{2} - \eta$. In conjunction with our expression for η we have $\beta = \frac{\pi}{2} - \eta = \frac{\pi}{2} - (\pi - (2\pi - (x + y)))/2 = \pi - \frac{x+y}{2}$. The same follows for γ with the equality of the base angles.

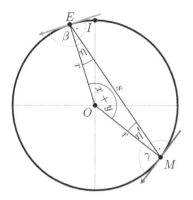 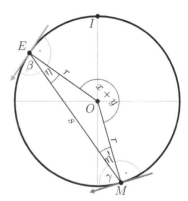

Fig. 4. Illustration of the cases $x + y < \pi$ (left) and $x + y \geq \pi$ (right) in the proof of Proposition 1.

We also make a straight-forward observation (about algorithms that follow the mentioned idea) that follows from monotonicity of the trajectories of both robots along the perimeter of the disk.

Observation 1. *For any two exit positions E and $E^{\sim}$ with the corresponding meeting points M and $M^{\sim}$ lying on the boundary and with $|\widehat{IE}| > |\widehat{IE^{\sim}}|$, we have $|\widehat{IM}| > |\widehat{IM^{\sim}}|$.*

3   Our Algorithm

In this section we present a class $\mathcal{A}(p_1, \alpha_1, d_1, p_2, \alpha_2, d_2)$ of parameterized algorithms with two cuts. The position of the first cut is specified by the parameter p_1, the parameter α_1 specifies the angle of the first cut and d_1 describes the depth of the first cut. Correspondingly, p_2 refers to the position, α_2 refers to the angle and d_2 refers to the depth of the second cut. The trajectories of the robots are described below. For an illustration, refer to Fig. 5.

Algorithm $\mathcal{A}(p_1, \alpha_1, d_1, p_2, \alpha_2, d_2)$:

If a robot finds the exit at any point, it immediately performs the meeting protocol and picks the other robot up at the calculated meeting point M. Both robots reach M at the same time and evacuate via the now known exit on a straight line together. Until this happens the trajectories of the robots are as follows:

1. Both robots move on a straight line to an arbitrary point I on the boundary.

2. At I the robots start their search on the boundary in opposite directions: robot R_1 moves counter-clockwise and robot R_2 moves clockwise.
3. After the robots each covered a distance of p_1 on the boundary, robot R_1 is at C_1 and robot R_2 is at C_1'. There both robots perform their first cut. They move on a straight line at angle α_1 towards the interior (see Figure 5).
4. After they each covered a distance of d_1 on the straight line, robot R_1 is at P_1 and robot R_2 is at P_1'.
5. Now both robots return on the straight line to C_1 and C_1', respectively.
6. After reaching C_1 and C_1', respectively, they continue their search on the boundary, proceeding as in step 2.
7. After the robots each covered a distance of p_2 on the boundary in total, robot R_1 is at C_2 and robot R_2 is at C_2'. There both robots perform their second cut. They move on a straight line at angle α_2 towards the interior (see Figure 5).
8. After they each covered a distance of d_2 on the straight line, robot R_1 is at P_2 and robot R_2 is at P_2'.
9. Now both robots return on the straight line to C_2 and C_2' respectively.
10. After reaching C_2 and C_2' respectively, they continue their search on the boundary, proceeding as in step 2.

---

4   Analysis

To achieve the stated bound on the evacuation time of 5.6234 we used local search to computationally determine the parameters $p_1 = 2.62666582851$, $\alpha_1 = 2\pi/9$, $d_1 = 0.490011696287$, $p_2 = 2.97374843355$, $\alpha_2 = 0.05523991\pi$ and $d_2 = 0.1670474016$ for our algorithm. In the following we rigorously prove that our algorithm indeed needs time at most 5.6234 to evacuate two robots from the unit disk, irrespective of the location of the exit. In order to improve readability, we refer to the above values of our parameters as p_1^*, α_1^*, d_1^*, p_2^*, α_2^* and d_2^*.

Because the presented algorithm is symmetric, it is sufficient to analyze only one half of the disk. We assume that robot R_1 finds the exit and robot R_2 gets picked up. For the analysis of our algorithm we proceed along the same

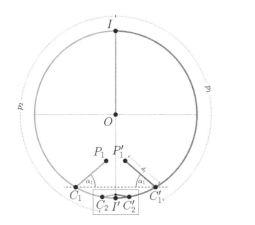
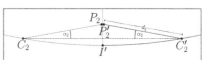

Fig. 5. Illustration of Algorithm $\mathcal{A}$. Left: trajectories of the two robots for the exit located at I'. Right: magnification of the region surrounding I'.

arguments as the authors in [5], but apply these to our approach with two cuts. First we partition the arc $\widehat{II'}$, into the following parts:

$\widehat{IE_1}$ For all exits on this arc robot R_2 is picked up before it leaves the boundary at point C_1'. In particular, if the exit is located at E_1, the meeting of the robots will take place at C_1'.

$\widehat{E_1E_2}$ For all exits on this arc, robot R_2 is picked up while performing its first cut. In particular, if the exit is located at E_2, robot R_2 will be picked up at C_1' after completing its first cut.

$\widehat{E_2E_3}$ If the exit lies on this arc, robot R_1 finds the exit before performing its first cut, but robot R_2 is picked up after performing its first cut.

$\widehat{E_3E_4}$ This part contains all exits for those robot R_2 is picked up while performing its second cut. In particular, if the exit is located at E_4, robot R_2 will be picked up at C_2' after completing its second cut. We exclude the point E_3 from this (half-open) part.

$\widehat{E_4E_5}$ If the exit lies on this arc, robot R_1 finds the exit before performing its second cut, but robot R_2 is picked up after performing its second cut.

$\widehat{E_5I'}$ This part contains all exits that robot R_1 finds after performing its second cut. We exclude the point E_5 from this (half-open) part.

We note that all points are well-defined for our parameter values. In particular, if the exit is found after the first cut, the meeting of the robots cannot take place before the other robot performs its second cut. We obtain (for our parameter values): $|\widehat{IE_1}| \approx 0.629973871925$, $|\widehat{IE_2}| \approx 2.590020657077$, $|\widehat{IE_3}| = 2.62666582851$, $|\widehat{IE_4}| \approx 2.972352082515$ and $|\widehat{IE_5}| = 2.97374843355$.

We now analyze each of these parts in detail and determine possible worst-case candidates.

Lemma 1. *If there is a (global) worst-case exit position on the arc $\widehat{IE_1}$, then it is at E_1.*

Proof. To prove the above statement, we will use Theorem 1 to exclude all interior points of the arc $\widehat{IE_1}$ as worst-case candidates.

Note that the movement of both robots is not differentiable at I and the movement of robot R_2 is not differentiable at C_1'. Therefore we cannot use Theorem 1 to exclude I and E_1 as worst-case candidates, as the corresponding meeting point for an exit at E_1 is at C_1'.

Recall that, for all possible exits on the arc $\widehat{IE_1}$, robot R_2 is picked up before it performs its first cut. In particular, for all these possible exits their corresponding meeting points lie on the boundary. Therefore Proposition 1 is applicable. Also note that we are in the case of conform movement.

With Proposition 1 we obtain $\beta = \gamma \approx 1.51327$ for the exit at E_1 and the corresponding meeting point at C_1', assuming that robot R_2 would continue its search on the boundary at point C_1'. This yields $2\cos(1.51327) + \cos(1.51327) < 1$. Note that by Proposition 1 and Observation 1 the obtained value of approximately 1.51327 is a lower bound on $\beta = \gamma$ for any exit on the interior of the arc $\widehat{IE_1}$ and its corresponding meeting point. In conjunction with the monotonicity of the cosine function on the interval $[0, \pi]$ the statement $2\cos(\beta) + \cos(\gamma) < 1$ holds for any exit on the interior of the arc $\widehat{IE_1}$ and its corresponding meeting point. Therefore we can use Theorem 1 to exclude any exit on the interior of the arc $\widehat{IE_1}$ as possible worst-case candidate.

It remains to show that the exit at I is not a worst-case candidate. This is obvious since the evacuation time for this placement would only be 1 because the robots would find the exit together after initially moving to point I.

Before we start with the next arc, we recall a statement by the authors of [5], that also applies to the second cut. We decouple the exit and meeting point for a moment and let X be an exit for which robot R_2 would be picked up while it is returning from the tip of a cut to the boundary. Now instead of the corresponding meeting point, we consider Y as the meeting point for the exit at X, where Y is on the corresponding cut. Let β' be the angle between the direction of movement of robot R_1 at X and $\overline{XY}$ and let γ' be the angle between $\overline{XY}$ and the direction of movement of robot R_2 while returning to the boundary along the corresponding cut. They state:

Lemma 2. *If $2\sin(\beta') - \sin(\gamma') > 0$, then the value of $2\cos(\beta') + \cos(\gamma')$ increases by moving Y by a small ϵ along the corresponding cut towards the boundary (but not on to the boundary).*

Proof sketch. To prove the statement we observe that moving Y in the aforementioned way decreases β' by the same amount by which γ' is increased. Because the cosine function is differentiable, if the ϵ is small enough the mentioned decrease of β' gets arbitrarily close to a value that is proportional to the derivative of the function $-\cos(\theta)$ at β'. Similarly the increase in γ' gets arbitrarily close to a value that is proportional to the derivative of the function $\cos(\theta)$ at γ'.

Therefore, if $2\sin(\beta') - \sin(\gamma') > 0$ the value of $2\cos(\beta') + \cos(\gamma')$ increases by moving Y in the aforementioned way.

We now can start with the analysis of the second arc.[1]

Lemma 3. *If there is a (global) worst-case exit position on the arc $\widehat{E_1 E_2}$, then it is at E_1 or E_2.*

For the worst-case candidates on the third arc we get an analogous result. (see Footnote 1)

Lemma 4. *If there is a (global) worst-case exit position on the arc $\widehat{E_2 E_3}$, then it is at E_2 or E_3.*

Before we start with the next arc, we recall that we explicitly specified that E_3 does not belong to the arc. To still have a closed arc we add an artificial point $E_3^{\sim}$, that coincides with E_3. However, for an exit at $E_3^{\sim}$ robot R_1 finds the exit immediately after performing its first cut. Without the addition of the artificial point $E_3^{\sim}$ our arc would be half open and there could be a sequence of points that converges towards E_3 with increasing evacuation times but the exit with the largest evacuation time would not belong to the arc. In [5] the authors observe that the evacuation time for the artificial exit at $E_3^{\sim}$ cannot be smaller than the evacuation time for the exit at E_3. This applies, since otherwise robot R_1 could improve the evacuation time for the exit at E_3 by simulating the movement for the exit at $E_3^{\sim}$.

Lemma 5. *If there is a (global) worst-case exit position on the arc $\widehat{E_3^{\sim} E_4}$, then it is at $E_3^{\sim}$ or E_4.*

Proof. To prove the statement, we further divide the arc into three parts. For this, we calculate the points Q_2 and S_2, where Q_2 is the second intersection of the boundary and the line through C_2' and P_2'. If the exit is placed at S_2, robot R_2 will be picked up at P_2'. We obtain: $E_3^{\sim} = E_3 \approx (-0.492471164; -0.870328761)$, $E_4 \approx (-0.16843; -0.98571)$, $Q_2 \approx (-0.492471152; -0.870328768)$ and also $S_2 \approx (-0.26963; -0.96296)$, $C_2' \approx (0.16706; -0.98595)$ and $P_2' \approx (0.00252; -0.95710)$. Note that Q_2 has a smaller second coordinate than $E_3^{\sim}$ and hence robot R_1 reaches the point Q_2 after $E_3^{\sim}$.

Now we analyze the three arcs $\widehat{E_3^{\sim} Q_2}$, $\widehat{Q_2 S_2}$ and $\widehat{S_2 E_4}$ separately.

$\widehat{E_3^{\sim} Q_2}$ We first note that our parameters are chosen in such a way that the meeting point $M_3^{\sim}$ for the exit at $E_3^{\sim}$ is on the line segment $\overline{C_2' P_2'}$. Once again we limit the respective β for any exit on the arc $\widehat{E_3^{\sim} Q_2}$ from above. To do this we decouple exit and meeting point. Instead of the β for the exit at $E_3^{\sim}$ and the corresponding meeting point at $M_3^{\sim}$, we calculate the angle β' for the exit at $E_3^{\sim}$ and consider C_2' as the meeting point. We obtain $\beta' \approx 0.34139$. Note that because of the definition of Q_2 the angle β' is greater than the respective β for the exit at $E_3^{\sim}$ and the corresponding meeting point at $M_3^{\sim}$.

---

[1] This and all other missing proofs can be found in the full version.

It remains to show that the angle β' is an upper bound for the respective angle β of any exit on the arc $\widehat{E_3^{\sim} Q_2}$. Therefore let E be an exit on the arc $\widehat{E_3^{\sim} Q_2}$ and M be the corresponding meeting point on the line segment $\overline{C_2' P_2'}$. If the line segment $\overline{E_3^{\sim} C_2'}$ is rotated around the origin so far that the point E is reached, denote the point on which C_2' is rotated by Y. Observe in particular that the line segment $\overline{C_2' P_2'}$ is not intersected during the rotation due to the definition of Q_2. We further observe, that the angle between the tangent at point E and the line segment $\overline{EY}$ is equal to the angle between the tangent at point $E_3^{\sim}$ and the line segment $\overline{E_3^{\sim} C_2'}$ (this angle is β'). Because the line segment $\overline{EY}$ is above the line segment $\overline{EM}$, where M is the corresponding meeting point of the exit at E, the angle β for an exit at E is smaller than the angle β'. Therefore, we can limit the respective β for any possible exit on $\widehat{E_3^{\sim} Q_2}$ by $\beta' \approx 0.34139$.

On the other hand, the angle γ at the corresponding meeting point for each possible exit of the arc $\widehat{E_3^{\sim} Q_2}$ can be limited from above by the angle $\angle E_3^{\sim} P_2' Q_2 \approx 0.00000001$.

Therefore with the monotonicity of the cosine function, the following statement $2\cos(\beta) + \cos(\gamma) > 2\cos(0.34139) + \cos(0.00000001) > 1.88458 + 0.99999999 > 1$ holds for each possible exit on the arc $\widehat{E_3^{\sim} Q_2}$ and the corresponding meeting point. With Theorem 1 we can exclude all exits but $E_3^{\sim}$ (because the movement of robot R_1 is not differentiable at $E_3^{\sim}$) on the arc $\widehat{E_3^{\sim} Q_2}$ as worst-case candidates.

$\widehat{Q_2 S_2}$ We first note that in this case the movement of the robots is converse (because of the definition of Q_2) and the movement of robot R_2 is not differentiable around P_2' and therefore we cannot exclude the exit at S_2 as the worst-case candidate with Theorem 1. With a similar reasoning as in the first case we can limit respective β of all other points on the arc $\widehat{Q_2 S_2}$ from above by the β for an exit placed at Q_2 and its corresponding meeting point. This statement holds because if we rotate the line segment $\overline{Q_2 C_2'}$ around the origin until we reach a point E on the arc $\widehat{Q_2 S_2}$ and denote the point on which C_2' is rotated on by Y, we observe the following: the line segment $\overline{EY}$ is above the line segment $\overline{EM}$, where M is the corresponding meeting point of an exit at E. The one thing left to argue is that while we rotated the line segment $\overline{Q_2 C_2'}$ we did not intersect the line segment $\overline{C_2' P_2'}$. This is because the intersection of the line segment $\overline{Q_2 C_2'}$ and the perpendicular line that contains the origin O is above the point P_2'. Therefore we can limit the respective β by approximately 0.34138552, which is the obtained value of the angle for an exit placed at Q_2.

On the other, hand it is easy to verify that the angle γ at the corresponding meeting point for each possible exit on the arc $\widehat{Q_2 S_2}$ (but S_2, because the movement of robot R_2 is not differentiable at S_2) can be limited from above by the angle $\angle Q_2 P_2' S_2 \approx 0.195072$.

Combining these two observations and with the monotonicity of the cosine function on the interval $[0, \pi]$ we can state that for any possible exit (but S_2) on the arc $\widehat{Q_2 S_2}$ the term $2\cos(\beta) + \cos(\gamma)$ is greater than $2\cos(0.34138552)$

$+ \cos{(0.195072)} > 1.884583 + 0.98103 > 1$. Therefore with Theorem 1 we can exclude all possible exits (but S_2) on the arc $\widehat{Q_2 S_2}$ as possible worst-case candidates.

$\widehat{S_2 E_4}$ Here we want to show that $2 \cos{(\beta)} + \cos{(\gamma)} < 1$ holds for all possible exits on the interior of the arc $\widehat{S_2 E_4}$. We first observe that the movement of robot R_2 is not differentiable at point C_2', which is the corresponding meeting point for E_4 and that the movement of the robots is conform. To prove the statement we decouple the exit and the meeting point and use Lemma 2: Let X be an arbitrary point on the arc $\widehat{S_2 E_4}$ and Y be an arbitrary point on the line segment $\overline{P_2' C_2'}$. First we will show that $2 \sin{(\beta')} - \sin{(\gamma')} > 0$ holds for all possible combinations of X and Y. To do so we verify the combination of X and Y that minimizes $2 \sin{(\beta')}$ and the combination that maximizes $\sin{(\gamma')}$ separately. It is easy to understand that $2 \sin{(\beta')}$ is minimized for $X = E_4$ and $Y = C_2'$ (note that we allow Y to be on the boundary as it is a lower bound for $2 \sin{(\beta')}$) and $\sin{(\gamma')}$ is maximized for $X = E_4$ and $Y = P_2'$. Calculating the respective angles we have: $2 \sin{(\beta')} - \sin{(\gamma')} > 0.33549 - -0.33289 > 0$. Therefore, we can use Lemma 2 and it is enough to show that for every X on the arc $\widehat{S_2 E_4}$ and $Y = C_2'$ the statement $2 \cos{(\beta')} + \cos{(\gamma')} < 1$ holds. This is because Lemma 2 says that moving Y closer to C_2' only increases the value of $2 \cos{(\beta')} + \cos{(\gamma')}$. Therefore, we need to verify the combinations X and $Y = C_2'$ that maximize $2 \cos{(\beta')}$ and $\cos{(\gamma')}$. We once again do this separately to obtain an upper bound. It is straight forward to verify that for $X = E_4$ the respective expressions are maximized. We calculate the respective angles (under the assumption that robot R_2 would continue his movement at C_2' as he did while returning from the tip of its cut) and obtain $2 \cos{(\beta')} + \cos{(\gamma')} < 1.971661 - 0.985099 < 1$. Therefore we can use Theorem 1 to exclude all exits on the interior of the arc $\widehat{S_2 E_4}$ as possible worst-case candidates. It remains to show that S_2 is not a worst-case candidate either. We do this by comparing the evacuation times for an exit at S_2 and E_4. Recall that the corresponding meeting point for an exit at S_2 is at P_2'. For an exit at S_2 we obtain an evacuation time of approximately 5.39304 and for an exit at E_4 we obtain an evacuation time of approximately 5.62335779. Comparing these evacuation times we see that S_2 is not a worst-case candidate.

Altogether we have shown that for all possible exits (but $E_{\tilde{3}}$ and E_4) on the arc $\widehat{E_{\tilde{3}} E_4}$ either $2 \cos{(\beta)} + \cos{(\gamma)} > 1$ or $2 \cos{(\beta)} + \cos{(\gamma)} < 1$ holds. With Theorem 1 we excluded all possible exits (but $E_{\tilde{3}}$ and E_4) on the arc $\widehat{E_{\tilde{3}} E_4}$ as worst-case candidates.

We continue with the analysis of the fifth segment.

Lemma 6. *If there is a (global) worst-case exit position on the arc $\widehat{E_4 E_5}$, then it is at E_4 or E_5.*

Proof. We first note that the movement of robot R_2 is not differentiable at the corresponding meeting point C_2' of an exit placed at E_4. Furthermore, the movement of robot R_1 is not differentiable at E_5. Therefore we cannot use Theorem 1 to exclude E_4 and E_5 as worst-case candidates.

Recall that for all possible exits on the arc $\widehat{E_4 E_5}$ robot R_2 is picked up after performing its second cut and robot R_1 finds the exit before performing its second cut (but after continuing its search on the boundary after completing its first cut). In particular, for all these possible exits their corresponding meeting points lie on the boundary. Therefore Proposition 1 is applicable. Also note that we are in the case of conform movement.

With Proposition 1 and Observation 1 we can obtain an upper bound for the angles $\beta = \gamma$ of any exit on the interior of the arc $\widehat{E_4 E_5}$ and its corresponding meeting point. We do so by calculating the value of $\beta = \gamma$ for the exit placed at E_4 and its corresponding meeting point C_2' (assuming that robot R_2 would continue its search on the boundary at point C_2'). For this exit and meeting point we obtain $\beta = \gamma \approx 0.16855$. With the monotonicity of the cosine function on the interval $[0, \pi]$ the statement $2 \cos{(\beta)} + \cos{(\gamma)} > 2 \cos{(0.16855)} + \cos{(0.16855)} > 2.95748 > 1$ holds for any exit on the interior of the arc $\widehat{E_4 E_5}$ and its corresponding meeting point.

Therefore we can use Theorem 1 to exclude any exit on the interior of the arc $\widehat{E_4 E_5}$ as a possible worst-case candidate.

Analogous to our preparation of Lemma 5 we again recall that we explicitly specified that E_5 does not belong to the final arc. To still have a closed arc we add an artificial point $E_5^\sim$, that coincides with E_5. For an exit at $E_5^\sim$ robot R_1 finds the exit immediately after performing its second cut. Again the evacuation time for the artificial exit at $E_5^\sim$ cannot be smaller than the evacuation time for the exit at E_5. This applies, since otherwise robot R_1 could improve the evacuation time for the exit at E_5 by simulating the movement for the exit at $E_5^\sim$. (see Footnote 1)

Lemma 7. *If there is a (global) worst-case exit position on the arc $\widehat{E_5^\sim I'}$, then it is at $E_5^\sim$.*

We summarize the statements of the Lemmas 1 through 7 as follows.

Theorem 2. *For Algorithm $\mathcal{A}(p_1^*, \alpha_1^*, d_1^*, p_2^*, \alpha_2^*, d_2^*)$ the worst-case exit position is at E_1, E_2, $E_3^\sim$, E_4 or $E_5^\sim$.*

To determine the evacuation time for our algorithm, we simply take the maximum of the evacuation times for these candidates. This leads to our stated upper bound and main result.

Theorem 3. *Algorithm $\mathcal{A}(p_1^*, \alpha_1^*, d_1^*, p_2^*, \alpha_2^*, d_2^*)$ needs time at most 5.6234 to evacuate two robots via an unknown exit on the boundary of the closed unit disk.*

Proof. For exit E_1 we have $|\widehat{IE_1}| \approx 0.629973871925$ and for the corresponding meeting point M_1 we have $|\widehat{IM_1}| = 2.62666582851$, which results in an evacuation time of less than 5.62335779. For the second worst-case candidate E_2 we have $|\widehat{IE_2}| \approx 2.590020657077$ and for the corresponding meeting point M_2

we have $|\widehat{IM_2}| = 2.62666582851$, which results in an evacuation time of less than 5.62335779. For exit $E_3^{\sim}$ we have $|\widehat{IE_3^{\sim}}| = 2.62666582851$ and for the corresponding meeting point $M_3^{\sim}$ we have $|\widehat{C_2'M_3^{\sim}}| \approx 0.161251676967$, which results in an evacuation time of less than 5.62335779. For the fourth worst-case candidate E_4 we have $|\widehat{IE_4}| \approx 2.972352082515$ and for the corresponding meeting point M_4 we have $|\widehat{IM_4}| = 2.97374843355$, which results in an evacuation time of less than 5.62335779. For exit $E_5^{\sim}$ we have $|\widehat{IE_5^{\sim}}| = 2.97374843355$ and for the corresponding meeting point $M_5^{\sim}$ we have $|\widehat{IM_5^{\sim}}| = 3.141494005121$, which results in an evacuation time of less than 5.62335778. The maximum of these evacuation times is limited from above by 5.6234 and we have proved our stated upper bound on the problem of the two robot evacuation from the closed disk with face-to-face communication.

References

1. Alpern, S.: The rendezvous search problem. SIAM J. Control Optim. **33**(3), 673–683 (1995)
2. Baezayates, R., Culberson, J., Rawlins, G.: Searching in the plane. Inf. Comput. **106**(2), 234–252 (1993)
3. Beck, A., Newman, D.J.: Yet more on the linear search problem. Isr. J. Math. **8**(4), 419–429 (1970). https://doi.org/10.1007/BF02798690
4. Borowiecki, P., Das, S., Dereniowski, D., Kuszner, L.: Distributed evacuation in graphs with multiple exits. In: Suomela, J. (ed.) SIROCCO 2016. LNCS, vol. 9988, pp. 228–241. Springer, Cham (2016). https://doi.org/10.1007/978-3-319-48314-6_15
5. Brandt, S., Laufenberg, F., Lv, Y., Stolz, D., Wattenhofer, R.: Collaboration without communication: evacuating two robots from a disk. In: Fotakis, D., Pagourtzis, A., Paschos, V.T. (eds.) CIAC 2017. LNCS, vol. 10236, pp. 104–115. Springer, Cham (2017). https://doi.org/10.1007/978-3-319-57586-5_10
6. Chalopin, J., Das, S., Disser, Y., Mihalák, M., Widmayer, P.: Mapping simple polygons: how robots benefit from looking back. Algorithmica **65**, 43–59 (2013). https://doi.org/10.1007/s00453-011-9572-8
7. Chalopin, J., Das, S., Disser, Y., Mihalák, M., Widmayer, P.: Mapping simple polygons: the power of telling convex from reflex. ACM Trans. Algorithms **11**, 1–16 (2015). Article No. 33
8. Czyzowicz, J., et al.: Priority evacuation from a disk using mobile robots. In: Lotker, Z., Patt-Shamir, B. (eds.) SIROCCO 2018. LNCS, vol. 11085, pp. 392–407. Springer, Cham (2018). https://doi.org/10.1007/978-3-030-01325-7_32
9. Czyzowicz, J., Georgiou, K., Kranakis, E., Narayanan, L., Opatrny, J., Vogtenhuber, B.: Evacuating robots from a disk using face-to-face communication (extended abstract). In: Paschos, V.T., Widmayer, P. (eds.) CIAC 2015. LNCS, vol. 9079, pp. 140–152. Springer, Cham (2015). https://doi.org/10.1007/978-3-319-18173-8_10
10. Czyzowicz, J., Kranakis, E., Krizanc, D., Narayanan, L., Opatrny, J., Shende, S.: Wireless autonomous robot evacuation from equilateral triangles and squares. In: Papavassiliou, S., Ruehrup, S. (eds.) ADHOC-NOW 2015. LNCS, vol. 9143, pp. 181–194. Springer, Cham (2015). https://doi.org/10.1007/978-3-319-19662-6_13

11. Czyzowicz, J., Dobrev, S., Georgiou, K., Kranakis, E., MacQuarrie, F.: Evacuating two robots from multiple unknown exits in a circle. In: Proceedings of the 17th International Conference on Distributed Computing and Networking (IDCN). p. 8 (2016). Article No. 28
12. Czyzowicz, J., Gasieniec, L., Gorry, T., Kranakis, E., Martin, R., Pajak, D.: Evacuating robots via unknown exit in a disk. In: Kuhn, F. (ed.) DISC 2014. LNCS, vol. 8784, pp. 122–136. Springer, Heidelberg (2014). https://doi.org/10.1007/978-3-662-45174-8_9
13. Czyzowicz, J., et al.: Evacuation from a disc in the presence of a faulty robot. In: Das, S., Tixeuil, S. (eds.) SIROCCO 2017. LNCS, vol. 10641, pp. 158–173. Springer, Cham (2017). https://doi.org/10.1007/978-3-319-72050-0_10
14. Czyzowicz, J., et al.: God save the queen. In: Proceedings of the 9th International Conference on Fun with Algorithms (FUN), pp. 16:1–16:20 (2018). Article No. 16
15. Feinerman, O., Korman, A.: The ants problem. Distrib. Comput. **30**(3), 149–168 (2017)
16. Kao, M.Y., Reif, J.H., Tate, S.R.: Searching in an unknown environment: an optimal randomized algorithm for the cow-path problem. Inf. Comput. **131**(1), 63–79 (1996)
17. Kranakis, E., Krizanc, D., Rajsbaum, S.: Mobile agent rendezvous: a survey. In: Flocchini, P., Gasieniec, L. (eds.) SIROCCO 2006. LNCS, vol. 4056, pp. 1–9. Springer, Heidelberg (2006). https://doi.org/10.1007/11780823_1
18. Lamprou, I., Martin, R., Schewe, S.: Fast two-robot disk evacuation with wireless communication. In: Gavoille, C., Ilcinkas, D. (eds.) DISC 2016. LNCS, vol. 9888, pp. 1–15. Springer, Heidelberg (2016). https://doi.org/10.1007/978-3-662-53426-7_1
19. Littlewood, J.: A Mathematician's Miscellany. Methuen & Co. Ltd., London (1953)
20. Pattanayak, D., Ramesh, H., Mandal, P.S., Schmid, S.: Evacuating two robots from two unknown exits on the perimeter of a disk with wireless communication. In: Proceedings of the 19th International Conference on Distributed Computing and Networking (ICDCN). p. 4 (2018). Article No. 20
21. Ta-Shma, A., Zwick, U.: Deterministic rendezvous, treasure hunts, and strongly universal exploration sequences. ACM Trans. Algorithms **10**(3), 12:1–12:15 (2014)

Distributed Pattern Formation in a Ring

Anne-Laure Ehresmann[1], Manuel Lafond[2], Lata Narayanan[1],
and Jaroslav Opatrny[1(✉)]

[1] Department of Computer Science and Software Engineering,
Concordia University, Montreal, Canada
alehresmann@gmail.com, {lata,opatrny}@cs.concordia.ca
[2] Department of Computer Science, University of Sherbrooke, Sherbrooke, Canada
Manuel.Lafond@USherbrooke.ca

Abstract. Motivated by concerns about diversity in social networks, we consider the following pattern formation problems in rings. Assume n mobile agents are located at the nodes of an n-node ring network. Each agent is assigned a colour from the set $\{c_1, c_2, \ldots, c_q\}$. The ring is divided into k contiguous *blocks* or neighbourhoods of length p. The agents are required to rearrange themselves in a distributed manner to satisfy given diversity requirements: in each block j and for each colour c_i, there must be exactly $n_i(j) > 0$ agents of colour c_i in block j. Agents are assumed to be able to see agents in adjacent blocks, and move to any position in adjacent blocks in one time step.

When the number of colours $q = 2$, we give an algorithm that terminates in time $N_1/n_1^* + k + 4$ where N_1 is the total number of agents of colour c_1 and n_1^* is the minimum number of agents of colour c_1 required in any block. When the diversity requirements are the same in every block, our algorithm requires $3k + 4$ steps, and is asymptotically optimal. Our algorithm generalizes for an arbitrary number of colours, and terminates in $O(nk)$ steps. We also show how to extend it to achieve arbitrary specific final patterns, provided there is at least one agent of every colour in every pattern.

1 Introduction

Recent research in sociology and network science indicates that diverse social connections have many benefits. For example, de Leon *et al.* [24] conclude that increasing diversity, and not just increasing size, of social networks may be essential for improving health and survival among the elderly. Eagle *et al.* find that social network diversity is at the very least a strong structural signature for the economic development of communities [12]. Reagans and Zuckerman [27] found that increased organizational tenure diversity in R&D teams correlated positively with higher creativity and productivity.

On the other hand, the pioneering work of Schelling [28,29] provided a model to describe how even small preferences for *locally* homogeneous neighbourhoods

Research supported by NSERC grants.

K. Censor-Hillel and M. Flammini (Eds.): SIROCCO 2019, LNCS 11639, pp. 215–229, 2019.
https://doi.org/10.1007/978-3-030-24922-9_15

result in globally segregated cities. In his model, individuals of two colours are situated on a path, ring or mesh (i.e. a one- or two-dimensional grid) and have a threshold for the minimum number of neighbors of the same colour they require in their local neighborhood. If this threshold is not met, they move to a random new location. Schelling showed via simulations that this process inevitably led to a globally segregated pattern. This model has been studied extensively and the results confirmed repeatedly; see for example [1,2,10,19,26,31,32]. Recent work in theoretical computer science [4,20] has shown that the expected size of these segregated communities can be exponential in the size of the local neighbourhoods in some social network graphs. The Schelling model has also been studied from a games perspective, where neighborhood formation is determined by agents that can be selfish, strategic [6,14] or form coalitions [5].

In this paper, we study an algorithmic approach to *seeking* diversity. Consider a social network formed by a finite number of individuals, henceforth called *agents* that can be classified into a set of categories, called *colours*. The group collectively seeks diversity in *local* neighborhoods. Is it possible to achieve a specified version of diversity? If so, how can the agents achieve this diversity? These questions were first raised and studied in a recent paper [22], in which the authors studied a model with red and blue agents specifying their local neighbourhood preferences in a ring network. Centralized algorithms to satisfy the preferences of all agents, when possible, were given in [22].

In this paper, we study distributed algorithms for achieving diversity in local neighborhoods (blocks) in a ring network.

1.1   Model and Problem Definition

We assume we are given a collection of n *agents* situated on the nodes of an n-node ring network. We fix a partition of the ring into k paths or *blocks* of length p, with $n = kp$. Each agent has a colour drawn from the set $\{c_1, c_2, \ldots, c_q\}$. For ease of exposition, we initially focus our analysis on the case $q = 2$, and call the colours c_1 and c_2 as *blue* and *red* respectively. Any specific clockwise ordering of these n coloured agents around the ring is called a *configuration* or *n-configuration*. Starting from an initial configuration, we are interested in distributed algorithms for the agents to rearrange themselves into final configurations that meet certain given constraints, for example, alternating red and blue agents, or each red agent has at least one blue agent as a neighbour.

A configuration can be represented by a string $\mathcal{C}$ of length n drawn from the alphabet $\{c_1, c_2, \ldots, c_q\}$. Following standard terminology, for any string u, we denote by $|u|$ the length of u; by u^i the string u repeated i times; by uv the string u concatenated with the string v. The number of occurrences of the colour c_i in a string u is denoted by $n_i(u)$. A *pattern* P is a string of length p drawn from the alphabet $\{c_1, c_2, \ldots, c_q\}$ with $n_i(P) > 0$ for all $1 \leq i \leq q$. Corresponding to the partition of the ring into blocks mentioned above, a configuration can be seen as a concatenation of k patterns $S_1 S_2 \ldots S_k$, in which S_i is the pattern of agents in the i-th block.

We are interested in distributed algorithms for the following problems:

P1: Given positive integers $n_i(j)$ for all $1 \leq i \leq q$ and $1 \leq j \leq k$ and a valid initial configuration $\mathcal{C}$, give a distributed algorithm that achieves a final configuration $S_1 S_2 \ldots S_k$ with $n_i(S_j) = n_i(j)$ for all $1 \leq i \leq q$ and $1 \leq j \leq k$.

P2: Given positive integers $n_i(j)$ for all $1 \leq i \leq q$ and $1 \leq j \leq k$ and a valid initial configuration $\mathcal{C}$, give a distributed algorithm that achieves a final configuration $S_1 S_2 \ldots S_k$ with $n_i(S_j) \geq n_i(j)$ for all $1 \leq i \leq q$ and $1 \leq j \leq k$.

P3: Given a sequence of patterns $P_1, P_2, \ldots, P_k$, and a valid initial configuration $\mathcal{C}$, give a distributed algorithm to achieve the final configuration $P_1 P_2 \cdots P_k$.

A given initial configuration $\mathcal{C}$ is *valid* for the three problems above (respectively) if and only if for all $1 \leq i \leq q$, we have (P1) $n_i(\mathcal{C}) = \sum_{j=1}^{k} n_i(j)$ (P2) $n_i(\mathcal{C}) \geq \sum_{j=1}^{k} n_i(j)$ and (P3) $n_i(\mathcal{C}) = \sum_{j=1}^{k} n_i(P_j)$. Clearly the problem is only solvable for valid initial configurations. Figure 1 shows an input configuration and an output configuration for problem P1 with the required number of agents of colours c_1 and c_2 being 2 in all the blocks.

Notice that an algorithm for P1 can be used to solve P3 with a few modifications. If specific patterns are required in blocks as in Problem P3, we can first apply an algorithm for problem P1 to obtain the correct number of agents of different colours in every block. Once a block has the correct number of agents of every colour, they can rearrange themselves in one additional step to form the required pattern.

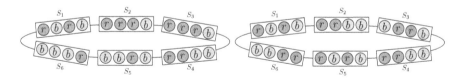

Fig. 1. An input configuration on the left, and a possible output on the right. (Color figure online)

Agent Model. Our agent model is similar to the widely studied *autonomous mobile agent* model [16], where initially the agents are assumed to occupy arbitrary positions in the plane, and each agent repeatedly performs a *Look-Compute-Move* cycle. First an agent looks at the positions of the other agents. Then, using the positions of other agents it found and its own, it computes its next position. Finally it moves to the newly computed position. The agents are generally assumed to have identical capabilities, they have no centralized coordination, and they do not communicate with each other. Indeed, their decisions are based only on their observations of their surroundings made during the look phase. Many different variations of the model have been studied, based on whether or not the agents are synchronized or not, how far they can see, whether or not they agree on a coordinate system, etc.

In this paper, as already stated, we assume that agents are initially placed at the nodes of a n-node ring network G. They belong to q colour classes, but agents within a colour class are identical, and they do not have or use any id. Agents are aware of their position in the ring, and may use this position in their calculations; for example agents at odd positions in the ring may behave differently from agents at even positions. They are completely synchronous, and time proceeds in discrete time steps. We assume that agents have a common *limited visibility range*: each agent can see all agents in the ring that are within the same block or in adjacent blocks. They also have a limited movement range: in one step, an agent can move to any position in the same or adjacent blocks[1]. It is the responsibility of the algorithm designer to ensure that there are no collisions, *i.e.*, we must never have two agents in the same position in the ring. Agents are memoryless, and do not communicate; their decisions are made solely on what they see during their "look" phase. We say that a distributed algorithm for pattern formation by agents *terminates* in a time step, if agents have formed the required pattern, and no agent moves after this time step.

1.2   Our Results

We give solutions for problems P1, P3, and a restricted case of P2. We first consider in detail the case when the number of blocks k is even, and the number of colours $q = 2$, that is, all agents are blue or red. We give an algorithm that terminates in time $N_b/n_b^* + k + 4$ where N_b is the total number of blue agents, and n_b^* is the minimum number of blue agents required in any block (Theorem 2). Note that this time is upper bounded by $n/2$ in the worst case where n is the total number of agents. When the requirements are homogeneous, that is, the same number of blue agents are required in every block, our algorithm terminates in $3k + 4$ steps; we show that this is optimal (Corollary 2).

We then give algorithms for the cases when:

1. k is odd.
2. the number of colours q is more than 2.
3. a specific final configuration $P_1 P_2 \ldots P_k$ is desired.

or any combination of the above (see Theorem 3). These algorithms terminate in $O(nk)$ time.

Finally, given $n_1(j) > 0$ for all $1 \leq j \leq k$, we give an $O(nk)$ algorithm to achieve a final configuration in which the number of agents of colour c_1 is at least $n_1(j)$ in every block S_j (see Theorem 4).

1.3   Related Work

There is a large literature on distributed computing by autonomous mobile agents, see the book [16] that provides a very complete survey of results in

---

[1] We note that if the visibility and movement ranges are smaller, our algorithms still work, though they take more time. The details are lengthy and uninteresting, and therefore omitted from this paper.

this area. Much of this work is done for mobile agents located in the plane or a continuous region, and the main problems that have been considered are gathering of robots [8,18], scattering and covering [9], and pattern formation [7,30] by autonomous mobile robots (see [[17], Sect. 1.3]). The solvability, i.e., termination or convergence, of these problems depends on the type of synchronization, visibility, agreement on orientation [16]. In [25], the authors introduce the problem of using a set of pairwise-communicating processes to constructing a given network.

In discrete spaces, typically graphs like grids or rings, similar problems have been studied, see for example: gathering [11,21]; exploration [23]; pattern formation [3]; and deployment [15]. Unlike in our paper, in all these studies, only a subset of the nodes of the graph is occupied by agents, and typically agents are identical, and not partitioned into colour classes.

The work closest to our work is [22], which considers a similar setting of red and blue agents on a ring of n nodes, and every node of a ring contains an agent. The agents are required to achieve final configurations on a ring, with specified diversity constraints similar to ours. However, unlike the present paper, in which the final configuration specifies the diversity constraints per block and agents are interchangeable, in [22], each agent has a preference for the colours of other agents in its w-neighborhood. The authors consider there three ways for agents to specify these preferences: each agent can specify (1) a preference list: the *sequence* of colours of agents in the neighborhood, (2) a preference type: the *exact* number of neighbors of its own colour in its neighborhood, or (3) a preference threshold: the *minimum* number of agents of its own colour in its neighborhood. The main result of [22] is that satisfying seating preferences is fixed-parameter tractable (FPT) with respect to parameter w for preference types and thresholds, while it can be solved in $O(n)$ time for preference lists. They also show a linear time algorithm for the case when all agents have homogeneous preference types. However, in [22], all algorithms are centralized unlike in the present paper, in which we are only interested in distributed algorithms.

2   Distributed Algorithm for Two Colours

In this section we consider problem P1 with $q = 2$, that is, all agents are blue or red. We denote the number of blue and red agents in a pattern u by $n_b(u)$ and $n_r(u)$ respectively. We first describe an algorithm for the first problem, *i.e.* when the exact number of blue and red agents in the i-th block, denoted by $n_b(i)$ and $n_r(i)$ respectively, is specified. For simplicity, we assume the number of blocks k is even. We will show in Sect. 2.2 that the algorithm also works when k is odd. Let N_b and N_r be the total number of blue and red agents in the initial configuration. Furthermore, let $n_b^* = \min_{1 \leq j \leq k} n_b(j)$ and $n_r^* = \min_{1 \leq j \leq k} n_r(j)$. We assume without loss of generality that $N_b/n_b^* \leq N_r/n_r^*$. If not, we reverse the roles of the blue and red agents in the algorithm and subsequent analysis.

We use S_i to denote both the i-th block and the pattern of agents in it at a particular time step; the meaning used will be clear from context. For every i,

$1 \leq i \leq k$ we denote by y_i the difference between the number of blue agents in S_i and the desired number of blue agents: i.e., $y_i = n_b(S_i) - n_b(i)$. We call y_i a *surplus* if positive, and a *deficit* if negative.

In each step, we partition the ring into *windows* of length $2p$, each window consisting of 2 adjacent blocks. If blocks S_i and S_{i+1} are paired into a window in this partitioning, we denote the window by $[S_i|S_{i+1}]$. Recall that in our agent model, agents can see and move to adjacent blocks, and know their positions in the ring. Therefore, agents in the same window can all see each other and can move to any position within the window (as S_i has access to S_{i+1} and vice-versa). Also, notice that this pairing is specific to our algorithm, and not a constraint of our model. Agents in the window $[S_i|S_{i+1}]$ execute the following algorithm

---

Algorithm 1. Algorithm for agents in the window $[S_i|S_{i+1}]$

---

if $y_i < 0$, i.e., S_i has a deficit of blue agents and an excess of red agents **then**

    $t \leftarrow \min(n_b^*, |y_i|, n_b(S_{i+1}))$

    Move blue agents before red in order-preserving manner in S_i and S_{i+1},

    t leftmost red agents of S_i swap places with t leftmost blue agents of S_{i+1}.

end if

Note that neither moving blue agents before red agents, nor the specification that it is the leftmost agents that swap places, is necessary for the algorithm to terminate, nor do these affect the termination time; they only simplify the analysis. Furthermore, agents can compute the net effect of the two moves mentioned above, and perform a single movement in a single Look-Compute-Move cycle; thus Algorithm 1 comprises a single time step.

Clearly agents in different windows can perform the above algorithm in parallel in the same time step without collisions. We give our pattern formation algorithm for agents of two colours below:

Algorithm 2. Pattern Formation Algorithm

$i \leftarrow 1$

loop

    Apply in parallel Algorithm 1 to windows $[S_i|S_{i+1}], \ldots, [S_{i-2}, S_{i-1}]$

    $i \leftarrow 1 + i \mod k$

end loop

We call one iteration of the loop in Algorithm 2 a *round*. It is straightforward to see that each round takes constant time and can be performed without collisions. Notice that if a block is paired with the block on its right in a round, then it will be paired with the block on its left in the next round. In one round, surplus red agents (if any) in the blocks $S_i, S_{i+2}, S_{i+4}, \ldots, S_{i-2}$ move to $S_{i+1}, S_{i+3}, S_{i+5}$, $\ldots, S_{i-1}$ respectively, and in the next round surplus red agents in the blocks

$S_{i+1}, S_{i+3}, S_{i+5}, \ldots, S_{i-1}$ move to $S_{i+2}, S_{i+4}, S_{i+6}, \ldots, S_i$, respectively, and this is repeated. In other words, there is a flow of red agents to the right and an equivalent flow of blue agents to the left. Since blue agents moving left from the block S_{i+1} to S_i are always the leftmost blue agents in S_{i+1}, and move to positions after the blue agents already present in S_i, and the blue agents move up to the beginning of each block before the next round, each round preserves the clockwise order of the blue agents as stated in the following lemma:

Lemma 1 (Order-preserving property of blue agents). *Let C and C' be the configurations before and after a round in Algorithm 2. Then the clockwise ordering of blue agents in C and C' is the same.*

2.1 Proof of Correctness and Termination

We now show that this flow of agents eventually stops, and the algorithm terminates with the desired number of blue and red agents in every block. We first introduce some definitions and observations about the properties of a configuration.

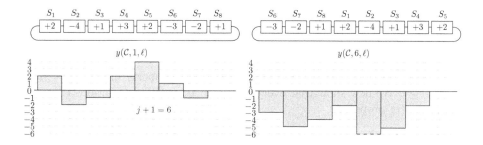

Fig. 2. The surplus of blue agents y_i is shown inside the box for slice S_i. The plot below shows the cumulative surplus $y(C, 1, \ell)$ on the left and $y(C, 6, \ell)$ on the right

Definition 1. *For any given valid configuration $C = S_1 S_2 \ldots S_k$, define $y(C, i, \ell)$ $= \sum_{j=0}^{\ell-1} y_{i+j}$, i.e., y_i is the cumulative surplus of blue agents in ℓ consecutive blocks starting from S_i.*

Definition 2. *For any given valid configuration $C = S_1 S_2 \ldots S_k$, define $y(C, i)$ to be the maximum in the set $\{y(C, i, 1), y(C, i, 2), \ldots, y(C, i, k)\}$.*

Figure 2 shows $y(C, 1, \ell)$ for a sample configuration. Notice that $y(C, 1) = y(C, 1, 5) = 4$.

Lemma 2. *In any given valid configuration $C = S_1 S_2 \ldots S_k$ there exists an integer i such that $y(C, i) = 0$.*

Proof. Let j be the integer such that $y(\mathcal{C}, 1, j) = y(\mathcal{C}, 1)$, i.e., the index of the block where the sum of differences of number of blue agents between the blocks and patterns, when counted from S_1, is maximized. See Fig. 2. Consider the value $y(\mathcal{C}, j+1, \ell)$ for some integer ℓ. Clearly, $y(\mathcal{C}, j+1, \ell) = y(\mathcal{C}, 1, \ell) - y(\mathcal{C}, 1, j) \leq 0$ by our choice of j. On the other hand in any valid configuration the total excess of red agents in the configuration is 0. Thus, $y(\mathcal{C}, j+1) = 0$. See Fig. 2; in this example, $j = 5$; the figure on the right shows $y(\mathcal{C}, 6, \ell)$, which is always ≤ 0 for every ℓ.   □

Since the input configuration is circular, and the algorithm runs in parallel in all blocks, we can rename the blocks. We will therefore assume in the sequel that $y(\mathcal{C}, 1) = 0$. This means that starting in block S_1, *the cumulative surplus of blue agents is never positive*. This implies the following useful non-positive prefix and non-negative suffix property.

Lemma 3 (Non-Negative Suffix Property). *For any valid input configuration $\mathcal{C}$ and for every integer j, $1 \leq j < k$ we have:*
$y(\mathcal{C}, 1, j) \leq 0$ *and* $y(\mathcal{C}, j+1, k-j) \geq 0$
In other words, the total surplus of blue agents in the first j blocks is at most 0 while the total surplus of blue agents in the last $k - j$ blocks is at least 0.

Proof. Since $y(\mathcal{C}, 1) = 0$, we have that $y(\mathcal{C}, 1, j) \leq 0$. Since $y(\mathcal{C}, 1) = y(\mathcal{C}, 1, j) + y(\mathcal{C}, j+1, k-j) = 0$, we get $y(\mathcal{C}, j+1, k-j) = -y(\mathcal{C}, 1, j) \geq 0$.   □

In particular, notice that the non-negative suffix property implies that $y_k = y(\mathcal{C}, k, 1) \geq 0$. We now show that every round of Algorithm 2 maintains the non-negative suffix property.

Lemma 4. *Let $\mathcal{C} = S_1 S_2 \ldots S_k$ be a configuration such that for any j, $1 \leq j \leq k$, we have $y(\mathcal{C}, 1, j) \leq 0$ and $y(\mathcal{C}, j+1, k-j) \geq 0$. Executing one round of Algorithm 2 gives a configuration $\mathcal{C}' = S_1' S_2' \ldots S_k'$ such that $y(\mathcal{C}', 1, j) \leq 0$ and $y(\mathcal{C}', j+1, k-j) \geq 0$ for any j, $1 \leq j \leq k$.*

Proof. For a block S_ℓ, define $z_\ell = n_r(S_\ell) - n_r(\ell)$ as the surplus of red agents in S_ℓ. Let ℓ be an integer, $1 \leq \ell < k$ such that Algorithm 1 is applied to the pair $[S_\ell | S_{\ell+1}]$. Assume for now that $\mathcal{C}'$ is the configuration obtained after only $[S_\ell | S_{\ell+1}]$ have made their exchanges and no other blocks. We will show that the conditions of the Lemma hold for $\mathcal{C}'$. This proves our statement, since we may repeat our analysis by applying the exchanges between pairs of blocks one at a time while maintaining the invariants of the Lemma.

Performing the exchanges between S_ℓ and $S_{\ell+1}$ can affect only the suffixes $y(\mathcal{C}', \ell, k - \ell + 1)$ and $y(\mathcal{C}', \ell+1, k - \ell)$; any movement of agents that happens cannot affect a suffix $y(\mathcal{C}', j, k - j)$ with $j < \ell$ or $j > \ell + 1$. Clearly, if $y(S_\ell) \geq 0$ or there are no blue agents in $S_{\ell+1}$ then there is no movement of agents between S_ℓ and $S_{\ell+1}$ and thus, obviously, $y(\mathcal{C}', \ell, k - \ell + 1) = y(\mathcal{C}, \ell, k - \ell + 1) \geq 0$, and $y(\mathcal{C}', 1, \ell - 1) = y(\mathcal{C}, 1, \ell - 1) \leq 0$.

Thus we only need to consider the case when S_ℓ has $z_\ell = -y_l > 0$ excess red agents and $S_{\ell+1}$ contains $n_b(S_{\ell+1}) > 0$ blue agents. Let $t =$

$min\{z_\ell, n_b(S_{\ell+1}), n_b^*\}$. Then t blue agents move from $S_{\ell+1}$ to S_ℓ and t move from S_ℓ to $S_{\ell+1}$. Clearly, the total surplus of blue agents in the ℓ-th suffix does not change, that is, $y(\mathcal{C}', \ell, k - \ell + 1) = y(\mathcal{C}, \ell, k - \ell + 1) \geq 0$. We only need to show that the movement of t blue agents to S_ℓ does not render the $(\ell+1)$-th suffix negative. Observe that $0 \leq y(\mathcal{C}, \ell, k - \ell + 1) = y(\mathcal{C}, \ell + 1, k - \ell) - z_\ell$.

In other words, since S_ℓ lacks z_ℓ blue agents and yet the ℓ-th suffix is non-negative, it must be that the $(\ell+1)$-st suffix is at least z_ℓ. Thus, $y(\mathcal{C}, \ell+1, k-\ell) \geq z_\ell \geq t$, and moving $t \leq z_\ell$ blue agents from $S_{\ell+1}$ to S_ℓ cannot make the $(\ell+1)$-st suffix negative. That is, $y(\mathcal{C}', \ell + 1, k - \ell) = y(\mathcal{C}, \ell + 1, k - \ell) - t \geq 0$.   □

Lemma 4 implies that the property $y_k \geq 0$ is maintained after every round. This means S_k *never* has a deficit of blue agents, and therefore when it participates in the window $[S_k|S_1]$, there is no swap of agents. Therefore we obtain the following corollary:

Corollary 1. *Algorithm 2 never does any swap of agents between the blocks S_k and S_1.*

We now develop a potential function argument to show that Algorithm 2 eventually reaches the desired final configuration, after which no agents ever move. Given a configuration $\mathcal{C} = S_1 S_2 \cdots S_k$, we define the *destination block* of the i-th blue agent (from position 1 in the ring) to be
$$dest(i) = \operatorname{argmin}_\ell \sum_{i=1}^\ell n_i(\ell) \geq i$$
That is, the destination block of the leftmost $n_b(1)$ blue agents in the input configuration is S_1, the next $n_b(2)$ blue agents in the input configuration is S_2, and so on. Next we define the *displacement* of the i-th blue agent (from position 1 in the ring) that is located in block S_j to be the difference between the index of its present block and the index of its destination block:
$$displacement(i, j) = j - dest(i).$$
Given a configuration $\mathcal{C} = S_1 S_2 \cdots S_k$, we define the *distance* of the configuration $d(\mathcal{C})$ to be the sum of displacements of all blue agents in the configuration. We first prove some properties of the distance function.

Lemma 5. *Let $\mathcal{C}$ be the configuration after i rounds in Algorithm 2. Then:*

1. $d(\mathcal{C}) \geq 0$
2. $d(\mathcal{C}) \leq d(\mathcal{C}')$ *where $\mathcal{C}'$ was the configuration before the i-th round.*
3. *If $d(\mathcal{C}) = 0$, no agent ever moves again.*

Proof. It follows from Lemma 4 that the displacement of any blue agent is non-negative, as otherwise, there would be a deficit of blue agents in some suffix. Thus $d(\mathcal{C})$ must always be non-negative. Second, since blue agents only move left in an order-preserving manner (Lemma 1), and from Corollary 1, they never move left past S_1 to S_k, the displacement of any blue agent cannot increase, and therefore $d(\mathcal{C})$ cannot increase. Finally, if $d(\mathcal{C}) = 0$, then every block has its desired number of blue agents, and no agent will ever move again.   □

We now show that the distance function is guaranteed to decrease after at most 2 rounds.

Lemma 6. *Assume that we are given a configuration* $\mathcal{C} = S_1 S_2 \ldots, S_k$ *and positive integers* $n_b(i) > 0$ *for* $1 \leq j \leq k$ *such that* $d(\mathcal{C}) > 0$. *After at most two rounds in Algorithm 2 to* $\mathcal{C}$ *we obtain configuration* $\mathcal{C}'$ *such that* $d(\mathcal{C}') \leq d(\mathcal{C}) - 1$.

Proof. Since $d(\mathcal{C}) > 0$, we know that $\mathcal{C}$ is not a valid final configuration, and Lemma 3 and the assumption $n_b(i) > 0$ for all $1 \leq j \leq k$ implies that there exists a window $[S_j | S_{j+1}]$ such that $n_b(S_j) < n_b(j)$ and $n_b(S_{j+1}) > 0$, i.e., there is a surplus of red agents in S_j, and S_{j+1} contains some blue agents. Then $t = \min\{|y_j|, n_b(S_{j+1}), n_b^*\}$ as defined in Algorithm 1 is positive. Now, if the window $[S_j | S_{j+1}]$ is active in the next round in Algorithm 2, then t blue agents from S_{j+1} are swapped with t red agents from S_j. Clearly, the displacement of these blue agents goes down by 1. Thus after this round, we obtain configuration $\mathcal{C}'$ with $d(\mathcal{C}') < d(\mathcal{C})$.

Suppose instead the window $[S_j | S_{j+1}]$ is not active in the next round of Algorithm 2. Let $\mathcal{C}_1$ be the configuration obtained after the next round of Algorithm 2. By Lemma 5, $d(\mathcal{C}_1) \leq d(\mathcal{C})$. In the following round of Algorithm 2 the window $[S_j | S_{j+1}]$ is active and we transfer t blue agents from S_{j+1} to S_j, and the displacement of these t agent goes down by 1. Thus after two rounds, we obtain the configuration $\mathcal{C}'$ with $d(\mathcal{C}') \leq d(\mathcal{C}_1) - 1 \leq d(\mathcal{C}) - 1$.   □

Notice that the condition $n_b(P_j) > 0$ for $1 \leq j \leq k$ is necessary. For example, if there is a block S_j with $n_b(P_j) = 0$ and $y(S_j) = 0$ then this block will never receive a blue agent, and this would prevent any flow of blue agents to the left of S_j, even when blue agents are needed there, and any flow of red agents across S_j.

Theorem 1. *Given* $n_b(j) > 0$ *and* $n_r(j) \geq 0$ *for all all* $1 \leq j \leq k$, *where* k *is even, and a valid input configuration, Algorithm 2 reaches a configuration where* $n_b(S_j) = n_b(j)$ *and* $n_r(S_j) = n_r(j)$ *in every block* S_j. *Furthermore, this takes* $O(nk)$ *steps, and after this configuration is reached, no agent ever moves again.*

Proof. Since the initial value of the distance function is at most nk, the result follows from Lemma 6 and Lemma 5.   □

In this section, we show an upper bound on the number of iterations required for Algorithm 2 to terminate. Recall that $n_b^* = p - \max_{1 \leq j \leq k}(n_r(j))$ and N_b is the total number of blue agents in the input configuration $\mathcal{C}$. Let B be the set of blue agents, and assume (for the purpose of the analysis) that they are numbered from left to right as $b_1, b_2, \ldots b_{N_b}$. Partition the members of B into subsets $B_1, \ldots, B_L$ as follows. Let $B_1 = \{b_1, \ldots, b_{n_b^*}\}$, $B_2 = \{b_{n_b^*+1}, \ldots, b_{2n_b^*}\}$, and so on, so that $L = \lceil N_b / n_b^* \rceil$ and each subset has n_b^* elements, except perhaps B_L. We say that an agent b_i is (t)-*cooperative* if, in every round $t' \geq t$, either (1) b_i moves left; or (2) b_i has reached its destination block as defined in the previous section.

The key idea is that the first n_b^* blue agents (i.e. B_1) start travelling every round at the latest by round 2 until they reach their destination blocks. After this round, they "free" the path for the blue agents of B_2, which then start travelling every round from round 4 until they reach their destination blocks.

This unblocks B_3 from round 6, and so on. In the end, the blue agents of B_L can freely travel from round $2L + 2$, and they only need to travel at most k slices. Note that agents in blocks B_l may indeed move before round $2l+2$ but they may be subsequently blocked; however from round $2l + 2$ onward, they will continue to travel every round until they reach their destination block. These ideas are formalized in the next lemma.

Lemma 7. *For every $l \in [L]$, each blue agent in B_l is $(2l + 2)$-cooperative.*

Proof. For convenience, define $B_0 = \emptyset$. We show by induction on $l \in \{0, 1, \dots, L\}$ that each blue agent in each B_l is $(2l + 2)$-cooperative. Using $l = 0$ as a base case, this statement is vacuously true: each blue agent in B_0 is (0)-cooperative. Now suppose that the lemma holds for all $B_{l'}$ with $l' < l$. Assume that there is $b_i \in B_l$ such that in some round $w \geq 2l + 2$, the agent b_i does not move left and it is not in its destination block. Let S_a be the block that contains b_i before (and after) the execution of round w. It follows from the earlier discussions that $S_a \neq S_1$, as this would mean the destination block of b_i requires it to move left from S_1 to S_k. There are 2 cases to consider: either, in round w,
(*) S_a is in the active window $[S_{a-1}|S_a]$;
(**) S_a is in the active window $[S_a|S_{a+1}]$.
Details of these two cases do not fit within the page limit, but they can be found in [13]. □

Theorem 2. *Given even k and positive integers $n_r(j)$ and $n_b(j)$ for all $1 \leq j \leq k$ and a valid input configuration $\mathcal{C}$, Algorithm 2 terminates in at most $\frac{2N_b}{n_b^*} + k + 4$ time steps, where $n_b^* = \min_{1 \leq j \leq k}(n_b(j))$ and N_b is the total number of blue agents in $\mathcal{C}$.*

Proof. Consider the partition $B_1, \dots, B_L$ of the blue agents as defined above. We have $L = \lceil N_b/n_b^* \rceil \leq N_b/n_b^* + 1$. By Lemma 7, each b_i is $(2L+2)$-cooperative, implying that each b_i is also $(2N_b/n_b^* + 4)$-cooperative. This means that after at most $2N_b/n_b^* + 4$ rounds, every blue agent that is not yet at its destination block advances a block in every round. Since there are k blocks, such a b_i will reach its destination after at most k additional rounds. Since this holds for any b_i, and recalling that every round can be executed in one step, this proves the statement. □

Corollary 2. *If the diversity requirements are homogeneous, that is $n_b(i) = m$ for all $1 \leq j \leq k$, then Algorithm 2 terminates in at most $3k + 4$ steps, and this is optimal.*

Proof. Suppose $n_b(j) = m$ for all $1 \leq j \leq k$. Then $n_b^* = m$ and $N_b = km$, and the bound given by Theorem 2 is $\frac{2km}{m} + k + 4 = 3k + 4$.

We now show that there are inputs for which $\Omega(k)$ is a lower bound on the number of steps required for *any* algorithm in our agent model. In particular consider an input with the size of the blocks p even, and $n_b(i) = p/2$ for every block i, and the input configuration such that the first $k/2$ blocks have only red agents and the last $k/2$ blocks have only blue agents. Then $pk/4$ blue agents need to move to blocks S_1 to $S_{k/2}$. Since agents can only move to the next block in a single time step, all these agents would have to move from $S_{k/2+1}$ to $S_{k/2}$ or from S_k to S_1 in some step. However, at most p blue agents can move from $S_{k/2+1}$ to S_k and similarly at most p blue agents can move from S_k to S_1 in one time step. It follows that the number of time steps required is at least $pk/(4(2p)) = k/8$. $\square$

Notice that when the diversity patterns are identical, given by a string u of length p, the solution obtained by our algorithm is of type u^k. The results in [22] imply that in the final ring network every neighbourhood of length p starting at any position of the ring contains the same number of blue and red agents as u. Thus, our algorithm solves in distributed manner the *homogeneous preference type problem* of [22].

2.2 When the Number of Blocks k is Odd

Consider now the situation when the number of blocks k is odd. Notice that in one round of Algorithm 2, the surplus red agents from the blocks $S_i, S_{i+2}, S_{i+4}, \ldots, S_{i-3}$ move to $S_{i+1}, S_{i+3}, S_{i+5}, \ldots, S_{i-2}$, respectively, and in the next round, surplus red agents from the blocks $S_{i+1}, S_{i+3}, S_{i+5}, \ldots S_{i-2}$ move to $S_{i+2}, S_{i+4}, S_{i+6}, \ldots, S_{i-1}$ respectively. In this case surplus red agents from S_{i-1} can move only in the subsequent third round, but this delay happens to S_{i-1} only once in k rounds since the value of i is incremented after every round. Then in the proof of Lemma 6 the window $[S_i, S_{i+1}]$ in which there is swap of agents between S_i and S_{i+1} is only in the third round of the algorithm. Thus, the value of $d(\mathcal{C})$ is only guaranteed to decrease after 3 rounds. However, due to the way the windows are shifted in the algorithm after each round, this can occur to S_{i-1} only once in k rounds. It is clear that Theorem 1 still holds.

3 Distributed Algorithm for q Colours

In this section we consider the pattern formation problem when the given set of colours $\{c_1, c_2, \ldots, c_q\}$ is of size $q \geq 3$. We show that Algorithm 2 can be generalized to solve the pattern formation problem for any $q \geq 3$.

The main idea of the generalization is to iteratively "correct" the number of agents of colour i, starting with colour c_1, while treating the remaining colors $c_{i+1}, c_{i+2}, \ldots, c_q$ as a single "new" color c'. Thus, essentially, by repeating $q-1$ times Algorithm 2 we get a solution for q colors. An optimized version is given below in Algorithms 3 and 4.

Clearly, the termination of this algorithm follows directly from the termination of Algorithm 4 as shown in Theorem 1. Notice that $n_i(c_j) \geq 1$ is needed for every $1 \leq i \leq q-1$ and $1 \leq j \leq k$ to guarantee termination.

Theorem 3. *Algorithm 4 solves the pattern formation problems P1 and P3 and terminates in time $O(nk)$ where n is the number of agents and k is the number of blocks.*

Algorithm 3. Algorithm for a Window $[S_j|S_{j+1}]$ for Agents of q colours

---

$i \leftarrow 1$
while $i < q$ & $(n_i(S_j) = n_i(j)$ & $n_i(S_{j+1}) = n_i(j+1))$ **do**
    $i \leftarrow i + 1$
end while
if $i < q$ **then**
    if $n_i(S_j) < n_i(j)$, i.e., S_j needs additional agents of colour c_i **then**
        $t \leftarrow \min(n_i(j) - n_i(S_j), n_i(S_{j+1}))$
        t agents of colour c_i in S_{j+1} swap places with t agents of colour greater than
i in S_j
    end if
    else Rearrange S_i into specific pattern P_i if required in the problem specification.
end if

Algorithm 4. Pattern Formation Algorithm for Agents of q colours

---

$i \leftarrow 1$
loop
    Apply in parallel Algorithm 3 to windows $[S_i|S_{i+1}], [S_{i+2}|S_{i+3}] \ldots$
    $i \leftarrow 1 + i \mod k$
end loop

3.1 When There Is a Lower Bound on the Number of Agents

In this section we address a restricted version of the second problem P2: given $n_1(j) > 0$ and $n_i(j) = 0$ for all $2 \leq i \leq q$ and $1 \leq j \leq k$, achieve a final configuration $\mathcal{C}' = S_1 S_2 \ldots S_k$ with $n_i(S_j) \geq n_i(j)$. In other words, there is a positive lower bound required on the number of agents of colour c_1 in every block, but no lower bound on agents of any other colour. Observe that in this case a valid initial configuration w must satisfy the condition $n_1(\mathcal{C}) \geq \sum_{i=1}^{k} n_1(S_i)$.

Let $d = n_1(\mathcal{C}) - \sum_{i=1}^{k} n_1(S_i)$, that is, d is the number of extra agents of colour c_1 present in the input configuration. Consider now applying Algorithm 2 (for 2 colours) on the given input instance, treating agents of all colours except for colour c_1 as agents of colour c'_2, a new colour. Given a valid block configuration $\mathcal{C}$ we use the definitions of $y(\mathcal{C}, i, \ell)$ and $y(\mathcal{C}, i, \ell)$ from the preceding section. It is easy to see that by replacing every 0 by d in the statements of Lemmas 2, 3, 4, and Corollary 1, they remain valid for this problem. This implies that Lemma 6 and Theorem 1 remain valid exactly as stated in the previous section. Thus we get the following:

Theorem 4. *Algorithm 4 solves the following restricted version of the pattern formation problem P2 in time $O(nk)$: Given a valid input configuration $\mathcal{C}$ and $n_1(j) > 0$ and $n_i(j) = 0$ for all $2 \leq i \leq q$ and $1 \leq j \leq k$, achieve a final configuration $\mathcal{C}' = S_1 S_2 \ldots S_k$ with $n_i(S_j) \geq n_i(j)$.*

4   Discussion

Given n agents of q different colours situated on the nodes of a ring network that has been partitioned into k blocks, we gave distributed algorithms for the agents to move to new locations where they satisfy specified patterns or diversity constraints in every block. Our analysis for the case of even k and two colours is tight, but for the other problems, it would be interesting to obtain tight bounds. Our algorithms need a positive number of agents of at least the first $q-1$ colours to be required in the final configuration in every block; it is unclear to what extent this restriction is necessary for the existence of distributed algorithms for the problem. It would also be interesting to consider how bounds on the number of agents moved between bloks or bandwidth would influence the number of steps needed in the algorithms.

References

1. Benard, S., Willer, R.: A wealth and status-based model of residential segregation. Math. Sociol. **31**(2), 149–174 (2007)
2. Benenson, I., Hatna, E., Or, E.: From Schelling to spatially explicit modeling of urban ethnic and economic residential dynamics. Sociol. Methods Res. **37**(4), 463–497 (2009)
3. Bose, K., Adhikary, R., Kundu, M.K., Sau, B.: Arbitrary pattern formation on infinite grid by asynchronous oblivious robots. In: Das, G.K., Mandal, P.S., Mukhopadhyaya, K., Nakano, S. (eds.) WALCOM 2019. LNCS, vol. 11355, pp. 354–366. Springer, Cham (2019). https://doi.org/10.1007/978-3-030-10564-8_28
4. Brandt, C., Immorlica, N., Kamath, G., Kleinberg, R.: An analysis of one-dimensional Schelling segregation. In: Proceedings of the 44th STOC, pp. 789–804. ACM (2012)
5. Bredereck, R., Elkind, E., Igarashi, A.: Hedonic diversity games (2019). arXiv preprint arXiv:1903.00303
6. Chauhan, A., Lenzner, P., Molitor, L.: Schelling segregation with strategic agents. In: Deng, X. (ed.) SAGT 2018. LNCS, vol. 11059, pp. 137–149. Springer, Cham (2018). https://doi.org/10.1007/978-3-319-99660-8_13
7. Cicerone, S., Di Stefano, G., Navarra, A.: Asynchronous embedded pattern formation without orientation. In: Gavoille, C., Ilcinkas, D. (eds.) DISC 2016. LNCS, vol. 9888, pp. 85–98. Springer, Heidelberg (2016). https://doi.org/10.1007/978-3-662-53426-7_7
8. Cohen, R., Peleg, D.: Convergence properties of the gravitational algorithm in asynchronous robot systems. SIAM J. Comput. **34**, 1516–1528 (2005)
9. Cohen, R., Peleg, D.: Local spreading algorithms for autonomous robot systems. Theor. Comput. Sci. **399**, 71–82 (2008)
10. Dall'Asta, L., Castellano, C., Marsili, M.: Statistical physics of the Schelling model of segregation. J. Stat. Mech.: Theory Exp. **2008**(07), L07002 (2008)
11. D'Angelo, G., Di Stefano, G., Klasing, R., Navarra, A.: Gathering of robots on anonymous grids without multiplicity detection. In: Even, G., Halldórsson, M.M. (eds.) SIROCCO 2012. LNCS, vol. 7355, pp. 327–338. Springer, Heidelberg (2012). https://doi.org/10.1007/978-3-642-31104-8_28

12. Eagle, N., Macy, M., Claxton, R.: Network diversity and economic development. Science **328**(5981), 1029–1031 (2010)
13. Ehresmann, A.L., Lafond, M., Narayanan. L., Opatrny, J.: Distributed pattern formation in a ring (2019). arXiv:1905.08856
14. Elkind, E., Gan, J., Igarashi, A., Suksompong, W., Voudouris, A.A.: Schelling games on graphs (2019). arXiv preprint arXiv:1902.07937
15. Elor, Y., Bruckstein, A.M.: Uniform multi-agent deployment on a ring. Theor. Comput. Sci. **412**(8–10), 783–795 (2011)
16. Flocchini, P., Prencipe, G., Santoro, N.: Distributed Computing by Oblivious Mobile Robots: Synthesis Lectures on Distributed Computing Theory. Morgan and Claypool Publishers, San Rafael (2012)
17. Flocchini, P., Prencipe, G., Santoro, N.: Distributed Computing by Mobile Entities: Current Research in Moving and Computing. LNCS, vol. 11340. Springer, Cham (2019). https://doi.org/10.1007/978-3-030-11072-7
18. Flocchini, P., Prencipe, G., Santoro, N., Widmayer, P.: Gathering of asynchronous mobile robots with limited visibility. Theor. Comput. Sci. **337**(1–2), 147–168 (2005)
19. Henry, A.D., Pralat, P., Zhang, C.-Q.: Emergence of segregation in evolving social networks. Proc. Nat. Acad. Sci. **108**(21), 8605–8610 (2011)
20. Immorlica, N., Kleinberg, R., Lucier, B., Zadomighaddam, M.: Exponential segregation in a two-dimensional Schelling model with tolerant individuals. In: Proceedings of the 20th SODA Symposium, pp. 984–993 (2017)
21. Klasing, R., Markou, E., Pelc, A.: Gathering asynchronous oblivious mobile robots in a ring. Theor. Comput. Sci. **390**(1), 27–39 (2008)
22. Krizanc, D., Lafond, M., Narayanan, L., Opatrny, J., Shende, S.: Satisfying neighbor preferences on a circle. In: Bender, M.A., Farach-Colton, M., Mosteiro, M.A. (eds.) LATIN 2018. LNCS, vol. 10807, pp. 727–740. Springer, Cham (2018). https://doi.org/10.1007/978-3-319-77404-6_53
23. Lamani, A., Potop-Butucaru, M.G., Tixeuil, S.: Optimal deterministic ring exploration with oblivious asynchronous robots. In: Patt-Shamir, B., Ekim, T. (eds.) SIROCCO 2010. LNCS, vol. 6058, pp. 183–196. Springer, Heidelberg (2010). https://doi.org/10.1007/978-3-642-13284-1_15
24. Mendes de Leon, C.F., Ali, T., Nilsson, C., Weuve, J., Rajan, K.B.: Longitudinal effects of social network diversity on mortality and disability among elderly. Innovation Aging **1**(suppl 1), 431 (2017)
25. Michail, O., Spirakis, P.G.: Simple and efficient local codes for distributed stable network construction. Distrib. Comput. **29**(3), 207–237 (2016). https://doi.org/10.1007/s00446-015-0257-4
26. Pancs, R., Vriend, N.J.: Schelling's spatial proximity model of segregation revisited. J. Public Econ. **91**(1), 1–24 (2007)
27. Reagans, R., Zuckerman, E.W.: Networks, diversity, and productivity: the social capital of corporate R&D teams. Organ. Sci. **12**(4), 502–517 (2001)
28. Schelling, T.C.: Models of segregation. Am. Econ. Rev. **59**(2), 488–493 (1969)
29. Schelling, T.C.: Dynamic models of segregation. J. Math. Sociol. **1**(2), 143–186 (1971)
30. Suzuki, I., Yamashita, M.: Distributed anonymous mobile robots: formation of geometric patterns. SIAM J. Comput. **28**(4), 1347–1363 (1999)
31. Young, H.P.: Individual Strategy and Social Structure: An Evolutionary Theory of Institutions. Princeton University Press, Princeton (2001)
32. Zhang, J.: A dynamic model of residential segregation. J. Math. Sociol. **28**(3), 147–170 (2004)

On Distributed Merlin-Arthur Decision Protocols

Pierre Fraigniaud[1], Pedro Montealegre[2(✉)], Rotem Oshman[3], Ivan Rapaport[4], and Ioan Todinca[5]

[1] CNRS and Université de Paris, Paris, France
[2] Universidad Adolfo Ibáñez, Santiago, Chile
p.montealegre@uai.cl
[3] Tel-Aviv University, Tel Aviv-Yafo, Israel
[4] DIM-CMM (UMI 2807 CNRS), Universidad de Chile, Santiago, Chile
[5] Université d'Orléans, Orléans, France

Abstract. In a distributed locally-checkable proof, we are interested in checking the legality of a given network configuration with respect to some Boolean predicate. To do so, the network enlists the help of a *prover*—a computationally-unbounded oracle that aims at convincing the network that its state is legal, by providing the nodes with certificates that form a distributed proof of legality. The nodes then verify the proof by examining their certificate, their local neighborhood and the certificates of their neighbors.

In this paper we examine the power of a *randomized* form of locally-checkable proof, called *distributed Merlin-Arthur protocols*, or dMA for short. In a dMA protocol, the prover assigns each node a short certificate, and the nodes then exchange *random messages* with their neighbors. We show that while there exist problems for which dMA protocols are more efficient than protocols that do not use randomness, for several natural problems, including Leader Election, Diameter, Symmetry, and Counting Distinct Elements, dMA protocols are no more efficient than standard nondeterministic protocols. This is in contrast with Arthur-Merlin (dAM) protocols and Randomized Proof Labeling Schemes (RPLS), which are known to provide improvements in certificate size, at least for some of the aforementioned properties.

Keywords: Distributed verification · Nondeterminism · Interactive computation · Interactive proof systems

1 Introduction

Nondeterminism is a fundamental concept in computer science. In particular, the class NP, introduced almost half a century ago [6], lies at the heart of computational complexity theory. Moreover, the P versus NP question is the largest unsolved problem in theoretical computer science.

© Springer Nature Switzerland AG 2019
K. Censor-Hillel and M. Flammini (Eds.): SIROCCO 2019, LNCS 11639, pp. 230–245, 2019.
https://doi.org/10.1007/978-3-030-24922-9_16

One way to define the class NP is as a computationally-efficient *proof system*: a language $\mathcal{L}$ is in NP if for any input x, a powerful but untrusted *prover* can convince a polynomial time *verifier* to accept whenever $x \in \mathcal{L}$, by providing the verifier with a *certificate* (a proof). However, if $x \notin \mathcal{L}$, not certificate will cause the verifier to accept.

This fundamental notion of nondeterminism (or polynomial time verification) was extended in the 90s to *interactive proof systems* [10,11], a model that allows *back-and-forth* interaction between the prover (*Merlin*) and the verifier (*Arthur*). This interaction gave the model tremendous power, equivalent to PSPACE [18,22].

Different *distributed* counterparts of the class NP have been introduced: locally checkable labelings [20], proof labeling schemes [16], non-deterministic local decision [8], and others. In all these models, roughly speaking, a powerful prover gives to every node $v \in V$ a certificate $c(v)$. This provides $G = (V, E)$ with a global distributed certificate. Then, every node v performs a *local verification* using its local information together with $c(v)$. Typically, the goal is to verify whether G belongs to a particular class of graphs (planar, bipartite, connected, k-colorable, etc.).

Very recently, these *distributed NP* models evolved—as already happened in the centralized setting almost thirty years ago—towards the study of *distributed interactive proofs* [14,19]. To state our results, let us recall some basic notions.

Distributed Languages. Let G be a simple connected n-node graph, let $x : V(G) \rightarrow \{0,1\}^*$ be a function assigning a label to every node of G, and let id $: V(G) \rightarrow \{1,\ldots,\text{poly}(n)\}$ be a one-to-one function assigning identifiers to the nodes. (The identifiers are $O(\log n)$-bit natural numbers.)

A *distributed language* is a Turing-Machine-decidable collection of triples (G, x, id), called *configurations*. In this paper, we are interested in the following distributed languages:

- LEADER $= \{(G, x, \text{id}) \mid x : V(G) \rightarrow \{0,1\} \text{ and } |\{v \in V(G) : x(v) = 1\}| = 1\}$, the language of graphs where every node is marked with a bit $x \in \{0,1\}$, and we require that exactly one node be marked 1.
- AMOS $= \{(G, x, \text{id})) \mid x : V(G) \rightarrow \{0,1\} \text{ and } |\{v \in V(G) : x(v) = 1\}| \leq 1\}$, the language of graphs where nodes are marked with a bit, and we require that *at most* one node be marked (AMOS stands for "at most one selected", and was introduced in [8]).
- DIAMETER$_{\leq k}$ $= \{(G, x, \text{id}) \mid diam(G) \leq k\}$, the language of graphs with diameter at most k.
- SYMMETRY $= \{(G, x, \text{id}) \mid G \text{ has a non-trivial automorphism}\}$. (An *automorphism* of a graph G is a one-to-one mapping $\phi : V(G) \rightarrow V(G)$ such that $\{u, v\} \in E(G) \iff \{\phi(u), \phi(v)\} \in E(G)$. It is *not-trivial* if it is not the identity function.)
- COUNT$_k$ $= \{(G, x, \text{id}) \mid x : V(G) \rightarrow \{0,1\}^* \text{ and } |\{x(u) : u \in V\}| = k\}$, the language of graphs where every node has an input $x(v) \in \{0,1\}^*$, and there are exactly k distinct inputs.

None of these languages refer to the node identifiers, but languages like

SPANNING TREE $= \Big\{(G, x, \text{id})) \mid \{\{\text{id}(v), x(v)\}, v \in V(G)\} \text{ forms a spanning tree of } G\Big\}$

do refer to the identifiers (here, $x(v)$ refers to the id of the parent of v in the tree).

In a locally-checkable proof, we ask a prover to provide the network nodes with a *certificate* that should convince them that $(G, x, \text{id}) \in \mathcal{L}$. The certificate is a function $c : V \to \{0, 1\}^*$ assigning to each $v \in V$ a label $c(v)$. The nodes exchange their certificates with their neighbors, examine their own input, and then decide whether to accept or reject; we require that $(G, x, \text{id}) \in \mathcal{L}$ iff there is some certificate c that causes all nodes to accept.

Formally, a deterministic distributed verification algorithm is specified as a collection of *decision functions*, $\mathsf{A} = \{\text{acc}_v\}_v$, where each function acc_v takes the ids, inputs and certificates of v and its neighbors, and outputs a decision whether to accept (1) or reject (0). We say that a $(G = (V, E), x, \text{id}, c)$ is *accepted by* A if for all $v \in V$ we have $\text{acc}_v(\{(\text{id}(u), c(u), x(v)) | u \in N[v]\}) = 1$.

A decision algorithm A *verifies* a distributed language $\mathcal{L}$ if, for every configuration (G, x, id),

$$(G, x, \text{id}) \in \mathcal{L} \iff \exists c : V(G) \to \{0, 1\}^* \mid (G, x, \text{id}, c) \text{ is accepted by } \mathsf{A}.$$

The *cost* of the algorithm A is the maximum number of bits assigned to any node in a certificate accepted by A, that is,

$$\max_{(G, x, \text{id}, c) \text{ accepted by } \mathsf{A}} \max_{v \in V} |c(v)|.$$

The class $\mathsf{LCP}(k)$, defined in [12], is the class of all distributed languages that have a distributed verification protocol with cost k. Other variants exist in the literature: *proof labeling schemes* [16] are defined similarly, except that at every node v, the verification algorithm does not take as input the data $x(u)$ of neighbors $u \in N(v)$, only the neighbors' certificates; *non-deterministic local decision*, defined in [8], is also similar, but the certificate c may not depend on the identifiers of the nodes (i.e., it is not used by the decision function).

Merlin-Arthur Protocols. *Merlin-Arthur (MA)* protocols extend locally-checkable proofs by allowing the nodes to use *randomness* when deciding whether to accept or reject. The prover remains nondeterministic, and it does not see the randomness of the nodes when choosing a certificate. After the prover assigns certificates to the nodes, each node randomly chooses a message, from a distribution specified by the protocol. This message is broadcast to all neighbors of the node, and then each node decides whether to accept or reject, based on its input and neighbors (including their ids), its certificate, and the messages it received from its neighbors.

Formally, an MA protocol is specified by two collections of functions, $\mathsf{A} = (\{\text{msg}_v\}_v, \{\text{acc}_v\}_v)$. After receiving a certificate assignment $c : V \to \{0, 1\}^*$, the protocol executes in two stages:

(1) Each node v generates a message $m(v)$, by calling the function msg_v, which takes as input $\text{id}(v)$, $\{\text{id}(u) : u \in N(v)\}$, $x(v), c(v)$, and a random string $r(v)$. The message $m(v)$ is broadcast to v's neighbors.

(2) Each node v uses the function acc_v to decide whether to accept or reject; acc_v takes as input $\mathsf{id}(v)$, $\{(\mathsf{id}(u), m(u)) : u \in N(v)\}$, $x(v), c(v), r(v)$.

For a given protocol A, the *acceptance probability* of (G, x, id, c) under A is the probability that all nodes accept the configuration (G, x, id) with certificate c. The probability here is taken over the nodes' internal randomness (the random strings $r(v)$).

A *Merlin-Arthur protocol* verifies a distributed language $\mathcal{L}$ with success probability $p \in (0, 1/2)$ if, for every configuration (G, x, id),

$$\begin{cases} (G, x, \mathsf{id}) \in \mathcal{L} \implies \exists c : V(G) \to \{0,1\}^* \mid \Pr[\mathsf{A} \text{ accepts } (G, x, \mathsf{id}, c)] \geq p \\ (G, x, \mathsf{id}) \notin \mathcal{L} \implies \forall c : V(G) \to \{0,1\}^*, \ \Pr[\mathsf{A} \text{ accepts } (G, x, \mathsf{id}, c))] \leq 1 - p. \end{cases}$$

A Merlin-Arthur protocol can be viewed as the *non-deterministic version* of randomized decision. It can also be viewed as the *randomized version* of locally checkable proofs (the randomized version of proof-labeling schemes has been considered in [3]).

The *cost* of an MA protocol is defined as the size of the longest certificate $c(v)$ accepted by a node v in any configuration on n nodes (the size may grow with n). (The standard definition of two-party MA protocols also charges for the communication between the players, which in our case corresponds to the messages $m(v)$. However, the lower bounds we prove apply even if the messages have unbounded length, as they depend more on the *local knowledge* of the nodes even after seeing the certificates.)

Given a distributed language $\mathcal{L}$, we define its Merlin-Arthur complexity, denoted $\mathsf{dMA}_p(\mathcal{L})$, as the minimum cost of a Merlin-Arthur protocol that decides $\mathcal{L}$ with success probability p.

Note that our definition above does not provide node v with the inputs and neighborhoods of its neighbors; this is similar to proof-labeling schemes (although we also provide ids), and dissimilar to locally-checkable proofs. However, it is easy to modify our lower bounds so that the view of a node is the same as it would be in a locally-checkable proof, except that instead of seeing the certificates of its neighbors, it only sees the messages they generated.

Comparison with Other Randomized One-Round Models of Verification. Let us point out how dMA protocols relate to two other models.

In an *Arthur-Merlin* distributed decision protocol (or dAM for short) [14], each node v sends a random string to the prover, and the prover responds by providing each node with a certificate (which can depend on the random strings of all the nodes). Each node then makes its decision based on its own randomness, its neighborhood, and its neighbors' certificates. The order of interaction is the opposite of dMA schemes, where the prover first commits to the certificates, and then the nodes send random messages. As we show in this paper, this reverse order gives dAM protocols more power than dMA protocols, at least in some scenarios.

Another related model is *randomized proof labeling schemes* (RPLS) [3]. These are very similar to dMA protocols, except that the certificate size is *unbounded*, and the protocol is only charged for the randomized messages the players send to each other. It was shown in [3] that any property admits an RPLS that is exponentially cheaper than the best proof labeling scheme; however, the construction in [3] not only does not reduce the certificate size, it in fact blows it up, by a factor of up to n. We show in this paper that this is inherent: if we *do* care about the certificate size, then randomness does not always help.

1.1   Our Results

Both AMOS and LEADER have proof-labeling schemes using certificates on $O(\log n)$ bits. (A tree rooted at the leader if any, or at an arbitrary node otherwise, suffices.) The next result shows that one cannot do better, even using randomization for the verification part.

Theorem 2.2. *Any 2-sided error* dMA *protocol for* AMOS *with success probability larger than* 4/5 *requires certificates on* $\Omega(\log n)$ *bits. Any 1-sided error* dMA *protocol for* AMOS *requires certificates on* $\Omega(\log n)$ *bits. The same result holds for* LEADER.

In contrast, whenever randomization is used *before* interacting with the prover, AMOS can be decided with certificates on $O(1)$ bits.

Theorem 2.1. *For every* $k \geq 1$, *there exists a* dAM *protocol for* AMOS *with success probability* $1 - 1/2^k$, *using* $(k+1)$-*bit certificates at each node.*

This shows that the gap between dAM and dMA (with success probability $\geq 4/5$) is potentially unbounded. Next, we show that a certain class of reductions from 2-party communication complexity can be adapted to show dMA lower bounds as well. As a consequence, we obtain lower bounds on DIAMETER, SYMMETRY, and COUNT.

Corollary 3.1. *Let* $0 \leq \varepsilon < 1/3$. *Then,* $\mathsf{dMA}_{1-\varepsilon}(\mathrm{DIAMETER}_{\leq 6}) = \Omega(n/\log n)$. *That is, every Merlin-Arthur protocol with success probability at least* $1 - \varepsilon$ *that is able to decide whether the diameter of the input graph is at most 6 requires certificates on* $\Omega(n/\log n)$ *bits.*

Corollary 3.2. *Let* $0 \leq \varepsilon < 1/3$. *Then,* $\mathsf{dMA}_{1-\varepsilon}(\mathrm{SYMMETRY}) = \Omega(n^2)$.

Corollary 3.3. *Let* $0 \leq \varepsilon < 1/3$. *Then,* $\mathsf{dMA}_{1-\varepsilon}\left(\mathrm{COUNT}_{n/2+1}\right) = \Omega(n)$.

Our lower bounds are shown by adapting existing tools for proving lower bounds on locally-checkable proofs and in CONGEST, thus showing that some types of lower bounds extend easily to dMA.

1.2   Related Work

This paper is very much related to two recent contributions on distributed interactive proofs. The concept of distributed interactive proofs was introduced in [14]. Among other results, [14] proves that SYMMETRY admits a dMAM protocol with $O(\log n)$-bit certificates, and a dAM protocol with $O(n \log n)$-bit certificates. Moreover, it is also proved that any dAM protocol for SYMMETRY requires certificates on $\Omega(\log \log n)$ bits. Graph non-isomorphism has also been studied in [14]—every node is given the adjacency list of a node in some graph H, and the nodes have to collectively decide whether the actual network G is isomorphic to H. It is proved that this problem admits a dAMAM protocol with certificates on $O(n \log n)$ bits.

The recent paper [19] carried on the investigations in [14]. In particular, [19] proves that NON-SYMMETRY can be decided by a dAMAM protocol with $O(\log n)$-bit certificates. It is also proved, using general reductions from circuit computation, that graph non-isomorphism can be decided by an interactive protocol with a constant number of interaction rounds between Arthur and Merlin, and certificates on $O(\log n)$ bits. Another variant of graph non-isomorphism is also considered in [19]—every node is given two subsets of incident edges, and the nodes have to collectively decide whether the resulting subgraphs of the actual network G are isomorphic. It is proved that this problem admits a dAMAM protocol with certificates on $O(\log n)$ bits.

Problem DIAMETER$_{\leq k}$ has been studied, in the framework of distributed verification algorithms, in [5]. More precisely, in the proof-labeling scheme model, the authors show, for the certificate size, an upper bound of $O(n \log n)$ and a lower bounds of $\Omega(n/k)$. They manage to improve the previous upper bound by introducing approximation ([5] defines *approximate* proof-labeling schemes).

2   Warmup: Deciding AMOS and LEADER

As a warm-up, let us consider the distributed language AMOS, for "at most one selected", introduced in [8]. Recall that for every configuration (G, x, id), we have $(G, x, \mathsf{id}) \in$ AMOS if and only if $x(v) \in \{0, 1\}$ for every $v \in V(G)$ and $|\{v \in V(G) : x(v) = 1\}| \leq 1$. A node v with $x(v) = 1$ is said to be *selected*. This language is therefore similar to LEADER, apart from the fact that having no leader is a legal configuration.

It is shown in [8] that AMOS cannot be decided *deterministically* in sublinear time without a prover, as a configuration with two selected nodes that are at distance $n - 1$ from one another cannot be detected. On the other hand, using randomization (but still without a prover), AMOS can be decided in zero rounds with success probability $p = (\sqrt{5} - 1)/2$: every selected node accepts with probability p, and the non-selected nodes all accept. A legal configuration is accepted with probability exactly p, while an illegal one is accepted with probability at most $p^2 = 1 - p$. In fact, [8] shows that p is the best success probability possible for a sublinear-time randomized algorithm.

A locally checkable proof for AMOS can simply be designed using certificates on $O(\log n)$ bits. On a legal instance, every node is given a pointer to a neighbor, on $O(\log n)$ bits, such that the set of all pointers encodes a spanning tree T rooted at an arbitrary node if there are no selected nodes, and rooted at the selected node otherwise. The certificate also includes $O(\log n)$ bits forming a distributed proof that T is indeed a spanning tree (see [16]). The verification algorithm consists, for every node v, to check that T is indeed a spanning tree. In addition, a node with $x(v) = 1$ that is not the root of T rejects. It was shown in [12] that $O(\log n)$-bit certificates is the best that can be achieved, that is, there is no locally checkable proofs for AMOS with certificates on $o(\log n)$ bits.

Remark 2.1. *With the previous example we can see the power of the* dMA *model in comparison with proof labelling schemes and randomized local decision. Suppose that we want to decide* AMOS∩BIPARTITE *(i.e., whether the input is a bipartite graph with at most one selected node). We can combine a one-bit certificate (for bipartiteness) with local randomness (for at-most-one-selected) in order to get a one-bit Merlin-Arthur protocol for* AMOS ∩ BIPARTITE *with probability of success at least* $\frac{\sqrt{5}-1}{2}$.

The following result is a simple illustration of the power of Arthur-Merlin protocols, by showing that one can design an Arthur-Merlin protocol for AMOS with success probability as close to 1 as desired, with certificates on $O(1)$ bits. For LEADER, we refer to [19] which describes a dMAM protocol using $O(1)$-bit certificates, but with one more interaction between Arthur and Merlin.

Theorem 2.1. *For every* $k \geq 1$, *there exists a* dAM *protocol for* AMOS *with success probability* $1 - 1/2^k$, *using* $(k + 1)$-*bit certificates at each node.*

Proof. Let $k \geq 1$. Every node picks k bits at random. On a legal instance, and given these k random bits at each node, Merlin sends -1 to every node if there are no selected nodes, and otherwise sends the bit string randomly selected by the selected node. The verification algorithm is as follows. Every node checks that the certificate given by Merlin is the same as the one given to its neighbors. If this test is passed, then a non-selected node systematically accepts, and a selected node accepts only if the bit string sent by Merlin is identical to the one it randomly generated. If there are more than one selected nodes, the probability that they all pick the same random string is at most $1/2^k$, thus the verification succeeds with probability at least $1 - 1/2^k$.   □

In contrast, the following results illustrates the limitation of Merlin-Arthur protocols, by showing that such protocols cannot achieve success probability much larger than $\frac{\sqrt{5}-1}{2} = 0.61\ldots$ whenever using certificates on $o(\log n)$ bits.

Theorem 2.2. *Any 2-sided error* dMA *protocol for* AMOS *with success probability larger than* 4/5 *requires certificates on* $\Omega(\log n)$ *bits. Any 1-sided error* dMA *protocol for* AMOS *requires certificates on* $\Omega(\log n)$ *bits. The same result holds for* LEADER.

Proof. The intuition of the proof is simple. Consider a configuration $I_1 \in$ AMOS consisting of an n-node cycle with a unique selected node v. Let us then take two copies of I_1, remove the edge e opposite to v in both, and create a cycle with $2n$ nodes by glueing the two resulting paths. Let us call this latter configuration I_2. We have $I_2 \notin$ AMOS. Let us consider a dMA protocol $\mathcal{P}$ for AMOS with success probability larger than $2/3$. We have $\Pr[\mathcal{P}$ accepts $I_1] > 2/3$ with the appropriate certificate assignment c to the nodes of I_1, and $\Pr[\mathcal{P}$ rejects $I_2] > 2/3$ for every certificate assignment to the nodes of I_2. On the other hand, for the certificate assignment c, since the nodes have the same view in I_1 and I_2, as far as the certificates are concerned, we get, by the union bound, that $\Pr[\mathcal{P}$ rejects $I_2] < 1/3 + 1/3 = 2/3$, yielding a contradiction. There is however a gap between this intuition and a correct proof. In particular, as nodes have identities, one cannot claim that the extremities of the removed edge e do not "see" the difference between I_1 and I_2. glueing legal instances to create illegal instances in which the nodes cannot distinguish which one they belong to requires some more work.

The sophisticated glueing technique introduced in [12] allowed Göös and Suomela to show that there is no locally checkable proof for AMOS and LEADER with certificates of size $o(\log n)$ bits. This glueing technique can also be used to prove that the same result holds for dMA protocols with success probability larger than $4/5$. To see why, let us first briefly summarize the construction in [12].

Let n be even, and let us consider an arbitrary partition of $\{1, \ldots, n^2\}$ of the form $(A_i, B_i)_{i \in \{1, \ldots, n\}}$ such that $\{1, \ldots, n^2\} = (\cup_{i=1}^n A_i) \cup (\cup_{i=1}^n B_i)$, where $|A_i| = |B_i| = n/2$ for every $i \in \{1, \ldots, n\}$. The elements of A_i are enumerated as $A_i[1], \ldots, A_i[n/2]$ for every $i \in \{1, \ldots, n\}$, and the same for every B_i. Let $\mathbf{A} = \{A_i, i = 1, \ldots, n\}$ and $\mathbf{B} = \{B_i, i = 1, \ldots, n\}$.

Given $(A, B) \in \mathbf{A} \times \mathbf{B}$, let $R_{A,B}$ be the n-node ring $(v_1, \ldots, v_n)$, where $\mathsf{id}(v_i) = A[i]$ for $i = 1, \ldots, n/2$, and $\mathsf{id}(v_{n-i+1}) = B[i]$ for $i = 1, \ldots, n/2$. For every node v in the ring, let $\ell_{A,B}(v) \in \{0, 1\}$ be its input label, specifying whether v is selected or not. Assume that only one node is selected in each R_{A_i, B_j} for $i, j \in \{1, \ldots, n\}$, and that this node is at distance at least 2 from the nodes v_{n-1}, v_n, v_1, v_2, with respective identities $B_j[2], B_j[1], A_i[1], A_i[2]$, which form a path of length 4 in R_{A_i, B_j}.

For $(A, B) \in \mathbf{A} \times \mathbf{B}$, let $c_{A,B}(v)$ be the certificates assigned to the nodes of $R_{A,B}$ with such a unique selected node, leading all nodes to accept, with probability $> 4/5$. Finally, for every node v, let $L_{A,B}(v) = (\ell_{A,B}(v), c_{A,B}(v))$, and set

$$L_{A,B} = (L_{A,B}(v_{n-2}), L_{A,B}(v_{n-1}), L_{A,B}(v_n), L_{A,B}(v_1), L_{A,B}(v_2), L_{A,B}(v_3)).$$

Let us consider the complete bipartite graph $K_{n,n}$ with bipartitions $\mathbf{A}$ and $\mathbf{B}$, and let us color every edge $\{A, B\}$, $(A, B) \in \mathbf{A} \times \mathbf{B}$, with $L_{A,B}$. Since $L_{A,B}$ is on $o(\log n)$ bits, it can be shown that the colored $K_{n,n}$ contains a monochromatic 4-cycle. Let (A_1, B_1, A_2, B_2) be such a cycle. The two n-node rings R_{A_1,B_1} and R_{A_2,B_2} are then glued to form a $2n$-node ring S by removing the edge $\{v_n, v_1\}$ in both n-node rings, and connecting the copy of v_1 in one ring to the copy of v_n in the other ring. Note that there are two selected nodes in the ring S.

Since $L_{A_1,B_1} = L_{A_2,B_1} = L_{A_2,B_2} = L_{A_1,B_2}$, no nodes can distinguish whether they are in one of the four small (legal) rings R_{A_i,B_j}, $i,j \in \{1,2\}$, or in the large (illegal) ring S.

We are now ready to apply the intuition provided at the beginning of the proof to the construction in [12]. Let us consider a dMA protocol $\mathcal{P}$ for AMOS with success probability larger than $4/5$. For every $i,j \in \{1,2\}$, we have $\Pr[\mathcal{P}$ accepts $R_{A_i,B_j}] > 4/5$, with the appropriate certificate assignment $c_{i,j}$ given to the nodes of R_{A_i,B_j}. Also, $\Pr[\mathcal{P}$ rejects $S] > 4/5$ for every certificate assignment to the nodes of S. However, consider S with the certificate assignment c consisting in giving the certificates defined by $c_{i,i}$ to the nodes coming from R_{A_i,B_i} in S, for $i = 1, 2$. By union bound, we have

$\Pr[\exists v \in S : \mathcal{P}$ rejects at v with certificate $c(v)]$

$\qquad \leq \Pr[\exists v \in \{v_4, \ldots, v_{n-3}\} : \mathcal{P}$ rejects at v in R_{A_1,B_1} with certificate $c_{1,1}]$

$\qquad + \Pr[\exists v \in \{v_4, \ldots, v_{n-3}\} : \mathcal{P}$ rejects at v in R_{A_2,B_2} with certificate $c_{2,2}]$

$\qquad + \Pr[\exists v \in \{v_{n-2}, v_{n-1}, v_n, v_1, v_2, v_3\} : \mathcal{P}$ rejects at v in R_{A_2,B_1} with certificate $c_{2,1}]$

$\qquad + \Pr[\exists v \in \{v_{n-2}, v_{n-1}, v_n, v_1, v_2, v_3\} : \mathcal{P}$ rejects at v in R_{A_1,B_2} with certificate $c_{1,2}]$.

Each of the four terms on the right hand side of the equation above is smaller than $1/5$. It follows that, with the certificate assignment c, we have $\Pr[\exists v \in S : \mathcal{P}$ rejects at $v] < 4/5$, which contradicts the fact that the success probability of $\mathcal{P}$ is larger than $4/5$.

The proof above applies to LEADER as well since all legal configurations considered in the proof have exactly one selected node, and all illegal configurations have exactly two selected nodes. For both AMOS and LEADER, the proof also applies to 1-sider error protocols, since, for such protocols, the union bound yields $\Pr[\exists v \in S : \mathcal{P}$ rejects at $v] = 0$, that is, $\mathcal{P}$ is incorrect with probability 1 for S with certificate c. $\qquad \square$

Remark 2.2. *Both* LEADER *and* AMOS *have locally checkable proofs with 2-bit certificates, whenever restricted to trees. Indeed, for* LEADER, *the certificate at every node v in a legal instance consists of the distance of v to the leader in the tree, modulo 3. The same for* AMOS, *apart that, if there is no leader, then the distance is from an arbitrary node of the tree. Such certificates enable to identify a unique root of the tree, which is the only node allowed to be leader (it must be selected in* LEADER, *but do not need to be selected in* AMOS).

3 The Canonical 2-Party Reduction

In this section we show that a widely-used class of reductions from 2-party communication complexity, which is typically used to prove lower bounds in CONGEST, also yields lower bounds on dMA. These reductions are typically used to relate the round complexity of a deterministic or randomized algorithm in CONGEST to the deterministic or randomized communication complexity of some 2-party problem, but here we use them as reductions from *nondeterministic* communication complexity.

Let $\mathcal{L}_{comm}$ be a two player communication complexity language with instances of the form $(x, y) \in X \times Y$, where both X and Y are finite sets. Let $\mathcal{L}_{dist}$ be a distributed language. We consider in this section distributed languages that represent "pure graph properties". Therefore, the instances are of the form (G, id), where G is a graph and id is the list of the identifiers of the nodes. In fact, for simplicity, we are going to consider the instances as being just graphs (and the ids will be fixed). In other words, a distributed interactive protocol $\mathcal{P}_{dist}$ that solves $\mathcal{L}_{dist}$, needs to implicitly answer whether $G \in \mathcal{L}_{dist}$.

A reduction from $\mathcal{L}_{comm}$ to $\mathcal{L}_{dist}$ is an explicit transformation of instances (x, y) of $\mathcal{L}_{comm}$ into instances $G_{x,y}$ of $\mathcal{L}_{dist}$ such that $(x, y) \in \mathcal{L}_{comm}$ if and only if $G_{x,y} \in \mathcal{L}_{dist}$. If the reduction is such that $G_{x,y} \in \mathcal{L}_{dist}$ has the specific structure we are going to define in the sequel, we say that the reduction is *canonical*. We consider here only reductions that generate graphs over a fixed set $V = \{1, \ldots, n\}$ of nodes, for any specific n.

The definition below captures "clean-cut" reductions where each player "owns" part of the graph, with a fixed cut between the two parts. Many reductions in the literature have this structure, or can be easily modified to have it.

Definition 3.1. *Let $s : \mathbb{N} \to \mathbb{N}$ be a computable function. A reduction from $\mathcal{L}_{comm}$ to $\mathcal{L}_{dist}$ is said to be s-canonical if there is some fixed partition $V = (V_1, V_2)$ of the node set of the graph, such that for all $(x, y) \in X \times Y$,*

- *The neighborhood of any node in V_1 in $G_{x,y}$ does not depend on y, and the neighborhood of any node in V_2 in $G_{x,y}$ does not depend on x.*
- *Consider the cut $E(V_1, V_2) = \{\{u, v\} \in E(G_{x,y}) : u \in V_1, v \in V_2\}$. Let V_c be the vertices of the cut (i.e., endpoints of edges in the cut). Then V_c does not depend on either x or y, and $|V_c| \leq s(n)$.*

Nondeterministic Communication Complexity. A 2-party nondeterministic protocol Π is modelled as a collection $\Pi = \{\Pi_c\}_{c \in \{0,1\}^\ell}$ of *deterministic* protocols. On inputs x, y, the protocol begins with the prover presenting Alice and Bob with a *proof* $c \in \{0,1\}^\ell$; the players then execute the protocol Π_c corresponding to the proof c. The *cost* of Π is defined to be $\ell + \max_c |\Pi_c|$, where $|\Pi_c|$ is the worst-case number of bits sent by Π_c on any input.

The protocol Π *solves* $\mathcal{L}_{comm}$ if, for any input (x, y), we have $(x, y) \in \mathcal{L}_{comm}$ iff there exists a proof $c \in \{0,1\}^\ell$ such that Π_c accepts (x, y).

We denote by $\mathsf{N}(\mathcal{L}_{comm})$ the nondeterministic cost of solving $\mathcal{L}_{comm}$, i.e., the cost of the best nondeterministic protocol that solves $\mathcal{L}_{comm}$. It is known, for example, that DISJOINTNESS has nondeterministic cost $\Omega(n)$.

Theorem 3.1. *If there exists an s-canonical reduction from $\mathcal{L}_{comm}$ to $\mathcal{L}_{dist}$, then, for every $\varepsilon < 1/3$,*

$$\mathsf{dMA}_{1-\varepsilon}\left(\mathcal{L}_{dist}\right) = \Omega\left(\frac{\mathsf{N}(\mathcal{L}_{comm})}{s(n)}\right).$$

Proof. Consider a dMA protocol $\mathcal{P}$ that solves $\mathcal{L}_{dist}$ with success probability at least $1 - \varepsilon$ and using $p(n)$-bit certificates. Our goal is to show that $p(n) = \Omega\left(\frac{N(\mathcal{L}_{comm})}{s(n)}\right)$, by constructing a nondeterministic protocol Π for $\mathcal{L}_{comm}$ with communication cost $O(p(n) \cdot s(n))$.

On input (x, y), the protocol Π proceeds as follows:

(1) Alice (resp. Bob) locally constructs $G_{x,y}[V_1]$ from x (resp., $G_{x,y}[V_2]$ from y). Note that both players agree on the neighborhoods of the cut nodes V_c, because the reduction is canonical: these nodes' neighborhoods do not depend on either x or y.
(2) The prover presents Alice and Bob with a proof $\pi \in \{0, 1\}^{p(n) \cdot s(n)}$, which the players interpret as an assignment of certificates to the cut nodes V_c.
(3) Alice (resp. Bob) enumerates over all possible assignments of $p(n)$-bit certificates to the nodes in $V_1 \setminus V_c$ (resp. $V_2 \setminus V_c$), and checks whether there is an assignment that, together with the certificates π of the cut nodes, causes all nodes of V_1 (resp. V_2) to jointly accept with probability at least $1 - \varepsilon$.
(4) The players inform each other whether they can find such an assignment. The players accept iff both were able to find some assignment that makes all nodes in V_1 (resp. V_2) accept.

Note that both Alice and Bob can perform step (3) above without need of communication: after fixing the certificates π of the nodes V_c on both sides of the cut, the acceptance probability of any node in V_1 does not depend on y, and vice-versa. This is because the neighborhood of any node in V_1 does not depend on y, and vice-versa.

Clearly, the cost of Π is $s(n) \cdot p(n) + 2$. It remains to prove its correctness:

- Suppose that (x, y) is a YES-instance of $\mathcal{L}_{comm}$. We are going to show the existence of a certificate $\tilde{c}$ that causes both Alice and Bob to accept.

  By definition of the reduction, $G_{x,y}$ is a YES-instance of $\mathcal{L}_{dist}$, so there exist certificates C to the nodes of $G_{x,y}$ such that, with probability at least $1 - \varepsilon$, all nodes accept. Let π be the restriction of the certificates to the nodes of V_c. In Π, the prover can give π to the players, causing them to accept: when enumerating over all possible certificates, Alice and Bob will each find the restriction of C to the nodes on their side of the graph (V_1 and V_2, respectively), and since C causes *all* nodes to accept w.p. $\geq 1 - \varepsilon$, in particular it causes all nodes of V_1 (resp. V_2) to accept w.p. $\geq 1 - \varepsilon$.
- Suppose that (x, y) is a NO-instance of $\mathcal{L}_{comm}$. We need to show that there is no certificate π that can be given to Alice and Bob to cause them to accept.

  Suppose for the sake of contradiction that there is such a certificate π, and let C_x, C_y be the extensions of π to the nodes of V_1 (resp. V_2) that cause them all to accept with probability at least $1 - \varepsilon$. Now consider the global certificate assignment $C = (C_x, C_y)$ where in the distributed dMA protocol $\mathcal{P}$, the prover assigns C_x to the nodes of V_1 and C_y to the nodes of V_2. By the union bound, when assigned C, the probability that either some node in V_1 or some node in V_2 (or both) reject is at most 2ε. Overall, we see that the proof is accepted by all nodes with probability at least $1 - 2\varepsilon > 1 - 2 \cdot (1/3) = 1/3$, which is a contradiction, because $G_{x,y} \notin \mathcal{L}_{comm}$. □

3.1 Lower Bound on DIAMETER

It is known that, for every $k \geq 1$, DIAMETER$_{\leq k} \in$ LCP$(O(n \log n))$, i.e., has a locally checkable proof—actually, a proof-labeling scheme—using certificates on $O(n \log n)$ bits [5]. (For this certificate, the prover constructs a BFS tree from every node of the graph.) We show that allowing randomization in the verification of the proof does not help.

Let DISJ be the two-player problem the players receive sets $x, y \subseteq [n]$, and their goal is to accept iff $x \cap y = \emptyset$.

Our canonical reduction from DISJ to DIAMETER$_{\leq 6}$ is a simple modification of a reduction of Censor-Hillel, Khoury and Paz [4]. The reduction of [4] mostly has the static structure required for a canonical reduction, and it has a sparse cut, of size $s(n) = O(\log n)$; however, it is not $O(\log n)$-canonical, only because the neighborhoods of the cut nodes may depend on x or on y. This is easily solved by replacing each edge in the cut by a path of length 3 (subdividing the edge by inserting two auxiliary nodes). Let $G_{x,y}$ be the resulting graph. After this modification, (x, y) are disjoint iff the diameter of $G_{x,y}$ is at most 6, and the new reduction is $O(\log n)$-canonical. Thus, we obtain:

Lemma 3.1. *There exists an* $O(\log(n))$-*canonical reduction from* DISJ *to* DIAMETER$_{\leq 6}$.

By Theorem 3.1, we have:

Corollary 3.1. *Let* $0 \leq \varepsilon < 1/3$. *Then,* dMA$_{1-\varepsilon}$(DIAMETER$_{\leq 6}$) $= \Omega(n/\log n)$.

The proof uses the fact that N(DISJ) $= \Omega(n)$ (see, e.g., the textbook [13]).

3.2 Lower Bound on SYMMETRY

It is known that SYMMETRY is among the most difficult graph properties to verify in a distributed manner, in the sense that every locally checkable proof for SYMMETRY requires certificates on $\Omega(n^2)$ bits [12], while *all* distributed languages on n-node graphs can be verified using a certificate on $O(n^2)$ bits at each node [16]. We show that allowing randomization in the verification of the proofs does not help.

We extend the SYMMETRY lower bound of [12] to dMA. The lower bound in [12] is not formally stated as a reduction; it essentially "re-proves" the 2-party nondeterministic lower bound for EQUALITY. By observing that this lower bound *is* in fact a canonical reduction, we obtain a dMA lower bound.

Let EQ$_\mathcal{D}$ be the two-player communication language where the players receive inputs $x, y \in \mathcal{D}$, and their goal is to output 1 iff $x = y$. Here, $\mathcal{D}$ is some domain of size N, which, following [12], we take to be a set of equivalence classes of all n-node asymmetric graphs, under the isomorphism equivalence relation. It is known that $|\mathcal{D}| = 2^{\Theta(n^2)}$ [7].

Let SYMMETRY be the distributed language defined on the set of all graphs, where the YES-instances are graphs having non-trivial automorphisms. Obviously, all the graphs in $\mathcal{D}$ are NO-instances of SYMMETRY.

Theorem 3.2 ([12], re-phrased). *There exists a 2-canonical reduction from* EQ$_\mathcal{D}$ *to* SYMMETRY *that transforms instances* $(G_x, G_y) \in \mathcal{D}^2$ *into graphs* $G_{x,y}$ *of size* $2n + 2$.

For completeness, we repeat the argument of [12], and show that it is a 2-canonical reduction:

Proof. Let $V_1 = \{1, \ldots, n+1\}$, $V_2 = \{n+2, \ldots, 2n+2\}$. On inputs G_x, G_y, Alice and Bob construct the following graph: Alice constructs a copy of some graph in the equivalence class G_x over the nodes $\{1, \ldots, n\}$, and Bob constructs a copy of some graph in G_y over nodes $\{n+3, \ldots, 2n+2\}$. In addition, Alice connects node $n+1$ to node n, Bob connects node $n+2$ to $n+3$, and "both players" add the edge $\{n+1, n+2\}$.

The reduction is 2-canonical because there is only one edge in the cut. Correctness follows from the fact that, since G_x and G_y contain only asymmetric graphs, and they are equivalence classes of the isomorphism relation, the resulting graph $G_{x,y}$ is symmetric iff $G_x = G_y$. $\square$

Since the nondeterministic cost of EQ$_\mathcal{D}$ is $|\mathcal{D}|$ [13], we obtain:

Corollary 3.2. *Let* $0 \le \varepsilon < 1/3$. *Then,* dMA$_{1-\varepsilon}$(SYMMETRY) $= \Omega(n^2)$.

3.3   Lower Bound on COUNT

Finally, we observe that the notion of a canonical reduction is easily extended to languages where the nodes have input in addition to the graph: to do this, we require a transformation from the communication problem $\mathcal{L}_{comm}$ to configurations $(G_{x,y}, d)$ (keeping the ids fixed, as before), such that $(x, y) \in \mathcal{L}_{comm}$ iff $(G_{x,y}, d) \in \mathcal{L}_{dist}$. We require all the conditions from the previous section; moreover, the *input* $d(u)$ of any neighbor $u \in N(v)$ for $v \in V_1$ (resp. V_2) may not depend on y (resp. x). With this additional restriction, Theorem 3.1 continues to hold.

For example, consider the problem of counting the number of distinct elements in the input. Cast as a decision problem, we define it as COUNT$_k = \{(G = (V, E), d) : |\{d(u) : u \in V\}| = k\}$.

In [21], Patt-Shamir showed by reduction from DISJ that counting the number of distinct elements in the input of an n-node network requires $\tilde{\Omega}(n)$ rounds in CONGEST, even if randomization is allowed. A similar argument was used in [2] to show that streaming algorithms for counting the number of distinct elements require linear memory (indeed, [2] shows that this holds either for randomized exact algorithms, or for deterministic approximate algorithms). Implicitly, the argument of [2] shows that the nondeterministic cost of DISJ with input sets of size $n/4$, and with the promise that either $x \cap y = \emptyset$ or $|x \cap y| \geq n/100$, is $\Omega(n)$.

The reduction of [21] is "almost" 2-canonical. We modify it slightly to make it 2-canonical; this involves restricting the size of the input sets, and fixing the input of the cut nodes.

Lemma 3.2. *There is a 2-canonical reduction from* DISJ *with sets of size $n/4$ to* COUNT$_{n/2+1}$ *in networks of size $n/2 + 2$.*

Proof. The modified reduction features a line network of $n/2 + 2$ nodes, $1, \ldots, n/2 + 2$, with Alice controlling nodes $1, \ldots, n/2 + 1$ and Bob controlling nodes $n/2 + 2, \ldots, n/2 + 2$. Nodes $n/2 + 1, n/2 + 2$, which are the cut nodes, always receive $\perp$ as their input (where $\perp$ is some fixed element that is not in the universe of DISJ). Let $x = \{x_1, \ldots, x_{n/4}\}, y = \{y_1, \ldots, y_{n/4}\}$ be the inputs of Alice and bob. Alice assigns each node i the input x_i, and Bob assigns each node $n/2 + 1 + j$ the input y_j.

If $x \cap y = \emptyset$, then the total number of distinct elements in the input is $|x| + |y| + 1 = n/2 + 1$, whereas if $x \cap y \neq \emptyset$, the number of distinct elements is smaller. □

For deterministic algorithms, using the argument from [2], this can be extended to a sufficiently small constant approximation (e.g., $1 \pm 1/100$).

We obtain:

Corollary 3.3. *Let $0 \leq \varepsilon < 1/3$. Then,* dMA$_{1-\varepsilon}$ $(\text{COUNT}_{n/2+1}) = \Omega(n)$.

Let us make two further remarks about verifying the approximate number of distinct elements in line networks. First, there is an $O(\log n)$-bit dAM scheme for this problem: we can simulate the execution of the streaming algorithm from [2], which uses $O(\log n)$ bits of randomness and $O(\log n)$ bits of memory, and gives a constant approximation. In the simulation, the first node in the line sends the prover $O(\log n)$ bits of randomness r, which serve to specify a pairwise-independent hash function in [2]. The prover responds by sending r to all the nodes, and also, it tell each node i the state of the streaming algorithm of [2] after processing the inputs of the first i nodes, using the hash function indicated by r. The nodes verify that they all received the same value of r, and also that, if node i received state s_i and node $i + 1$ received state s_{i+1}, then indeed, with randomness r, the algorithm of [2] transitions from state s_i to state s_{i+1} upon processing the input of node i. This idea can be extended to arbitrary networks, by using *mergeable sketches* [1], asking the prover to specify a spanning tree, and "summing" the sketches up the tree.

Next, we observe that $\Omega(\log n)$ is in fact a lower bound on the dAM-cost of computing the *exact* number of elements in a network. This can be shown by the following argument, which is very similar to a recent $\Omega(\log n)$ lower bound for the dAM-cost of SYMMETRY [15].

Theorem 3.3. *We have* dAM$(\text{COUNT}_{n/2}) = \Omega(\log n)$.

Proof. Given an ℓ-bit dAM protocol for COUNT$_{n/2}$, we construct a $2^{O(\ell)}$-bit, private-coin, randomized two-party protocol for DISJ with sets of size $n/4$ (without a prover). Since DISJ requires $\Omega(n)$ bits of communication, we conclude that $\ell = \Omega(\log n)$.

The protocol proceeds as follows: given inputs (x, y), the players construct the network from Lemma 3.2, but they use only $n/2$ nodes in total (with each player responsible for $n/4$ nodes), and omit the input $\bot$ (it is not necessary here). Then, Alice and Bob each sample a private random string r_A, r_B (respectively). We say that a pair of ℓ-bit certificates c, c' is r_A -*good* if there is an assignment of certificates to all the nodes in Alice's side, where the cut nodes receive the certificates c and c' (respectively), such that when their randomness is r_A (here, r_A represents a list of the random string of each node on Alice's side), all nodes on Alice's side accept when their randomness is r_A. Similarly, we say that c, c' is r_B -*good* if the same holds for Bob's side with randomness r_B. Alice and Bob announce to each other the list of pairs c, c' that are r_A-good and r_B-good, respectively. This requires $2^{2\ell}$ bits. Finally, the players accept iff there is some pair c, c' that is good for both players.

It is easy to verify (see, e.g., [14]) that the probability that the players accept is exactly the probability that a prover has of convincing all nodes of the network to accept. Therefore, the protocol correctly solves DISJ. □

As a final remark, the argument above also yields an $\Omega(\log \log n)$ lower bound on the dAM cost of deciding whether the number of distinct elements is $(1 \pm 1/100)k$, for $k = \Theta(n)$. We use the same reduction, but reduce from the gap version of DISJ, where it is promised that either $x \cap y = \emptyset$ or $|x \cap y| \geq n/100$. This problem has randomized private-coin communication complexity $\Omega(\log n)$ (as its deterministic cost is $\Omega(n)$ [2], and the private-coin randomized cost of a problem is never exponentially better than its deterministic cost [13]). Interestingly, our *upper bound* of $O(\log n)$ on approximating the number of distinct elements could be improved to $O(\log \log n)$, if the nodes had shared randomness. We could then simulate the famous Flajolet-Martin streaming algorithm [9], which assumes perfectly random hash functions, and requires $O(\log \log n)$ bits of memory.

Acknowledgements. Partially supported by CONICYT PIA/Apoyo a Centros Científicos y Tecnológicos de Excelencia AFB 170001 (P.M. and I.R.), Fondecyt 1170021 (I.R.) and CONICYT via PAI + Convocatoria Nacional Subvención a la Incorporación en la Academia Año 2017 + PAI77170068 (P.M.). Rotem Oshman is supported by ISF i-core Center for Excellence, No. 4/11.

References

1. Agarwal, P.K., Cormode, G., Huang, Z., Phillips, J.M., Wei, Z., Yi, K.: Mergeable summaries. ACM Trans. Database Syst. **38**(4), 26:1–26:28 (2013)
2. Alon, N., Matias, Y., Szegedy, M.: The space complexity of approximating the frequency moments. J. Comput. Syst. Sci. **58**(1), 137–147 (1999)

3. Baruch, M., Fraigniaud, P., Patt-Shamir, B.: Randomized proof-labeling schemes. In: 34th ACM Symposium on Principles of Distributed Computing (PODC), pp. 315–324 (2015)
4. Censor-Hillel, K., Khoury, S., Paz, A.: Quadratic and near-quadratic lower bounds for the CONGEST model (2017). arXiv preprint arXiv:1705.05646
5. Censor-Hillel, K., Paz, A., Perry, M.: Approximate proof-labeling schemes. In: Das, S., Tixeuil, S. (eds.) SIROCCO 2017. LNCS, vol. 10641, pp. 71–89. Springer, Cham (2017). https://doi.org/10.1007/978-3-319-72050-0_5
6. Cook, S.A.: The Complexity of theorem-proving procedures. In: Proceedings of the Third Annual ACM Symposium on Theory of Computing (STOC 1971), New York, NY, USA, pp. 151–158. ACM (1971)
7. Erdős, P., Rényi, A.: Asymmetric graphs. Acta Math. Hungar. **14**(3–4), 295–315 (1963)
8. Fraigniaud, P., Korman, A., Peleg, D.: Towards a complexity theory for local distributed computing. J. ACM **60**(5), 35:1–35:26 (2013)
9. Flajolet, P., Martin, G.N.: Probabilistic counting algorithms for data base applications. J. Comput. Syst. Sci. **31**(2), 182–209 (1985)
10. Goldreich, O., Micali, S., Wigderson, A.: Proofs that yield nothing but their validity or all languages in NP have zero-knowledge proof systems. J. ACM (JACM) **38**(3), 690–728 (1991)
11. Goldwasser, S., Micali, S., Rackoff, C.: The knowledge complexity of interactive proof systems. SIAM J. Comput. **18**(1), 186–208 (1989)
12. Göös, M., Suomela, J.: Locally checkable proofs in distributed computing. Theory Comput. **12**(1), 1–33 (2016)
13. Kushilevitz, E., Nisan, N.: Communication Complexity, pp. 1–189. Cambridge University Press, New York (1997). ISBN 978-0-521-56067-2
14. Kol, G., Oshman, R., Saxena, R.R.: Interactive distributed proofs. In: 37th ACM Symposium on Principles of Distributed Computing (PODC), pp. 255–264 (2018)
15. Kol, G., Oshman, R., Saxena, R.R.: AM Lower Bound for Symmetry. Private communication (2019)
16. Korman, A., Kutten, S., Peleg, D.: Proof labeling schemes. Distrib. Comput. **22**(4), 215–233 (2010). https://doi.org/10.1007/s00446-010-0095-3
17. Kushilevitz, E., Nissan, N.: Communication Complexity. Cambridge University Press, Cambridge (2006)
18. Lund, C., Fortnow, L., Karloff, H., Nisan, N.: Algebraic methods for interactive proof systems. J. ACM **39**(4), 859–868 (1992)
19. Naor, M., Parter, M., Yogev, E.: The power of distributed verifiers in interactive proofs (2018). CoRR abs/1812.10917
20. Naor, M., Stockmeyer, L.J.: What can be computed locally? SIAM J. Comput. **24**(6), 1259–1277 (1995)
21. Patt-Shamir, B.: A note on efficient aggregate queries in sensor networks. Theor. Comput. Sci. **370**(1–3), 254–264 (2007)
22. Shamir, A.: IP = PSPACE. J. ACM **39**(4), 869–877 (1992)

Anonymous Read/Write Memory: Leader Election and De-anonymization

Emmanuel Godard[1], Damien Imbs[1],
Michel Raynal[2,3(✉)], and Gadi Taubenfeld[4]

[1] LIS, Université d'Aix-Marseille, Marseille, France
[2] Univ Rennes IRISA, Rennes, France
raynal@irisa.fr
[3] Department of Computing, Polytechnic University, Kowloon, Hong Kong
[4] The Interdisciplinary Center, 46150 Herzliya, Israel

Abstract. Anonymity has mostly been studied in the context where processes have no identity. A new notion of anonymity was recently introduced at PODC 2017, namely, this notion considers that the processes have distinct identities but disagree on the names of the read/write registers that define the shared memory. As an example, a register named A by a process p and a shared register named B by another process q may correspond to the very same register X, while the same name C may correspond to different registers for p and q.

Recently, a memory-anonymous deadlock-free mutual exclusion algorithm has been proposed by some of the authors. This article addresses two different problems, namely election and memory de-anonymization. Election consists of electing a single process as a leader that is known by every process. Considering the shared memory as an array of atomic read/write registers $SM[1..m]$, memory de-anonymization consists in providing each process p_i with a mapping function $\mathsf{map}_i()$ such that, for any two processes p_i and p_j and any integer $x \in [1..m]$, $\mathsf{map}_i(x)$ and $\mathsf{map}_j(x)$ allow them to address the same register.

Let n be the number of processes and α a positive integer. The article presents election and de-anonymization algorithms for $m = \alpha\,n + \beta$ registers, where β is equal to 1, $n - 1$, or belongs to a set denoted $M(n)$ (which characterizes the values for which mutual exclusion can be solved despite anonymity). The de-anonymization algorithms are based on the use of election algorithms. The article also shows that the size of the permanent control information that, due to de-anonymization, a register must save forever, can be reduced to a single bit.

Keywords: Anonymous registers · Asynchronous system ·
Atomic read/write registers · Concurrent algorithm · Leader election ·
Local memory · Mapping · Memory de-anonymization ·
Mutual exclusion · Synchronization

K. Censor-Hillel and M. Flammini (Eds.): SIROCCO 2019, LNCS 11639, pp. 246–261, 2019.
https://doi.org/10.1007/978-3-030-24922-9_17

1   Anonymous Memory, Model, and Aim of the Article

1.1   Anonymous Memory

Memory Anonymity. While the notion of *process anonymity* has been studied for a long time from an algorithmic and computability point of view, both in message-passing systems (e.g., [1,4,17]) and shared memory systems (e.g., [3,5, 8]), the notion of *memory anonymity* has been introduced only very recently by [15]. (See also [11] for an introductory survey on process and memory anonymity.)

Let us consider a shared memory SM made up of m atomic read/write registers. Such a memory can be seen as an array with m entries, namely $SM[1..m]$. In a non-anonymous memory system, for each index x, the name $SM[x]$ denotes the same register whatever the process that invokes the address $SM[x]$. As stated in [15], in the classical system model, there is an a priori agreement on the names of the shared registers. This a priori agreement facilitates the implementation of the coordination rules the processes have to follow to progress without violating the safety (consistency) properties associated with the application they solve [10,14].

This a priori agreement does no longer exist in a memory-anonymous system. In such a system the very same identifier $SM[x]$ invoked by a process p_i and invoked by a different process p_j does not necessarily refer to the same atomic read/write register. More precisely, a memory-anonymous system is such that:

- prior the execution, an adversary defined, for each process p_i, a permutation $f_i()$ over the set $\{1, 2, \cdots, m\}$, such that when p_i uses the address $SM[x]$, it actually accesses $SM[f_i(x)]$, and
- no process knows the permutations.

The read/write registers of a memory-anonymous system are necessarily MWMR.

Results on Memory Anonymity in Mutual Exclusion. The work described in [15] on anonymous read/write memory addressed mutual exclusion, consensus, election and renaming, problems for which it presented algorithms and impossibility results. The consensus, election and renaming algorithms in [15] satisfy the starvation-freedom progress condition, namely, if a process executes alone during a long enough period, it eventually decides. This progress condition is different from the one considered in this article.

Among the results from [15], one states a condition on the size m of the anonymous memory which is necessary for any symmetric deadlock-free algorithm, where *symmetric* means that process identities can only be compared with equality (hence, there is no notion of a total order on process identities). More precisely, given an n-process system where $n \geq 2$, there is no deadlock-free mutual exclusion algorithm if the size m does not belong to the set $M(n) = \{ m \text{ such that } \forall \ell : 1 < \ell \leq n\colon \gcd(\ell, m) = 1 \} \setminus \{1\}$.

Recently, it has been shown in [2] that the condition $m \in M(n)$ is also a sufficient condition for symmetric deadlock-free mutual exclusion in read/write anonymous memory systems.

1.2   Computing Model

Processes. The system is composed of a finite set of $n \geq 2$ asynchronous processes denoted $p_1, .., p_n$. The subscript i in p_i is only a notation convenience, which is not known by the processes. *Asynchronous* means that each process proceeds to its own speed, which can vary with time and remains always unknown to the other processes. Initially, each process p_i knows only its identity id_i, the total number of processes n, and the fact that no two processes have the same identity. It is assumed that there are no process failures. Furthermore, unlike the mutual exclusion model where a process may never leave its remainder region, it is assumed that all the processes must participate in the algorithm.

Anonymous Shared Memory. The shared memory is made up of m atomic anonymous read/write registers denoted $SM[1...m]$. As a system composed of a single atomic register is not anonymous, it is assumed that $m > 1$. Hence, *all* registers are anonymous. As already indicated, when a process p_i invokes the address $SM[x]$, it actually accesses $SM[f_i(x)]$, where $f_i()$ is a permutation statically defined once and for all by an external adversary. We will use the notation $SM_i[x]$ to denote $SM[f_i(x)]$, to stress the fact that no process knows the permutations. It is assumed that all the registers are initialized to the same value. Otherwise, thanks to their different initial values, it would have been possible to distinguish different registers, which consequently will no longer be fully anonymous.

Symmetry Constraint on the Algorithms. A *symmetric algorithm* is an "algorithm in which the processes are executing exactly the same code and the only way for distinguishing processes is by comparing identifiers. Identifiers can be written, read, and compared, but there is no way of looking inside an identifier. Thus it is not possible to know whether an identifier is odd or even" [15]. Furthermore, the only comparison that can be applied to identifiers is equality. There is no order structuring the identifier name space. (Other notions of symmetry are described in [6,9]). Let us notice that as all the processes have the same code and all the registers are initialized to the same value, process identities become a key element when one has to design an algorithm in such a constrained context.

1.3   Problems Addressed in This Article

Leader Election. In this problem, the input of each process p_i is its identity id_i. Its output will be deposited in a write-once local variable $leader_i$. The aim is to design an algorithm that provides the local variable $leader_i$ of each process p_i with the same process identity. The only process such that $leader_i = id_i$ is the elected process.

Anonymous Memory De-anonymization. In this problem, as before, the input of each process p_i is its identity id_i. The aim is for each process p_i to compute an addressing function $\mathsf{map}_i()$, which is a permutation over the set of the memory indexes $\{1, \cdots, m\}$, such that the two following properties are satisfied.

- Safety. Let $y \in \{1, \cdots, m\}$. For any process p_i: $SM_i[\mathsf{map}_i(y)] = SM[y]$.
- Liveness. There is a finite time after which all the processes have computed their addressing function $\mathsf{map}_i()$.

The safety property states that once a process p_i has computed $\mathsf{map}_i()$, its local anonymous memory address $SM_i[x]$, where $x = \mathsf{map}_i(y)$, denotes the shared register $SM[y]$.

1.4   Content

This article presents first an impossibility result. Then, it presents symmetric algorithms solving the two previous problems in a system where the process cooperate through m atomic anonymous read/write registers. As already indicated, it is assumed that all the processes participate in the algorithms, and the size of the memory is $m = \alpha\, n + \beta$, where α is a positive integer and β can take the following values:

- $\beta = 1$. The size of the anonymous memory is then $m = \alpha\, n + 1$.
- $\beta = n - 1$. The size of the anonymous memory is then $m = \alpha\, n + (n - 1)$.
- $\beta \in M(n)$ where $M(n)$ is as defined above. Namely, $M(n)$ is the set of values for which deadlock-free mutual exclusion can be solved [2,15]. This is due to the fact that when $\beta \in M(n)$, the algorithms use a deadlock-free mutual exclusion algorithm to solve conflicts - which do not exist when $\beta = 1$ or $\beta = n - 1$). In this specific case, α can also be 0.

Find a characterization of the set of the values of m for which leader election can be solved in a memory anonymous system remains an open problem (see the Conclusion section).

2   An Impossibility Result

Theorem 1. *There is neither a de-anonymizing algorithm nor an election algorithm for n processes using m anonymous registers, where $m = \alpha\, n$ and α is a positive integer.*

Proof. First, we observe that once de-anonymizing is solved using $m = \alpha\, n$ registers, it is straightforward to solve election using $m = \alpha\, n$ registers. First, run the de-anonymizing algorithm to get $m = \alpha\, n$ non-anonymous registers. Then, using these registers, simply run the symmetric mutual exclusion algorithm from [13] which uses exactly n registers, and let the first process to enter its critical section be the leader. Thus, to prove the theorem, we only need to prove that it is impossible to solve election using $m = \alpha n$ registers.

Assume to the contrary, that there is a symmetric election algorithm for n processes using $m = \alpha\, n$ registers where α is a positive integer. Let us arrange the m registers on a ring with m nodes where each register is placed on a different node. Let us call the n processes $p_0, ..., p_{n-1}$. To each one of the n processes, we assign an initial register (namely, the first register that the process accesses) such that for every two processes p_i and $p_{i+1 \ (mod \ n)}$, the distance between their initial registers is exactly α when walking on the ring in a clockwise direction. Here we use the assumption that $m = \alpha n$.

The lack of global names allows us to assign for each process an initial register and an ordering which determines how the process scans the registers. An execution in which the n processes are running in *lock steps*, is an execution where we let each process take one step (in the order $p_0, ..., p_{n-1}$), and then let each process take another step, and so on. For process p_i and integer k, let $order(p_i, k)$ denote the kth new register that p_i accesses during an execution where the n processes are running in lock steps, and assume that we arrange that $order(p_i, k)$ is the register whose distance from p_i's initial register is exactly $(k - 1)$, when walking on the ring in a clockwise direction.

We notice that $order(p_i, 1)$ is p_i's initial register, $order(p_i, 2)$ is the next new register that p_i accesses and so on. That is, p_i does not access $order(p_i, k + 1)$ before accessing $order(p_i, k)$ at least once, but for every $j \leq k$, p_i may access $order(p_i, j)$ several times before accessing $order(p_i, k + 1)$ for the first time.[1]

With this arrangement of registers, we run the n processes in lock steps. Since only comparisons for equality are allowed, and all registers are initialized to the same value –which (to preserve anonymity) is not a process identity– processes that take the same number of steps will be at the same state, and thus it is not possible to break symmetry. It follows that either all the processes will be elected, or no process will be elected. A contradiction.   □$_{Theorem 1}$

3   Memory Anonymous Leader Election When $m = \alpha n + 1$

3.1   Algorithm

Local Variables. In addition to $leader_i$, each process p_i manages the following local variables: $towrite_i$, $overwritten_i$, $written_i$, which contain sets of memory indexes, $last_i$ which is a memory index, and nb_i which is a non-negative integer. The meaning of these variables will appear clearly in the text of Algorithm 1.

First Part of the Algorithm: Lines 1–12. Each anonymous register $SM[x]$ is initialized to $\langle start, \bot \rangle$, where $\bot$ is default value, which can be compared (with equality) with any process identity.

When it invokes election(id_i), a process p_i first writes the pair $\langle start, id_i \rangle$ in the first (from its point of view) α registers, namely, $SM_i[1], ..., SM_i[\alpha]$ (line 3). Then, it waits until all the registers (except one) are tagged $start$, or a register in which it wrote $\langle start, id_i \rangle$ has been overwritten. There are consequently two cases.

---

[1] Once a process accesses a register for the first time, say register x, we may map x to any (physical) register that it hasn't accessed yet. However, when it accesses x again, it must access the same register it has accessed before when referring to x.

– If registers in which p_i wrote $\langle start, id_i \rangle$ have been overwritten (the first part of the predicate of line 5 is then satisfied), p_i updates its local variables $overwritten_i$, nb_i, $towrite_i$ and $last_i$, and re-enters the repeat loop, the goal being to have α registers containing $\langle start, id_i \rangle$.
– If all the registers except one (i.e., exactly $m - 1 = \alpha\, n$ registers) are tagged start, p_i exits the loop.

As we will see in the proof, it follows from this collective behavior of the processes that there is time at which exactly one register still contains its initial value $\langle start, \perp \rangle$, while for each $j \in \{1, \cdots, n\}$, exactly α registers contain $\langle start, id_j \rangle$ (this property is named P1 in the algorithm).

---

init: each $SM[x]$ is initialized to $\langle start, \perp \rangle$; $m = \alpha\, n + 1$.

operation election(id_i) **is** % code for process p_i, $i \in \{1, \cdots, n\}$
(01) $towrite_i \leftarrow \{1, ..., \alpha\}$; $overwritten_i \leftarrow \emptyset$; $written_i \leftarrow \emptyset$; $last_i \leftarrow \alpha$;
(02)   **repeat**
(03)   **for each** $x \in towrite_i$ **do** $SM_i[x] \leftarrow \langle start, id_i \rangle$ **end do**;
(04) $written_i \leftarrow (written_i \setminus overwritten_i) \cup towrite_i$;
(05)   **wait until**$\big((\exists\, x \in written_i : SM_i[x] \neq \langle start, id_i \rangle)$
                        $\vee\ (|\{\ell \text{ such that } SM_i[\ell] \neq \langle start, \perp \rangle\}| = \alpha\, n)\big)$;
(06)   **if** $(|\{\ell \text{ such that } SM_i[\ell] \neq \langle start, \perp \rangle\}| = \alpha\, n)$
(07)   **then** exit repeat loop
(08)   **else** $overwritten_i \leftarrow \{\, x \in written_i \text{ such that } SM_i[x] \neq \langle start, id_i \rangle\}$;
(09) $nb_i \leftarrow |overwritten_i|$;
(10) $towrite_i \quad \leftarrow \{last_i + 1, ..., last_i + nb_i\}$; $last_i \leftarrow last_i + nb_i$;
(11)   **end if**
(12)   **end repeat**;
        % Property P1: There is a time at which exactly one register contains $\langle start, \perp \rangle$
        % and, for each $j \in \{1, \cdots, n\}$, α registers contain $\langle start, id_j \rangle$
(13)   **let** ℓ_i be such that $SM_i[\ell_i] = \langle start, \perp \rangle$ or $SM_i[\ell_i] = \langle leader, - \rangle$;
(14) $SM_i[\ell_i] \leftarrow \langle leader, id_i \rangle$;
(15)   **wait until**$\big((SM_i[\ell_i] \neq \langle leader, id_i \rangle)$
                        $\vee\ (SM_i[1..m]$ has exactly $\alpha + 1$ entries not tagged done)$\big)$;
(16)   **for each** x such that $SM_i[x] = \langle start, id_i \rangle$ **do** $SM_i[x] \leftarrow \langle done, id_i \rangle$ **end for**;
        % Property P2: There is a time from which there is exactly one
        % index $\ell \in \{1, \cdots, n\}$ such that a register contains $\langle leader, id_\ell \rangle$, and
        % for each $j \in \{1, \cdots, n\}$, there are α registers containing $\langle done, id_j \rangle$
(17)   **if** $(SM_i[\ell_i] \neq \langle leader, id_i \rangle)$ **then**
                    wait until$\big(SM_i[1..m]$ has only one entry not tagged done$\big)$ **end if**;
(18) $\langle -, id \rangle \leftarrow SM_i[\ell_i]$; $leader_i \leftarrow id$.
        % Here, one register is tagged leader, all the others are tagged done.

Algorithm 1: n-process election with $m = \alpha\, n + 1$ anonymous read/write registers

Second Part of the Algorithm: Lines 13–18. As just seen, the previous part of the algorithm has identified a single register of the anonymous memory, namely the only one containing $\langle start, \perp \rangle$. This register is known by all the processes, more precisely, it is known as $SM_i[\ell_i]$ by p_i, $SM_j[\ell_j]$ by p_j, etc.

So, to become the leader, each process p_i writes the pair $\langle \text{leader}, id_i \rangle$ in this register (known as $SM_i[\ell_i]$ by p_i, line 14). It follows that the last process that will write this register will be the leader. There are then two cases.

- If p_i discovers it has not been elected (we have then $SM_i[\ell_i] \neq \langle \text{leader}, id_i \rangle$, first predicate of line 15), it resets all the registers containing its tagged identity ($\langle \text{start}, id_i \rangle$) to the value $\langle \text{done}, id_i \rangle$ (line 16). Then, p_i waits until all registers except one are tagged $\langle \text{done}, - \rangle$.
- If p_i is the last process to write in the single register locally known as $SM_i[\ell_i]$, it waits until all the other processes have written $\langle \text{done}, - \rangle$ in the registers containing their identity (second part of the predicate of line 15). When this is done, the elected process p_i writes $\langle \text{done}, id_i \rangle$ in all the registers containing its identity (line 16), which allows each other process not to remain blocked at line 17 and progress to the last line of the algorithm. When this occurs, each process can assign the identity of the leader to its local variable $leader_i$ (line 18).

As before, we will see in the proof, that there is a time from which there is exactly one index $\ell \in \{1, \cdots, n\}$ such that a register contains $\langle \text{leader}, id_\ell \rangle$, and, for each $j \in \{1, \cdots, n\}$, there are α registers containing $\langle \text{done}, id_j \rangle$ (This property is named P2 in the algorithm).

3.2 Proof of Algorithm 1

Lemma 1 (Property P1). *Before a process executes line 14, there is a finite time at which one register contains $\langle \text{start}, \bot \rangle$, and, for each $j \in \{1, \cdots, n\}$, α registers contain $\langle \text{start}, id_j \rangle$.*

Proof. Considering time instants before a process executes line 14, we have the following.

- Let us first observe that the order on the entries of $SM[1..m]$ in which p_i writes them has been statically predefined by the adversary (namely, according to the –unknown– permutation $f_i()$: $SM_i[x]$ is actually $SM[f_i(x)]$). The important point is that a process p_i never backtracks while scanning $SM[1..m]$, and its successive accesses are $SM[f_i(1)]$, $SM[f_i(2)]$, etc.
- The first writes of a process p_i involve the registers $SM_i[1]$, ..., until $SM_i[\alpha]$ (lines 1 and 3). Then, as indicated above, its next writes in SM follows a statically predefined order. The process p_i issues a write of $\langle \text{start}, id_i \rangle$ in a register it has not yet written, for each of its previous writes that have been overwritten by another process (line 4). These writes by p_i concern entries of $SM_i[1..n]$ in which it has not yet written (management of the local variables $towrite_i$, $overwritten_i$, $written_i$, and $last_i$, at lines 1,4, and 8–10). As p_i writes only in new registers, it follows that, for any p_i we have $|\{x \text{ such that } SM[x] = \langle \text{start}, id_i \rangle\}| \leq \alpha$, and from a global point of view we have

$$\sum_{i=1}^{n} \left(|\{x \text{ such that } SM[x] = \langle \text{start}, id_i \rangle\}| \right) \leq n\alpha.$$

– It follows from $m = \alpha\, n + 1$ and the previous inequality, that there is enough room in the array $SM[1..m]$ for each process p_i to write n times the pair $\langle \mathtt{start}, id_i \rangle$. Consequently, there is time after which the first predicate of line 5 is false for each process p_i, and as $m = n\alpha + 1$, the remaining entry of $SM[1..m]$ has still its initial value, namely $\langle \mathtt{start}, \bot \rangle$, from which we conclude that a process neither remains forever blocked at line 4, nor forever executes the "repeat" loop (lines 2–12).

It follows from the previous observations that before a process executes line 14, there is a time at which, for each identity id_i, the pair $\langle \mathtt{start}, id_i \rangle$ is present in α entries of $SM[1..m]$, and an entry of $SM[1..m]$ has still its initial value, which concludes the proof of the lemma. $\hfill \square_{Lemma1}$

The Number of Write Accesses Between Line 3 and Line 12. When considering the proof of Lemma 1, it is easy to count the number of writes in the anonymous memory. In the best case, the (unknown) permutations assigned by the adversary to the processes are such that no process overwrites the pairs written by the other processes. In this case, line 2 generates $\alpha\, n$ writes into the shared memory.

In the worst case, the permutations assigned by the adversary, and the asynchrony among the processes are such that the first α writes of a process are overwritten $(n-1)$ times, the first α writes of another process are overwritten $(n-2)$ times, etc., until a last process whose none of its first α writes are overwritten. In this case, line 2 generates $\alpha\frac{n(n+1)}{2}$ writes into the anonymous shared memory.

Lemma 2 (Property P2). *There is a finite time from which there is $\ell \in \{1, \cdots, n\}$ such that exactly one register contains $\langle \mathtt{leader}, id_\ell \rangle$, and, for each $j \in \{1, \cdots, n\}$, there are α registers containing $\langle \mathtt{done}, id_j \rangle$.*

Proof. It follows from Lemma 1 that no process blocks or loops forever in the "repeat" loop (2–12). Hence, each process eventually executes lines 13–14. Let p_ℓ the last process that executes line 14. This means that after it executed this line, we have $SM_i[\ell_i] = \langle \mathtt{leader}, id_\ell \rangle$ for any process p_i (namely, p_ℓ is the process that has been elected). There are two cases.

– A process p_i that is not the leader, is such that $SM_i[\ell_i] \neq \langle \mathtt{leader}, id_i \rangle$. Consequently, it cannot be blocked at line 15. So, such a process p_i eventually writes $\langle \mathtt{done}, id_i \rangle$ in the α registers containing $\langle \mathtt{start}, id_i \rangle$ (line 16). Let us recall that, due to Property $P1$, these exactly α registers do exist. When the $(n-1)$ processes that are not leader have executed line 16, there are $\alpha(n-1)$ registers containing $\langle \mathtt{done}, - \rangle$, α registers containing $\langle \mathtt{start}, id_\ell \rangle$, and one register containing $\langle \mathtt{leader}, id_\ell \rangle$.
– As far as the leader process p_ℓ is concerned, we have the following. Due to the previous item, the second predicate of line 15 is eventually satisfied. When this occurs, p_ℓ writes $\langle \mathtt{done}, id_\ell \rangle$ in the α registers containing $\langle \mathtt{start}, id_\ell \rangle$ (line 16) and, from then on, a single register is not tagged $\langle \mathtt{done}, - \rangle$, namely the one containing $\langle \mathtt{leader}, id_\ell \rangle$.

The lemma follows directly from the two previous items. $\square_{Lemma2}$

Theorem 2. *Algorithm 1 solves the election problem.*

Proof. Once Property P2 is satisfied, no non-leader process is blocked at line 17, and each process eventually execute line 18. When this occurs, they all agree on the very same leader, namely the only process p_ℓ whose identity is tagged leader. $\square_{Theorem2}$

4   From Leader Election to De-anonymization When $m = \alpha\, n + 1$

4.1   A Simple Leader-Based De-anonymization Algorithm

As soon as a process has been elected, it is easy to de-anonymize the anonymous memory. To this end, the elected process p_ℓ imposes its mapping function to all the processes.

Algorithm 2 is such a de-anonymization algorithm, which relies on Property P2. Each process p_i invokes the operation election(id_i) (line 1). Then for each register $SM_\ell[x]$, the elected process p_ℓ writes the pair $\langle$desa, $x\rangle$ in $SM_\ell[x]$ (line 3). Hence, its mapping function is $\forall\, x \in \{1, \cdots, m\}$: map$_i(x) = x$. On the other side, any non-leader process p_i waits until all the registers are tagged desa (line 4). When this occurs, p_i computes its own mapping function (line 5), which is such that map$_i(y) = x$, where $SM_i[x] = \langle$desa, $y\rangle$. The proof of this algorithm is easy and left to the reader.

---

operation SM_i.scan() **returns** ($[SM_i[1], \cdots, SM_i[m]]$).

operation de-anonymize(id_i) **is**
(01)   election(id_i);
      % in the following ℓ_i has the value computed in election(id_i); moreover, if p_i is the
      % first process that exits from election(id_i):
      % one register is tagged leader, all the others are tagged done
(02)   **if** ($SM_i[\ell_i] = \langle$leader, $id_i\rangle$)   % this predicate is equivalent to $leader_i = id_i$
(03)   **then** **for each** $x \in \{1, \cdots, m\}$ **do** $SM_i[x] \leftarrow \langle$desa, $x\rangle$ **end for**
             % the permutation for p_i is: $\forall\, y \in \{1, \cdots, m\}$: map$_i(y) = y$ %
(04)   **else** **repeat** $sm_i \leftarrow SM_i$.scan() **until** ($\forall\, x :\ sm_i[x]$ is tagged desa) **end repeat**;
(05)   **for each** $x \in \{1, \cdots, m\}$ **do** map$_i(y) \leftarrow x$ where $sm_i[x] = \langle$desa, $y\rangle$ **end for**
             % perm. of p_i is: $\forall\, y \in \{1, \cdots, m\}$: map$_i(y) = x$, where $sm_i[x] = \langle$desa, $y\rangle$
(06)   **end if.**   % Here, each register $SM_i[x]$ is tagged desa.

Algorithm 2: Election-based de-anonymization (code for p_i, $m = \alpha\, n + 1$)

As a simple example see Fig. 1, where p_ℓ has been elected as leader, and $f_\ell()$ is the permutation defined by the adversary for p_ℓ (this permutation remains always unknown to the processes). $SM_i[x] = \langle$desa, $y\rangle$, and $SM_j[z] = \langle$desa, $y\rangle$ address the same register, which is $SM_\ell[y]$. Hence, this register is locally known as $SM_i[$map$_i(y)]$ by p_i, $SM_j[$map$_j(y)]$ by p_j, and $SM_\ell[$map$_\ell(y)] = SM_\ell[y]$ by p_ℓ.

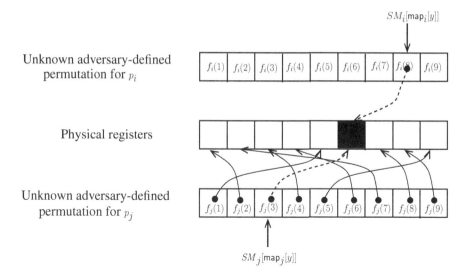

Fig. 1. An example of de-anonymization, $n = 4$ and $m = 2n + 1$

4.2   Using the De-anonymized Memory

When a process p_i returns from Algorithm 2, it knows that all the processes will share the same index for the same register (i.e., if $SM_i[x] = \langle \text{desa}, y \rangle$, then $SM_i[\text{map}_i(y)]$ is $SM_i[x]$). When this occurs, process p_i could start executing its local algorithm defined by the upper layer application, but if it writes an application-related value in some of these registers, this value can overwrite a pair tagged **desa** stored in a register not yet read by other processes. A way to prevent this problem from occurring consists in tagging all the values written by a process at the application level by the tag **apply**, and include a field containing the common index y associated with this register. Hence, at the application level, a register will contain $\langle \text{apply}(y), v \rangle$. In this way, despite asynchrony, any process p_j will be able to compute its local mapping function $\text{map}_j()$, and start its upper layer application part, as soon as it has computed $\text{map}_j()$.

Let us notice that one bit is needed to distinguish the tag **desa** and the tag **apply**. Hence each of a pair $\langle \text{desa}, y \rangle$ and a pair $\langle \text{apply}(y), - \rangle$ requires $(1 + \log_2 m)$ control bits.

4.3   Reducing the Size of the Permanent Control Information

Aim and Additional Assumption. This section shows that, at the price of an additional synchronization phase, the control information that each register must forever contain can be reduced from $(1 + \log_2 m)$ to a single bit.

To this end, we assume now that each atomic read/write register $SM[x]$ is composed of two parts $SM[x].BIT$ and $SM[x].RM$ (i.e., $SM[x] = \langle SM[x].BIT, SM[x].RM \rangle$). $SM[x].BIT$ is for example the leftmost bit of $SM[x]$, and

$SM[x].RM$ the other bits. The meaning and use of $SM[x].RM$ are exactly the same as $SM[x]$ in Algorithms 1 and 2. For each x, $SM[x].BIT$ is initialized to 0, while $SM[x].RM$ is initialized to $\langle \text{start}, \bot \rangle$. We assume that the previous algorithms are appropriately updated so that they do not modify the bits $SM[x].BIT$.

---

operation $SM_i.\text{scan}()$ **returns** $([SM_i[1], \cdots, SM_i[m]])$.

operation efficient_de-anonymize(id_i) **is**
(01) de-anonymize(id_i);
    % As all reg. are tagged desa when the first process returns from de-anonymize()
    % the tags start and done disappeared from the system and can be re-used
(02) execute lines 1-10 of Algorithm 1 where start is replaced by desa;
    % in the following, ℓ_i has the value obtained in de-anonymize(id_i)
(03) $SM_i[\ell_i] \leftarrow \langle \text{done}, id_i \rangle$;
(04) wait until$\big((SM_i[\ell_i] \neq \langle \text{done}, id_i \rangle)$
            $\vee\ (SM_i[1..m]$ has exactly $\alpha + 1$ entries not tagged done$)\big)$;
(05) **for each** x such that $SM_i[x] = \langle \text{desa}, id_i \rangle$ **do** $SM_i[x] \leftarrow \langle \text{done}, id_i \rangle$ **end for**;
    % <u>Property P1'</u>: There is a time at which exactly one register contains $\langle \text{start}, z \rangle$
    % where z is an integer and for each $j \in \{1, \cdots, n\}$, α registers contain $\langle \text{start}, id_j \rangle$
(06) **if** $(leader_i = id_i)$
    % Here the leader knows that every process p_j knows its mapping function $\text{map}_j()$
(07)   **then for each** $x \in \{1, \cdots, m\}$ **do** $BIT_i[x] \leftarrow 1$ **end for**
(08)   **else repeat** $bit_i \leftarrow BIT_i.\text{scan}()$ **until** $(\forall x :\ bit_i[x] = 1)$ **end repeat**
(09) **end if**.

Algorithm 3: Reduction to a single bit of control information per register (code for p_i)

Not to overload the presentation, the following notation shortcuts are used in Algorithm 3.

- The read of $SM_i[x]$ at lines 3 and 4 concerns the field $SM_i[x].RM$.
- The write of $SM_i[x]$ at lines 2 and 4 writes 0 in its leftmost bit (which actually is not modified).
- The statement "$BIT_i[x] \leftarrow 1$" at line 6, means that only the leftmost bit of $SM_i[x]$ is modified. As this statement is issued by the leader process only, this process can first read $SM_i[x]$, prefix it by 1, and rewrite this new value so that only the leftmost bit $SM_i[x]$ is modified.
- The statement "$BIT_i.\text{scan}()$" stands for "$SM_i.\text{scan}()$" from which only the leftmost bits are extracted.

After they return from de-anonymize(), the processes execute the same synchronization pattern as lines 14–17 of Algorithm 1 where the tag start is replaced by the tag desa. As the reader can see, at this time the tag done is no longer present in a register, so it can be re-used. Moreover, as the type "process identity" and the type "integer" are different, any integer x is considered as a synonym of $\bot$ when looking at a pair $\langle \text{desa}, x \rangle$ (which now is a synonym of $\langle \text{start}, x \rangle$).

It follows that we have then the property P1': there is a time at which exactly one register contains $\langle \mathtt{start}, z \rangle$ where z is an integer and, for each $j \in \{1, \cdots, n\}$, α registers contain $\langle \mathtt{start}, id_j \rangle$. Here, the important point is that the process previously elected as a leader knows that any process p_j knows its mapping function $\mathsf{map}_j()$. So, it can inform of it the other processes. This is done at lines 6–9 of Algorithm 3. As soon as a process p_j sees the leftmost bit of all the registers equal to 1, it knows that each process knows its mapping function, and p_j can consequently start writing application-related values in the other bits of the registers.

The lines 2–9 of Algorithm 3 and the code of Algorithm 1 are nearly the same. More precisely, they differ in the fact that Algorithm 1 *elects* a leader at lines 13–14, while Algorithm 3 uses at line 3 the leader that has been *previously been elected*. It follows that the proof of Algorithm 3 is very close to the proof of Algorithm 1, and is left to the reader.

5 Memory Anonymous Leader Election When $m = \alpha\, n + (n - 1)$

Leader Election. Algorithm 1, which solves the election problem for a system of $m = \alpha\, n + 1$ anonymous registers, is based on the fact that each process can write its identity in α registers that – after some finite time – will not be overwritten, and when this occurred, the single remaining not yet written anonymous register is used to elect the leader (which will be the last process that writes its identity in this register well-identified by each process).

The principle that underlies the election when there are $m = \alpha\, n + (n - 1)$ anonymous registers is dual in the sense that each of the n processes can write its identity in $\alpha + 1$ anonymous registers, except one which can write its identity in only α registers. When this occurs, the corresponding process becomes elected.

Algorithm. The operational view of this idea is captured by Algorithm 4, obtained from a simple adaptation of Algorithm 1 to the fact that the leader is selected from a memory occupation criterion (instead of a competition on a single read/write register, where the last writer is the winner). The main difference lies in the management of the local variables $towrite_i$, $overwritten_i$, $written_i$, $last_i$, and nb_i. Property P1" captures the result of the algorithm, namely, there is a time at which α registers contain the same pair $\langle start, id_\ell \rangle$, and for each $j \in \{1, \cdots, n\} \setminus \{\ell\}$, $\alpha + 1$ registers contain $\langle \mathtt{start}, id_j \rangle$. Its proof is a simple adaptation of the proof of Algorithm 1.

---

init: each $SM[x]$ is initialized to $\langle$start$, \perp\rangle$. $\langle$START$, \perp\rangle$; $m = (\alpha + 1)n - 1$.

operation election(id_i) **is** % code for process p_i, $i \in \{1, \cdots, n\}$
(01) $towrite_i \leftarrow \{1, ..., \alpha + 1\}$; $overwritten_i \leftarrow \emptyset$; $written_i \leftarrow \emptyset$; $last_i \leftarrow \alpha + 1$;
(02)   **repeat**
(03)   **for each** $x \in towrite_i$ **do** $SM_i[x] \leftarrow \langle$start$, id_i\rangle$ **end do**;
(04) $written_i \leftarrow (written_i \setminus overwritten_i) \cup towrite_i$;
(05)   **wait until** $((\exists\, x \in written_i : SM_i[x] \neq \langle$start$, id_i\rangle)$
                                   $\vee\, (|\{\ell$ such that $SM_i[\ell] \neq \langle$start$, \perp\rangle\}| = m))$;
(06)   **if** $(|\{\ell$ such that $SM_i[\ell] \neq \langle$start$, \perp\rangle\}| = m)$
(07)   **then** exit repeat loop
(08)   **else** $overwritten_i \leftarrow \{\, x \in written_i$ such that $SM_i[x] \neq \langle$start$, id_i\rangle\}$;
(09) $nb_i \leftarrow |overwritten_i|$;
(10) $towrite_i \leftarrow \{last_i + 1, ..., \min(last_i + nb_i, m)\}$;
(11) $last_i \leftarrow \min(last_i + nb_i, m)$
(12)   **end if**
(13)   **end repeat**;
              % Property P1": There is a time at which α reg. contain the same pair $\langle$start$, id_\ell\rangle$,
              % and for each $j \in \{1, \cdots, n\} \setminus \{\ell\}$, $\alpha + 1$ registers contain $\langle$start$, id_j\rangle$
(14)   leader$_i \leftarrow id$ **where** id
(15)   **is such that** α registers exactly contain the same pair $\langle$start$, id\rangle$.

Algorithm 4: n-process election for $m = \alpha\, n + (n - 1)$ anonymous registers

6 Election and De-anonymization for $m = \alpha\, n + \beta$, $\beta \in M(n)$

This section considers the case where an underlying mutex algorithm, suited to an anonymous memory, is used to elect a leader.

Mutual Exclusion in an Anonymous System. Mutual exclusion in memory anonymous systems was introduced in [15], which presents a symmetric deadlock-free mutex algorithm for *two* processes only, and a theorem stating that there no symmetric deadlock-free mutual exclusion algorithm if the size m does not belong to the set $M(n) = \{\, m$ such that $\forall\, \ell : 1 < \ell \leq n$: $\gcd(\ell, m) = 1\} \setminus \{1\}$. Recently, a symmetric deadlock-free mutual exclusion algorithm has been proposed, which works any number of processes and for any value $m \in M(n)$ [2], from which follows that $m \in M(n)$ is a necessary and sufficient condition for anonymous mutual exclusion.

Leader Election in a System of $m = \alpha\, n + \beta$ Anonymous Registers. The idea is to rely on the underlying mutex algorithm to elect a leader. But, to this end, the processes have first to isolate a set of β anonymous registers in order to be thereafter able to use a symmetric deadlock-free mutex algorithm accessing this subset of registers.

init: each $SM[x]$ is initialized to $\langle \text{start}, \perp \rangle$. $\langle \text{START}, \perp \rangle$; $m = \alpha\, n + \beta$, $\beta \in M(n)$.

operation election(id_i) **is** % code for process p_i, $i \in \{1, \cdots, n\}$
(01) $towrite_i \leftarrow \{1, ..., \alpha\}$; $overwritten_i \leftarrow \emptyset$; $written_i \leftarrow \emptyset$; $last_i \leftarrow \alpha$;
(02)   **repeat**
(03)   **for each** $x \in towrite_i$ **do** $SM_i[x] \leftarrow \langle \text{start}, id_i \rangle$ **end do**;
(04) $written_i \leftarrow (written_i \setminus overwritten_i) \cup towrite_i$;
(05)   **wait until** $((\exists\, x \in written_i : SM_i[x] \neq \langle \text{start}, id_i \rangle)$
                        $\vee \,(|\{\ell \text{ such that } SM_i[\ell] = \langle \text{start}, \perp \rangle\}| = \beta))$;
(06)   **if** $(|\{\ell \text{ such that } SM_i[\ell] = \langle \text{start}, \perp \rangle\}| = \beta)$
(07)   **then** exit repeat loop
(08)   **else** $overwritten_i \leftarrow \{\, x \in written_i \text{ such that } SM_i[x] \neq \langle \text{start}, id_i \rangle\}$;
(09) $nb_i \leftarrow |overwritten_i|$;
(10) $towrite_i \leftarrow \{last_i + 1, ..., last_i + nb_i\}$; $last_i \leftarrow last_i + nb_i$
(11)   **end if**
(12)   **end repeat**;
        % Property P1"': There is a time at which β registers contain the pair $\langle \text{start}, \perp \rangle$,
        % and for each $j \in \{1, \cdots, n\}$, α registers contain $\langle \text{start}, id_j \rangle$
(13)   **let** $SM\beta_i[1..\beta]$ be the sub-array of the β registers
                    that do not contain $\langle \text{start}, id \rangle$, for any process identity id;
(14)   Now, using the previous sub-array (locally knows as $SM\beta_i[1..\beta]$ by p_i) the processes
        processes execute a symmetric deadlock-free mutex algorithm at the end of which the
        last process to enter the critical section is elected. While it is in the critical section,
        the elected process p_ℓ write $\langle \text{leader}, id_\ell \rangle$ in all the registers of $SM\beta_\ell[1..\beta]$,
        which allows the other processes to know which is the leader.

Algorithm 5: Election in a system of $m = \alpha\, n + \beta$, $\beta \in M(n)$ anonymous reg.

Algorithm 5 realizes this at lines 1–12, which are a simple adaptation of the same line numbers in Algorithms 1 and 4. When the processes exit the repeat loop (line 12), we have property P1", namely, there is a time at which β registers contain the pair $\langle \text{start}, \perp \rangle$ and, for each $j \in \{1, \cdots, n\}$, α registers contain $\langle \text{start}, id_j \rangle$. Hence, the set of β registers define a common anonymous memory on top of which the n processes can execute a symmetric deadlock-free mutex algorithm. As $\beta \in M(n)$, such mutex algorithms do exist (e.g., [2]). Moreover, as the mutex algorithm is deadlock-free and each process invokes it once, each process eventually enters the critical. It is shown in [7] how a symmetric deadlock-free mutual exclusion algorithm can be used to allow a process to know it is the last that entered the critical section. Finally, the last process to enter is the elected process.

We point out that a memory de-anonymization algorithm is described in [7]. However, as it is based on an underlying mutual exclusion algorithm, it is a specific algorithm that works only for $m \in M(n)$, which is not the general case addressed here, namely $m = \alpha\, n + \beta$.

Memory De-Anonymization in a System of $m = \alpha\, n + \beta$ Anonymous regis-ters. The previous algorithm can be modified in order to solve de-anonymization. When the last process is inside the critical section, it can impose its mapping function to all the processes by executing lines 4–5 of Algorithm 2, while all the other processes execute lines 5–6 of this algorithm.

7   Conclusion

This article is on synchronization problems in an n-process system in which the communication is through m anonymous read/write registers only. In such a system there is no a priori agreement on the names of the registers: the same register name A used by several processes can head them to different registers. In such a context, the article addressed the following problems: leader election and memory de-anonymization. It was first shown that these problems are impossible to solve if $m = \alpha \, n$, where α is a positive integer. Then, considering $m = \alpha \, n + \beta$, it has presented election algorithms for $\beta = 1$, $\beta = n - 1$, and $\beta \in M(n)$ where $M(n)$ is the set of the memory anonymous sizes for which symmetric deadlock-free mutual exclusion can be solved in n-process systems. De-Anonymization algorithms have also been presented, each based on an underlying election algorithm.

As stated in [15], the memory-anonymous communication model "enables us to better understand the intrinsic limits for coordinating the actions of asynchronous processes". It consequently enriches our knowledge of what can be (or cannot be) done when an adversary replaced a common addressing function, by individual and independent addressing functions, one per process. Additional results regarding the computational power of anonymous and non-anonymous objects can be found in [16]. On a more practical side, it appears that the concept of an anonymous memory allows us to model epigenetic cell modifications [12].

On the open problems side, it seems that finding a characterization of all the values of m (the size of the read/write anonymous memory) for which leader election (and de-anonymization) can be solved in an n-process system is particularly important as soon as we want to understand the power and the limits of n-process memory anonymous systems. Finally, since we assume a model where participation is required, in the case where the mutex algorithm from [2] (which also works when participation is not required) is used, it might be possible to replace the algorithm from [2] with a simpler algorithm. In such a case we might not need to assume that $\beta \in M(n)$, but something weaker.

Acknowledgments. This work was partially supported by the French ANR project DESCARTES (16-CE40-0023-03) devoted to layered and modular structures in distributed computing. The authors want to thank the referees for their constructive comments.

References

1. Angluin, D.: Local and global properties in networks of processes. In: Proceedings of 12th Symposium on Theory of Computing (STOC 1980), pp. 82–93. ACM Press (1980)
2. Aghazadeh, Z., Imbs, D., Raynal, M., Taubenfeld, G., Woelfel, Ph.: Optimal memory-anonymous symmetric deadlock-free mutual exclusion. In: Proceedings of 38th ACM Symposium on Principles of Distributed Computing (PODC 2019), 10 pages. ACM Press (2019)

3. Attiya, H., Gorbach, A., Moran, S.: Computing in totally anonymous asynchronous shared-memory systems. Inf. Comput. **173**(2), 162–183 (2002)
4. Bonnet, F., Raynal, M.: Anonymous asynchronous systems: the case of failure detectors. Distrib. Comput. **26**(3), 141–158 (2013)
5. Bouzid, Z., Raynal, M., Sutra, P.: Anonymous obstruction-free (n, k)-set agreement with $(n-k+1)$ atomic read/write registers. Distrib. Comput. **31**(2), 99–117 (2018)
6. Garg, V.K., Ghosh, J.: Symmetry in spite of hierarchy. In: Proceedings of 10th International Conference on Distributed Computing Systems (ICDCS 1990), pp. 4–11. IEEE Computer Press (1990)
7. Godard E., Imbs D., Raynal M., Taubenfeld G.: Mutex-based de-anonymization of an anonymous read/write memory. In: Proceedings of 7th International Conference on Networked Systems (NETYS 2018). LNCS, 15 pages. Springer (2019, to appear)
8. Guerraoui, R., Ruppert, E.: Anonymous and fault-tolerant shared-memory computations. Distrib. Comput. **20**, 165–177 (2007)
9. Johnson, R.E., Schneider, F.B.: Symmetry and similarity in distributed systems. In: Proceedings of 4th ACM Symposium on Principles of Distributed Computing (PODC 1985), pp. 13–22. ACM Press (1985)
10. Raynal, M.: Concurrent Programming: Algorithms, Principles and Foundations. Springer, Heidelberg (2013). https://doi.org/10.1007/978-3-642-32027-9. ISBN 978-3-642-32026-2
11. Raynal, M., Cao, J.: Anonymity in distributed read/write systems: an introductory survey. In: Podelski, A., Taïani, F. (eds.) NETYS 2018. LNCS, vol. 11028, pp. 122–140. Springer, Cham (2019). https://doi.org/10.1007/978-3-030-05529-5_9
12. Rashid, S., Taubenfeld, G., Bar-Joseph, Z.: Genome wide epigenetic modifications as a shared memory consensus. In: 6th Workshop on Biological Distributed Algorithms (BDA 2018), London (2018)
13. Styer, E., Peterson, G.L.: Tight bounds for shared memory symmetric mutual exclusion problems. In: Proceedings of 8th ACM Symposium on Principles of Distributed Computing, pp. 177–191. ACM Press (1989)
14. Taubenfeld, G.: Synchronization Algorithms and Concurrent Programming. Pearson Education/Prentice Hall, 423 pages (2006). ISBN 0-131-97259-6
15. Taubenfeld G., Coordination without prior agreement. In: Proceedings of 36th ACM Symposium on Principles of Distributed Computing (PODC 2017), pp. 325–334. ACM Press (2017)
16. Taubenfeld, G.: Set agreement power is not a precise characterization for oblivious deterministic anonymous objects. In: Censor-Hillel, K., Flammini, M (eds.) SIROCCO 2019. LNCS, pp 293–308 (2019)
17. Yamashita, M., Kameda, T.: Computing on anonymous networks: Part I-characterizing the solvable cases. IEEE Trans. Parallel Distrib. Syst. **7**(1), 69–89 (1996)

Faster Construction of Overlay Networks

Thorsten Götte, Kristian Hinnenthal$^{(\boxtimes)}$, and Christian Scheideler

Paderborn University, Paderborn, Germany
{thgoette,krijan,scheidel}@mail.upb.de

Abstract. We consider the problem of transforming any weakly connected overlay network of polylogarithmic degree into a topology of logarithmic diameter. The overlay network is modeled as a directed graph, in which messages are sent in synchronous rounds, and new edges can be established by sending node identifiers. However, every node can only send and receive a polylogarithmic number of bits in each round, which makes the naive approach of introducing all neighbors to each other until the network forms a clique infeasible. We present an algorithm that takes time $O(\log^{3/2} n)$, w.h.p. At the heart of our algorithm lies a deterministic strategy to group and merge large components of nodes, but we make use of randomized load-balancing techniques to keep the communication load of each node low. To the best of our knowledge, this is the first algorithm to improve upon the algorithm by Angluin et al. [SPAA 2005], which solves the problem in time $O(\log^2 n)$, and comes closer to the $\Omega(\log n)$ lower bound.

Keywords: Overlay networks · Peer-to-peer · Pointer jumping

In order to exchange information in a distributed system that is established over the Internet, IP addresses of its members have to be known. If member v knows the address of member w, we can interpret this as a directed edge (v, w). The set of these directed edges is commonly referred to as an *overlay network*. There exists a vast amount of papers that have already dealt with the question of how to best maintain an overlay network between the members of a distributed system. An important and particularly challenging threat to overlay networks lies in *churn* and *adversarial behavior*. Such dynamics may push an overlay network into a state where it does not function correctly anymore, or, in the worst case, even becomes disconnected. Recovering an overlay network from an undesired state is quite difficult. One approach that has been pursued in theory is to come up with *self-stabilizing* overlay networks, which are overlay networks that can recover from *any* illegal state (as long as its members are still weakly connected). In order to keep the maintenance overhead low, the members of a self-stabilizing overlay network should ideally only initiate a repair process if they discover some problem locally. However, almost none of the self-stabilizing

This work is partially supported by the German Research Foundation (DFG) within the CRC 901 "On-The-Fly Computing" (project number 160364472-SFB901).

K. Censor-Hillel and M. Flammini (Eds.): SIROCCO 2019, LNCS 11639, pp. 262–276, 2019.
https://doi.org/10.1007/978-3-030-24922-9_18

overlay networks that follow this approach have been shown to converge quickly to a legal state, and for the few for which a fast recovery time has been shown, no reasonable bound on the communication work (i.e., the number of exchanged messages) is known.

An alternative approach is to start rebuilding an overlay from scratch every T steps, for some suitable time span T, irrespective of whether its members noticed a problem with the overlay or not. Once an overlay has been transformed into a low-degree overlay of logarithmic diameter, one can quickly reconstruct various overlay topologies from there. In fact, many well-known overlays, e.g., [2, 9,10,15,18–20], can be built in $O(\log n)$ rounds if the nodes form a sorted ring. Given an overlay of logarithmic diameter, a sorted ring can be constructed by first performing a BFS and then applying the algorithm of [3]. In this paper, we focus on the problem of transforming an arbitrary overlay of polylogarithmic degree into a constant degree tree of depth $O(\log n)$. Similar problems have been studied before, but, to the best of our knowledge, the best time bound known to our problem for more than a decade is $O(\log^2 n)$. We improve this to $O(\log^{3/2} n)$.

Note that there is a fundamental lower bound of $\Omega(\log n)$ time steps to construct an overlay of logarithmic diameter, so we make a significant step towards this lower bound. To illustrate this, consider an overlay that forms a linked list. Even if all nodes exhaustively exchange their neighborhoods through pointer jumping in every round, this can only halve the diameter in every iteration. Hence, it takes $\Omega(\log n)$ rounds until the diameter is logarithmic.

Model and Problem Statement. We consider overlay networks with a fixed node set V. Each node u has a unique *identifier* id(u), which is a bit string of length $O(\log n)$, where $n = |V|$. Time proceeds in *synchronous rounds*. Let $E_i(u)$ be the set of identifiers stored by a node u at the beginning of round i. We define the set of overlay edges in round i as $E_i = \{(u, v) \mid u \in V \text{and} v \in E_i(u)\}$ and the overlay in round i as $G_i = (V, E_i)$.

In each round i, every node u can send and receive at most a polylogarithmic number of messages of size $O(\log n)$ (i.e., such a message can carry only a constant number of identifiers), and messages from node u can only be sent to nodes in $E_i(u)$. If more than a polylogarithmic number of messages is sent to a node, it receives an *arbitrary* subset (and the rest is simply dropped by the network). A message sent in round i arrives at the beginning of round $i+1$. We assume that every node has sufficient memory for our protocol to work correctly and every node is sufficiently fast so that it can process all messages that arrived at the beginning of round i within that round[1].

Before we formally define our problem, we first review some basic concepts from graph theory. A node's *outdegree* denotes the number of outgoing edges, i.e., the number of identifiers it stores. Analogously, its *indegree* denotes the number of incoming edges, i.e., the number of nodes that store its identifier. A node's *degree* is the sum of its in- and outdegree, and a graph's degree is the

---

[1] Note that for our algorithm polylogarithmic memory and a small number of local computations is sufficient.

maximum degree of any node. We say that a graph is *weakly connected* if there is a (not necessarily directed) path between all pairs of nodes. A graph's *diameter* is the maximum over all node pairs v, w of the length of a shortest path between v and w (where we ignore the edges' directions).

Using these definitions, we define the *Overlay Construction Problem* as follows. Given a weakly connected graph $G_0 = (V, E_0)$ of polylogarithmic degree, the goal is to arrange the nodes into a bidirected tree of constant degree and depth $O(\log n)$ that is rooted at the node with highest identifier as quickly as possible.

Related Work. General research on overlay networks started in the early 2000s. Some popular examples for these early overlays are Chord [20], Pastry [19], and skip graphs [2]. Most of these solutions focus on the problem of efficiently joining and leaving such an overlay, or keeping it in some legal state despite some potentially heavy churn (see, e.g., [5,6,12] for some recent works that even allow adversarial churn). Still, adversarial nodes and churn beyond the limits prescribed in these papers may push an overlay into an corrupted state. Here, solutions for self-stabilizing overlays can be used.

There is a rich collection of papers on self-stabilizing overlays, e.g., [9,10, 15,17,18]. Note that a self-stabilizing overlay must be able to recover from *any* initial state, including nodes with arbitrarily corrupted local memory. Many of them do not provide time or work bounds, but rather focus on showing self-stabilization in a very general context like the asynchronous message passing model. Some also present bounds on the convergence time, but these are often much higher than polylogarithmic, like the $O(n)$ time bound for self-stabilizing lists [17] and the $O(n^3)$ time bound for self-stabilizing Delaunay graphs [14]. Notable exceptions are [8,13]: In [13], the authors show a convergence time of $O(\log^2 n)$ rounds for the SKIP+ graph, and in [8] the authors present a general framework for the self-stabilizing construction of a large class of overlays that can be used, for example, to achieve a convergence time of just $O(\log n)$ for SKIP+ graphs. However, no low bounds for the communication work are known; in fact, the work required for the constructions in [8,13] can be prohibitively large.

The problem of recovering from an arbitrary topology seems to become much easier once we assume a *well-initialized* overlay, i.e., the nodes are in a well-defined initial state. To the best of our knowledge, the first paper that considered the problem of constructing a low-diameter overlay from an arbitrary well-initialized overlay is by Angluin et al. [1]. Given an initial graph of degree d and identifiers of size W, their algorithm constructs an overlay of logarithmic diameter in time $O((d + W) \log n)$, w.h.p.[2]. Note, however, that their communication model slightly differs from ours as nodes are only allowed to send $O(1)$ messages via an edge each round. If we allow polylogarithmic communication on every edge (and assume that $W = O(\log n)$), the algorithm yields a convergence time of $O(\log^2 n)$. Aspnes and Wu [3] came up with an improved time bound if

---

[2] We say an event holds *with high probability* (w.h.p.), if it holds with probability at least $1 - 1/n^c$ for any fixed constant $c > 0$.

the initial graph has outdegree at most 1. In particular, for that case they gave a construction that only requires $O(\log n)$ time. Gmyr et al. [11] considered the problem for arbitrary overlays in our model. Incorporating ideas from both [1] and [3], they were able to present a deterministic algorithm that converges in $O(\log^2 n)$ rounds.

The main difficulty of designing efficient algorithms in our model lies in the fact that, irrespective of its degree, a node is only allowed to send and receive polylogarithmically many bits in each round. Recently, the impact of this restriction has been studied in the so-called *Node-Capacitated Clique* model [4]. In fact, if we restrict our initial network to be of logarithmic degree only, with little effort our algorithm would be adaptable to work in the Node-Capacitated Clique as well. We reuse some of the techniques presented in [4] to quickly aggregate and disseminate data without violating the communication constraints (see Sect. 1 for more details).

Our Contribution. We present an algorithm that solves the Overlay Construction Problem in time $O(\log^{3/2} n)$. The algorithm requires *polylogarithmic communication work* (i.e., in every round, each node sends and receives a polylogarithmic number of messages), w.h.p., and thereby ensures no message will ever be dropped. Similar to the algorithms in [1,3,11], we incrementally construct our desired topology by alternatingly grouping and merging *supernodes*. Our improved time bound comes from the fact that we are able to always merge large clusters of supernodes by first exploring each supernode's neighborhood up to a certain distance using *pointer jumping*. To do so efficiently, we utilize the communication power of a large subset of the nodes that make up each supernode. As this requires a high amount of coordination between the nodes, we carefully organize supernodes internally and use a variety of algorithmic techniques.

1   Algorithmic Primitives

Our algorithm relies on a set of primitives, which are briefly described in this section. Similar to the approach of [4], we distribute the communication load of a supernode by performing routing strategies in a simulated *butterfly network*. Formally, for $d \in \mathbb{N}$, the *d-dimensional wrapped butterfly* is a graph with node set $[d+1] \times [2^d]$, where we denote $[k] = \{0, \ldots, k-1\}$, and an edge set $E_1 \cup E_2 \cup E_3$ with

$$E_1 = \{\{(i, \alpha), (i+1, \alpha)\} \mid i \in [d],\ \alpha \in [2^d]\},$$
$$E_2 = \{\{(0, \alpha), (d, \alpha)\} \mid \alpha \in [2^d]\},$$
$$E_3 = \{\{(i, \alpha), (i+1, \beta)\} \mid i \in [d], \alpha, \beta \in [2^d],$$
$$\alpha \text{ and } \beta \text{ differ only at the } i\text{-th bit}\}.$$

The node set $\{(i, j) \mid j \in [2^d]\}$ represents *level i* of the butterfly, and node set $\{(i, j) \mid i \in [d+1]\}$ represents *column j* of the butterfly. As an example, the columns of the nodes $x_0, \ldots, x_7$ in Fig. 1 form a 3-dimensional wrapped butterfly network. Given a d-dimensional butterfly, we introduce the following primitives:

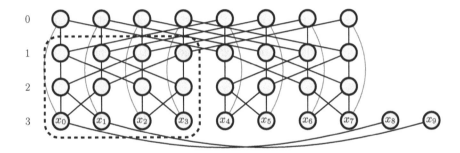

Fig. 1. Internal butterfly B_v of a supernode v with $|v| = 10$. The nodes $x_0, \ldots, x_9$ at the bottom are members of v. The first eight nodes $x_0, \ldots, x_7$ construct the 3-dimensional wrapped butterfly by simulating one column each. The other two nodes x_8 and x_9 connect to their corresponding helper nodes. Indicated by the dashed outline, the first four nodes $x_0, \ldots, x_3$ simulate the small butterfly b_v, which is described in Sect. 3.2

- **Routing.** Every node of the butterfly's top level has at most polylogarithmically many messages that it wishes to send to nodes of the bottom level such that every node is the target of at most polylogarithmically many messages. Using the algorithm from [16], which we refer to as the *Routing Primitive*, every message can be routed to its destination within $2d + 2$ rounds and with polylogarithmic communication work, w.h.p.
- **Aggregate-and-Broadcast.** Every node of the butterfly's top level stores a subset of input values $A := \{a_0, \ldots, a_{2^d}\}$ to a *distributive aggregate function*[3] f. The *Aggregate-and-Broadcast Primitive* allows every node to compute $f(A)$ within $O(d)$ rounds and with polylogarithmic communication work, w.h.p.
- **Pipelined Sampling.** Every node of the butterfly's top level has an infinite supply of data items that the nodes of the bottom level wish to sample from. The *Pipelined Sampling Primitive* ensures that every butterfly node of the bottom level constantly receives up to a polylogarithmic number of data items selected uniformly at random at every round in a pipelined fashion (after an initial delay of $O(d)$) and with polylogarithmic communication work, w.h.p.
- **Filtering.** The nodes of the butterfly's top level store data items (m, g) each consisting of some payload m of size $O(\log n)$ and some group identifier $g \in \mathcal{G}$ of size $O(\log n)$. The data items are arbitrarily distributed among the nodes of the butterfly's top level such that every node stores at most polylogarithmically many. All data items with the same group identifier have the same target, which is a node of the butterfly's bottom level, but every node is only target of at most polylogarithmically many groups. The *Filtering Primitive* passes the data items down in the butterfly such that eventually exactly one (arbitrary) data item of each group reaches its target within $O(d)$ rounds and with polylogarithmic communication work, w.h.p.

---

[3] An aggregate function f is called distributive if there is an aggregate function g such that for any partition $A_1, \ldots, A_k \subset A$, it holds $f(A) = g(f(A_1), \ldots, f(A_k))$ (e.g., MAX, MIN, and SUM are distributive).

All of our primitives can be performed using the butterfly's path system and techniques from [16] and [4]. At a high level, for the Routing and Filtering Primitive, we first route messages to random intermediate targets, and finally forward them to their actual targets. In contrast to classical routing protocols, we do not wait to send messages. Thereby, we obtain guaranteed runtimes, but are only able to show that the communication bounds are met w.h.p.

2   High-Level Algorithm

Our algorithm to solve the Overlay Construction Problem is divided into consecutive *phases*, where each phase relies on a set of invariants maintained after the previous phase. We first present the algorithm from a high level, and then give the details of a single phase in Sect. 3. Note that as we assume G_0 to have polylogarithmic degree, we can easily bidirect all edges of G_0; therefore, we will refer to them as if they were undirected.

We organize sets of nodes into *supernodes*, where initially each node makes up a supernode on its own, and repeatedly merge supernodes into larger supernodes until only a single supernode containing all nodes of V remains. We internally organize each supernode v that consists of $|v|$ nodes (we say v is of *size* $|v|$) as a constant degree tree of depth $O(\log |v|)$. Whenever a set of supernodes merges to form a larger supernode u, we maintain this invariant by making sure that u becomes organized as a constant degree tree of depth $O(\log |u|)$ as well. Maintaining the low-diameter internal structures allows us to merge any set of supernodes in time $O(\log m)$, where m is the size of the resulting supernode. Most importantly, we always merge *large* sets of supernodes in a highly coordinated fashion, which, compared to previous approaches, results in a faster growth of supernodes, and fewer rounds until only a single supernode remains. Clearly, once a single supernode remains, all nodes are organized in a constant degree tree of depth $O(\log n)$, and thus the Overlay Construction Problem is solved.

More precisely, our algorithm proceeds in phases $0, \ldots, \lceil \log \log n \rceil$, where the goal of phase i is to grow every supernode of size $|v| \in [2^{2^i}, 2^{2^{i+1}} - 1]$ to a supernode of size at least $2^{2^{i+1}}$. To optimally balance the number of phases with the required runtime of each phase, we further divide each phase i into subphases $0, \ldots, \lceil \sqrt{2^i} \rceil - 1$. Correspondingly, the goal of subphase j of phase i is to grow every *active* supernode, which is a supernode whose size lies in the interval $I = [2^{2^i + j \lceil \sqrt{2^i} \rceil}, 2^{2^i + (j+1) \lceil \sqrt{2^i} \rceil} - 1]$, to a supernode of size at least $2^{2^i + (j+1) \lceil \sqrt{2^i} \rceil}$. If a supernode is of size at least $2^{2^i + (j+1) \lceil \sqrt{2^i} \rceil}$ already at the beginning of the subphase, we call it *inactive*. As we will later show, our algorithm ensures that there are no supernodes of smaller size than $2^{2^i + j \lceil \sqrt{2^i} \rceil}$ at the beginning of the subphase[4].

---

[4] For the first subphase, we have to assume that every supernode is of size at least 2 already, which can, e.g., be ensured by letting each node simulate two virtual nodes.

Lemma 1. *If in subphase j of phase i every active supernode grows to a supernode of size at least $2^{2^i+(j+1)\lceil\sqrt{2^i}\rceil}$ in time $O(2^i)$, then the algorithm merges all nodes into a single supernode in time $O(\log^{3/2} n)$.*

Proof. We only have to prove the overall runtime. Summing up over all phases results in $\sum_{i=0}^{T:=\lceil\log\log n\rceil} O(\sqrt{2^i})\cdot O(2^i) = O(\sqrt{\log n})\cdot\sum_{i=0}^{T} O(2^i) = O(\log^{3/2} n)$. $\square$

3   A Single Subphase

We now describe the details of a single subphase of the algorithm; more specifically, in the following we consider a subphase j of phase i. The subphase is divided into three stages: In the *Expansion Stage*, the goal of each active supernode is to get to know at least $2^{\lceil\sqrt{2^i}\rceil}$ other active supernodes. In the *Grouping Stage*, every active supernode chooses to merge with at most one supernode among the active and inactive supernodes it has learned. It remains to reconfigure each resulting component into a constant degree tree of logarithmic depth, which is the goal of the *Merging Stage*. Due to space reasons, most proofs of this section are deferred to the full version of this paper.

3.1   Beginning of the Subphase

Before we present the three stages of the subphase in detail, we first describe the situation at its beginning. We show that our algorithm ensures the corresponding situation to hold at the beginning of the next subphase, which inductively establishes the correctness of the overall algorithm. As already mentioned, we ensure that every supernode is of size at least $2^{2^i+j\lceil\sqrt{2^i}\rceil}$. We further make sure that the *members* of each supernode (i.e., the nodes of which it consists) know whether it is active or inactive in this subphase. Let v be an active supernode, i.e., $|v| \in [2^{2^i+j\lceil\sqrt{2^i}\rceil}, 2^{2^i+(j+1)\lceil\sqrt{2^i}\rceil} - 1]$, and define ℓ_v to be v's member of highest identifier. We ensure that v's members know v's *identifier* $\text{id}(v) = \text{id}(\ell_v)$ and v's size $|v|$, and are internally organized in two overlay topologies:

- **Internal Tree.** The *internal tree* T_v of v is a tree of constant degree and depth $O(\log|v|)$ that consists of all members of v and is rooted at ℓ_v.
- **Internal Butterfly.** The *internal butterfly* B_v of v is a $\lfloor\log|v|\rfloor$-dimensional wrapped butterfly network consisting of virtual nodes. More specifically, let k be the largest power of 2 such that $k \leq |v|$, and let $x_0,\ldots,x_{|v|-1} = \ell_v$ be the members of v in some arbitrary order. For $0 \leq l \leq k-1$, x_l simulates the nodes of column l of the butterfly. To do so, x_l knows the identifier of every other member of v that simulates a butterfly node that is adjacent to a butterfly node of column l. If $|v|$ is not a power of 2, then for every $k \leq l \leq |v| - 1$, the node x_l, which does not simulate a column of the butterfly, is connected to the butterfly via a bidirected edge to its *helper node* x_{l-k}. An example of an internal butterfly can be found in Fig. 1.

For an inactive supernode u we cannot retain such strong invariants; particularly, v's members might not know $id(v)$ nor $|v|$. Still, we ensure that u is internally organized as a rooted tree R_v of polylogarithmic degree.

3.2 Expansion Stage

In the Expansion Stage, every active supernode v maintains a set of *neighbors* N_v, which initially consists of all active supernodes v is adjacent to in G_0 (we say two supernodes v, u are *adjacent in* G_0 if there exist members x and y of v and u, respectively, such that $\{x, y\} \in E_0$). v's goal is to expand its set of neighbors to be of size $2^{\sqrt{2^i}}$; this will allow us to find large clusters of supernodes to be merged in the Grouping Stage. To do so, v performs $\lceil \sqrt{2^i} \rceil$ *introduction steps*, in each of which it introduces all of its neighbors known so far to each other.

To carry out the required communication, v utilizes its $(2\lceil \sqrt{2^i} \rceil)$-dimensional wrapped butterfly that results from taking only the leftmost $2^{2\lceil \sqrt{2^i} \rceil}$ columns and bottommost $2\lceil \sqrt{2^i} \rceil + 1$ rows of B_v, and which we call v's *small butterfly* b_v (see dashed outline in Fig. 1). Note that b_v must exist as $|v| \geq 2^{2^i + j\lceil \sqrt{2^i} \rceil} \geq 2^{2\lceil \sqrt{2^i} \rceil}$ for $i \geq 2$ (for $i \leq 1$ we simply choose $b_v = B_v$). Our idea is to store N_v exclusively within b_v, which allows us to perform each introduction step efficiently. However, we have to make sure that N_v never grows too large; specifically, we ensure that N_v contains at most $2^{2\lceil \sqrt{2^i} \rceil}$ many supernodes.

Computing $|N_v|$. Before the first introduction step, v has to determine whether it is already adjacent to at least $2^{\lceil \sqrt{2^i} \rceil}$ many active supernodes in G_0. To do so, every member x of v collects all of its neighbors in G_0 that are members of a different active supernode. In the following, we refer to a node that simulates a column of B_v as a *node of B_v*. If x is not a node of B_v, its sends its collected neighbors to its helper node. To correctly determine $|N_v|$, v needs to filter all members corresponding to the same supernodes and keeps only one *representative* for each of its adjacent supernodes. Then, by using the Aggregate-and-Broadcast Primitive to count the number of representatives, v learns $|N_v|$.

More specifically, filtering the members is performed by using the Filtering Primitive in the following way: Every member x of v that is a node of B_v and that stores some member y of a supernode u (either because y is a neighbor of x in G_0, or because y has been sent to x by some member of v of which x is the helper node) sends y from the topmost node of x's column in B_v to the bottommost node of column $h(id(u))$, where $h : [n] \rightarrow [k]$ is a common (pseudo-) random hash function[5], and k is the number of columns of B_v. Note that as G_0 has polylogarithmic degree, every node is source of at most polylogarithmically many messages; furthermore, by using Chernoff bounds, it can be shown that every node is target of at most polylogarithmically many messages, w.h.p. Therefore, the Filtering Primitive ensures that exactly one member of u reaches the bottommost node of column $h(id(u))$ in time $O(2^i)$.

---

[5] To agree on a suitable random hash function h, ℓ_v needs to broadcast $O(\log^2 n)$ random bits to all nodes in B_v (see [4,7] for details).

Preparation Steps. If $|N_v| \geq 2^{\lceil \sqrt{2^i} \rceil}$, v becomes *successful*, in which case v does not participate in the Expansion Stage any further. For the following description we thus assume that v was *unsuccessful*. Every member of v that stores a representative a asks a whether its supernode u has been successful. In this case, v must refrain from introducing u any further, and marks a as *finished*. Then, all representatives that have not been marked as finished are moved into b_v by using the Routing Primitive in the following way: Every member x of v that stores a representative y of some supernode u sends y from the topmost node of x's column in B_v (note that x must be a node of B_v) to the bottommost node of column $h'(\mathrm{id}(y))$, where h' is h restricted to range $[2^{2\lceil \sqrt{2^i} \rceil}]$. By using Chernoff bounds and the fact that there are at most $2^{\lceil \sqrt{2^i} \rceil}$ messages, it can easily be shown that every node is source and target of at most polylogarithmically many messages, which allows us to route all messages in time $O(2^i)$.

Before v can finally begin every representative a of some supernode u that is stored at a member x of v needs to be replaced by a randomly chosen node of b_u. Note that by using the Pipelined Sampling Primitive, we can steadily provide all nodes of b_v with polylogarithmically many other random nodes of b_v. Thus, x simply requests a random node of b_u from a, which becomes v's new representative of u. Then, v is ready to begin the first introduction step. It will continually perform introduction steps until it either declares itself successful at the end of an introduction step, or after having performed $\lceil \sqrt{2^i} \rceil$ steps. As each introduction step takes time $O(\sqrt{2^i})$, the total introduction will take time $O(2^i)$.

Introduction Step. At the beginning of each introduction step, v stores exactly one representative of each neighbor u it has learned so far, which is a random node of b_u stored at a random node of b_v. To be able to introduce all other neighbors of v to u, of which there can be up to $2^{\lceil \sqrt{2^i} \rceil} - 1$, v's goal is to first obtain $2^{\lceil \sqrt{2^i} \rceil}$ many representatives of u. Furthermore, v needs to find an assignment of representatives to one another such that every neighbor receives exactly one representative of each other neighbor of v, and each representative is sent to at most one supernode. It does so by enumerating its neighbors from 0 to $|N_v| - 1$, and enumerating the representatives of each neighbor from 0 to $2^{\lceil \sqrt{2^i} \rceil} - 1$; then, it introduces the p-th representative of its q-th neighbor to the q-th representative of its p-th neighbor (if both exist).

To achieve that, v assigns each neighbor u a unique label $r(u) \in [|N_v|]$ as follows. Consider the subtree T of b_v that results from taking all paths connecting the topmost node of column 0 with all nodes on the bottom level of b_v. Every inner node of T can easily compute the number of representatives stored at its leaves by performing an aggregation in T, and inform its parent about this value. This allows the root of T to assign intervals of labels to its children in T, which further divide the interval according to the values received from their children, until every leaf of T that stores a representative receives unique labels for all representatives stored at it.

We now show how v obtains $2^{\lceil \sqrt{2^i} \rceil}$ representatives of each neighbor u. Alongside, the process will also assign each representative a of u obtained in that way a unique label $r(a) \in [2^{\lceil\sqrt{2^i}\rceil}]$. Let x be the member of u that stores the first representative y of u. First, x sends a request that contains two random nodes of b_v to y. To be able to later locally associate representatives with their supernodes, the message contains the value $r(u)$; furthermore, it contains an empty bitstring l from which the representative's labels will be constructed. When y receives the request, it responds with two random nodes of b_u, each sent to either of the two nodes contained in the request. One of the messages gets associated with label $0 \circ l$ (where $\circ$ denotes the concatenation of two binary strings), the other with $1 \circ l$, and both contain $r(u)$. Whenever a node of b_v receives a response from u associated with label l', it sends a new request containing $r(u)$ and l' to the node of b_u that was contained in the response. In the $\lceil\sqrt{2^i}\rceil$-th iteration, v receives $2^{\lceil\sqrt{2^i}\rceil}$ many representatives of u. Each representative a gets stored together with its associated label $r(a)$ and u's label $r(u)$ by its recipient in b_v.

To finally perform the actual introduction, every node x of b_v that stores a representative a of some neighbor u aims to send a to the representative b of the supernode w such that $r(a) = r(w)$ and $r(u) = r(b)$, if that node exists (in which case we say b is a's *counterpart*). However, x does not know b, therefore it first has to send a to the node of b_v that currently stores b, and let that node take care of the introduction. To do so, we first relocate all representatives in b_v to allow for an easy retrieval: x moves a to the node y of b_v that simulates column $h(r(u), r(a))$ using the Routing Primitive and a (pseudo-)random hash function $h : [2^{\lceil\sqrt{2^i}\rceil}]^2 \to [2^{2\lceil\sqrt{2^i}\rceil}]$. Afterwards, y simply routes a message $(a, r(a), r(u))$ to the node z that simulates column $h(r(a), r(u))$. If a's counterpart b, which is a member of a supernode w, exists, then it must be stored by z, as $h(r(a), r(u)) = h(r(w), r(b))$. Consequently, z can send a to b. Note that z may receive and store multiple representatives and counterparts, respectively, but is able to match them accordingly as their labels have been sent along.

After the introduction, the nodes of b_v store many representatives, which need to be filtered such that exactly one representative is stored for each new neighbor of v. This can easily be done by performing the Filtering Primitive as in the beginning of the Expansion Stage, but in b_v only. v then counts the number of representatives stored in b_v; if this value is at least $2^{\lceil\sqrt{2^i}\rceil}$, it becomes *successful*. Irrespectively, v also declares itself successful if one of its neighbors has become successful in the previous introduction step (i.e., if v has marked one of its stored representatives as finished at the end of the previous step). Finally, every member of v that stores a representative a asks a whether its supernode has been successful; if so, a is marked as finished and does not partake in the next introduction step.

After v has performed $\lceil\sqrt{2^i}\rceil$ introduction steps, it stops. If v is unsuccessful until the end of the last introduction step, but determines that one of its neighbors has become successful in this step, it also declares itself successful. Note that at the end of the Expansion Stage, v may have been unsuccessful, and even if is has been successful, it might have learned fewer than $2^{\lceil\sqrt{2^i}\rceil}$ many supernodes (i.e., when it became successful because of one of its neighbors). The following lemma establishes the correctness of each introduction step.

Lemma 2. *Let v be an active and not yet successful supernode and consider an introduction step in which none of v's neighbors have been successful already. For each $u \in N_v$, v sends exactly one representative for each supernode $w \in N_v$ to u. Further, each representative stored at v is sent at most once, and is the recipient of at most one representative of another supernode.*

Note that the last claim of Lemma 2 also holds when v has already marked some representative as finished, after which v does not participate in any further introduction step. We conclude the presentation of the Expansion Stage with the following lemma.

Lemma 3. *Let $H = (A, R)$ be the undirected graph whose node set is the set of all active supernodes A, and whose edge set R contains an edge $\{u, v\}$ for all supernodes u, v that are adjacent in G_0. Let C be a connected component of H. If $|C| \geq 2^{\lceil \sqrt{2^i} \rceil} + 1$, then every supernode in C is successful at the end of the Expansion Stage. Otherwise, no supernode in C is successful. The Expansion Stage takes time $O(2^i)$.*

3.3   Grouping Stage

After having collected potentially large sets of neighbors in the Expansion Stage, the goal of each active supernode in the Grouping Stage is to establish directed *merge edges* between members of supernodes that ensure the following: the total number of members contained in each component of supernodes connected by merge edges (which we call *merge component*) amounts to at least $2^{2^i+(j+1)\lceil \sqrt{2^i} \rceil}$.

Successful Supernodes. Let v be a successful supernode. We define $t(v)$ to be the introduction step in which v became successful; $t(v) = 0$, if v was successful at the beginning already. We further define $\phi(v) = (t(v), \mathrm{id}(v))$, and let $\phi(u) < \phi(v)$ if and only if $t(u) < t(v)$, or $t(u) = t(v)$ and $\mathrm{id}(u) < \mathrm{id}(v)$. To ensure that v becomes part of a large merge component, it chooses a neighbor to merge with according to the following rules:

1. If v has learned a neighbor u in the Expansion Stage such that $\phi(u) < \phi(v)$, and there is no neighbor w of v such that $\phi(w) < \phi(u) < \phi(v)$, v chooses u, randomly picks a member x of v and a member y of u, and selects (x, y) as its merge edge.
2. If otherwise v has learned a neighbor u whose neighbor w of smallest ϕ-value is such that $\phi(w) < \phi(v)$ (and which, consequently, is not a neighbor of v itself), and none of v's neighbors has a neighbor with even smaller ϕ-value, v chooses u, randomly picks a member x of v and a member y of u, and selects (x, y) as its merge edge.

Note that v might be unable to determine a supernode to merge with, i.e., if v has the minimum ϕ-value among all neighbors of v and their respective neighbors (we say v is a *local minimum*).

It remains to show how v can retrieve the information required for its choice. First, note that v can easily determine its own ϕ-value $\phi(v)$, which it sends to all of its neighbors via their representatives. By using the Aggregate-and-Broadcast Primitive, v learns the smallest of all ϕ-values of its neighbors and can determine whether it can select a merge edge according to the first rule. In the same way, we let each supernode inform its neighbors about the smallest ϕ-value it knows; this allows v to select a merge edge according to the second rule. If v has selected a neighbor u to merge with, then the member of v that stores the representative a of u requests a randomly selected member y of u from a; upon receipt, it sends y to a randomly selected member x of v, which learns the merge edge (x, y).

Unsuccessful Supernodes. If v was unsuccessful in the Expansion Stage, then by Lemma 3, the component C of adjacent active supernodes that contains v consists of fewer than $2^{\lceil \sqrt{2^i} \rceil} + 1$ supernodes, all of which are unsuccessful. It can easily be seen that the neighborhood of each supernode in C is the set of all other supernodes in C. Furthermore, if there still exists an inactive supernode, one of the supernodes of C must be adjacent to an inactive supernode in G_0. Conversely, if none is adjacent to an inactive supernode anymore, C contains all remaining supernodes. In the first case, we let the supernode u of C that has lowest identifier merge with an inactive supernode u is adjacent to in G_0, if such a node exists, and let all the others merge with u. In the second case, all supernodes simply merge with the supernode in C that has lowest identifier.

More specifically, v does the following: it first determines whether it is adjacent to an inactive supernode in G_0 by performing the Aggregate-and-Broadcast Algorithm, and, if so, sends its own identifier to all of its neighbors via their representatives in v. By performing the algorithm a second time, all members of v learn the supernode in C that has lowest identifier among all that are adjacent to an inactive supernode. If $u = v$, then v determines the edge $\{x, y\} \in E_0$ such that x is a member of u and y is a member an inactive supernode adjacent to u, and $\mathrm{id}(x) \circ \mathrm{id}(y)$ is minimal among all such edges. The edge (x, y) is selected as v's merge edge. If otherwise $u \neq v$, then v creates a merge edge (x, y) between v and u by randomly choosing a member x of v and replacing its representative of u by a randomly chosen member y of u as described in the previous paragraph. Finally, if v did not receive any identifier, and thus no supernode in C is adjacent to an inactive supernode, v determines its neighbor u that has the lowest identifier, and, if $\mathrm{id}(u) < \mathrm{id}(v)$, creates a random merge to u in the same way.

Lemma 4. *After the Grouping Stage, every merge component C is a tree that (1) entirely consists of at least $2^{\lceil \sqrt{2^i} \rceil}$ successful supernodes, (2) consists of unsuccessful supernodes, only its root being an inactive supernode, (3) entirely consists of unsuccessful supernodes and contains all nodes of V. The Grouping Stage takes time $O(2^i)$.*

3.4 Merging Stage

Let C be a merge component, and let v' be the resulting supernode after merging C. As we show at the end of this section, $|v'| \geq 2^{2^i + (j+1)\lceil \sqrt{2^i} \rceil}$. If

$|v'| \leq 2^{2^i + (j+2)\lceil\sqrt{2^i}\rceil} - 1$, then $|v'|$ will be active in the next subphase; otherwise, it is too large and will be inactive. In the Merging Stage, we attempt to construct the internal tree $T_{v'}$ of v' for $O(2^i)$ rounds. If we succeed, v' uses its internal tree to compute its size, and determines whether it is actually active in the next subphase. Otherwise, v' is clearly too large, in which case every member of v' will immediately learn that v' will be inactive. We first describe how $T_{v'}$ is constructed and finally show how to obtain $B_{v'}$ as a by-product.

For now, assume that the root of C is an active node, and let v be an inner node of C (i.e., v has selected a merge edge (x, y) from a member of v to a member of its parent u). We first establish x as the new root of T_v by performing a broadcast in T_v. The graph that results from taking the internal trees together with all merge edges is a rooted tree that contains all nodes of C (i.e., all members of v'). To transform this tree into a constant degree tree of diameter $O(\log|v'|)$, we perform the merging step of the algorithm of [11]: given a rooted tree of size m that has polylogarithmic degree, the algorithm transforms it into a tree of constant degree and diameter $O(\log m)$ within $O(\log m)$ rounds. By letting the algorithm run for only $O(2^i)$ rounds, v' can therefore successfully construct $T_{v'}$, if $|v'| \leq 2^{2^i + (j+2)\lceil\sqrt{2^i}\rceil} - 1$. Note that the algorithm can be modified to also detect whether it has been successful or not.

Now assume that the root u of C is inactive. By our assumption in Sect. 3.1, u is internally organized as a rooted tree R_v of polylogarithmic degree. Note that R_v might have large diameter, and thus only the members of u which have been selected as the endpoint of a merge edge know whether an active supernode intends to merge with u. However, the members of u irrespectively participate in the execution of the algorithm of [11], and eventually learn whether it has been successful or not. By doing so, every inactive supernode essentially attempts to construct its internal tree in each subphase, and finally succeeds in doing when its size permits it to become active in the subsequent subphase.

Note that irrespective of whether or not the algorithm is successful, the internal trees of the supernodes of C (which are not the root u of C), together with T_u, if u is active, or R_u, otherwise, connected by the merge edges of C, form a rooted tree of polylogarithmic degree. Therefore, the nodes know $R_{v'}$ in case v' in inactive in the next subphase.

Finally, we show how to construct $B_{v'}$ alongside the above execution of the algorithm of [11] (we remark that if the algorithm is unsuccessful, then the construction of $B_{v'}$ must also be terminated). As an intermediate step, the algorithm constructs a *ring of virtual nodes* in which every member of v' simulates some node. This ring can easily be transformed into a line that consists of all members of v' in an arbitrary order, enumerated from 0 to $|v'| - 1$, in time $O(\log|v'|)$. Our goal is to let the l-th node of that line simulate column l of the butterfly. To do so, every node only has to learn the identifier of each node to which its distance on the line is a power of 2. This can easily be achieved by performing pointer jumping: Every node with degree 2 introduces its two neighbors to each other; in each subsequent round, every node that receives two nodes introduces them to each other. After $O(\log|v'|)$ rounds, every node knows the identifiers of all nodes simulating its neighbors in the butterfly. Furthermore, if merging is successful, every member of v' learns $id(v')$ and $|v'|$.

The following lemma follows from [11], the fact that every node is chosen as the endpoint of a merge edge at most polylogarithmically often, and Lemma 4.

Lemma 5. *After the Merging Stage, every resulting supernode v is of size at least $2^{2^i+(j+1)\lceil\sqrt{2^i}\rceil}$ or contains all nodes of V. If $|v| \leq 2^{2^i+(j+2)\lceil\sqrt{2^i}\rceil} - 1$, then every member of v learns $id(v)$ and $|v|$, and the algorithm succeeds in constructing T_v and B_v. Otherwise, the members of v know of their adjacent edges in R_v. The Merging Stage takes time $O(2^i)$.*

By taking all lemmas of this section together with Lemma 1, and proving that the random load-balancing of our algorithm only causes polylogarithmic communication work, w.h.p., we conclude the following theorem.

Theorem 1. *If G_0 is a weakly connected graph with polylogarithmic degree, then the algorithm transforms G_0 into a constant degree tree of depth $O(\log n)$ in time $O(\log^{3/2} n)$ and with polylogarithmic communication work, w.h.p.*

4   Conclusion

We have shown how to construct overlays from arbitrary initial topologies efficiently by leveraging each supernode's capability to perform massive amounts of communication. To the best of our knowledge, this is the first algorithm that achieves an $O(\log^{3/2} n)$ runtime for this problem, and which is a further step towards closing the gap to the lower bound of $\Omega(\log n)$. However, we do not see any evidence why our algorithm should be asymptotically optimal; for example, our algorithm is not able to exploit the potential communicational power of a supernode to its limit. Furthermore, the algorithm does not take into account the initial graph topology, which might lead to more efficient solutions.

Finally, we point out that our algorithm is a *Monte Carlo algorithm*: whereas its runtime is guaranteed, it may require a node to perform more than polylogarithmic communication work with very small probability. However, we strongly believe that one can derive a *Las Vegas algorithm* from our algorithm with little effort, i.e., an algorithm that is guaranteed to work, but whose runtime only holds w.h.p. It may also be possible to come up with a fully deterministic algorithm.

References

1. Angluin, D., Aspnes, J., Chen, J., Wu, Y., Yin, Y.: Fast construction of overlay networks. In: Proceedings of the 17th Annual ACM Symposium on Parallelism in Algorithms and Architectures (SPAA), pp. 145–154 (2005)
2. Aspnes, J., Shah, G.: Skip graphs. In: Proceedings of the 14th ACM-SIAM Symposium on Discrete Algorithms (SODA), pp. 384–393 (2003)
3. Aspnes, J., Wu, Y.: $O(\log n)$-time overlay network construction from graphs with out-degree 1. In: Tovar, E., Tsigas, P., Fouchal, H. (eds.) OPODIS 2007. LNCS, vol. 4878, pp. 286–300. Springer, Heidelberg (2007). https://doi.org/10.1007/978-3-540-77096-1_21

4. Augustine, J., et al.: Distributed computation in node-capacitated networks. In: Proceedings of the 31st Annual ACM Symposium on Parallelism in Algorithms and Architectures (SPAA) (2019, to appear). http://arxiv.org/abs/1805.07294

5. Augustine, J., Pandurangan, G., Robinson, P., Roche, S.T., Upfal, E.: Enabling robust and efficient distributed computation in dynamic peer-to-peer networks. In: Proceedings of 56th IEEE Annual Symposium on Foundations of Computer Science (FOCS), pp. 350–369 (2015)

6. Augustine, J., Sivasubramaniam, S.: Spartan: a framework for sparse robust addressable networks. In: Proceedings of the 32nd IEEE International Parallel and Distributed Processing Symposium (IPDPS), pp. 1060–1069 (2018)

7. Barenboim, L., Elkin, M., Pettie, S., Schneider, J.: The locality of distributed symmetry breaking. J. ACM **63**(3), 20:1–20:45 (2016)

8. Berns, A., Ghosh, S., Pemmaraju, S.V.: Building self-stabilizing overlay networks with the transitive closure framework. Theor. Comput. Sci. **512**, 2–14 (2013)

9. Feldmann, M., Scheideler, C.: A self-stabilizing general De Bruijn graph. In: Spirakis, P., Tsigas, P. (eds.) SSS 2017. LNCS, vol. 10616, pp. 250–264. Springer, Cham (2017). https://doi.org/10.1007/978-3-319-69084-1_17

10. Feldotto, M., Scheideler, C., Graffi, K.: HSkip+: a self-stabilizing overlay network for nodes with heterogeneous bandwidths. In: Proceedings of the 14th IEEE International Conference on Peer-to-Peer Computing (P2P), pp. 1–10 (2014)

11. Gmyr, R., Hinnenthal, K., Scheideler, C., Sohler, C.: Distributed monitoring of network properties: the power of hybrid networks. In: Proceedings of the 44th International Colloquium on Automata, Languages, and Programming (ICALP), pp. 137:1–137:15 (2017)

12. Götte, T., Vijayalakshmi, V.R., Scheideler, C.: Always be two steps ahead of your enemy. In: Proceedings of the 33rd IEEE International Parallel and Distributed Processing Symposium (IPDPS) (2019, to appear)

13. Jacob, R., Richa, A.W., Scheideler, C., Schmid, S., Täubig, H.: Skip+: a self-stabilizing skip graph. J. ACM **61**(6), 36:1–36:26 (2014)

14. Jacob, R., Ritscher, S., Scheideler, C., Schmid, S.: Towards higher-dimensional topological self-stabilization: a distributed algorithm for delaunay graphs. Theor. Comput. Sci. **457**, 137–148 (2012)

15. Kniesburges, S., Koutsopoulos, A., Scheideler, C.: Re-chord: a self-stabilizing chord overlay network. In: Proceedings of the 23rd Annual ACM Symposium on Parallelism in Algorithms and Architectures (SPAA), pp. 235–244 (2011)

16. Mitzenmacher, M., Upfal, E.: Probability and Computing. Cambridge University Press, Cambridge (2005)

17. Onus, M., Richa, A., Scheideler, C.: Linearization: locally self-stabilizing sorting in graphs. In: Proceedings of the Meeting on Algorithm Engineering and Experiments (ALENEX), pp. 99–108 (2007)

18. Richa, A., Scheideler, C., Stevens, P.: Self-stabilizing De Bruijn networks. In: Défago, X., Petit, F., Villain, V. (eds.) SSS 2011. LNCS, vol. 6976, pp. 416–430. Springer, Heidelberg (2011). https://doi.org/10.1007/978-3-642-24550-3_31

19. Rowstron, A., Druschel, P.: Pastry: scalable, decentralized object location, and routing for large-scale peer-to-peer systems. In: Guerraoui, R. (ed.) Middleware 2001. LNCS, vol. 2218, pp. 329–350. Springer, Heidelberg (2001). https://doi.org/10.1007/3-540-45518-3_18

20. Stoica, I., Morris, R.T., Karger, D.R., Kaashoek, M.F., Balakrishnan, H.: Chord: a scalable peer-to-peer lookup service for internet applications. In: Proceedings of the 2018 Conference of the ACM Special Interest Group on Data Communication (SIGCOMM), pp. 149–160 (2001)

Partial Gathering of Mobile Agents Without Identifiers or Global Knowledge in Asynchronous Unidirectional Rings

Masahiro Shibata[1]([⊠]), Norikazu Kawata[2], Yuichi Sudo[2], Fukuhito Ooshita[3], Hirotsugu Kakugawa[4], and Toshimitsu Masuzawa[2]

[1] Graduate School of Computer Science and Systems Engineering, Kyushu Institute of Technology, Iizuka, Japan
shibata@cse.kyutech.ac.jp
[2] Graduate School of Information Science and Technology, Osaka University, Suita, Japan
{n-kawata,y-sudou,masuzawa}@ist.osaka-u.ac.jp
[3] Graduate School of Science and Technology, NAIST, Ikoma, Japan
f-oosita@is.naist.jp
[4] Faculty of Science and Technology, Ryukoku University, Ootsu, Japan
kakugawa@rins.ryukoku.ac.jp

Abstract. In this paper, we consider the partial gathering problem of mobile agents in asynchronous unidirectional rings. This problem requires that, for a given positive integer g, all the agents terminate in a configuration such that at least g agents or no agent exist at each node. While the previous work achieves move-optimal partial gathering using distinct IDs or knowledge of the number of agents, in this paper we aim to achieve this without such information. We consider deterministic and randomized cases. First, in the deterministic case, we show that unsolvable initial configurations exist. In addition, we propose an algorithm to solve the problem from any solvable initial configuration in $O(gn)$ total number of moves, where n is the number of nodes. Next, in the randomized case, we propose an algorithm to solve the problem in $O(gn)$ expected total number of moves from any initial configuration. Since agents require $\Omega(gn)$ total number of moves to solve the partial gathering problem, our algorithms can solve the problem in asymptotically optimal total number of moves without global knowledge.

Keywords: Distributed system · Mobile agent · Gathering problem · Partial gathering

1 Introduction

1.1 Background and Related Works

The *total gathering problem* is a fundamental problem for (mobile) agents' coordination. Agents are software object that can traverse the distributed system with

© Springer Nature Switzerland AG 2019
K. Censor-Hillel and M. Flammini (Eds.): SIROCCO 2019, LNCS 11639, pp. 277–292, 2019.
https://doi.org/10.1007/978-3-030-24922-9_19

carrying information collected at visited nodes [1,2]. This problem requires that all the k agents distributed in the system terminate at a single node in finite time. By meeting at a single node, all agents can share information or synchronize behaviors. The total gathering problem has been considered in various kinds of networks such as rings [3–5], trees [6], tori [7], and arbitrary networks [8].

Recently, a variant of the total gathering problem, called the g-partial gathering problem [9], has been considered. This problem does not require all the agents to meet at a single node, but allows the agents to meet partially at several nodes. Concretely, the problem requires that, for a given positive integer $g\,(<k)$, all the agents terminate in a configuration such that at least g agents or no agent exist at each node. The g-partial gathering problem is still useful especially in large-scale networks. That is, after achieving g-partial gathering, each agent can share information and tasks with at least g agents (or a group) staying at the same node, and each group can partition the network and patrol its area that they should monitor efficiently. The g-partial gathering problem is interesting also from a theoretical point of view. If $k/2 < g < k$ holds, the g-partial gathering problem is clearly equivalent to the total gathering problem. On the other hand, if $2 \leq g \leq k/2$ holds, the requirement for the g-partial gathering problem is no stronger than that for the total gathering problem. Thus, there exists possibility that the g-partial gathering problem can be solved with strictly fewer total number of moves (i.e., lower costs) compared to the total gathering problem.

Our previous works considered the g-partial gathering problem in rings [9], trees [10], and arbitrary networks [11]. In [9], we considered it in unidirectional rings with whiteboards (or memory spaces) at nodes. We considered two problem settings about agents: distinct agents (i.e., agents with distinct IDs) and anonymous agents (i.e., agents without IDs) with knowledge of k. For distinct agents, we gave a deterministic algorithm to solve the problem in $O(gn)$ total number of moves, where n is the number of nodes. For anonymous agents with knowledge of k, we considered deterministic and randomized cases. In the deterministic case, we showed that unsolvable initial configurations exist. In addition, we gave an algorithm to solve the problem from any solvable initial configuration in $O(kn)$ total number of moves. In the randomized case, we gave an algorithm to solve the problem in $O(gn)$ expected total number of moves. The g-partial (resp., the total) gathering problem in rings requires $\Omega(gn)$ (resp., $\Omega(kn)$) total number of moves. Thus, the first and third results are asymptotically optimal in terms of total number of moves, and the total number $O(gn)$ of moves is strictly fewer than that for the total gathering problem when $g = o(k)$. In tree and arbitrary networks, we also proposed algorithms to solve the g-partial gathering problem with strictly fewer total number of moves compared to the total gathering problem for some settings, but we omit to explain the results due to page limit.

1.2   Our Contribution

In this paper, for the case of $2 \leq g \leq k/2$ we consider the g-partial gathering problem in asynchronous unidirectional ring networks with whiteboards at nodes as in [9]. While the previous work [9] achieved move-optimal g-partial gathering using distinct IDs or knowledge of k, in this paper we aim to achieve this without such information.

As contributions, we consider deterministic and randomized cases. First, in the deterministic case, we show that the set of unsolvable initial configurations is the same as that for agents with knowledge of k. We also show (unlike the case of [9]) that agents cannot detect whether the initial configuration is an unsolvable one or not. In addition, we propose an algorithm to solve the g-partial gathering problem from any solvable initial configuration in $O(gn)$ total number of moves. Next, in the randomized case, we propose an algorithm to solve the problem with probability 1 in $O(gn)$ expected total number of moves from any initial configuration. Thus, our algorithms can solve the g-partial gathering problem in asymptotically optimal total number of moves without distinct IDs or global knowledge. These results improve our previous results in [9]. Due to page limit, we omit several pseudocodes and proofs of lemmas and theorems.

2   Preliminaries

2.1   System Model

A *unidirectional ring network* R is defined as 2-tuple $R = (V, E)$, where $V = \{v_0, v_1, \ldots, v_{n-1}\}$ is a set of n nodes and $E = \{e_0, e_1, \ldots, e_{n-1}\}$ ($e_i = (v_i, v_{(i+1) \bmod n})$) is a set of directed links. For simplicity, we denote $v_{(i+j) \bmod n}$ by v_{i+j} for any integers i and j. The *distance* from node v_i to v_j is defined to be $(j - i) \bmod n$. We define the direction from v_i to v_{i+1} (resp., v_{i-1}) as the *forward* (resp., *backward*) direction. We assume that nodes are anonymous. Every node $v_i \in V$ has a whiteboard that agents at node v_i can read from and write on.

Let $A = \{a_0, a_1, \ldots, a_{k-1}\}$ be a set of k ($\leq n$) anonymous agents. Agents can move through directed links, that is, they can move from v_i to v_{i+1} for any i. Agents do not have knowledge of k or n, and they cannot detect whether other agents exist at the current node or not.

We consider two models. In the first model, we consider agents executing a deterministic algorithm. An agent a_i is a finite automaton $(S, W, \delta, s_{initial}, s_{final}, w_{initial}, w'_{initial})$. The first element S is the set of all states of an agent, including two special states, initial state $s_{initial}$ and final state s_{final}. The second element W is the set of all states (contents) of a whiteboard, including two special initial states $w_{initial}$ and $w'_{initial}$. We explain the meanings of $w_{initial}$ and $w'_{initial}$ later. The third element $\delta : S \times W \mapsto S \times W \times M$ is the state transition function that decides the next states of a_i and the current node's whiteboard, and whether a_i moves to the next node or not based on the current states of a_i and the whiteboard. The variable $M = \{1, 0\}$ in δ represents whether a_i makes a movement or not. The value 1 represents movement to the next node and 0 represents stay at the current node. We assume that $\delta\left(s_{final}, w_i\right) = \left(s_{final}, w_i, 0\right)$ holds for any state $w_i \in W$, which means that a_i never changes its state, updates the contents of a whiteboard, or leaves the current node once it reaches state s_{final}.

In the second model, we consider agents executing a randomized algorithm. An agent a_i in this model is a probabilistic automaton $(S, W, R, \delta, s_{initial}, s_{final}, w_{initial}, w'_{initial})$. The third element R is a set of random numbers. Since we treat a randomized algorithm, δ is a mapping $S \times W \times R \mapsto S \times W \times M$. If the state of a_i is s_{final}, then $\delta\left(s_{final}, w_i, r\right) = \left(s_{final}, w_i, 0\right)$ holds for any state $w_i \in W$ and any random number $r \in R$. The other elements in the automaton are the same as those in the deterministic model. Note that for both the models all the agents are modeled by the same state machine since they are anonymous.

In an agent system, (global) *configuration* c is defined as a product of the states of all the agents, the states (whiteboards' contents) of all the nodes, and the locations (i.e., the current nodes) of all the agents. We define C as a set of all configurations. In an initial configuration $c_0 \in C$, we assume that the states of all the agents are $s_{initial}$, agents are arbitrary placed at nodes so that no two agents stay at the same node, and the states of whiteboards are $w_{initial}$ or $w'_{initial}$ depending on existence of an agent. That is, when an agent exists at node v, the state of v's whiteboard is $w_{initial}$. Otherwise, the state is $w'_{initial}$.

During execution of the algorithm, we assume that agents move instantaneously, that is, agents always exist at nodes (do not exist on links). Each agent at node v executes the following four operations in an *atomic step*: (1) reads the contents of v's whiteboard, (2) executes local computation (or changes its state), (3) updates the contents of v's whiteboard, and (4) moves to the next node or stays at v. A configuration changes to the next one when a *scheduler* activates some agents and the agents take atomic steps as mentioned before. Concretely, letting A_i be a non-empty set of agents, configuration c_i changes to c_{i+1} when the scheduler activates each agent $a_j \in A_i$ and a_j takes an atomic step. If multiple agents at the same node are included in A_i, the agents take atomic steps interleavingly in an arbitrary order. We denote the transition by $c_i \xrightarrow{A_i} c_{i+1}$.

A sequence of configurations $E = c_0, c_1, \ldots$ is called an *execution* starting from c_0 if there exists a sequence $A_0, A_1, \ldots$ of non-empty agent sets such that $c_i \xrightarrow{A_i} c_{i+1}$ holds for every $i \geq 0$. Execution E is infinite, or ends in final configuration c_{final} where every agent's state is s_{final}. We assume that the scheduler is *fair*, that is, each agent is activated infinitely often. When the scheduler activates all the agents for every transition $c_i \xrightarrow{A_i} c_{i+1}$, that is, $c_i \xrightarrow{A} c_{i+1}$ holds for every i, the execution is called a *synchronous execution*. Otherwise, i.e., if $c_i \xrightarrow{A} c_{i+1}$ does not always hold, the execution is called an *asynchronous execution*. In this paper, we consider the asynchronous execution.

2.2   Partial Gathering Problem

The requirement of the partial gathering problem is that, for a given integer g $(2 \leq g \leq k/2)$, agents terminate in a configuration such that at least g agents or no agent exist at each node. Formally, we define the problem as follows.

Definition 1. *Execution E solves the g-partial gathering problem when the following conditions hold:*

- *Execution E is finite (i.e., all agents terminate).*
- *In the final configuration, all agents are in the finial state, and for any node v_j where an agent exists, at least g agents exist at v_j.* □

Definition 2. *A deterministic algorithm $\mathcal{A}$ solves the g-partial gathering problem if any fair execution of $\mathcal{A}$ solves the problem.* □

Definition 3. *A randomized Algorithm $\mathcal{A}$ solves the g-partial gathering problem with probability 1 if an execution of $\mathcal{A}$ that solves the problem occurs with probability 1.* □

In [9], the following lower bound on the total number of agent moves for the g-partial gathering problem in ring networks is shown.

Theorem 1 [9]. *The total number of agent moves required to solve the g-partial gathering problem is $\Omega(gn)$ even if the algorithm is randomized.*

3   Deterministic g-partial Gathering

In this section, we consider the deterministic case. First, we show a sufficient condition of initial configurations from which agents cannot solve the problem, and then propose an algorithm to solve the problem from any initial configuration other than unsolvable initial configurations mentioned above. Thus, the sufficient condition of unsolvable initial configurations is also a necessary condition.

3.1   Unsolvable Initial Configurations

In this section, we show a sufficient condition of unsolvable initial configurations and show that agents cannot detect whether the initial configuration is an unsolvable one or not. The set of unsolvable configurations is the same as that for agents with knowledge of k shown in [9]. To prove this, we define *periodic initial configurations* as follows. At first, we define the i-th $(i \neq 0)$ forward (resp., backward) agent a' of agent a as the agent such that $i - 1$ agents exist between a and a' in a's forward (resp., backward) direction. We call the a's 1-st forward and backward agents *neighboring agents* of a. In initial configuration c_0, we assume that agents $a_0, a_1, \ldots, a_{k-1}$ exist in this order, that is, a_i is the i-th forward agent of a_0. Then, we define the *distance sequence* of agent a_i in c_0 as $D_i(c_0) = (d_0^i(c_0), \ldots d_{k-1}^i(c_0))$, where $d_j^i(c_0)$ is the distance from the j-th forward agent of a_i to the $(j + 1)$-st forward agent of a_i in c_0. In addition, we define the distance sequence $D(c_0)$ of configuration c_0 as the lexicograph-ically minimum sequence among $\{D_i(c_0)|a_i \in A\}$. Moreover, let $shift(D, x) = (d_x, d_{x+1}, \ldots, d_{k-1}, d_0, d_1, \ldots, d_{x-1})$ for sequence $D = (d_0, d_1, \ldots, d_{k-1})$. Then, we say that configuration c_0 is *periodic* if $D(c_0) = shift(D(c_0), x)$ holds for some x $(0 < x < k)$. Otherwise, we say c_0 is *aperiodic*. For an initial configuration, we define the period *peri* of the ring as the minimum positive integer x satisfying $D(C_0) = shift(D(C_0), x)$. Then, we have the following theorem.

Theorem 2. *No algorithm exists such that all its executions starting from a periodic initial configuration c_0 with peri less than g solve the g-partial gathering problem. In addition, no algorithm exists such that, in each of its executions, all agents can detect whether the initial configuration has peri less than g or not.*

Proof Sketch. The former argument can be proved by considering a synchronous execution. For each q $(0 \leq q \leq peri - 1)$, agents $a_q, a_{q+peri}, \ldots, a_{q+\ell \times peri}$ $(\ell = k/peri)$ always execute the same action simultaneously and they cannot break the symmetry. The latter argument holds because some agent a cannot distinguish a periodic initial configuration c_0 with $peri \leq g - 1$ from an aperiodic initial configuration c_0' with $peri = k$ $(>g)$ that consecutively includes c_0 sufficiently many times as a part of c_0'. Thus, a's execution starting from c_0' is the exactly same as that starting from c_0.   □

3.2   Proposed Algorithm

In this section, we propose a deterministic algorithm to solve the problem from any solvable initial configuration (i.e., any initial configuration with $peri \geq g$) in $O(gn)$ total number of moves. The algorithm consists of two parts. In the first part, several agents are elected as leader agents by executing the leader election algorithm to the middle. In the second part, the leader agents instruct the other agents which node they should meet at, and the other agents move to the node.

3.2.1   The First Part: Leader Election

The aim of this part is similar to [9], that is, to elect several leaders and satisfy the following properties: (1) At least one agent is elected as a leader, and (2) at least $g - 1$ non-leader agents exist between two leaders. Each agent takes a status from the following three statuses:

- active: The agent is performing leader election as a candidate for leaders.
- inactive: The agent has dropped out from the set of the leader candidates.
- leader: The agent has been elected as a leader.

Initially, all agents are active. At first, we explain the idea of leader election in [9] to adopt it to this paper. In [9], the network is a unidirectional ring, agents have distinct IDs, and each node has a whiteboard. For intuitively understanding, we firstly explain the idea for the case of bidirectional rings. Then, we apply the idea to the unidirectional ring. The algorithm consists of several phases. In each phase, each active agent a_i compares its own ID with IDs of its forward and backward neighboring active agents. Concretely, a_i writes its ID on the current node's whiteboard, and then moves forward and backward to observe IDs of its forward and backward active agents. Then, if its own ID is the smallest among the three IDs, a_i remains active (as a candidate for leaders) in the next phase. Otherwise, a_i becomes inactive and drops out from the set of leader candidates. Note that, in each phase, neighboring active agents never remain active because of distinct IDs. Hence, the number of inactive agents between two active agents at least doubles in each phase. Then, from [12], after executing j phases at least $2^j - 1$ inactive agents exist between two active agents. Thus, after executing $\lceil \log g \rceil$ phases, the following properties hold: (1) At least one agent remains active, and (2) the number of inactive agents between two active agents is at least $g - 1$. Therefore, all the remaining active agents become leaders.

Next, we implement the above algorithm in unidirectional rings using a traditional approach [12]. In unidirectional rings, active agent a_i cannot move backward or observe the ID of its backward active agent. Instead, a_i moves forward until it observes IDs of two active agents. Thus, a_i observes IDs of three successive active agents including a_i itself, say id_1, id_2, id_3 in this order. Note that id_1 is the ID of a_i. Here, this situation is similar to that in which the active agent with ID id_2 observes id_1 of its backward active agent and id_3 of its forward active agent in a bidirectional ring. For this reason, a_i behaves as if it would be an active agent with ID id_2 in the bidirectional ring. That is, if id_2 is the smallest among the three IDs, a_i remains active. Otherwise, a_i becomes inactive.

In the following, we explain how to apply the above leader election to anonymous agents. Let *active nodes* (resp., *inactive nodes*) be nodes where active agents (resp., inactive agents) start some phase. In this section, agents use *virtual IDs*. A virtual ID is given in the form of $(disArray[], nInactive)$, where $disArray[]$ and $nInactive$ are a distance sequence and the number of inactive nodes between active nodes, respectively. Concretely, we assume that active agent a_i starts some phase at node v_j and $v_{j'}$ is v_j's forward active node. In addition, let $v_{ina}^1, v_{ina}^2, \ldots, v_{ina}^\ell$ be inactive nodes between v_j and $v_{j'}$. That is, nodes $v_j (=v_{ina}^0), v_{ina}^1, v_{ina}^2, \ldots, v_{ina}^\ell, v_{j'} (=v_{ina}^{\ell+1})$ exist in this order. Then, when a_i moves from v_j to $v_{j'}$, it observes a distance sequence $(d_1, d_2, \ldots, d_{\ell+1})$, where $d_m (1 \le m \le \ell + 1)$ is the distance from v_{ina}^{m-1} to v_{ina}^m. Then, a_i gets $disArray[] = (d_1, d_2, \ldots d_{\ell+1})$ and $nInactive = \ell$ as its virtual ID. An example is given in Fig. 1(a). Note that each active agent can detect whether the current node is an active node, an inactive node, or another node using whiteboards. Each active agent moves until it observes such three virtual IDs. Note that, multiple agents may have the same virtual ID, and we explain this case next.

After observing three virtual IDs id_1, id_2, id_3, each active agent a_i compares the virtual IDs by the lexicographical order and decides whether it remains active in the next phase or not. Different from [9], multiple agents may have the same virtual ID. To treat this, when at least one virtual ID differs from other IDs, if $id_2 < \min(id_1, id_3)$ or $id_2 = id_3 < id_1$, a_i remains active. Otherwise, a_i becomes inactive. When all the three virtual IDs are identical (i.e., $id_1 = id_2 = id_3$ holds), a_i compares the value of $nInactive$ with the value of g. If $nInactive \ge g-1$, it still

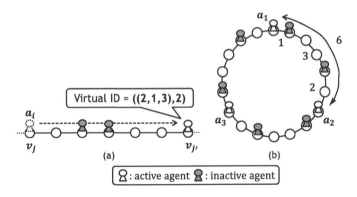

Fig. 1. (a): An example of a virtual ID. (b): An example of comparison of virtual IDs. Active agents a_1, a_2, and a_3 observe the same virtual IDs $((1,3,2),2)$. Then, they remain active if the value 2 of *nInactive* is $g-1$ or larger, and become inactive if the value is smaller than $g-1$.

remains active. Otherwise, it becomes inactive[1]. After executing such a phase $\lceil \log g \rceil$ times, agents complete leader election and all the remaining active agents become leaders. At the end of leader election, we can show that at least $g-1$ inactive agents exist between two leaders (Lemma 1). Intuitively, this is because (1) when three IDs have different values, neighboring active agents never remain active, and (2) when at least two IDs have the same value and two neighboring active agents a_i and a_j remain active, at least $g-1$ inactive agents already exist between a_i and a_j. An example is given in Fig. 1(b).

The pseudocode of active agents in the first part is described in Algorithm 1. Variable $a_i.phase$ (resp., $v_j.phase$) represents the phase number of agent a_i (resp., node v_j). Variable $v_j.initial$ represents existence of an agent in the initial configuration c_0. That is, $v_j.initial = true$ holds if an agent exists at node v_j in c_0 and $v_j.initial = false$ otherwise. In addition, variable $v_j.inactive$ represents existence of an inactive agent. That is, $v_j.inactive = true$ holds if an inactive agent exists at node v_j and $v_j.inactive = false$ otherwise. Initially, $v_j.inactive = false$ holds for any v_j. Moreover, agents use procedure *NextActive*() to move to the next active node and get a virtual ID (we omit the detailed description). Concerning leader election, we have the following lemmas.

Lemma 1. *Algorithm 1 eventually terminates and satisfies the following two properties when it terminates.*

- *At least one leader agent exists.*
- *At least $g-1$ inactive agents exist between two leader agents.* □

Lemma 2. *Algorithm 1 requires $O(n \log g)$ total number of moves.* □

---

[1] Even if a_i becomes inactive, it does not happen that all the active agents become inactive because we consider the case of $peri \geq g$.

Algorithm 1. Behavior of active agent a_i in the first part (v_j is the current node of a_i.)

Variables for Agent a_i
int $a_i.phase = 0$;
int $nInactive_1, nInactive_2, nInactive_3$;
array $disArray_1[], disArray_2[], disArray_3[]$;
Variables for Node v_j
int $v_j.phase = 0$;
boolean $v_j.initial$;
boolean $v_j.inactive = false$;
Main Routine of Agent a_i
1: $a_i.phase++$, $v_j.phase := a_i.phase$
2: **for** $l = 1, l \leq 3, l++$ **do**
   $id_l := NextActive(disArray_l[], nInactive_l)$
3: **if** $(id_2 < \min(id_1, id_3)) \vee (id_2 = id_3 < id_1) \vee ((id_1 = id_2 = id_3) \wedge (nInactive_2 \geq g-1))$
   then
4:    **if** $a_i.phase = \lceil \log g \rceil$ **then** terminate the first part and enter the second part
      with a leader status
5:    **else** go to line 1
6: **else**
7:    $v_j.inactive := true$
8:    terminate the first part and enter the second part with an inactive status
9: **end if**

3.2.2 The Second Part: Leaders' Instructions and Agents' Movement

After leader agent election, agents achieve g-partial gathering by leader agents' instructions. Let *leader nodes* be nodes where leader agents start this part. At the beginning of this part, a node where an agent exists is either a leader node or an inactive node. We use the same technique as in [9]. We assume that a leader agent a_i starts this part at leader node v_j, $v_{j'}$ is v_j's next leader node, and inactive nodes $v_{ina}^1, v_{ina}^2, \ldots v_{ina}^\ell$ exist between v_j and $v_{j'}$. That is, nodes $v_j, v_{ina}^1, v_{ina}^2, \ldots, v_{ina}^\ell, v_{j'}$ exist in this order. Then, briefly speaking, leader agent a_i at v_j moves to the next leader node $v_{j'}$. During the movement, a_i marks node v_{ina}^t as a non-gathering node (resp., a gathering node) if $(t+1) \mod g \neq 0$ (resp., $(t+1) \mod g = 0$). That is, each leader continues to mark some consecutive $g-1$ inactive nodes as non-gathering nodes and mark the next inactive node as a gathering node until it visits the next leader node. Then, agents are partitioned into groups each of which has at least g agents. After the instruction, each agent moves to the nearest gathering node and they achieve g-partial gathering. This part can be achieved in $O(gn)$ total number of moves since each link is passed by at most $2g$ times. By this fact and Lemma 2, we have the following theorem.

Theorem 3. *From any solvable initial configuration, the proposed algorithm solves the g-partial gathering problem in $O(gn)$ total number of moves.* □

4   Randomized g-partial Gathering

In this section, for the case of $2 \leq g \leq k/2$ we propose a randomized algorithm to solve the problem with probability 1 in $O(gn)$ (i.e., optimal) expected total number of moves from any initial configuration. The basic idea is the same as that of the previous section, that is, agents elect multiple leaders by comparing (virtual) IDs. In this section, since we consider a randomized algorithm, agents can use random numbers as IDs. In addition, each agent compares $2g-1$ random IDs at one time instead of comparing three IDs as in the previous section. By this behavior, if there exists a unique minimum ID among any consecutive $2g-1$ IDs, agents can make a configuration such that at least $g - 1$ non-leader agents exist between two leaders. However, if several IDs have the same minimum value among the $2g - 1$ IDs, agents cannot make such a configuration. Agents treat this case by additional behaviors explained in the following subsections.

The algorithm consists of two parts. In the first part, agents determine several candidate nodes for g-partial gathering using random IDs. In the second part, agents determine gathering nodes from the candidate nodes and achieve g-partial gathering. In the following, we refer to "candidate nodes for (g-partial) gathering" as "gathering candidate nodes".

4.1   The First Part: Determination of Gathering Candidate Nodes

Each agent takes a status from the following three statuses:

- active: The agent moves in the ring to determine gathering candidate nodes.
- leader: The agent elects the current node as a gathering candidate node.
- waiting: The agent is staying at a gathering candidate node and waiting for the next instruction.

Initially, all agents are active. This part consists of several phases. At the beginning of each phase, each active agent a_i creates a $\lceil 7 \log g \rceil$-bit random ID, and writes the ID and the current phase number on the current whiteboard. Thereafter, a_i moves until it observes $2g - 1$ random IDs (including the one it creates the current phase. Let $id_1, id_2, \ldots, id_{2g-1}$ be the IDs. Then, this situation is similar to that in which the active agent with ID id_g observes $g - 1$ IDs $id_1, \ldots, id_{g-1}$ (resp., $id_{g+1}, \ldots, id_{2g-1}$) of its backward (resp., forward) $g - 1$ active agents in a bidirectional ring. Hence, a_i behaves as if it would be an active agent with ID id_g (or the ID at the middle of the ID sequence) in the bidirectional ring. That is, if id_g is the uniquely minimum among the $2g - 1$ IDs, it becomes a leader. Different from the behavior of a leader in the previous section, a leader in this section sets the flag $v_j.candi$ to declare that the current node v_j is a gathering candidate node and waits at v_j until at least g agents gather at v_j. The reason why leaders execute such a behavior is that, even if some two agents a_i and $a_{i'}$ become leaders, it is possible that less than $g - 1$ non-leader agents exist between a_i and $a_{i'}$ (such a case does not happen in Sect. 3.2), and another leader a_h such that at least $g - 1$ non-leader agents exist between a_h and a_h's backward leader agent treats this situation (the detail is explained in the next part). If id_g is not the uniquely minimum, that is, id_g is not the minimum or id_g is the minimum but another ID has the same value as id_g, a_i additionally moves until it observes g IDs to check whether a gathering candidate node exists within this range or not. If a_i visits a gathering candidate node v_j (or $v_j.candi = true$) during the movement, it enters a waiting status there. Otherwise, a_i proceeds to the next phase. Each active agent repeats such a behavior until it becomes a leader and sets $v_j.candi = true$, or visits some gathering candidate node and enters a waiting status. An example is given in Fig. 2.

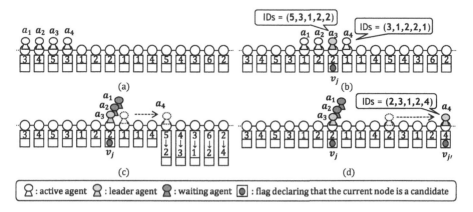

Fig. 2. Behavior outlines of the first part for the case of $g = 3$. Each number represents a random ID written on the whiteboard. For simplicity, we assume that an agent exists at each node in the initial configuration and random IDs are already written in (a). We consider behaviors of agents a_1, a_2, a_3, and a_4. From (a) to (b), each agent moves until it observes five ($=2g-1$) random IDs. Since agent a_3 observes random IDs $(5, 3, 1, 2, 2)$ and the middle ID 1 is the uniquely minimum, it becomes a leader and sets a candidate flag at v_j. On the other hand, the other agents continue to move in the ring. From (b) to (c), agents a_1, a_2, and a_4 move to observe additional three ($=g$) random IDs. During the movement, since agents a_1 and a_2 observe a candidate flag at v_j, they enter a waiting status there. On the other hand, since a_4 does not observe a flag, it updates a random ID and proceeds to the next phase (the four IDs from the rightmost node are updated similarly). From (c) to (d), since a_4 observes five random IDs $(2, 3, 1, 3, 4)$ and the middle ID 1 is the uniquely minimum, it becomes a leader and sets a candidate flag at $v_{j'}$. However, no agent exists between v_j and $v_{j'}$. This situation is handled in the second part.

The Pseudocode of active agents in the first part is described in Algorithm 2. As in Algorithm 1, variables $a_i.phase$ and $v_j.phase$ are used to maintain phase numbers. In addition, agent a_i has variables $a_i.id[\ ]$ and $a_i.nIDs$ to store the IDs and its number that a_i observed during the current phase, respectively. Node v_j has variable $v_j.id$ to store the random ID created at v_j. Variables $v_j.nVisited$ and $v_j.nAgents$ represent the number of visited times by active agents during the current phase and the number of agents staying at v_j, respectively. Moreover, in Algorithm 2 agents use procedure $NextActive2()$ to move to the next active node, and enter a waiting status when visiting a gathering candidate node (we omit the detailed description). We have the following lemma concerning Algorithm 2.

Lemma 3. *Algorithm 2 eventually terminates with probability 1 and all agents enter a leader status or a waiting status.* □

4.2   The Second Part: Achievement of g-partial Gathering

In this part, agents achieve g-partial gathering based on the gathering candidate nodes. Each agent takes a status from the following three statuses:

- leader: The agent checks whether the current candidate node finally becomes a gathering node or not. If the node becomes a gathering node, the agent instructs waiting agents where they should move.
- waiting: The agent is waiting for the leader's instruction.
- moving: The agent moves to its gathering node.

We consider the situation such that all agents complete the first part, stay at some gathering candidate nodes, and never move from the beginning of this part[2]. Then, there exist several (possibly one) gathering candidate nodes, and at least one candidate node has at least g agents and some candidate nodes have less than g agents each. Note that at each gathering candidate node one leader agent exists and the other agents are waiting agents. We denote a set of candidate nodes with at least g (resp., less than g) agents in the above situation by V_{candi}^{more} (resp., V_{candi}^{less}). Then, the basic movement of this part is as follows. Each leader agent a_i at a node in V_{candi}^{more} moves to the next candidate node $v_j \in V_{candi}^{more}$. During the movement, when a_i visits a candidate node $v_{j'} \in V_{candi}^{less}$, it sets a flag $v_{j'}.lVisited$ to declare that $v_{j'}$ is visited by a leader. Then, all the agents at $v_{j'}$ move to the nearest candidate node $v_{j''}$ such that the number of agents existing between $v_{j'}$ and $v_{j''}$ is at least g. After the movement, agents achieve g-partial gathering.

First, we explain the behavior of leader agents. Each leader agent a_i firstly waits at the current node v_j until at least g agents gather at v_j. Thereafter, a_i moves to the next candidate node $v_{j'}$ and sets $v_{j'}.lViisted = true$. Then, if at least $g - 1$ waiting agents exist at $v_{j'}$ (i.e., $v_{j'} \in V_{candi}^{more}$), a_i terminates the algorithm at $v_{j'}$ because all the agents at $v_{j'}$ eventually terminate the algorithm there and this guarantees that at least g agents gather at $v_{j'}$. If less than $g - 1$ waiting agents exist at $v_{j'}$ (i.e., $v_{j'} \in V_{candi}^{less}$), a_i stores the number of the waiting agents at $v_{j'}$ to a variable $a_i.nAgentsTemp$, moves to the next candidate node $v_{j''}$, and updates the number of agents staying at $v_{j''}$ (or agents that eventually gather at $v_{j''}$) using $a_i.nAgentsTemp$. Then, if the updated number of agents at $v_{j''}$ is

---

[2] We consider the situation for explanation, and it is possible that some agents execute the second part and the other agents still execute the first part.

Algorithm 2. Behavior of active agent a_i in the first part (v_j is the current node of a_i.)

Variables for Agent a_i
int $a_i.phase = 1$;
int $a_i.nIDs = 0$;
array $a_i.id[\,]$;
Variables for Agent v_j
int $v_j.phase = 1$;
int $v_j.nAgents = 0$;
int $v_j.id = \perp$;
int $v_j.nVisited = 0$;
boolean $v_j.candi = false$;
Main Routine of Agent a_i
1: **while** true **do**
2:    $a_i.id[0] := random(\lceil 7 \log g \rceil)$,
     $v_j.id := a_i.id[0]$
3:    $a_i.nIDs := 1, v_j.nVisited := 1$
4:    **while** $a_i.nIDs < 2g - 1$ **do**
5:       $NextActive2()$
6:       $a_i.id[a_i.nIDs] := v_j.id$
7:       $a_i.nIDs$++$, v_j.nVisited$++
8:    **end while**
9:    **if** $\forall h \in [0, 2g - 2] \setminus \{g - 1\}; a_i.id[g - 1] < a_i.id[h]$ **then**
10:      $v_j.candi := true$
11:      $v_j.nAgents := 1$
12:      terminate the first part and enter the second part with a leader status
13:    **else**
14:      **while** $a_i.nIDs < 3g - 1$ **do**
15:        $NextActive2()$
16:        $a_i.nIDs$++$, v_j.nVisited$++
17:      **end while**
18:    **end if**
19:    $a_i.phase$++$, v_j.phase := a_i.phase$
20: **end while**

---

at least g, $v_{j''}$ eventually becomes a gathering node regardless $v_{j''}$ is in V^{more}_{candi} or V^{less}_{candi}. In this case, a_i resets $a_i.nAgentsTemp$ to 0. If the updated number is less than g, $v_{j''}$ becomes a non-gathering node and a_i stores the number of agents at $v_{j''}$ to $a_i.nAgentsTemp$. This operation means that $a_i.nAgentsTemp$ (possibly 0) agents move to the next candidate node. After updating the value of $a_i.nAgentsTemp$, a_i moves to the next candidate node. Each leader agent repeats such a behavior until it visits a candidate node v_j in V^{more}_{candi}. However, when a leader agent is staying at node v_j until at least g agents gather at v_j, it is possible that $v_j.lViisted$ is set to true. This means that another leader agent $a_{i'}$ visits v_j and v_j is in V^{less}_{candi}. In this case, a_i enters a waiting status, whose behavior is described next.

Next, we explain the behaviors of waiting agents and moving agents. Each waiting agent a_i stays at the current node v_j until some leader agent visits v_j and sets $v_j.lVisited = true$. Then, it checks (from the whiteboard content) whether at least g agents eventually gather at v_j or not. If at least g agents gather, a_i terminates the algorithm there. Otherwise, a_i enters a moving status. Each moving agent a_i moves to the next candidate node $v_{j'}$ and enters a waiting status there. Then, the number of agents that eventually gather at $v_{j'}$ is the sum of agents that visit $v_{j'}$ from v_j and agents that already stay at $v_{j'}$. Thus, if the number of agents that eventually gather at $v_{j'}$ is at least g, a_i terminates the algorithm there. Otherwise, a_i enters a moving status again. Each moving agent repeats such a behavior until it visits a candidate node v_j such that at least g agents eventually gather at v_j. When all agents terminate the algorithm, the final configuration is a solution of the g-partial gathering problem. An example is given in Fig. 3.

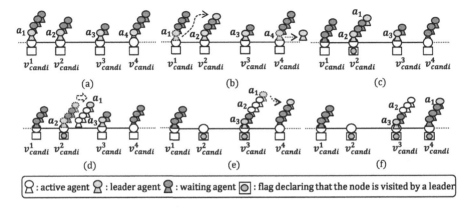

(a)   (b)   (c)

(d)   (e)   (f)

🧍 : active agent 🧍 : leader agent 🧍 : waiting agent ⊙ : flag declaring that the node is visited by a leader

Fig. 3. Behavior outlines of the second part for the case of $g = 4$. For simplicity, we omit nodes where no agent exists. From (a) to (b), since the number of agents at nodes v_{candi}^1 and v_{candi}^4 are respectively $4 (\geq g)$, leader agents a_1 and a_4 move to their next candidate nodes, respectively. On the other hand, since the number of agents at v_{candi}^2 and v_{candi}^3 are respectively less than g, leader agents a_2 and a_3 stay at the current node. Then, the system reaches the configuration of (c) and a flag declaring that the node is visited by a leader is set at v_{candi}^2. Since the number of agents at v_{candi}^2 except for a_1 is $3 < g$, all the agents (including a leader) at v_{candi}^2 move to the next candidate node v_{candi}^3 (d). Then, the system reaches the configuration of (e) and a flag is set by a_1. Since the number of agents at v_{candi}^3 except for a_1 is $5 > g$, a_1 moves to the next candidate node v_{candi}^4 and the other agents at v_{candi}^3 terminate the algorithm there. In (f), a_1 sets a flag at v_{candi}^4. Since the number of agents that sets the flag (i.e., a_1) and agents that already stay at v_{candi}^4 is $4 \geq g$, all the agent at v_{candi}^4 terminate the algorithm there.

We have the following lemmas for the proposed algorithm.

Lemma 4. *The proposed algorithm solves the g-partial gathering problem with probability 1 from any initial configuration.*   □

Lemma 5. *The expected total number of moves of the proposed algorithm is* $O(gn)$. □

By Lemmas 4 and 5, we have the following theorem.

Theorem 4. *The proposed algorithm solves the g-partial gathering with probability 1 in* $O(gn)$ *expected total number of moves from any initial configuration.*

5 Conclusion

In this paper, we considered the g-partial gathering problem for anonymous agents without global knowledge in asynchronous unidirectional rings. We considered deterministic and randomized cases. In the deterministic case, we showed that unsolvable initial configurations exist and agents cannot detect whether the initial configuration is an unsolvable one or not. In addition, we proposed an algorithm to solve the problem from any solvable initial configuration in $O(gn)$ total number of moves. In the randomized case, we proposed an algorithm to solve the problem with probability 1 in $O(gn)$ expected total number of moves from any initial configuration. Thus, our algorithms can solve the problem in asymptotically optimal total number of moves without global knowledge.

Acknowledgement. This work was partially supported by JSPS KAKENHI Grant Number 17K19977, 18K18000, 18K11167, 18K18031, and 19K11826, and Japan Science and Technology Agency (JST) SICORP.

References

1. Gray, R.S., Kotz, D., Cybenko, G., Rus, D.: D'Agents: applications and performance of a mobile-agent system. Softw. Pract. Exper. **32**(6), 543–573 (2002)
2. Lange, D.B., Oshima, M.: Seven good reasons for mobile agents. CACM **42**(3), 88–89 (1999)
3. Kranakis, E., Krozanc, D., Markou, E.: The Mobile Agent Rendezvous Problem in the Ring. Synthesis Lectures on Distributed Computing Theory, vol. 1. Morgan & Claypool, San Rafael (2010)
4. Kranakis, E., Santoro, N., Sawchuk, C., Krizanc, D.: Mobile agent rendezvous in a ring. In: Proceedings of ICDCS, pp. 592–599 (2003)
5. Flocchini, P., Kranakis, E., Krizanc, D., Santoro, N., Sawchuk, C.: Multiple mobile agent rendezvous in a ring. In: Farach-Colton, M. (ed.) LATIN 2004. LNCS, vol. 2976, pp. 599–608. Springer, Heidelberg (2004). https://doi.org/10.1007/978-3-540-24698-5_62
6. Fraigniaud, P., Pelc, A.: Deterministic rendezvous in trees with little memory. In: Taubenfeld, G. (ed.) DISC 2008. LNCS, vol. 5218, pp. 242–256. Springer, Heidelberg (2008). https://doi.org/10.1007/978-3-540-87779-0_17
7. Kranakis, E., Krizanc, D., Markou, E.: Mobile agent rendezvous in a synchronous torus. In: Correa, J.R., Hevia, A., Kiwi, M. (eds.) LATIN 2006. LNCS, vol. 3887, pp. 653–664. Springer, Heidelberg (2006). https://doi.org/10.1007/11682462_60
8. Dieudonné, Y., Pelc, A.: Anonymous meeting in networks. Algorithmica **74**(2), 908–946 (2016)

9. Shibata, M., Kawai, S., Ooshita, F., Kakugawa, H., Masuzawa, T.: Partial gathering of mobile agents in asynchronous unidirectional rings. Theor. Comput. Sci. **617**, 1–11 (2016)
10. Shibata, M., Ooshita, F., Kakugawa, H., Masuzawa, T.: Move-optimal partial gathering of mobile agents in asynchronous trees. Theor. Comput. Sci. **705**, 9–30 (2018)
11. Shibata, M., Nakamura, D., Ooshita, F., Kakugawa, H., Masuzawa, T.: Partial gathering of mobile agents in arbitrary networks. IEICE Trans. Inf. Syst. **102**(3), 444–453 (2019)
12. Peterson, G.L.: An $O(n \log n)$ unidirectional algorithm for the circular extrema problem. ACM Trans. Program. Lang. Syst. **4**(4), 758–762 (1982)

Set Agreement Power Is Not a Precise Characterization for Oblivious Deterministic Anonymous Objects

Gadi Taubenfeld$^{(\boxtimes)}$

The Interdisciplinary Center, P.O. Box 167, 46150 Herzliya, Israel
tgadi@idc.ac.il

Abstract. Anonymous shared memory systems, recently introduced in [36], are composed of objects for which there is no a priori agreement between processes on their names. We resolve the following foundational open problems in theoretical distributed computing, for a model which includes both non-anonymous and anonymous shared objects: (1) Are non-trivial oblivious deterministic objects with the same set agreement power have the same computational power? (2) Is there a non-trivial oblivious deterministic object which is strictly weaker than an atomic read/write register? We prove that the answer to the first problem is negative, while the answer to the second problem is positive. The positive answer to the second problem implies that the common belief that every non-trivial deterministic object of consensus number one is at least as strong as atomic read/write registers is false. A noteworthy property of the proofs of our results lies in their simplicity.

Keywords: Anonymous shared memory · Anonymous objects ·
Set agreement · Consensus · Read/write registers · RMW registers

1 Introduction

1.1 Set Agreement, Oblivious Objects, Anonymous Objects

Among the most fundamental problems in distributed computing are agreement and its generalization, k-set agreement. The k-set agreement problem is to design an algorithm for n processes, where each process starts with an input value from some domain and must choose some participating process input as its output. All n processes together may choose no more than k distinct output values [10]. The 1-set agreement problem is the familiar consensus problem [27]. The k-set agreement number of an object is the largest integer m such that using any number of instances of that object and registers k-set agreement can be solved in a wait-free manner among m processes, or the number is ∞ if k-set agreement can be solved among any number of processes. The 1-set agreement number is also called the consensus number [18]. The set agreement *power* of an object is

© Springer Nature Switzerland AG 2019
K. Censor-Hillel and M. Flammini (Eds.): SIROCCO 2019, LNCS 11639, pp. 293–308, 2019.
https://doi.org/10.1007/978-3-030-24922-9_20

the infinite sequence $(n_1, n_2, ..., n_k, ...)$ where n_k is the k-set agreement number of that object for all $k \geq 1$.

A shared object type is defined using a sequential specification, which describes the operations that may be performed on the object and the responses, if the operations are performed sequentially. We consider objects that are *linearizable* (with respect to their sequential specification): they behave as if all operations, including concurrent ones, are applied sequentially, so that each operation appears to take effect instantaneously at some distinct point between its invocation and response [21].

Each operation causes a state transition and may return a response. If the state transition and response are uniquely determined by the current state of the object and the operation applied, then the object is *deterministic*. If the state transition and the response for any operation do not depend on the process that invokes the operation and every process can invoke every operation, then the object is *oblivious*. All common deterministic object types which are supported by modern multiprocessor architectures (such as, bits, registers, test&set, fetch&add, swap, compare&swap, queues, and stacks) are oblivious objects.

Anonymous objects, recently introduced in [36], are objects for which there is no a priori agreement between processes on their names. That is, anonymous objects do not have global names. The lack of global names makes it convenient to think of each process as being assigned an initial object and an ordering of the objects which determines how the process scans the objects. Thus, algorithms which use only anonymous objects should be correct assuming a very powerful adversary, which can determine the order in which processes access the objects.

In addition to its usefulness in modeling biologically inspired distributed computing methods, especially those that are based on ideas from molecular biology [30], the anonymous shared memory model enables to understand better the intrinsic limits for coordinating the actions of asynchronous processes.

1.2   Is the Set Agreement Power a Precise Characterization?

A characterization of objects is *precise* if it can always indicate when two objects are able to implement each other. In the last thirty years, researchers have tried to find a precise characterization of an object's ability to implement other objects in a wait-free manner, in the shared memory model. The first suggestion for such a characterization was the object's consensus number [18]. However, it was shown that this characterization is not precise. That is, some objects have the same consensus number but do not have the same computational power (i.e., cannot implement each other). This was first shown for oblivious non-deterministic objects [29] and later for oblivious deterministic objects as well [1].

Since the consensus number of an object does not fully characterize its ability to implement other objects, the next natural question to ask is whether the set agreement power of an object a precise characterization of its ability to implement other objects [6,12]? In [6], it was shown that the set agreement characterization is not a precise characterization for non-deterministic objects, leaving open the question of what happens when the universe of objects is restricted to deterministic objects. That is, are any two deterministic objects with the same set agreement power equivalent (i.e., can they implement each other)?

In [8], it is shown that the answer is *negative*, for *non-oblivious* deterministic objects with consensus numbers greater than 1. Non-oblivious objects, as defined in [8], have ports, each operation is invoked on a specific port, and the response received *may depend* on the port number chosen. Processes can choose to invoke an operation on any port of any object, but no two operations may be applied on the same port of an object concurrently. The number of ports of an object effectively limits the number of processes that may access it concurrently.

It is not possible to simulate non-oblivious objects using oblivious objects by simply including the port number as part of the process' input when invoking an operation. Since the total number of processes may be larger than the number of ports, in such a naive simulation two or more processes may end up using the same port number concurrently. So, this leaves open the following question.

> *Is the set agreement power of a non-trivial oblivious deterministic object a precise characterization of its ability to implement other oblivious deterministic objects? That is, are any two non-trivial oblivious deterministic objects with the same set agreement power have the same computational power?*

Non-trivial objects are objects that can be used to solve problems whose solutions require communication. We prove that for a universe of objects which includes both non-anonymous and anonymous objects, the answer to the above problem is *negative*. That is, there are two non-trivial oblivious deterministic objects (both with consensus number 1), one of which is an anonymous object, that have the same set agreement power, yet one of the two is strictly weaker than the other.

1.3   Is an Atomic Read/Write Register the Weakest Object?

A related question that attracted the attention of researchers investigating the relative computational power of shared objects, is the following open problem [24],

> *Is an atomic read/write register computationally the weakest possible non-trivial object? Put another way, is there a non-trivial deterministic object strictly weaker than an atomic read/write register?*

We prove that for a universe of objects which includes both non-anonymous and anonymous objects, atomic register is *not* the weakest non-trivial object. That is, we show that there is a non-trivial oblivious deterministic anonymous object which is strictly weaker than an atomic register. The answer to the above open problem implies that the common belief that every non-trivial deterministic object of consensus number one is computationally equivalent to or stronger than atomic read/write registers is *false*.

We managed to resolve the above open problems, by showing that,

1. Anonymous read/write bits are strictly weaker than both anonymous and non-anonymous read/write registers (Sects. 3 and 4); and
2. Anonymous read/write bits are non-trivial objects, even when assuming that processes may fail (Sect. 5).

2   Preliminaries

We consider an asynchronous shared memory system that consists of a collection of n deterministic processes with unique identifiers which communicate via anonymous atomic objects that do not have global names and via standard non-anonymous objects. For m *anonymous* objects, $o_1, ..., o_m$, the adversary can fix, for each process p, a permutation $\pi_p : \{o_1, ..., o_m\} \rightarrow \{o_1, ..., o_m\}$ of the objects such that, for process p, the j'th anonymous object is $\pi_p(o_j)$. In particular, when process p accesses its j'th anonymous object, it accesses $\pi_p(o_j)$. Algorithms designed for such a system must be correct regardless of the permutations chosen by the adversary.

With an *atomic* object, it is assumed that operations on the object occur in some definite order. That is, each operation is an indivisible action. All objects are assumed to be deterministic, that is, invoking an operation on an object may have only one possible result. Asynchrony means that there is no assumption on the relative speeds of the processes. Processes may fail by crashing, that is, they fail only by never entering the algorithm or by leaving the algorithm at some point and after that permanently refraining from accessing the shared objects. A process that crashes is said to be *faulty*; otherwise, it is *correct*.

A read/write register (register for short) is a shared object that supports (atomic) read and write operations. The fact that anonymous registers do not have global names implies that only multi-writer multi-reader anonymous registers are possible. Such registers can both be written and read by all the processes. A read-modify-write register (RMW register for short) is a shared object that supports read-modify-write operation in which a process can atomically read a value of a shared register and based on the value read, compute some new value and assign it back to the register.

An atomic bit is an object that supports atomic read and write operations, and can store only two values (0 or 1). Through the paper, by a register we will mean a multi-valued register, that is, a register which can store many different values (but only one value at any given time).

Several progress conditions have been proposed for algorithms in which processes may fail. The strongest, and most extensively studied condition, is wait-freedom. *Wait-freedom* guarantees that *every* active process will always be able to complete its pending operations in a finite number of steps [18]. *Obstruction-freedom* guarantees that an active process will be able to complete its pending operations in a finite number of steps, if all the other processes "hold still" long enough [19]. In a model where participation is required, every correct process must eventually become active and execute its code. A more common and practical situation is one in which participation is not required. Unless explicitly stated otherwise, we assume that participation is *not* required.

The *sequential specification* of an object describes its behavior when operations are applied sequentially. We consider objects that are *linearizable* (w. r. t. their sequential specification): they behave as if all operations, including concurrent ones, are applied sequentially, so that each operation appears to take effect instantaneously at some distinct point between its invocation and response [21].

Two objects with the same consensus number are *equivalent* if and only if,

1. Their consensus number is 1, and each object can be implemented by instances of the other object in a wait-free manner; or
2. Their consensus number is more than 1, and each object can be implemented by instances of the other object and registers in a wait-free manner.

In the above definition, for objects with consensus number 1, the use of registers is forbidden. Otherwise, we would get that, by definition, no object is weaker than a register.

3   An Impossibility Result for Anonymous RMW Bits

An object of type A is *strictly weaker* than an object of type B if using objects of type B it is possible to implement, in a wait-free manner, an object of type A, but not vice versa. We show that there is a non-trivial deterministic object, namely anonymous read/write bit, which is strictly weaker than an (anonymous or non-anonymous) read/write register, and that there are non-trivial deterministic objects with the same set agreement power which have different computational power. This implies that *not* every deterministic object of consensus number one is computationally equivalent to or stronger than a non-anonymous read/write register.

3.1   Basic Notions and Notations

An *event* corresponds to an atomic step performed by a process. A (global) *state* of an algorithm is completely described by the values of the (local and shared) objects and the values of the location counters of all the processes. A *run* is defined as a sequence of alternating states and events (also referred to as steps). It is convenient to define a run as a sequence of events omitting all the states except the initial state. Since the events and the initial state uniquely determine the states in a run, no information is lost by omitting the states.

We use x, y and z to denote runs. When x is a prefix of y (and y is an *extension* of x), we denote by $(y - x)$ the suffix of y obtained by removing x from y. We denote by $x; seq$ the sequence obtained by extending x with the sequence of events seq. Saying that an extension y of x involves only process p means that all events in $(y - x)$ are only by process p.

Runs x and y are *indistinguishable* for process p, denoted $x[p]y$, if the subsequence of all events by p in x is the same as in y, the initial values of the local registers of p in x are the same as in y, and the values of all the shared objects in x are the same as in y. Notice that the indistinguishability relation is an equivalence relation. We assume that the processes are *deterministic*, that is, for every two runs $x; e$ and $x; e'$ if e and e' are events by the same process then $e = e'$. We notice that if two runs are indistinguishable to a given process, then the next step by that process in both runs is the same.

3.2   The Impossibility Result

The consensus problem is defined as follows: There are n processes where each process $i \in \{1, ..., n\}$ has an input value in_i. The requirements are that there exists a decision value v such that, (1) *Agreement & termination*: each non-faulty process eventually decides on v; and (2) *Validity*: $v \in \{in_1, ..., in_n\}$. When the only possible input values are 0 and 1, the problem is called *binary* consensus.

Theorem 1. *For any $m \geq 1$, there is no obstruction-free binary consensus algorithm for two (or more) processes using m anonymous RMW bits.*

Proof. We assume to the contrary that there is an obstruction-free consensus algorithm for two processes using m anonymous RMW bits, and show how this leads to a contradiction. Let p and q be the identifiers of the processes, let S be the set of all the RMW bits used by the algorithm, and assume that the initial values of all the RMW bits in S are 0.

Let x_0 be a run of the algorithm in which p with input 0 runs alone until it decides on 0 and terminates. Let x_1 be a run of the algorithm in which p with input 1 runs alone until it decides on 1 and terminates. Clearly, by the agreement requirement, in any extension of x_0 (resp. of x_1) in which q runs alone and decides, q must also decide on 0 (resp. on 1). For an arbitrary run z, let $number(z)$ be the number of all the RMW bits that, at the end of z, have value 1. Assume w.l.o.g. that $number(x_0) \leq number(x_1)$.

Since the anonymous RMW bits do not have global names, each process independently names each of one of them with a unique name. For simplicity, assume that the names are natural numbers. The consensus algorithm, assumed at the beginning of the proof, is correct only if it always reaches agreement regardless of how the RMW bits are numbered by the different processes. Thus, it follows from the existence of the run x_0 that, for every set of RMW bits $R \subseteq S$ such that $|R| = number(x_0)$, there must exist a run x_0^R of the algorithm such that: (1) in x_0^R, p with input 0 runs alone until it decides on 0 and terminates, (2) at the end of x_0^R the values of all the RMW bits in R are 1, and (3) $|R| = number(x_0) = number(x_0^R)$.

Let x_1' be a prefix of x_1 such that $number(x_1') = number(x_0)$. Notice that the input of p in x_1' is 1. Let W be the set of all the RMW bits that, at the end of x_1', have value 1. We notice that $|W| = number(x_0)$. As explained above there exists a run x_0^W of the algorithm in which (1) p with input 0 runs alone until it decides on 0 and terminates, (2) at the end of x_0^W the values of all the RMW bits in W are 1, (3) $|W| = number(x_0) = number(x_0^W)$.

Let y be an extension of x_0^W in which q decides and terminates such that (1) in $(y - x_0^W)$ only q takes steps, and (2) the input of q is 1 (such an extension exists by the obstruction-freedom assumption). What is the value that q decides on in y? There are two possibilities both lead to a contradiction:

1. Process q decides on 0 in y. Since $x_0^W[q]x_1'$, it follows that $z = x_1'; (y - x_0^W)$ is a legal run. However, in z process q decides on 0 while the inputs of both p and q are 1. This contradicts the requirement that the decision value must be the input value of one of the processes.
2. Process q decides on 1 in y. By assumption, p decides on 0 in x_0^W, and since x_0^W is a prefix of y, it follows that p decides on 0 in y. Thus p and q decide on different values in y. A contradiction.   □

An interesting open problem is to determine what are the smallest anonymous registers with which obstruction-free consensus and set agreement can be solved.

4   Implications of the Impossibility Result

It is easy to design a wait-free consensus algorithm for two processes using three non-anonymous RMW bits. Assume that the initial values of all the three bits are 0. Each process uses one bit to announce its input, and then tries to set the last bit to 1. The decision value is the input of the process that was the first to access the third bit, changing it from 0 to 1. Thus, by Theorem 1,

Corollary 1. *An anonymous RMW bit is strictly weaker than a non-anonymous RMW bit.*

Also, for any $n \geq 1$ and $m \geq 1$, it is easy to design a wait-free consensus algorithm for n processes using m anonymous RMW (multi-valued) registers. Assume that the initial values of all the m registers are 0. Each process first scans the m registers and only if the value of a register is 0 the process writes its identifier and input value into that register. The decision value is the input of the process with the maximum identifier among all the identifiers found in the m registers. Thus, by Theorem 1,

Corollary 2. *An anonymous RMW bit is strictly weaker than an anonymous RMW register.*

We observe that an anonymous RMW register is *not* necessarily weaker than a non-anonymous RMW bit since the consensus number of non-anonymous RMW bits is only two [26]. Although both anonymous RMW bits and anonymous read/write registers have consensus number one, it is an open question whether one can implement the other. A RMW bit supports read and write operations, and it also can be used as a read/write bit. Thus, Theorem 1 implies a similar result for read/write bits.

Corollary 3 (An impossibility result for anonymous read/write bits). *There is no obstruction-free binary consensus algorithm for two (or more) processes using anonymous read/write bits.*

Anonymous bits are *non-trivial* objects – they can be used to solve problems whose solutions require communication. An interesting *wait-free* consensus algorithm that makes use of anonymous read/write bits together with anonymous RMW bits in a general anonymous shared memory model, is presented in Sect. 5. We describe below a simple consensus algorithm for a failure-free model.

Proposition 1. *There is a binary consensus algorithm for two processes using two anonymous read/write bits, assuming participation is required and processes never fail.*

Proof. We assume that the initial values of both bits are 0. The two processes are called the *sender* and the *receiver*. When the sender starts, it first sets one of the bits to 1, then it spins on that bit until its value is changed (by the receiver) back to 0. When this happens, it writes its input value into the other bit, sets again to 1 the bit it has previously set to 1, decides on its input and terminates. The receiver, when it starts, keeps on checking the two bits until it notices that the value of one of them is 1. Then, it changes this bit back to 0, and spins on that bit until its value is changed back to 1. When this happens, it decides on the value of the other bit and terminates. Clearly, the algorithm guarantees that both processes eventually decides on the input value of the sender.   □

Theorem 2 (main result). *In a system of two or more processes:*

1. *There is a non-trivial oblivious deterministic object which is strictly weaker than an anonymous (and hence also non-anonymous) read/write register, for two or more processes;*
2. *There are non-trivial oblivious deterministic objects with the same set agreement power which have different computational power;*
3. *Not every non-trivial oblivious deterministic object of consensus number one is computationally equivalent to or stronger than a non-anonymous (or anonymous) read/write register.*

Proof. An anonymous read/write register trivially implements an anonymous read/write bit. It was shown in [36], that there is an obstruction-free consensus algorithm for two (or more) processes using anonymous read/write registers. Since, by Corollary 3, there is no obstruction-free consensus algorithm for two (or more) processes using anonymous read/write bits, it follows that anonymous read/write bits cannot implement an anonymous register. Thus, an anonymous read/write bit is strictly weaker than an anonymous read/write register.

The k-set agreement problem can trivially be solved for k processes, by simply letting each process decides on its own input. Thus, the set agreement power of any object is at least $(1, 2, 3, ...)$. It was proven in [4,20,32], that the set agreement power of a non-anonymous read/write register is exactly $(1, 2, 3, ...)$. Thus, also the set agreement power of an anonymous register is exactly $(1, 2, 3, ...)$. Since anonymous bit is strictly weaker than an anonymous register, its set agreement power is also $(1, 2, 3, ...)$. Thus, anonymous bits and anonymous (or non-anonymous) registers are deterministic objects with the same set agreement power but with different computational power. From the fact that an anonymous bit is strictly weaker than anonymous (or non-anonymous) register, it immediately follows that *not* every non-trivial deterministic object of consensus number one is computationally equivalent to non-anonymous register. □

5 Mixing Objects: Wait-Free Consensus for Two Processes

So far we have considered a model wherein each algorithm processes communicate via anonymous objects all of which are of the same type. We now consider a more general setting in which, in a given algorithm, processes may access different types of anonymous objects. In the more general model, there are different *groups* of objects. All the objects in the same group must all be of the same type. Objects from different groups may be different (but are not required to be different). All the objects which reside in the same group are anonymous, but the groups themselves are not anonymous. Thus, when a process needs to access an object, it can specify in which group the object resides, but cannot point at a specific object within the group (unless the group is a singleton). We can now think of a non-anonymous object as an object which resides in a group with exactly one element (i.e., a singleton).

We have already shown that it is not possible to solve obstruction-free consensus for two processes using only *one group* of anonymous RMW bits regardless of the size of that group (Theorem 1). This result immediately implies that it is not possible to solve obstruction-free consensus for two processes using only *one group* of anonymous read/write bits regardless of the size of that group (Corollary 3). We now prove that it is possible to solve wait-free consensus for two processes using *two groups*, where the elements of the first group are (anonymous) RMW bits, and the elements of the second group are (anonymous) read/write bits, regardless of the size of the groups. At first sight, this result seems counterintuitive since RMW registers are strictly stronger the read/write bits, so how adding read/write bits can make a difference? What makes the difference is that we now have *two* groups and, although the objects within each group are anonymous, the groups are *not* anonymous.

Theorem 3. *For every $\ell \geq 1$ and $m \geq 1$, there is a wait-free binary consensus algorithm for two processes using a group of ℓ anonymous RMW bits, and a group of m anonymous read/write bits.*

It follows immediately from Theorem 3 that,

Corollary 4. *Anonymous read/write bits are non-trivial objects, also when assuming that participation is not required and that processes may fail.*

For $\ell = 1$ and $m = 1$, the following result for non-anonymous objects follows immediately from Theorem 3.

Corollary 5. *There is a wait-free binary consensus algorithm for two processes using a single (non-anonymous) RMW bit and single (non-anon.) read/write bit.*

What is the point of considering the cases when ℓ and m are greater than 1? In general, the fact that a problem is solvable using m anonymous objects, does not imply that it is solvable also using $m + 1$ anonymous objects [36].

5.1 The Algorithm

The code of the algorithm is given in Fig. 1. The algorithm makes use of two group of objects called X and Y. The group X includes ℓ RMW bits, and the group Y includes m read/write bits. As the objects within each group do not have global names, each process independently numbers them. We use the following notations: $X.i[j]$ denotes the jth RMW bit according to process i numbering, for $1 \le j \le \ell$, and $Y.i[j]$ denotes the jth read/write bit according to process i numbering, for $1 \le j \le m$.

ALGORITHM 1: CODE OF PROCESS i WITH INPUT $in_i \in \{0, 1\}$

Constants:
  ℓ, m: positive integers   // # of shared objects in the two groups
Shared variables:
  $X.i[1..\ell]$: array of ℓ anonymous RMW bits, initially all 0   // X group
  $Y.i[1..m]$: array of m anonymous read/write bits, initially all 0   // Y group
Local variables:
  $j, index$: integer, the initial value of $index$ is 1

1 **if** $in_i = 0$ **then**
2      $X.i[1] \leftarrow 1$ // write operation
3   **for** $j \leftarrow 1$ **to** m **do if** $Y.i[j] = 1$ **then** $index \leftarrow j$ **fi od**   // read operations
4   **if** $Y.i[index] = 0$ **then** $decide(0)$ // no rival; read operation
5   **else**
6   **if** $X.i[1] = 0$ // lines 6–8 are one RMW operation
7   **then** $decide(0)$ // the rival changed $X.i[1]$
8   **else** $X.i[1] \leftarrow 0; decide(1)$ **fi**   // process i changed $X.i[1]$
9   **fi**
10 **else**   // $in_i = 1$
11     $Y.i[1] \leftarrow 1$ // write operation
12   **for** $j \leftarrow 1$ **to** ℓ **do if** $X.i[j] = 1$ **then** $index \leftarrow j$ **fi od**   // read operations
13   **if** $X.i[index] = 0$ // lines 13–15 are one RMW operation
14   **then** $decide(1)$ // no rival or the rival changed $X.i[index]$
15   **else** $X.i[index] \leftarrow 0; decide(0)$ **fi**   // process i changed $X.i[index]$
16 **fi**

Fig. 1. Wait-free binary consensus for two processes using a group of $\ell \ge 1$ RMW bits and a group of $m \ge 1$ read/write bits.

5.2   Correctness Proof

Lemma 1. *The algorithm is a correct wait-free binary consensus algorithm for two processes.*

Proof. The correctness proof is as follows:

- If both processes have input value 0, then clearly they will both decide on 0, as no read/write bit in group Y is ever updated and hence the values of all the bits in group Y are always 0. Thus, the condition in line 4 will be evaluated to true and both processes will decide 0.
- If both processes have input value 1, then clearly they will both decide on 1, as no RMW bit in group X is ever updated and hence the values of all the bits in group X are always 0. Thus, the condition in line 13 will be evaluated to true, and both processes will decide on 1.
- When the processes have different input values, and one of the two processes is faster and decides on a value without noticing that the other process "is around", the common decision value is that of the fast process, or
- When the processes have different input values, and both processes try to RMW the same bit in group X (i.e., $X.i[1]$ for the process with input 0, and $X.i[index]$ for the processes with input 1), the common decision value is the input of the *second* process that tried to RMW this bit.

This completes the proof.   □

6   Related Work

Anonymous Shared Memory. In [36], the notion of anonymous objects was defined, and several results were presented for a model where communication is only via anonymous (read/write) registers. In particular, it was shown that for a model where the number of processes is *not* a priori known (or is unbounded) anonymous registers are strictly weaker than non-anonymous registers. However, when the number of processes is not a priori known, it seems that anonymous registers are *trivial* objects – they cannot be used to solve any problem that requires communication. The question of whether anonymous registers are weaker than non-anonymous registers when the number of processes is known is open.[1]

   The work on anonymous objects was inspired by Michael O. Rabin's paper on solving the Choice Coordination Problem (k-CCP) [28]. In the k-CCP, n processes must choose between k alternatives. The agreement on a single choice is complicated by the fact that there is *no a priori* agreement on names for the

---

[1] In [36], it is mentioned that anonymous registers are non-trivial objects which are strictly weaker than non-anonymous registers, when the number of processes is not a priori known. This statement is misleading. Indeed, it was proved in [36] that anonymous registers are strictly weaker than non-anonymous registers when the number of processes is *not* a priori known (or unbounded). However, it was not proved that anonymous registers are non-trivial objects for such a model.

alternatives. Rabin has assumed that processes communicate by applying RMW operations to exactly k registers which do not have global names. The k different registers represent the k possible alternatives.

In [2], tight space bounds for solving the symmetric deadlock-free mutual exclusion problem using anonymous read/write registers and anonymous RMW registers, are presented. In [14], the election and the de-anonymization problems are studied in a model where processes may not fail. In the de-anonymization problem, processes must agree on unique names for the anonymous objects.

In [5], the naming problem of assigning unique names to initially identical processes is considered. It is assumed that each register is *owned* by some unique process which can write into it and that register is partially anonymous for the other processes that can only read it. For such a model, with single-writer registers, it is shown that wait-free naming is not solvable by a deterministic algorithm, while it is solvable by a randomized algorithm. According to our definition, the notion of an anonymous register is meaningful only when all the processes can both read and write the register.

In [30], it is shown how the process of genome wide epigenetic modifications, which allows cells to utilize the DNA, can be modeled as an anonymous shared memory system where, in addition to the shared memory, also the processes (that is, proteins modifiers) are anonymous. Epigenetic refers in part to post-translational modifications of the histone proteins on which the DNA is wrapped. Such modifications play an important role in the regulation of gene expression.

Consensus Numbers and the Consensus Hierarchy. The consensus problem was formally defined in [27]. The notion of a consensus number was defined in [18]. The consensus hierarchy, defined in [18], is an infinite hierarchy of objects such that the objects at level i of the hierarchy are exactly those objects with consensus number i. In the consensus hierarchy (1) no object at one level together with registers can wait-free implement any object at a higher level, and (2) each object at level i together with registers can wait-free implement any object at a lower level in a system of i processes.

In [1], it is shown that for every $n \geq 2$, there is an infinite sequence of deterministic objects of consensus number n with strictly increasing computational power in a system of more than n processes, leaving open the question of whether all deterministic objects with consensus number 1 are at least as strong as atomic registers. We resolve this question by showing that the answer is negative (Theorem 2(1)). In [15], it was shown that there is a non-deterministic object with consensus number 1 which cannot be wait-free implemented from atomic registers. Recently, it was shown that there are also deterministic objects with consensus number 1, but with different set consensus numbers than atomic registers, which are strictly stronger than atomic registers [11].

The consensus hierarchy is *robust* if no object in any level of the hierarchy can be implemented using a number of (possibly different) types of objects from lower levels [22]. It is shown in [9] that the consensus hierarchy is not robust, if non-oblivious non-deterministic objects are allowed. In [34] it is proved that the consensus hierarchy is not robust, even for oblivious objects, if objects with unbounded non-determinism are allowed. This last result is improved in [25], showing that the hierarchy is not robust even when restricted to oblivious objects when non-determinism is bounded.

The consensus hierarchy is known to be robust for deterministic one-shot objects [17] and deterministic read-modify-write and readable objects [31]. It is unknown whether the consensus hierarchy is robust for general deterministic objects and, in particular, for oblivious deterministic objects. Additional issues regarding the robustness question are discussed in [22,23]. For randomized computation, the consensus hierarchy collapses [3].

Set Agreement Power. The k-set agreement problem was defined in [10]. In [7], it was shown that a precise classification of linearizable objects must divide the objects into *uncountably* many classes. In [6], it is shown that for every $n \geq 2$, there exists a pair of non-deterministic objects with consensus number n that have the same set agreement power but are not computationally equivalent. In [8], it is shown that every level $n \geq 2$ of the consensus hierarchy has two deterministic objects, one of which is a non-oblivious object, with the same set agreement power that are not equivalent.

We show that in level one of the consensus hierarchy, there is such a pair of non-trivial oblivious deterministic objects, where one of the two objects is anonymous (Theorem 2(2)). That is, there exists a pair of non-trivial oblivious deterministic objects (i.e, an anonymous r/w bit and an atomic r/w register), that have the same set agreement power but are not computationally equivalent.

In [12], it is written: "We hope that this work will be a step towards proving a more general conjecture that our set-consensus numbers capture precisely the computing power of any 'natural' shared memory model." It follows from Theorem 2 that this hope cannot be realized.

Objects Weaker Than an Atomic Register. The investigation whether various objects are weaker than an atomic read/write register was initiated in [24], where three classes of shared registers are defined, which support read and write operations, called—safe, regular and atomic—depending on their properties when several reads and/or writes are executed concurrently. It was shown in [24] that an atomic register can be implemented from both safe bits and from regular bits.

In [33,35], relaxations of the notions of safe, regular and atomic registers called k-safe, k-regular and k-atomic registers, were considered and it was shown that they are all as strong as atomic registers. We have shown that an anonymous atomic bit is strictly weaker than an atomic non-anonymous register (Theorem 2(1)). Hence, an anonymous atomic bit is also strictly weaker than non-anonymous safe, regular and atomic bits (and registers). It is interesting to observe that the correctness of the algorithm in Fig. 1 is preserved even when the anonymous atomic bits are replaced with anonymous safe bits.

In [16] the authors introduce the family of d-solo models, where d processes may concurrently run solo, $1 \leq d \leq n$. The 1-solo model corresponds to the wait-free read/write model and the n-solo model corresponds to the wait-free message-passing model. Among other results, it is shown that, when the processes are anonymous any d-solo model with $d \geq 2$, is weaker than the wait-free read/write model, yet it is powerful enough to solve a non-trivial task, called the (d, ϵ)-solo approximate agreement task, which cannot be solved in the $(d + 1)$-solo model.

In [13], it is shown how n processes, with unique identifiers taken from a very large namespace, can emulate *single-write* multi-reader registers non-blocking using n multi-write multi-reader (MWMR) non-anonymous registers and wait-free using $2n-1$ MWMR non-anonymous registers. The emulations used to prove these interesting results would not work for anonymous registers.

7   Discussion

We have resolved important open problems, assuming a universe of objects which includes both non-anonymous and anonymous objects. In particular, we proved that anonymous bits are non-trivial objects which are strictly weaker than anonymous registers. It would be interesting to investigate the "mixed objects" model further. Finally, it would be interesting to investigate a model where both the processes and the objects are anonymous, as such a model seems to be suited for the study of "algorithms in nature", i.e., how collections of molecules, cells, and organisms process information and solve computational problems.

References

1. Afek, Y., Ellen, F., Gafni, E.: Deterministic objects: life beyond consensus. In: Proceedings of the ACM Symposium on Principles of Distributed Computing, PODC 2016, pp. 97–106 (2016)
2. Aghazadeh, Z., Imbs, D., Raynal, M., Taubenfeld, G., Woelfel, Ph.: Optimal memory-anonymous symmetric deadlock-free mutual exclusion. In: Proceedings of the ACM Symposium on Principles of Distributed Computing, PODC 2019 (2019)
3. Aspnes, J.: Randomized protocols for asynchronous consensus. Distrib. Comput. **16**(2–3), 165–175 (2003)
4. Borowsky, E., Gafni, E.: Generalized FLP impossibility result for t-resilient asynchronous computations. In: Proceedings of 25th ACM Symposium on Theory of Computing, pp. 91–100 (1993)
5. Buhrman, H., Panconesi, A., Silvestri, R., Vitanyi, P.: On the importance of having an identity or, is consensus really universal? Distrib. Comput. **18**(3), 167–176 (2006)
6. Chan, D.Y.C., Hadzilacos, V., Toueg, S.: Life beyond set agreement. In: Proceedings of the ACM Symposium on Principles of Distributed Computing, PODC 2017, pp. 345–354 (2017)
7. Chan, D.Y.C., Hadzilacos, V., Toueg, S.: On the number of objects with distinct power and the linearizability of set agreement objects. In: Proceedings of 31st International Symposium on Distributed Computing (DISC 2017), pp. 12:1–12:14 (2017)
8. Chan, D.Y.C., Hadzilacos, V., Toueg, S.: On the classification of deterministic objects via set agreement power. In: Proceedings of the ACM Symposium on Principles of Distributed Computing, PODC 2018, pp. 71–80 (2018)

9. Chandra, T., Hadzilacos, V., Jayanti, P., Toueg, S.: Wait-freedom vs. t-resiliency and the robustness of wait-free hierarchies. In: Proceedings of the 13th Annual ACM Symposium on Principles of Distributed Computing, PODC 1994, pp. 334–343 (1994)
10. Chaudhuri, S.: More choices allow more faults: set consensus problems in totally asynchronous systems. Inf. Comput. **105**(1), 132–158 (1993)
11. Daian, E., Losa, G., Afek, Y., Gafni, E.: A wealth of sub-consensus deterministic objects. In: 32nd International Symposium on Distributed Computing, DISC 2018, pp. 17:1–17:17 (2018)
12. Delporte-Gallet, C., Fauconnier, H., Gafni, E., Kuznetsov, P.: Set-consensus collections are decidable. In: 20th International Conference on Principles of Distributed Systems (OPODIS 2016), pp. 7:1–7:15 (2017)
13. Delporte-Gallet, C., Fauconnier, H., Gafni, E., Rajsbaum, S.: Linear space bootstrap communication schemes. Theoret. Comput. Sci. **561**, 122–133 (2015)
14. Godard, E., Imbs, D., Raynal, M., Taubenfeld, G.: Anonymous read/write memory: leader election and de-anonymization. In: Censor-Hillel, K., Flammini, M. (eds.) SIROCCO 2019. LNCS, vol. 11639, pp. 246–261. Springer, Cham (2019)
15. Herlihy, M.: Impossibility results for asynchronous pram. In: Proceedings of the 3rd Annual ACM Symposium on Parallel Algorithms and Architectures, pp. 327–336 (1991)
16. Herlihy, M., Rajsbaum, S., Raynal, M., Stainer, J.: From wait-free to arbitrary concurrent solo executions in colorless distributed computing. Theoret. Comput. Sci. **683**, 1–21 (2017)
17. Herlihy, M., Ruppert, E.: On the existence of booster types. In: Proceedings of 41st IEEE Symposium on Foundations of Computer Science, FOCS 2000, pp. 653–663 (2000)
18. Herlihy, M.P.: Wait-free synchronization. ACM Trans. Program. Lang. Syst. **13**(1), 124–149 (1991)
19. Herlihy, M.P., Luchangco, V., Moir, M.: Obstruction-free synchronization: double-ended queues as an example. In: Proceedings of the 23rd International Conference on Distributed Computing Systems, p. 522 (2003)
20. Herlihy, M.P., Shavit, N.: The topological structure of asynchronous computability. J. ACM **46**(6), 858–923 (1999)
21. Herlihy, M.P., Wing, J.M.: Linearizability: a correctness condition for concurrent objects. TOPLAS **12**(3), 463–492 (1990)
22. Jayanti, P.: On the robustness of Herlihy's hierarchy. In: Proceedings of 12th ACM Symposium on Principles of Distributed Computing, PODC 1993, pp. 145–157 (1993)
23. Jayanti, P.: Robust wait-free hierarchies. J. ACM **44**(4), 592–614 (1997)
24. Lamport, L.: On interprocess communication, parts I and II. Distrib. Comput. **1**(2), 77–101 (1986)
25. Lo, W., Hadzilacos, V.: All of us are smarter than any of us: nondeterministic wait-free hierarchies are not robust. SIAM J. Comput. **30**(3), 689–728 (2000)
26. Loui, M.C., Abu-Amara, H.: Memory requirements for agreement among unreliable asynchronous processes. Adv. Comput. Res. **4**, 163–183 (1987)
27. Pease, M., Shostak, R., Lamport, L.: Reaching agreement in the presence of faults. J. ACM **27**(2), 228–234 (1980)
28. Rabin, M.O.: The choice coordination problem. Acta Informatica **17**, 121–134 (1982)

29. Rachman, O.: Anomalies in the wait-free hierarchy. In: Tel, G., Vitányi, P. (eds.) WDAG 1994. LNCS, vol. 857, pp. 156–163. Springer, Heidelberg (1994). https://doi.org/10.1007/BFb0020431

30. Rashid, S., Taubenfeld, G., Bar-Joseph, Z.: Genome wide epigenetic modifications as a shared memory consensus problem. In: The 6th Workshop on Biological Distributed Algorithms (BDA 2018), London, July 2018

31. Ruppert, E.: Determining consensus numbers. SIAM J. Comput. **30**(4), 1156–1168 (2000)

32. Saks, M., Zaharoglou, F.: Wait-free k-set agreement is impossible: the topology of public knowledge. SIAM J. Comput. **29**, 1449–1483 (2000)

33. Shavit, N., Taubenfeld, G.: The computability of relaxed data structures: queues and stacks as examples. Distrib. Comput. **29**(5), 395–407 (2016)

34. Shenk, E.: The consensus hierarchy is not robust. In: Proceedings of 16th Annual ACM Symposium on Principles of Distributed Computing, PODC 1997, 279 p. (1997)

35. Taubenfeld, G.: Weak read/write registers. In: Frey, D., Raynal, M., Sarkar, S., Shyamasundar, R.K., Sinha, P. (eds.) ICDCN 2013. LNCS, vol. 7730, pp. 423–427. Springer, Heidelberg (2013). https://doi.org/10.1007/978-3-642-35668-1_29

36. Taubenfeld, G.: Coordination without prior agreement. In: Proceedings of ACM Symposium on Principles of Distributed Computing, PODC 2017, pp. 325–334 (2017)

Making Randomized Algorithms Self-stabilizing

Volker Turau[✉]

Institute of Telematics, Hamburg University of Technology, Hamburg, Germany
turau@tuhh.de

Abstract. It is well known that the areas of self-stabilizing algorithms and local algorithms are closely related. Using program transformation techniques local algorithms can be made self-stabilizing, albeit an increase in run-time or memory consumption is often unavoidable. Unfortunately these techniques often do not apply to randomized algorithms, which are often simpler and faster than deterministic algorithms. In this paper we demonstrate that it is possible to take over ideas from randomized distributed algorithms to self-stabilizing algorithms. We present two simple self-stabilizing algorithms computing a maximal independent set and a maximal matching and terminate in the synchronous model with high probability in $O(\log n)$ rounds. The algorithms outperform all existing algorithms that do not rely on unique identifiers.

1 Introduction

Self-stabilizing algorithms are a special type of distributed algorithms. Their distinctiveness is that they provide non-masking fault tolerance. That means that after a transient fault they return to a legal state without external intervention within a finite amount of time, i.e., within a finite number of rounds. An important class of distributed algorithms are local algorithms. They work on the assumption that each node only knows about its immediate neighborhood. The majority of the proposed local algorithms work in the synchronous model and do not care about fault tolerance. Despite some disparity local algorithms and self-stabilizing algorithms use a common set of models and complexity metrics. Nevertheless, the two communities do not pay much attention to each others results as already remarked by Lenzen et al. [18]. Another observation is that for many classical problems known local algorithms are faster by orders of magnitude. Whereas, the majority of self-stabilizing algorithms has prohibitively large stabilization time of $\Theta(n)$ or more, $O(\log n)$ or even $O(\log^* n)$ are common run-times for local algorithms. One reason is that most self-stabilizing algorithms are order invariant, i.e., they solve the symmetry breaking problem by utilizing only the ordering of identifiers.

This work is supported by the Deutsche Forschungsgemeinschaft (DFG) under grant DFG TU 221/6-3.

K. Censor-Hillel and M. Flammini (Eds.): SIROCCO 2019, LNCS 11639, pp. 309–324, 2019.
https://doi.org/10.1007/978-3-030-24922-9_21

A breakthrough with respect to efficient self-stabilizing algorithms is due to Barenboim et al. In 2018 they presented self-stabilizing algorithms with a sub-linear run-time for some classical problems [5]. In particular they proposed self-stabilizing algorithms for $(\Delta + 1)$–coloring, $(2\Delta - 1)$-edge-coloring, maximal independent set and maximal matching with $O(\Delta + log^*n)$ run-time. For symmetry breaking these algorithms still rely on unique identifiers but go beyond using only the ordering of identifiers.

Using program transformation techniques local algorithms can be made self-stabilizing. Several transformers were proposed using proof labeling schemes and self-stabilizing reset algorithms [1,3,4,18]. But they all come with some overhead in run-time or memory consumption. Lenzen et al. describe a transformer that also works in the asynchronous message passing model [18]. Each node stores the messages exchanged with each neighbor in a register. Each node simulates its own actions during a complete execution of the algorithm and writes its own outgoing messages into the register. After $T + 1$ asynchronous rounds, the initial state of the system has been replaced by the values the local algorithm would compute in a single run, assuming that the original algorithm completes in T rounds. Hence, the transformed algorithm is self-stabilizing and has the same time complexity, but it has a high memory overhead. The authors point out, that this transformation cannot be apply to randomized algorithms.

Using randomization for symmetry breaking has received only limited attention in self-stabilization [9,10,14,22]. On the other hand it is very well established in local algorithms. There are many efficient randomized algorithms that work in the synchronous model. Many make use of a phase concept, the realization of this concept in the self-stabilizing model requires an additional effort. The question we are concerned with in this work is how to transform phase-oriented distributed algorithms into self-stabilizing algorithms without increasing the run-time. Our main contribution are two randomized self-stabilizing algorithms for maximal independent set and maximal matching that do not rely on unique identifiers and stabilize w.h.p. in $O(\log n)$ rounds in the synchronous model. In contrast to the algorithms of Barenboim et al. which are rather difficult to implement our randomized algorithms have a very simple implementation.

We regard this work as a first step towards a transformation tool that automatically transforms phase-oriented synchronous distributed algorithms into self-stabilizing algorithms without introducing any overhead.

2 Phase-Oriented Distributed Algorithms

Many distributed algorithms using the synchronous model operate in phases. A phase consists of a fixed number of rounds. Phases are executed periodically and nodes perform a dedicated task in each round of a phase. In a phase-oriented algorithm the number of active nodes decreases in every phase. Initially all nodes are active. A typical phase is as follows. In the first round active nodes explore the states of their active neighbors. Next they validate if their state is consistent. If valid, they inform their neighbors and become passive. Otherwise they change their states according to the states of the active neighbors. Phase-oriented algorithms converge fast because in each phase a fair share of the nodes become passive. Now suppose that, because of some faults, the algorithm starts from an arbitrary configuration. Then nodes have a validated state, but this is not consistent with the states of the neighbors, some nodes may be still in the first step of a phase, some others already in the second step, etc. In such a scenario, phase-oriented algorithms will produce incorrect results.

The implementation of phases in a synchronous system is based on a synchronized variable counting rounds. This counter enables the nodes to decide in which round of a phase they are and when a new phase begins. This concept can be used in a self-stabilizing system if a phase consists only of a single round. Otherwise an algorithm must be prepared to handle transient faults that hit this counter. Thus, phase-oriented self-stabilizing algorithms require a self-stabilizing phase clock algorithm for synchronous networks, i.e., self-stabilizing synchronous unison. Such algorithms have been proposed but they require $\Omega(Diam(G))$ (the diameter of G) rounds to stabilize [7,15,17]. Since we aim at a run-time of $O(\log n)$ an approach that relinquishes the phase concept is required.

Instead of operating in globally synchronized phases, we propose that each node continuously and independently tries to perform its actions in the order as they would appear in a phase. In order for a node to know in which logical round of a phase it currently is, we introduce an additional state variable. The crucial point is that due to transient errors the order in which different nodes execute their actions is no longer synchronized but interleaved. Thus, we need additional actions that enforce the correct behavior even if neighboring nodes are desynchronised. One option is to perform a kind of *local reset* after detecting an inconsistency at a node. The peril of this option is that it may trigger a cascading reset leading to a kind of global reset that may last $Diam(G)$ rounds. We refrain from synchronizing nodes with respect to their position in a phase. To deal with the desynchronization effect we map the phase-dependent behavior to a state variable. This enables a node to determine from its own state and that of its neighbors its position within a phase and to act accordingly. The essential point is that when a fault hits the state variable, a node makes a local reset without triggering an avalanche of corrections.

In this paper we demonstrate this transformation for two phase-oriented algorithms solving classical problems. We show that the transformed algorithms are self-stabilizing and retain their run-time despite desynchronization. We like to point out that this transformation will not work in case the operations of the phases changes after some rounds. For example, when one algorithm runs for a fixed number of phases and then another algorithm takes over.

3   Notations and Computational Model

This paper uses the synchronous model of distributed computing as defined in the standard literature [21]. Extensions to asynchronous systems are discussed in Sect. 6. A distributed system is represented as an undirected graph $G(V, E)$

where V is the set of *nodes* and $E \subseteq V \times V$ is the set of *edges*. Let $n = |V|$, $m = |E|$, and let Δ denote the maximal degree of G. For any subgraph U of G and $v \in U$ denote by $d_U(v)$ the degree of v in U. The set of neighbors of node v is denoted by $N(v)$ and $N[v] = N(v) \cup \{v\}$. The diameter $Diam(G)$ of a graph G is the length of the longest shortest path between any two nodes.

Each node stores a set of variables. The values of all variables constitute the *local state* of a node. Let σ denote the set of possible local states of a node. The *configuration* of a system is the tuple of all local states of all nodes. $\Sigma = \sigma^n$ denotes the set of global states. A configuration is called *legitimate* if it conforms with the specification. Nodes communicate via locally shared memory. In this model each node executes a protocol consisting of a list of rules of the form *guard* $\longrightarrow$ *statement*. The guard is a Boolean expression over the node's variables and its neighbors' variables. The statement consists of a series of commands. A node is called *enabled* if one of its guards evaluates to true. The execution of a statement is called a *move*.

Execution of the statements is performed in a synchronous style, i.e., all enabled nodes execute their code in every round. An *execution* $\mathcal{E} = c_0, c_1, \ldots$, with $c_i \in \Sigma$ is a sequence of configurations, where c_0 is called the *initial configuration* and c_i is the configuration at the beginning of round i. Thus, if the current configuration is c_{i-1} and all enabled nodes make a move, then this yields c_i.

A distributed algorithm is called *self-stabilizing* if it satisfies two properties: *closure property* (every execution with a legitimate initial configuration remains in legitimate configurations) and *convergence property* (every execution reaches a legitimate configuration within finite time). We say that a randomized algorithm terminates w.h.p. (*with high probability*) within $O(f(n))$ time if it does so with probability at least $1 - 1/n^c$ for some $c \geq 1$. A randomized distributed algorithm is called *self-stabilizing* if it satisfies the closure property and if, starting from any configuration, reaches w.h.p. a legitimate configuration within finite time.

4   Maximal Matching

There is a bulk of literature on self-stabilizing algorithms for maximal matching [5, 8, 13]. These algorithms operate in different models (asynchronous, synchronous) and have different assumptions (anonymous, unique identifiers). Except for the algorithm in [5] – which stabilizes in $O(\Delta + \log^* n)$ rounds – all require $\Omega(n)$ rounds to stabilize. There are much stronger results for general distributed algorithms. For example Lotker et al. present a distributed algorithm to compute a maximal matching with approximation ratio $(1 - \epsilon)$ in $O(\log n)$ rounds [19]. Barenboim et al. proposed a maximal matching algorithm with runtime $O(\log \Delta + \log^4 \log n)$ [6]. Fischer recently presented an algorithm based on rounding requiring $O(\log^2 \Delta \log n)$ rounds [11]. In this section we show how to transform a randomized parallel maximal matching algorithm of Israeli and Itai [16]. The resulting self-stabilizing algorithm stabilizes w.h.p. in $O(\log n)$ rounds. If $\log n \in o(\Delta)$ it outperforms the algorithm of [5].

4.1   Algorithm $\mathcal{A}_{\mathsf{MAT}}$

The algorithm of Israeli and Itai can be transformed into a distributed algorithm that uses phases of length four rounds. In the first round each node invites a random neighbor. In the following round invited nodes randomly accept one invitation. At this point the accepted invitations form a subgraph U where the connected components are paths and cycles. In round three a matching of U is computed. For this purpose, nodes that accepted an invitation or nodes whose invitation was accepted randomly select either the edge towards the accepted or to the accepting neighbor, the corresponding edge is called a peer. In the last round edges that were selected by both end-nodes as peers join the matching. The corresponding nodes do not participate in the next phase. The algorithm terminates when none of the remaining nodes are connected by an edge.

To make this algorithm self-stabilizing the nodes mimic the phase-oriented behavior by storing information about their position within a phase in variable *state*. There are four positions: *none*, *invit*, *accept*, and *peer*. In contrast to the original algorithm, the positions do not necessarily follow a fixed pattern. A node may for example remain at *invit* for some rounds or may transit from *invit* to *peer* in one round. Also, the actions of the individual nodes are not synchronized, i.e., in one round some nodes invite a neighbor and at the same time other nodes may already select a peer. The invited, accepted or peered neighbor is stored in variable *partner*. $\mathcal{A}_{\mathsf{MAT}}$ makes use of the following three variables

- *match*: Indication whether the node is incident to an edge of the matching (*true*) or not (*false*). The fallback value is *false*.
- *partner*: The value of this variable is either a neighbor of the node or *null*. If *match* = *true* then the edge connecting the node with its *partner* belongs to the matching. Otherwise it indicates an invitation or the acceptance of an invitation. The fallback value is *null*.
- *state*: It describes the semantics of variable *partner*: *invit* (*partner* is invited), *accept* (node accepts invitation of *partner*), *peer* (edge connecting node and *partner* is proposed for or already belongs to the matching), and *none* (no *partner* selected, i.e., *partner* = *null*). The fallback value is *none*.

There are three rules RESET, MATCH, and RANDOM. The first rule is used to correct inconsistent states. After an inconsistent state is detected, rule RESET assigns the fallback values to the three variables. For example if *partner* = *null* but *state* $\neq$ *none* then RESET sets *state* to *none*. Each of the following three conditions is regarded as an inconsistent state (regardless of the value of *match*):

- $v.state = none$ but $v.partner \neq null$
- $v.state \neq none$ but $v.partner = null$
- $v.partner \notin N(v) \cup \{null\}$

For a node v with $match = true$ each of the following three conditions is regarded as an inconsistent state:

- $v.state \neq peer$
- $v.partner.partner \neq v$
- $v.partner.state \neq peer$

The rule MATCH promotes a node with $match = false$ to $match = true$ provided all of the following three conditions are all fulfilled:

- $v.state = peer$
- $v.partner.partner = v$
- $v.partner.state = peer$

Rule RANDOM only applies to nodes with $match = false$. It updates variables *partner* and *state* with the help of the following sets which contain the candidates for a peer, an acceptance, or for an invitation.

- $P(v) = \{w \in N(v) \mid w.state = accept \wedge w.partner = v \wedge w.match = false\} \cup \{v.partner \mid v.state = accept \wedge v.partner.match = false\}$
- $A(v) = \{w \in N(v) \mid w.state = invit \wedge w.partner = v \wedge w.match = false\}$
- $I(v) = \{w \in N(v) \mid w.state \in \{none, invit\} \wedge w.match = false\}$

The three sets are considered in this order and from the first non-empty set a random element w is selected; $w = null$ if all sets are empty. Let t be the corresponding *state*, i.e., *peer*, *accept*, or *invit* if $w \in P(v), w \in A(v)$, or $w \in I(v)$. The new values of *partner* and *state* are w and t respectively. The complete set of rules is as follows:

RESET: $(v.state = none) = (v.partner \neq null) \vee v.partner \notin N(v) \cup \{null\} \vee$
 $(v.match = true \wedge (v.state \neq peer \vee v.partner.partner \neq v \vee v.partner.state \neq peer))$
 $\longrightarrow \quad v.match := false; v.state := none; v.partner := null;$
MATCH: $v.state = peer \wedge v.partner.state = peer \wedge v.partner.partner = v$
 $\longrightarrow \quad v.match := true$
RANDOM: $v.match = false \wedge (v.partner, v.state) \neq (w, t)$
 $\longrightarrow \quad (v.partner, v.state) := (w, t)$

Figure 1 shows an execution of algorithm $\mathcal{A}_{\mathsf{MAT}}$. In the first round all nodes make a RANDOM move and invite a neighbor. The invited nodes v_1, v_2, and v_3 thereafter accept an invitation, while v_0 and v_4 make an new invitation. v_1, v_2, and v_3 select a peer in round three. None of these three nodes can make a MATCH move since the selected peers do not match. Thus, in round four v_1 and v_3 again make an invitation. Node v_2 cannot make an invitation in round four since none of its neighbors has $state \in \{none, invit\}$. In round five nodes v_0 and v_4 select each other as peer and therefore make a MATCH move in round five. After eight rounds a legitimate state is reached.

Let $\mathcal{E} = c_0, c_1, \ldots$ be an error-free execution of algorithm $\mathcal{A}_{\mathsf{MAT}}$. The following lemma summarizes basic properties of the $\mathcal{A}_{\mathsf{MAT}}$.

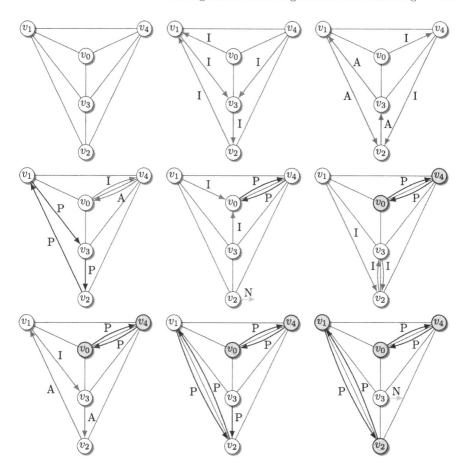

Fig. 1. An execution of Algorithm $\mathcal{A}_{\mathsf{MAT}}$ of length 8. Variable *partner* is pictured as an arrow. The attached label indicates the value of variable *state*. I, A, P, or N stand for *invit, accept, peer*, or *none*. Dark nodes have *match* = *true*. In the final configuration (bottom right) edges (v_0, v_4) and (v_1, v_2) form a maximal matching.

Lemma 1. *For $v \in V$ the following properties hold.*

(i) Node v executes rule MATCH *at most once in $\mathcal{E}$. If v executes* MATCH *then either $v.partner.match = true$ and $v.partner$ is disabled or $v.partner$ executes* MATCH *in the same round. The first case only occurs in round 0.*

(ii) Rule RESET *is only executed in round 0. In particular after round 0 either $v.state \neq none$ and $v.partner \in N(v)$ or $v.state = none$ and $v.partner = null$.*

(iii) After round 0 (round 1) v has at most one neighbor w with $w.partner = v$ and $w.state = accept$ ($w.state = peer$), in particular $|P(v)| \leq 2$ after round 0.

(iv) If v.state = peer after round 0 then either v.match = true and v is disabled for the remaining part of $\mathcal{E}$ or v has assigned a value to v.peer through RANDOM in the round before.

(v) If v executes rule RANDOM and sets v.state to peer and v.partner to w after round 0 then w will also have w.state = peer at the end of the round.

Lemma 2. *Let c_i with $i > 0$ be a configuration of $\mathcal{E}$ in which no node is enabled. Then the set*

$$M = \{(v, v.partner) \mid v.match = true \ in \ c_i\}$$

is a maximal matching.

Proof. If $v.match = true$ then $v.partner \in N(v)$, $v.partner.partner = v$, and $v.partner.match = true$ since rule RESET is disabled. Thus, M is a matching.

Let v be node with $v.match = false$. Assume $v.state = peer$ and $u = v.partner$. Since RANDOM is not enabled for v we have $u \in P(v)$. By definition of $P(v)$ this implies $u.partner = v$ and $u.state = accept$. Hence $v \in P(u)$. This yields $P(u) \neq \emptyset$ and $u.state = peer$. This contradiction shows $v.state \neq peer$. A similar argument shows that $v.state \neq accept$. Thus, $v.state \in \{invit, none\}$.

Assume there exists an edge $e = (v, u)$ of G such that e is not incident to an edge of M. Then $u.match = v.match = false$. Thus, the above implies that $I(v)$ and $I(u)$ are not empty and hence, $u.state = v.state = invit$. This yields that u and v are enabled for rule RANDOM. Contradiction. This proves that M is a maximal matching.   □

Lemma 3. *Let v be a node with $v.match = false$ in c_i with $i > 0$.*

(i) If $v.state = peer$ then the probability that $v.match = true$ in c_j for all $j \geq i + 1$ is at least 0.5.

(ii) If $v.state = accept$ then the probability that $v.match = true$ in c_j for all $j \geq i + 2$ is at least 0.5.

(iii) If $w.state = invit$ and $w.partner = v$ then the probability that $v.match = true$ in c_j for all $j \geq i + 3$ is at least 0.5.

Proof. (i) By Lemma 1(iv) v has set $v.partner$ to w by executing rule RANDOM before round i. By Lemma 1(v) w executed RANDOM in the same round and set $w.state$ to $peer$. By Lemma 1(iii) $|P(w)| \leq 2$. Hence, the probability that $w.partner = v$ is at least 0.5. Thus, v executes rule MATCH with probability at least 0.5. The result follows from Lemma 1(i).

(ii) When v changed it state to $accept$ and set $v.partner$ to w, w satisfied $w.state = invit$, $w.partner = v$ and $w.match = false$. Therefore, in c_i node w still satisfies $w.match = false$. Thus, $P(v) \neq \emptyset$ since $w \in P(v)$. Hence, v will execute rule RANDOM and set $v.state$ to $peer$ in round i. The lemma follows from (i).

(iii) If $v.state \in \{peer, accept\}$ then the result follows from (i), (ii), and Lemma 1(i). Otherwise, v will execute rule RANDOM and set $v.state$ to $accept$. The result again follows from (ii).   □

For $i > 0$ let $U_i = \{v \in V \mid v.match = false \text{ in } c_i\}$ and G_i the subgraph of G induced by U_i. Note that $G_i \subseteq G_{i-1}$ by Lemma 1(i). We will show that after expected $O(\log n)$ rounds graph G_i consists of isolated nodes only.

A node v of a graph G is called *good* if the degree of more than a third of v's neighbors is at most as large as the degree of v. An edge of G is called *good* if at least one of its endpoints is good.

Lemma 4. *Let $i > 0$ and v be a good node of G_i. Denote the expected number of edges incident to v in G_i that are not contained in G_{i+4} by $E_i(v)$. Then $E_i(v) \geq (1 - e^{-1/6})d_{G_i}(v)/12$ for $i > 1$.*

Proof. We can assume $d = d_{G_i}(v) > 0$. If $v.state \in \{peer, accept\}$ in c_i then by Lemma 3 v will have $match = true$ in c_{i+2} with probability at least 0.5. Thus, with probability 0.5 an edge incident to v in G_i will not be contained in G_{i+2}. Hence, in this case $E_i(v) \geq d_{G_i}(v)/2$.

Next consider the case that $v.state \in \{invit, none\}$. Since v is good in G_i it has k neighbors $v_1, \ldots, v_k$ in G_i with $d_{G_i}(v_i) \leq d$ and $k > d/3$. Let S be the set of those nodes v_i that set variable $state$ to $invit$ in round i. Thus, $v \in I(u)$ for all $u \in S$. Let $s = |S|$. First consider the case $s \geq k/2$. Then $s > d/6$. For $u \in S$ we have

$$Prob(u.partner \neq v \text{ at the end of round } i) \leq 1 - \frac{1}{d_{G_i}(u)}.$$

This yields

$$Prob(\{w \in U_{i+1} \mid w.partner = v\} = \emptyset) \leq \prod_{u \in S} \left(1 - \frac{1}{d_{G_i}(u)}\right)$$

$$\leq \left(1 - \frac{1}{d}\right)^{d/6} < e^{-1/6}.$$

Hence, the probability that there exists a neighbor $w \in U_{i+1}$ of v with $w.state = invit$, and $w.partner = v$ in c_{i+1} is at least $1 - e^{-1/6}$. Thus, Lemma 3 implies that $v.match = true$ at the end of round $i + 4$ is greater than $(1 - e^{-1/6})/2$. Thus, for $j = 1, \ldots, k$ the probability that edge (v, v_j) is not contained in G_{i+4} is at least $(1 - e^{-1/6})/2$. Hence, in this case $E_i(v) \geq (1 - e^{-1/6})d_{G_i}(v)/2$.

At last consider the case that $s < k/2$. Note that the neighbors of v in U_i that are not in S set their variable $state$ to $peer$ or $accept$. Lemma 3 imply that with probability 0.5 the edge (v, v_j) with $v_j \notin S$ is not contained in G_{i+2}. Thus, in this last case $E_i(v) \geq 0.5(k - s) \geq d_{G_i}(v)/12$. $\square$

Theorem 1. *Algorithm $\mathcal{A}_{\mathsf{MAT}}$ is self-stabilizing and computes w.h.p. in $O(\log n)$ rounds using $O(\log n)$ memory a maximal matching.*

Proof. Denote the number of edges of G_i by m_i. By Lemma 4.4 of [2] at least half of the edges of any graph are good. Thus,

$$\sum_{v \in V, v \text{ good}} d_{G_i}(v) \geq \frac{m_i}{2}.$$

Let E_i be the expected number of edges in G_i that are not contained in G_{i+4}. Then by Lemma 4

$$E_i \geq \sum_{v \in V, v \text{ good}} E_i(v)/2 \geq \sum_{v \in V, v \text{ good}} (1 - e^{-1/6})d_{G_i}(v)/24 \geq m_i(1 - e^{-1/6})/48.$$

Let $\alpha = 96/(1 - e^{-1/6})$. Hence, in expectation at least $2m_i/\alpha$ edges of G_i are not contained in G_{i+4}. Let ξ be the probability that at least $1/\alpha$ of the edges of G_i are not contained in G_{i+4}. Then $2/\alpha \leq E_i/m_i \leq \xi + (1 - \xi)/\alpha$. Hence, $\xi \geq 1/(\alpha - 1)$. This yields that with probability at least $1/(\alpha - 1)$ at least $1/\alpha$ of the edges of G_i are not contained in G_{i+4}. Thus, in expectation every $4(\alpha - 1)$ rounds $1/\alpha$ of the edges is removed from G_i. Hence, after $4(\alpha - 1)i$ rounds there are at most $m(1 - 1/\alpha)^i$ edges left. Thus, after $4(\alpha - 1)(\log m/\log(\alpha/(\alpha - 1)))$ rounds in expectation the algorithm terminates. Let $C > 1$ be a constant such that $C \log n > 4(\alpha - 1)(\log m/\log(\alpha/(\alpha - 1)))$.

For $i = 1, \ldots, 2(\alpha - 1)C \log n$ let Y_i be random variables such that Y_i is equal to 1 if during the rounds $4(\alpha - 1)i, \ldots, 4(\alpha - 1)(i + 1) - 1$ a fraction of $1/\alpha$ of the edges disappears and otherwise 0. Then $\Pr[Y_i = 1] = 1/(\alpha - 1)$. Let $Y = \sum_{i=1}^{2(\alpha-1)C \log n} Y_i$. Then $E[Y] = 2C \log n$. The Chernoff bound implies that $P[Y \leq (1 - \delta)E[Y]] \leq e^{-E[Y]\delta^2/2}$. Let $\delta = 1/2$. Then

$$e^{-E[Y]\delta^2/2} = e^{-C \log n/4} = 1/n^{C/4}$$

Thus, w.h.p. we have $2Y > E[Y] = 2C \log n$. This yields $Y > C \log n$. Therefore, we have proved that w.h.p. algorithm $\mathcal{A}_{\mathsf{MIS}}$ terminates in $O(\log n)$ rounds. Lemma 2 completes the proof. $\qquad\square$

5   Maximal Independent Sets

There exist several proposals for self-stabilizing algorithms to compute a maximal independent set (MIS), see [13] for a survey. Up to the work of Barenboim et al. none had a run-time in $o(n)$ [5]. Their algorithm has a run-time of $O(\Delta + \log^* n)$. A randomized MIS algorithm – albeit not self-stabilizing – with run-time $O(\log^2 \Delta + 2^{o(\sqrt{\log \log n})})$ is also due to Barenboim et al. [6]. This was later improved by Ghaffari to $O(\log \Delta + 2^{o(\sqrt{\log \log n})})$ [12]. In this section we transform a randomized MIS-algorithm of Métivier et al. into a self-stabilizing algorithm that terminates w.h.p. in $O(\log n)$ rounds [20].

5.1   Algorithm $\mathcal{A}_{\mathsf{MIS}}$

The algorithm of Métivier et al. uses the synchronous model and works in phases, where one phase lasts three rounds. In each phase some nodes join the MIS. These nodes and their neighbors become passive, i.e., they do not participate in future phases. In the first round of each phase the active nodes generate a random number from a range $0, \ldots D$, where D is a constant larger than n^3.

Then the nodes exchange their choice among their neighbors. A node is included in the independent set if its number is minimal within its local neighborhood. A node informs its neighbors in the second round of a phase in this case. In the third and final round the neighbors of the inserted nodes announce that they are passive.

Observe that the correctness does not rely on the phase-oriented execution, symmetry breaking is not affected. At the end of each phase the nodes that were included in the MIS together with their neighbors are no longer active. We use a state variable to indicate, whether a node is active or not. It may happen that a node that will be excluded in the next round (because one of its neighbors has joined the MIS in the current round) prevents a neighbor to join the MIS. But this will only delay convergence. For convenience we describe algorithm $\mathcal{A}_{\mathsf{MIS}}$ in the shared memory model. $\mathcal{A}_{\mathsf{MIS}}$ makes use of the following three variables:

- mis: It indicates whether a node belongs to the independent set ($true$) or not ($false$).
- $state$: The value $true$ indicates that the node has at least one neighbor with $mis = true$. Thus, only nodes with $state = false$ and $mis = false$ are active.
- r: The value is in the range $0, \ldots, D - 1$ and is used to break symmetries.

For $v \in V$ define $N^{act}(v) = \{w \in N(v) \mid w.state = false\}$ and $N^{mis}(v) = \{w \in N(v) \mid w.mis = true\}$. $\mathcal{A}_{\mathsf{MIS}}$ uses the following five simple rules.

LEAVE : $mis = true \wedge N^{mis}(v) \neq \emptyset$
$\longrightarrow$ $mis := false \wedge state := false$
FLAG : $mis = false \wedge state \neq (N^{mis}(v) \neq \emptyset)$
$\longrightarrow$ $state := (N^{mis}(v) \neq \emptyset)$
RESET : $mis = true \wedge state = true$
$\longrightarrow$ $state := false$
JOIN : $mis = false \wedge state = false \wedge \forall u \in N^{act}(v)\ v.r < u.r$
$\longrightarrow$ $mis := true$
RANDOM : $mis = false \wedge state = false$
$\longrightarrow$ $r := random(D)$

A node that is enabled for two or more rules only executes the first of these rules. Figure 2 shows an execution of algorithm $\mathcal{A}_{\mathsf{MIS}}$ with an illegal initial configuration.

Let $\mathcal{E} = c_0, c_1, \ldots$ be an error-free execution of algorithm $\mathcal{A}_{\mathsf{MIS}}$.

Lemma 5. *Throughout $\mathcal{E}$ each node $v \in V$*

(i) makes at most one JOIN move and all nodes in $N[v]$ become disabled after the following round and remain so in $\mathcal{E}$,
(ii) can only make a LEAVE move in round 0,
(iii) makes after round 1 at most one FLAG move, this move sets state to true,
(iv) can only make a RESET move in round 0.

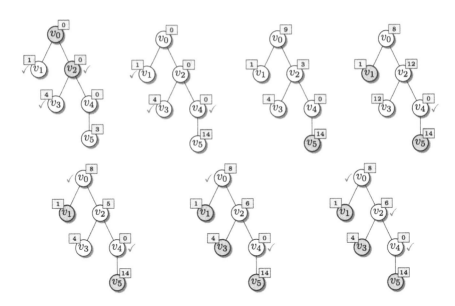

Fig. 2. An execution of Algorithm $\mathcal{A}_{MIS}$ for a tree. The numbers attached to the nodes correspond to variable r and the symbol $\checkmark$ indicates $state = true$. Dark nodes have $mis = true$. The initial state is illegal, since the neighboring nodes v_0 and v_2 are both in the MIS. Therefore, these two nodes make LEAVE moves in round 1. While v_1 and v_3 are disabled, node v_4 makes a FLAG move, and node v_5 a RANDOM move in this round. In round 2 nodes v_1, v_3, v_4 make a FLAG, v_0, v_2 a RANDOM, and v_5 a JOIN move. Node v_3 is the last node to join the MIS. After six rounds a legal configuration is reached.

Proof. (i) When v makes a JOIN move rule FLAG was not enabled. Thus, all neighbors have $mis = false$. The condition on variable r enforces that no neighbor can currently join. In the next round all nodes in $N^{act}(v)$ will execute FLAG and set $state$ to $true$. The nodes in $N(v) \setminus N^{act}(v)$ will be disabled for the rest of $\mathcal{E}$ since v will remain disabled. Hence, v and its neighbors will be disabled thereafter.

(ii) The condition $mis = true \wedge N^{mis}(v) \neq \emptyset$ can only be the result of a fault.

(iii) Suppose v makes a FLAG move in round $i \geq 2$ setting $state$ to $false$. Thus, $v.mis = false, v.state = true$, and $N^{mis}(v) = \emptyset$ in c_i. Assume that v was disabled in c_{i-1}. This yields that $N^{mis}(v) \neq \emptyset$ in c_{i-1}. Hence, a neighbor of v made a LEAVE move in round $i - 1$. Property (ii) implies that $i = 1$. This contradiction implies that v was enabled in round $i - 1 > 0$ for move FLAG. Thus, in c_{i-1} we had $v.mis = false, v.state = false$, and $N^{mis}(v) \neq \emptyset$. This requires that a neighbor of v made a LEAVE move in round $i - 1$. Contradiction. Hence, a FLAG move in round $i \geq 2$ can only set $state$ to $true$. Thus, $v.mis = false, v.state = false$, and $N^{mis}(v) \neq \emptyset$ in c_i. If a neighbor of v made a JOIN move in round $i - 1$ then the result follows from property (ii). Assume no neighbor of v made a JOIN move in round $i - 1$. Hence, $N^{mis}(v) \neq \emptyset$ in c_{i-1}. This yields that v was in round $i-1$ enabled for move FLAG, but then v would have $v.state = true$ in c_i. This contradiction proves property (iii).

(iv) Note that if v is enabled for RESET in round $i > 0$ it was also enabled for RESET in round $i - 1$. □

Lemma 6. *Let $c_i \in \mathcal{E}$ with $i > 1$ in which no node is enabled. Then*

$$I = \{v \in V \mid v.mis = true \text{ in } c_i\}$$

is a maximal independent set.

Theorem 2. *Algorithm $\mathcal{A}_{MIS}$ is self-stabilizing and computes w.h.p. in $O(\log n)$ rounds using $O(\log n)$ memory an independent set.*

Proof. For $i > 1$ let $R_i = \{v \in V \mid v.mis = false \wedge v.state = false \text{ in } c_i\}$ and G_i the subgraph of G induced by R_i. Property (iii) of Lemma 5 implies $R_{i+1} \subseteq R_i$ for all $i > 1$. Let $v, w \in R_i$. Then $P(v.r = w.r) = 1/D$. The union bound implies

$$P(\exists v, w \in R_i \text{ s.t. } v.r = w.r) \leq |R_i|^2/2D \leq n^2/2D.$$

Thus, with probability at least $1 - n^2/2D \geq 1 - 1/2n$ the values $v.r$ of all $v \in R_i$ are different. Consider the case that all values $v.r$ are different. The following part of the proof is based on ideas of Métivier et al. [20]. A vertex v making a JOIN move *preemptively removes* a neighbor u if $v.r$ is less than $u.r$ and $w.r$ for all other neighbours w of v and u. If this happens in round i then u will set $state$ to $true$ in round $i+1$. Furthermore, v and u will not be in R_j for $j \geq i+2$. We say that in this case the $d_{G_i}(u)$ edges (u, w) are *preemptively removed* from G_i. Note an edge (v, u) can only be preemptively removed twice, once when v preemptively removes u and once when u preemptively removes v.

The probability that v preemptively removes u is at least $1/(d_{G_i}(v) + d_{G_i}(u))$, since the r values of all nodes of G_i are different and this is the probability that $v.r = \min\{x.r \mid x \in N_{G_i}(v) \cup N_{G_i}(u)\}$. Let random variable X_i denote the number of edges preemptively removed from G_i. Then $E[X_i]$ is at least

$$\left(\sum_{\{v,u\} \in G_i} \frac{d_{G_i}(u)}{d_{G_i}(v) + d_{G_i}(u)} + \frac{d_{G_i}(v)}{d_{G_i}(v) + d_{G_i}(u)} \right) \bigg/ 2 = \frac{m_i}{2}$$

where m_i is the number of edges of G_i. Thus, with probability $1 - n^2/2D$ in expectation at least half of the edges of G_i are not contained in G_{i+2}.

Let ξ be the probability that at least a quarter of the edges of G_i are not contained in G_{i+2} provided all values of r are different. Then

$$1/2 \leq E[X_i]/m_i \leq \xi + (1 - \xi)/4 = (1 + 3\xi)/4.$$

Hence, $\xi \geq 1/3$. This yields that with probability at least $(1 - n^2/2D)/3 > 1/4$ at least a quarter of the edges of G_i are not contained in G_{i+2}. Thus, in expectation every eight rounds at least a quarter of the edges is removed. Hence, after $8i$

rounds there are at most $m(3/4)^i$ edges left. Thus, after $8\,(\log m/\log(4/3)+1)$ rounds in expectation the algorithm terminates. Let $C > 1$ be a constant such that $C\log n > 8\,(\log m/\log(4/3)+1)$.

For $i = 1,\ldots,8C\log n$ let Y_i be random variables such that Y_i is equal to 1 if during the rounds $8i,\ldots,8i+7$ a quarter of the edges disappears and otherwise 0. Let $Y = \sum_{i=1}^{8C\log n} Y_i$. Then $E[Y] = 8pC\log n$ with $p = (1 - n^2/2D)/3$. The Chernoff bound implies that $P[Y \leq (1-\delta)E[Y]] \leq e^{-E[Y]\delta^2/2}$. For $\delta = 1/2$

$$e^{-E[Y]\delta^2/2} = e^{-8pC\log n/8} = 1/n^{pC}$$

Hence, $\lim_{n\to\infty} e^{-E[Y]\delta^2/2} = 0$. Thus, w.h.p. $2Y > E[Y] = 8pC\log n > 2C\log n$ and $Y > C\log n$. Consequently we have proved that w.h.p. algorithm $\mathcal{A}_{\mathsf{MIS}}$ terminates in $O(\log n)$ rounds. Lemma 6 completes the proof.   □

6   Asynchronous Systems

Self-stabilizing algorithms are often designed for asynchronous systems. The degree of asynchronicity is controlled by a scheduler acting as an adversary. It can delay the execution of a move of an enabled node. A common scheduler is the unfair distributed scheduler. This scheduler can delay a move of a node as long as other nodes are enabled. Algorithm $\mathcal{A}_{\mathsf{MIS}}$ does not stabilize under this scheduler. Consider a graph with three nodes v_0, v_1, and v_2 that form a path such that $v_0.mis = true$ and $v_1.mis = v_2.mis = false$. Furthermore, $v_i.state = false$ for all i and $v_1.r = 0$. Then v_1 is enabled for rule FLAG and v_2 for rule RANDOM. An unfair distributed scheduler can continuously schedule v_2 for execution and withhold v_1 at the same time. A similar example can be constructed for algorithm $\mathcal{A}_{\mathsf{MAT}}$. In general rules that only select a random value such as rule RANDOM in algorithm $\mathcal{A}_{\mathsf{MIS}}$ do not work under an unfair distributed scheduler.

What about a fair distributed scheduler? With a fair scheduler each enabled node eventually is selected for execution, e.g. after at most k rounds. It is easy to prove that algorithm $\mathcal{A}_{\mathsf{MIS}}$ is self-stabilizing for the fair distributed scheduler. Note that nodes make at most one LEAVE, RESET, and JOIN move and at most three FLAG moves. Thus, the probability that nodes make RANDOM moves that lead to a configuration with all nodes disabled is positive. The expected number of rounds strongly depends on the fairness character of the scheduler. A similar argument shows that this statement also holds for the probabilistic scheduler.

7   Conclusion

In this paper we demonstrated that phase-oriented randomized distributed algorithms can be made self-stabilizing in the synchronous model while retaining their time complexity. We transformed two classical distributed randomized graph algorithms into self-stabilizing algorithms. The algorithms are next to those presented in [5] the only ones with a sublinear stabilization time for the corresponding problems. The ultimate goal of this work is to operationalize this transformation and to have a tool that automatically performs this program transformation for a rich class of randomized algorithms even in the asynchronous model.

References

1. Afek, Y., Kutten, S., Yung, M.: The local detection paradigm and its applications to self-stabilization. Theor. Comput. Sci. **186**(1), 199–229 (1997)
2. Alon, N., Babai, L., Itai, A.: A fast and simple randomized parallel algorithm for the maximal independent set problem. J. Algorithms **7**(4), 567–583 (1986)
3. Awerbuch, B., Varghese, G.: Distributed program checking: a paradigm for building self-stabilizing distributed protocols. In: Proceedings of 32nd Annual Symposium of Foundations of Computer Science, pp. 258–267, October 1991
4. Awerbuch, B., Patt-Shamir, B., Varghese, G., Dolev, S.: Self-stabilization by local checking and global reset. In: Tel, G., Vitányi, P. (eds.) WDAG 1994. LNCS, vol. 857, pp. 326–339. Springer, Heidelberg (1994). https://doi.org/10.1007/BFb0020443
5. Barenboim, L., Elkin, M., Goldenberg, U.: Locally-iterative distributed $(\Delta + 1)$-Coloring below szegedy-vishwanathan barrier, and applications to self-stabilization and to restricted-bandwidth models. In: Proceedings of ACM Symposium on Principles of Distributed Computing, pp. 437–446 (2018)
6. Barenboim, L., Elkin, M., Pettie, S., Schneider, J.: The locality of distributed symmetry breaking. J. ACM **63**(3), 20:1–20:45 (2016)
7. Boulinier, C., Petit, F., Villain, V.: Synchronous vs. asynchronous unison. In: Tixeuil, S., Herman, T. (eds.) SSS 2005. LNCS, vol. 3764, pp. 18–32. Springer, Heidelberg (2005). https://doi.org/10.1007/11577327_2
8. Cohen, J., Lefevre, J., Maamra, K., Pilard, L., Sohier, D.: A self-stabilizing algorithm for maximal matching in anonymous networks. PPL **26**(04), 1650016 (2016)
9. Devismes, S., Tixeuil, S., Yamashita, M.: Weak vs. self vs. probabilistic stabilization. Int. J. Found. Comput. Sci. **26**(3), 291–319 (2015)
10. Dolev, S., Israeli, A., Moran, S.: Analyzing expected time by scheduler-luck games. IEEE Trans. Softw. Eng. **21**(5), 429–439 (1995)
11. Fischer, M.: Improved deterministic distributed matching via rounding. In: Richa, A. (ed.) Distributed Computing, vol. 91, pp. 17:1–17:15. Springer, Heidelberg (2017). https://doi.org/10.1007/s00446-018-0344-4
12. Ghaffari, M.: An improved distributed algorithm for maximal independent set. In: Proceedings of 27th Annual ACM-SIAM Symposium on Discrete Algorithms, pp. 270–277 (2016)
13. Guellati, N., Kheddouci, H.: A survey on self-stabilizing algorithms for independence, domination, coloring, and matching in graphs. J. Parallel Distrib. Comput. **70**(4), 406–415 (2010)
14. Herman, T.: Probabilistic self-stabilization. IPL **35**(2), 63–67 (1990)
15. Herman, T., Ghosh, S.: Stabilizing phase-clocks. IPL **54**(5), 259–265 (1995)
16. Israeli, A., Itai, A.: A fast and simple randomized parallel algorithm for maximal matching. IPL **22**(2), 77–80 (1986)
17. Kravchik, A., Kutten, S.: Time optimal synchronous self stabilizing spanning tree. In: Afek, Y. (ed.) DISC 2013. LNCS, vol. 8205, pp. 91–105. Springer, Heidelberg (2013). https://doi.org/10.1007/978-3-642-41527-2_7

18. Lenzen, C., Suomela, J., Wattenhofer, R.: Local algorithms: self-stabilization on speed. In: Guerraoui, R., Petit, F. (eds.) SSS 2009. LNCS, vol. 5873, pp. 17–34. Springer, Heidelberg (2009). https://doi.org/10.1007/978-3-642-05118-0_2
19. Lotker, Z., Patt-Shamir, B., Pettie, S.: Improved distributed approximate matching. J. ACM **62**(5), 38:1–38:17 (2015)
20. Métivier, Y., Robson, J.M., Saheb-Djahromi, N., Zemmari, A.: An optimal bit complexity randomized distributed MIS algorithm. Distrib. Comput. **23**(5), 331–340 (2011)
21. Peleg, D.: Distributed Computing: A Locality-Sensitive Approach. SIAM Society for Industrial and Applied Mathematics, Philadelphia (2000)
22. Turau, V.: Computing fault-containment times of self-stabilizing algorithms using lumped Markov chains. Algorithms **11**(5), 58 (2018)

Brief Announcements

How to Color a French Flag

Biologically Inspired Algorithms for Scale-Invariant Patterning

Bertie Ancona, Ayesha Bajwa$^{(\boxtimes)}$, Nancy Lynch, and Frederik Mallmann-Trenn

Massachusetts Institute of Technology, Cambridge, MA 02139, USA
{bancona,abajwa,mallmann}@mit.edu, lynch@csail.mit.edu

Abstract. In the *French flag problem*, initially uncolored cells on a grid must differentiate to become blue, white or red. The goal is for the cells to color the grid as a French flag, i.e., a three-colored triband, in a distributed manner. To solve a generalized version of the problem in a distributed computational setting, we consider two models: a biologically-inspired version relying on morphogens (gradients of chemicals acting as signals) and a more abstract version based on reliable message passing between cellular agents. We show that both models easily achieve a French ribbon - a French flag in the 1D case. However, extending the ribbon to the 2D flag in the concentration model is somewhat difficult unless each agent has additional positional information. Assuming that cells are are identical, it is impossible to achieve a French flag or even a close approximation. In contrast, using a message-based approach in the 2D case only requires assuming that agents can be represented as constant size state machines. We hope our insights may lay some groundwork for what kind of message passing abstractions or guarantees are useful in analogy to cells communicating at long and short distances to solve patterning problems. In addition, we hope that our models and findings may be of interest in the design of nano-robots.

1 Introduction

In the *French flag problem*, initially uncolored cells on a grid must differentiate to become blue, white, or red, ultimately coloring the grid as three stripes without centralized decision-making. Wolpert's original French flag problem formulation using positional information gives a model of how organisms determine cell type, a question central to developmental biology [4]. Our work is loosely inspired by biological mechanisms; we relate a reliable message passing model with local cell-cell communication and a concentration-based model with chemical gradients over long distances. By analyzing a generalized French flag problem for k colors in these two computational models, we aim to understand the minimum set of assumptions required to solve the problem exactly or approximately. We also consider whether cells must know their absolute positions to solve the k-flag problem.

© Springer Nature Switzerland AG 2019
K. Censor-Hillel and M. Flammini (Eds.): SIROCCO 2019, LNCS 11639, pp. 327–331, 2019.
https://doi.org/10.1007/978-3-030-24922-9_22

We begin by studying the *French ribbon problem*, the 1D scenario in which both exact and approximate solutions are possible. While both models easily achieve a French ribbon, extending to the French flag is provably difficult in the concentration model. We hope this work may illuminate computational abstractions that may be useful in analogy to cells communicating to solve patterning problems. For the detailed algorithms and full proofs, see the full paper [1].

Related Work. Gradients of chemicals called morphogens are thought to underlie cell-cell communication over long distances, but exactly how they produce scale-invariant patterns in tissues of varying size is an interesting biological question. Mechanisms for local cell-cell communication include cell surface receptors and ligands, which we liken to message passing between neighboring agents. Wolpert focused the French flag problem and model on positional information and its generalization to patterning problems [4]. Subsequent papers validated the positional information paradigm in empirical studies [3]. We consider morphogens that form steady state gradients and trigger synchronized fate decisions.

Problem Statement and Notation. In the 1D French ribbon problem, we assume a line graph consisting of n nodes which we refer to as agents. In the French flag, the graph is a $a \times b$ grid on $n = a \cdot b$ agents. To solve the French ribbon problem, each agent must output a color so that the line is segmented into three colors: blue, white, and red from left to right. Formally, if b, w, and r denote the number of agents of each respective color, $\max\{|b - w|, |b - r|, |w - r|\} \leq 1$. In addition, each color should be in a single, contiguous sub-line of the line graph. We also define the more general 1D k-Ribbon problem in the same model, in which there are k distinct colors $\{1, ..., k\}$ which must form bands of approximately equal size, in increasing numerical order, along a line graph of n agents. We generally assume the number of colors k to be constant[1].

A solution to the French flag problem requires that every agent outputs a single color, so the grid is divided into three vertical blocks. Rows must abide by French ribbon requirements, with blue on the left and red on the right. An agent must be the same color as the agents above and below in its column. The 2D k-flag problem generalizes in the same way.

We say a k-colored flag of dimensions $a \times b$ is an *ε-approximate* flag if for every color $z \in \{1, ..., k\}$ the following hold. First, agents that are clearly within one stripe should have the corresponding color. Second, agents close to a color border (c_1, c_2) should have either color c_1 or c_2. Formally, we require that for each agent u with coordinates (x, y):

- If $x \in \left[\left(\frac{z-1}{k} + \varepsilon \right) \cdot a, \left(\frac{z}{k} - \varepsilon \right) \cdot a \right]$, then u has color z.
- If u has color z, then $x \in \left[\left(\frac{z-1}{k} - \varepsilon \right) \cdot a, \left(\frac{z}{k} + \varepsilon \right) \cdot a \right]$.

2   Concentration Model

For concentration-based solutions to the French flag problem, we assume that each agent receives concentration inputs from up to four source agents s_1, s_2, s_3, and s_4.

---

[1] However, for clarity, we sometimes highlight the dependency on k.

The *measured concentration* from source s_i at 2D coordinate $C = (x, y)$ is given by a *gradient function* $\lambda_i(C)$. We assume that the gradient function is (1) invertible and (2) monotonically decreasing in $dist(C, s_i)$, where $dist(C, s_i)$, denotes the distance between the agent at C and the source s_i. We can choose a gradient function such as a power-law (e.g., $\lambda_i(C) = 1/dist(C, s_i)^\alpha$) for convenience, but our bounds hold for general gradient functions satisfying (1) and (2).

We assume zero noise and therefore arbitrarily good precision in measuring concentration. Agents do not receive any other input such as knowledge of their coordinate or the total ribbon or flag size, and they perform the same algorithms. No messages are passed between agents, so we consider only local computation for time complexity. In the French ribbon, we assume sources s_1 and s_2 are positioned at the ends of the line. In the French flag, we assume the sources s_i are positioned at the corners. We make this assumption to probe whether the concentration model can solve the problem without additional communication.

On the positive side, we can solve the French ribbon problem exactly. Consider an n-process line of length a in the concentration model, with morphogens m_1 and m_2 (with concentrations c_1 and c_2) each secreted by an endpoint. We assume the noiseless, underlying gradient function given position x is the inverse power law with parameter α, as above.

Algorithm Exact Concentration Ribbon: Assume that m_1 is secreted at $x = 0$ and m_2 is secreted at $x = a$. We have $c_1 = 1/x^\alpha$ and $c_2 = 1/(a-x)^\alpha$. The ratio of c_2 to c_1 is then $(a-x)^\alpha/x^\alpha$. Each agent computes this ratio independently from the measured values of c_1 and c_2. Let $ratio = c_2/c_1$. Each agent computes the smallest color z such that $ratio \geq ((z-1)/(k-z))^\alpha$, decides color z, and halts. We prove in [1] that this algorithm is size-invariant and holds for any line graph of arbitrary finite length.

Theorem 1. *Algorithm Exact Concentration Ribbon solves the k-ribbon problem in the concentration model for an n-process line graph of arbitrary finite length a, with constant time and communication complexity, given that processes have knowledge of morphogen concentrations c_1 and c_2, which have reached steady states, as well as the gradient function.*

On the negative side, extending to the 2D French flag with just four corner sources is infeasible; symmetry prevents us from obtaining an ε-approximate algorithm. The following theorem shows that without absolute positional information, the concentration model cannot produce a correct French flag (or good approximation) regardless of the gradient function.

Theorem 2. *Consider the concentration model. Fix any $\varepsilon \in (0, 1/6)$. No algorithm can produce an ε-approximate French flag.*

Proof Sketch. Given an arbitrary flag G of dimensions $a \times b$, we show that we can construct a flag G' with dimensions $a' \times b'$ such that there are agents in both flags that (1) have the same distances from the respective sources and (2) must choose different colors. Since the two agents have the same respective distance to

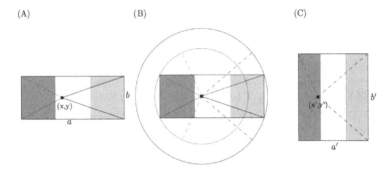

Fig. 1. (A) depicts an arbitrary original flag. The proof of Theorem 2 argues how to construct a new flag as in (C) such that there are agents in both flags with exactly the same distances from the respective sources that must also choose different colors, yielding impossibility. (B) shows construction of the new flag by modifying the aspect ratio to maintain the distances but change the correct color for the agent. (Color figure online)

every source, they receive the same concentration inputs and cannot distinguish between the settings, making it impossible to always color correctly. See Fig. 1 for an illustration. To show such a G' exists, we frame the constraints as a system of equations and show that there is a valid solution. The formal proof is in [1].

3   Message-Passing Model

Our message-passing model is similar to the standard LOCAL distributed model, with a few exceptions. Agents do not know their global position but have a common sense of direction $dir \in \{up, down, left, right\}$ and know which of their neighbors exist, meaning they know whether they are endpoints of rows or columns (or both, if corners). Initially, all but one arbitrary agent called the *starting agent*, representing the source of the communication signal, are *asleep* and thus perform no computation. Sleeping agents wake upon receiving a message.

We summarize results in Table 1, with exact statements in [1]. We would like to highlight the memory and message complexity of *Bubble Sort*, which is independent of n and constant assuming $k = O(1)$. The full version details all algorithms, which we summarize below, and shows extension to the 2D case.

Algorithm *Exact Count* uses a simple binary counter that measures the distance from the endpoints, from which agents learn their location in the ribbon. Algorithm *Exact Count* holds in an asynchronous model. Algorithm *Exact Silent Count* is based on the fact that the counters (measuring the distances from the endpoints) don't necessarily have to be communicated. Instead, relying on symmetry, each agent can count locally how many steps have been passed between receiving the different counters in order to color itself correctly. The beauty of Algorithm *Bubble Sort* lies in the small memory and message sizes. Initially, all

Table 1. Message passing algorithms.

Algorithm	Rounds	Memory per agent	# Msgs	Bits per Msg	Exact
Exact Count	$(2 - 1/k)n$	$3\log n + O(1)$	$O(n)$	$O(\log n)$	✓
Exact Silent Count	$3n$	$2\log n + O(1)$	$O(n)$	$O(1)$	✓
Bubble Sort	$3n$	$O(\log k)$	$O(n^2)$	$O(\log k)$	✓
Approx Count	$2n$	$2\log\log n + O(1)$	$O(n)$	$O(\log\log n)$	✗

agents are colored alternatingly, followed by a distributed execution of bubble sort. Algorithm *Approx Count* uses an approximate counting argument developed by Flajolet [2] to color agents approximately correctly.

References

1. Ancona, A., Bajwa, A., Lynch, N., Mallmann-Trenn, F.: How to color a French flag-biologically inspired algorithms for scale-invariant patterning. arXiv preprint arXiv:1905.00342 (2019)
2. Flajolet, P.: Approximate counting: a detailed analysis. BIT **25**(1), 113–134 (1985)
3. Nüsslein-Volhard, C., Wieschaus, E.: Mutations affecting segment number and polarity in drosophila. Nature **287**(5785), 795–801 (1980)
4. Wolpert, L.: Positional information and the spatial pattern of cellular differentiation. J. Theor. Biol. **25**(1), 1–47 (1969)

Self-adjusting Linear Networks

Chen Avin[1], Ingo van Duijn[2(✉)], and Stefan Schmid[3]

[1] Ben Gurion University of the Negev, Beersheba, Israel
[2] Aalborg University, Aalborg, Denmark
ingo@cs.aau.dk
[3] University of Vienna, Vienna, Austria

Abstract. Emerging networked systems become increasingly flexible and "reconfigurable". This introduces an opportunity to adjust networked systems in a demand-aware manner, leveraging spatial and temporal locality in the workload for *online* optimizations. However, it also introduces a tradeoff: while more frequent adjustments can improve performance, they also entail higher reconfiguration costs.

This paper initiates the formal study of *linear* networks which self-adjust to the demand in an online manner, striking a balance between the benefits and costs of reconfigurations. We show that the underlying algorithmic problem can be seen as a distributed generalization of the classic dynamic list update problem known from self-adjusting datastructures: in a network, requests can occur between *node pairs*. This distributed version turns out to be significantly harder than the classical problem in generalizes. Our main results are a $\Omega(\log n)$ lower bound on the competitive ratio, and a (distributed) online algorithm that is $\mathcal{O}(\log n)$-competitive if the communication requests are issued according to a linear order.

Keywords: Self-adjusting datastructures · Competitive analysis · Distributed algorithms · Communication networks

1 Introduction

Communication networks are becoming increasingly flexible, along three main dimensions: routing (enabler: software-defined networking), embedding (enabler: virtualization), and topology (enabler: reconfigurable optical technologies, for example [3]). In particular, the possibility to quickly reconfigure communication networks, e.g., by migrating (virtualized) communication endpoints [1] or by reconfiguring the (optical) topology [2], allows these networks to become *demand-aware*: i.e., to adapt to the traffic pattern they serve, in an online and *self-adjusting* manner. For example, in a self-adjusting network, frequently communicating node pairs can be moved *topologically closer*, saving communication costs (e.g., bandwidth, energy) and improving performance (e.g., latency, throughput).

© Springer Nature Switzerland AG 2019
K. Censor-Hillel and M. Flammini (Eds.): SIROCCO 2019, LNCS 11639, pp. 332–335, 2019.
https://doi.org/10.1007/978-3-030-24922-9_23

However, today, we still do not have a good understanding yet of the algorithmic problems underlying self-adjusting networks. The design of such algorithms faces several challenges. As the demand is often not known ahead of time, *online* algorithms are required to react to changes in the workload in a clever way; ideally, such online algorithms are "competitive" even when compared to an optimal offline algorithm which knows the demand ahead of time. Furthermore, online algorithms need to strike a balance between the benefits of adjustments (i.e., improved performance and/or reduced costs) and their costs (i.e., frequent adjustments can temporarily harm consistency and/or performance, or come at energy costs).

The vision of self-adjusting networks is reminiscent of self-adjusting datastructures such as *self-adjusting lists* and *splay trees*, which optimize themselves toward the workload. In particular, the *dynamic list update problem*, introduced already in the 1980s by Sleator and Tarjan in their seminal work [4], asks for an online algorithm to reconfigure an unordered linked list datastructure, such that a sequence of lookup requests is served optimally and at minimal reconfiguration costs (i.e., pointer rotations). It is well-known that a simple *move-to-front* strategy, which immediately promotes each accessed element to the front of the list, is *dynamically optimal*, that is, has a constant competitive ratio.

This paper initiates the study of pairwise communication problems in a dynamic network reconfiguration model. This model consists of a a set of communication nodes V, and a graph called the *host network*, denoted $H = (N, L)$ where $L \subseteq N \times N$. The communication nodes are 'hosted' on H, denoted with an injection $h : V \to N$ called a *configuration*. Without loss of generality, we assume that $|V| = |N|$, so that every configuration is a bijection. Two nodes $u, v \in V$ are *connected* on a configuration h if $(h(u), h(v)) \in L$.

Now, given a *communication sequence* $\sigma = \sigma_1, \sigma_2, ...$, where $\sigma_i \in V \times V$, the pairwise communication problem asks to serve all communication requests in order. A communication request is served if the two constituent nodes are connected. Additionally, the network is allowed to be *reconfigured*, by migrating any two nodes which are connected. That is, given a configuration h in which u, v are connected, migrating them corresponds to producing a new configuration h' where $h'(u) = h(v)$, $h'(v) = h(u)$, and $h'(w) = h(w)$ for $w \neq u, v$. For simplicity, we assume in this paper that both serving a communication request and reconfiguring the network are constant cost operations. Note that in general migrating two nodes is likely (a large constant factor) more expensive than serving a single communication. However, with the assumption that communicating and migrating are the same up to a (large) constant factor, one can think of the communication cost between two arbitrary node as simply the number of reconfigurations necessary to make them (temporarily) connected. With this simple cost model, we can thus phrase the pairwise communication problem as:

Definition 1 (Pairwise Communication Problem). *Given a host network with an initial configuration, give an algorithm that serves all communication requests from a sequence σ that minimises the total reconfiguration cost.*

In the *offline* version of this problem, σ is given in advance, whereas in an *online* setting, a communication request σ_i is only revealed after σ_{i-1} has been served. A *competitive analysis* compares an online algorithm ON to an offline algorithm OFF. The ultimate goal is to devise online algorithms ON for the pairwise communication problem which minimise the competitive ratio ρ:

$$\rho = \max_{\sigma} \frac{\text{cost}(ON(\sigma))}{\text{cost}(OFF(\sigma))}$$

2   Results and Open Problems

In this paper, we study host networks with the topology of a d-dimensional grid. The primary problem we investigate is therefore:

Definition 2 (Distributed Grid Update). *What is the competitive ratio for the pairwise communication problem for host networks with a d-dimensional grid topology?*

We show that for such networks there is a $\Omega(\log n)$ lower bound on the competitive ratio. Furthermore, in pursuit of a non-trivial upper bound, we show that for communication sequences with *linear demand*, there is an algorithm that is $O(\log n)$-competitive. A communication sequence is said to have linear demand if there is a configuration h that serves each request (without reconfigurations), or more formally:

Definition 3 (Linear Demand). *Let $R(\sigma)$ denote the set of communication requests interpreted as edges over V. A sequence σ has linear demand over a network $H = (N, L)$ if there exists a configuration h such that for all $(u, v) \in R(\sigma)$, it holds that $(h(u), h(v)) \in L$.*

From a lower bound perspective, our main result is that the competitive ratio is at least $\Omega(\log n)$. We show this by explicitly constructing a hard sequence:

Theorem 1. *For every online algorithm* ON *solving* DISTRIBUTED GRID UPDATE *on a grid of n nodes in total and for every $0 < \varepsilon \leq 1$, there is a sequence σ_{ON} of length $\mathcal{O}(\varepsilon n^{1+\varepsilon} \log n)$ such that $\text{cost}(ON(\sigma_{ON})) = \Omega(\varepsilon n^{1+1/d} \log n)$. The resulting request graph $R(\sigma_{ON})$ is a d-dimensional grid graph.*

Since $R(\sigma_{ON})$ is a grid graph, an offline algorithm can simply configure to the configuration h that serves all requests (i.e. σ_{ON} has linear demand). From an arbitrary starting configuration, it takes $\Omega(n^{1+1/d|})$ to configure h, which dominates the cost for serving all requests as long as $\varepsilon \leq 1/d$. Therefore, we find the following lower bound on ρ:

$$\rho \geq \frac{\text{cost}(ON(\sigma_{ON}))}{\text{cost}(OFF(\sigma_{ON}))} = \Omega(\log n)$$

Because we rely on linear demand, our technique does not leverage the full power available to an offline algorithm: namely reconfigurations in between communication requests. As future work, we pose the following open problem:

Problem 1. Construct a sequence σ with nonlinear demand for DISTRIBUTED GRID UPDATE, such that any online algorithm is a factor $\omega(\log n)$ off the optimal offline algorithm.

From an upper bound perspective, we investigate DISTRIBUTED GRID UPDATE restricted to line topologies, dubbed DISTRIBUTED LIST UPDATE. We show that in the very restricted case of linear demand, there is a matching upper bound to the competitive ratio:

Theorem 2. *There is an algorithm* GREAD *solving* DISTRIBUTED LIST UPDATE, *such that if σ has linear demand, then*

$$cost(\text{GREAD}(\sigma)) = \mathcal{O}(m + nk \log k)$$

where $m = |\sigma|$ and $k = |R(\sigma)|$

Again, it is the restriction to linear demand that allows for this tight upper bound. However, we see this as an initial step to finding any nontrivial upper bound for the general case:

Problem 2. Given a sequence σ for DISTRIBUTED LIST UPDATE with arbitrary request graph $R(\sigma)$, is there an algorithm with competitive ratio $o(n)$?

Note that a competitive ratio of $o(n)$ means *anything* better than a trivial algorithm. However, a better understanding of the behaviour of offline algorithms for DISTRIBUTED GRID UPDATE on nonlinear demand is required to answer this question.

References

1. Avin, C., Loukas, A., Pacut, M., Schmid, S.: Online balanced repartitioning. In: Gavoille, C., Ilcinkas, D. (eds.) DISC 2016. LNCS, vol. 9888, pp. 243–256. Springer, Heidelberg (2016). https://doi.org/10.1007/978-3-662-53426-7_18
2. Avin, C., Schmid, S.: Toward demand-aware networking: a theory for self-adjusting networks. In: ACM SIGCOMM Computer Communication Review (CCR) (2018)
3. Ghobadi, M., et al.: Projector: agile reconfigurable data center interconnect. In: Proceedings of ACM SIGCOMM, pp. 216–229 (2016)
4. Sleator, D.D., Tarjan, R.E.: Amortized efficiency of list update and paging rules. Commun. ACM **28**(2), 202–208 (1985)

Mutual Visibility for Asynchronous Robots

Subhash Bhagat[1], Sruti Gan Chaudhuri[2], and Krishnendu Mukhopadhyaya[3(✉)]

[1] University of Ottawa, Ottawa, Canada
subhash.bhagat.math@gmail.com
[2] Jadavpur University, Kolkata, India
sruti.ganchaudhuri@jadavpuruniversity.in
[3] Indian Statistical Institute, Kolkata, India
krishnendu@isical.ac.in

Abstract. This paper presents a study of the *mutual visibility* problem for a set of opaque asynchronous robots in the Euclidean plane. Due to opacity, if three robots lie on a line, the middle robot obstructs the visions of the two other robots. The mutual visibility problem requires the robots to coordinate their movements to form a configuration in which no three robots are collinear. This work presents a distributed algorithm which solves the *mutual visibility* problem for a set of $n \geq 2$ asynchronous robots under the FStatE computational model. The proposed algorithm assumes 1 bit of persistent memory and the knowledge of n. The proposed solution works under the *non-rigid* movements of the robots and also provides collision free movements for the robots.

1 Introduction

A *swarm* of robots is a distributed system of autonomous, homogeneous, anonymous small mobile robots which are capable of carrying out some task in a cooperative environment. An active robot repeatedly executes a computational cycle consisting of three phases, *look-compute-move*. In the *look phase*, a robot takes the snapshot of its surrounding. In the *compute phase*, it computes a destination point using the information obtained in the *look phase*. Finally, in the *move phase*, it moves towards the destination point. Robots may have some additional capabilities. Robots may be endowed with visible lights. These lights can assume a finite number of pre-defined colours. Each colour indicates a different state of the robots. The colours do not change automatically and they are persistent. The lights of the robots can be used in three different ways [1]: (i) the robots can set limited communications between themselves using visible lights and also retain some information about their previous states (FALL model) or (ii) only to remember information about their last states (FSTATE model) and this piece of information is not available to the other robots or (iii) the robots can use visible lights only to communicate with other robots and a robot does not remember the color of its light (FCOMM model). Thus, in FSTATE and FALL models, robots

K. Censor-Hillel and M. Flammini (Eds.): SIROCCO 2019, LNCS 11639, pp. 336–339, 2019.
https://doi.org/10.1007/978-3-030-24922-9_24

use lights as persistent memory to carry forward some information from thier previous computational cycles.

The problem considered in this work is defined as follows: *Consider a set of stationary robots occupying distinct positions in the Euclidean plane. The mutual visibility problem requires the robots to form a configuration, within finite time and without collision, in which no three robots are collinear.*

1.1 Earlier Works

The work in [2] proposed a solution to the problem under the *SSYNC* model for oblivious robots with the knowledge of n, the total number of robots in the system. The algorithm in [3] solves the problem under the *ASYNC* model for oblivious robots with one coordinate axis agreement and knowledge of n. Both the algorithms work for *non-rigid* movements of the robots. The algorithm proposed in [4] provides a solution to the problem for *fat* robots under the *ASYNC* model with the assumption of common *chirality*. Di Luna et al. proposed algorithms to solve the problem both for semi-synchronous and asynchronous robots under the FALL model [5]. The solution under the *SSYNC* model uses 3 colors and the solution in the *ASYNC* model also uses 3 colors with one coordinate axis agreement. In both the algorithms, the robots do not know the value of n. Bhagat and Mukhopadhyaya solved the problem for asynchronous robots under the FALL model using 7 colors [6]. During the execution of their algorithm, each robot moves exactly once. Recently, Sharma et al. proposed an algorithm for asynchronous robots under the FALL model using 47 colors which runs in $O(1)$ asynchronous rounds [7]. Their solution assumes *rigid* movements for the robots. The solution presented in [8] works for semi-synchronous robots under the FSTATE model and uses 1 bit of persistent memory.

1.2 Contribution of the Paper

A distributed algorithm for the *mutual visibility* problem has been proposed under *ASYNC* model. The proposed algorithm considers the FSTATE model which does not have communication overhead of the FALL model. All the existing solutions to the mutual visibility problem under the *ASYNC* model assume either (i) persistent memory for both communication and internal memory purpose or (ii) axis agreement or (iii) *chirality*. Our solution does not assume any axis agreement, *chirality* and it does not have any communication overhead. The proposed algorithm uses only 1 bit of persistent memory and the solution works under the *non-rigid* movements of the robots. It also provides collision free movements for the robots.

2 Assumptions and Notations

We consider a set $\mathcal{R} = \{r_1, \ldots, r_n\}$ of n point robots in the Euclidean plane under the *ASYNC* model. The robots are opaque. They know the value of n.

The movements of the robots are *non-rigid*. Each robot has 1 bit of internal persistent memory and the piece of information stored in this internal memory is not available to the other robots. Except this persistent memory, the robots are oblivious. Let $r_i(t)$ denote the position of robot r_i at time t. A configuration $\mathcal{R}(t) = \{r_1(t), \dots, r_n(t)\}$, denotes the set of all distinct positions occupied by the robots at time t. The class of all configurations, in which all the robot positions are collinear, is denoted by $\widetilde{C}_L$. The class of all configurations, in which at least three robot positions are non-collinear, is denoted by $\widetilde{C}_{NL}$. The vision of a robot r_i at the time t is defined as the set of all distinct positions occupied by the robots which are visible to r_i at time t ($r_i(t)$ is not included). This set is denoted by $\mathcal{V}_i(t)$. The robot positions in $\mathcal{V}_i(t)$ are sorted angularly in anti clockwise direction w.r.t. $r_i(t)$ (since robots do not have common orientation, this ordering may vary if computed by the other robots). Starting from any robot position in $\mathcal{V}_i(t)$, if we connect the other robot positions following that ordering, we get a simple polygon $\mathcal{VP}_i(t)$. A robot r_i is called an *internal robot* if it lies between two other robot positions on the line segment joining them. Otherwise, r_i is called a *terminal* robot. Let $\mathcal{L}_{ij}(t)$ denote the straight line passing through $r_i(t)$ and $r_j(t)$. The perpendicular distance of the line $\mathcal{L}_{ij}(t)$ from the point $r_k(t)$ is denoted by $d_{ij}^k(t)$. Let $D_i(t)$ denote the distance of $r_i(t)$ from the closest robot position in $\mathcal{V}_i(t)$.

3   Algorithm $MutualVisibility()$

A robot uses its persistent 1 bit memory to remember information about its last movement. Initially all robots have 0 in their persistent memory. A 0 in the internal bit means that the robot has not made any move yet and 1 means that it has moved at least once. A robot r_i acts according to the following:

(i) If r_i is a terminal robot and its internal bit is 0, it computes a destination point, changes its internal bit to 1 and moves straight towards the destination point.

(ii) If r_i is a terminal robot and its internal bit is 1, it does nothing.

(iii) If r_i is a non-terminal robot and $|\mathcal{V}_i(t)| < n - 1$, it does nothing.

(iv) If r_i is a non-terminal robot and $|\mathcal{V}_i(t)| = n - 1$, it computes a destination point, changes its internal bit to 1 and moves straight towards the destination point. Note that the movements of the non-terminal robots do not depend on their internal bits; it depends on the number of robots visible to them.

3.1   Computing Destination Point

Let r_i be a robot which finds itself eligible for movement at time t. For $\mathcal{R}(t) \in \widetilde{C}_{NL}$, let $\Gamma_i(t) = \{\angle r_j(t) r_i(t) r_k(t) : r_j(t), r_k(t) \text{ are two consecutive vertices on } \mathcal{VP}_i(t)\}$. Let $\alpha_i(t) = max\{\theta \in \Gamma_i(t) : \theta < \pi\}$ (tie, if any, is broken arbitrarily). The bisector of $\alpha_i(t)$ is denoted by $Bisec_i(t)$. Now suppose $\mathcal{R}(t) \in \widetilde{C}_L$. Let $\mathcal{L}^*$

be the perpendicular at $r_i(t)$ to the line of collinearity $\hat{\mathcal{L}}$. The robot r_i arbitrarily chooses a direction along $\mathcal{L}^*$ and let $\mathcal{L}^+$ denote the ray along this direction.

Let $d_i(t) = minimum\{d_{ij}^k(t), d_{ik}^j(t), d_{jk}^i(t) : \forall r_j(t), r_k(t) \in \mathcal{V}_i(t)\}$. The direction of movement of r_i is along $DIR_i(t)$ which is defined as follows:

$$DIR_i(t) = \begin{cases} Bisec_i(t) & \text{if } \mathcal{C}(t) \in \widetilde{C}_{NL} \\ \mathcal{L}^+ & \text{if } \mathcal{C}(t) \in \widetilde{C}_L \end{cases}$$

The amount of displacement $\sigma_i(t)$ of r_i at time t is defined as follows,

$$\sigma_i(t) = \begin{cases} \frac{1}{n}d_i(t) & \text{if } \mathcal{R}(t) \in \widetilde{C}_{NL} \\ \frac{1}{n}D_i(t) & \text{if } \mathcal{R}(t) \in \widetilde{C}_L \end{cases}$$

Let $p_i(t)$ be the point on $DIR_i(t)$ at a distance $\sigma_i(t)$ from $r_i(t)$. The destination point of $r_i(t)$ is $p_i(t)$.

Theorem 1. *The mutual visibility problem is deterministically solvable in finite time for a set of asynchronous robots under the FSTATE model with 1 bit of persistent memory and non-rigid movements.*

References

1. Flocchini, P., Santoro, N., Viglietta, G., Yamashita, M.: Rendezvous of two robots with constant memory. In: Moscibroda, T., Rescigno, A.A. (eds.) SIROCCO 2013. LNCS, vol. 8179, pp. 189–200. Springer, Cham (2013). https://doi.org/10.1007/978-3-319-03578-9_16
2. Di Luna, G.A., Flocchini, P., Poloni, F., Santoro, N., Viglietta, G.: The mutual visibility problem for oblivious robots. In: Proceedings of 26th Canadian Conference on Computational Geometry (CCCG) (2014)
3. Bhagat, S., Chaudhuri, S.G., Mukhopadhyaya, K.: Formation of general position by asynchronous mobile robots under one-axis agreement. In: Kaykobad, M., Petreschi, R. (eds.) WALCOM 2016. LNCS, vol. 9627, pp. 80–91. Springer, Cham (2016). https://doi.org/10.1007/978-3-319-30139-6_7
4. Agathangelou, C., Georgiou, C., Mavronicolas, M.: A distributed algorithm for gathering many fat mobile robots in the plane. In: Proceedings of the 32nd ACM Symposium on Principles of Distributed Computing (PODC), pp. 250–259 (2013)
5. Di Luna, G.A., Flocchini, P., Gan Chaudhuri, S., Poloni, F., Santoro, N., Viglietta, G.: Mutual visibility by luminous robots without collisions. Inf. Comput. **254**, 392–418 (2017)
6. Bhagat, S., Mukhopadhyaya, K.: Optimum algorithm for mutual visibility among asynchronous robots with lights. In: Spirakis, P., Tsigas, P. (eds.) SSS 2017. LNCS, vol. 10616, pp. 341–355. Springer, Cham (2017). https://doi.org/10.1007/978-3-319-69084-1_24
7. Sharma, G., Vaidyanathan, R., Trahan, J.L.: Constant-time complete visibility for asynchronous robots with lights. In: Spirakis, P., Tsigas, P. (eds.) SSS 2017. LNCS, vol. 10616, pp. 265–281. Springer, Cham (2017). https://doi.org/10.1007/978-3-319-69084-1_18
8. Bhagat, S., Mukhopadhyaya, K.: Mutual visibility by robots with persistent memory. In: Proceedings of 13th International Frontiers of Algorithmics Workshop (FAW), pp. 144–155 (2019)

Infinite Grid Exploration by Disoriented Robots

Quentin Bramas[1(✉)], Stéphane Devismes[2], and Pascal Lafourcade[3]

[1] University of Strasbourg, ICUBE, CNRS, Strasbourg, France
bramas@unistra.fr
[2] Université Grenoble Alpes, VERIMAG, Grenoble, France
[3] University Clermont Auvergne, CNRS, UMR 6158, LIMOS,
Clermont-Ferrand, France

Abstract. We deal with a set of autonomous robots moving on an infinite grid. Those robots are opaque, have limited visibility capabilities, and run using synchronous Look-Compute-Move cycles. They all agree on a common chirality, but have no global compass. Finally, they may use lights of different colors, but except from that, robots have neither persistent memories, nor communication mean. We consider the *infinite grid exploration* (IGE) problem. We first show that two robots are not sufficient in our settings to solve the problem, even when robots have a common coordinate system. We then show that if the robots' coordinate systems are not self-consistent, three or four robots are not sufficient to solve the problem neither. Finally, we present three algorithms that solve the IGE problem in various settings. The first algorithm uses six robots with constant colors and a visibility range of one. The second one uses the minimum number of robots, *i.e.*, five, as well as five modifiable colors, still under visibility one. The last algorithm requires seven oblivious anonymous robots, yet assuming visibility two. Notice that the two last algorithms also achieve exclusiveness.

1 Context and Motivation

We deal with a swarm of mobile robots having low computation and communication capabilities. The robots we consider are opaque (*i.e.*, a robot is able to see another robot if and only if no other robot lies in the line segment joining them) and run in synchronous Look-Compute-Move cycles, where they can sense their surroundings within a limited visibility range. All robots agree on a common chirality (*i.e.*, when a robot is located on an axis of symmetry in its surroundings, it is able to distinguish its two sides one from another), but have no global compass (they agree neither on a North-South, nor a East-West direction). However, they may use lights of different colors [13,18]. These lights can be seen by robots in their surroundings. However, except from those lights, robots have neither persistent memories nor communication capabilities.

This study has been partially supported by the ANR projects DESCARTES (ANR-16-CE40-0023) and ESTATE (ANR-16-CE25-0009).

K. Censor-Hillel and M. Flammini (Eds.): SIROCCO 2019, LNCS 11639, pp. 340–344, 2019.
https://doi.org/10.1007/978-3-030-24922-9_25

We are interested in coordinating such weak robots, endowed with both typically small visibility range (*i.e.*, one or two) and few light colors (only a constant number of them), to solve an infinite task in an infinite discrete environment. As an attempt to tackle this general problem, we consider the exploration of an infinite grid, where nodes represent locations that can be sensed by robots and edges represent the possibility for a robot to move from one location to another. Precisely, the exploration task consists in ensuring that each node of the infinite grid is visited within finite time by at least one robot. We refer to this problem as the *Infinite Grid Exploration* (IGE) problem.

2   Contribution

We present both negative and positive results. On the negative side, we show that if robots have a common chirality but a bounded visibility range, then the IGE problem is not solvable with:

- two robots, even if those robots agree on common North (the proof is essentially the adaptation to our context of the impossibility proof given in [14]);
- three or four robots with self-inconsistent compass (*i.e.*, the compass may change throughout the execution).

On the positive side, we provide three algorithms for solving the IGE problem using opaque robots equipped with self-inconsistent compass, yet agreeing on a common chirality. Two of them also satisfy *exclusiveness* [2], which requires any two robots to never simultaneously occupy the same position nor traverse the same edge. The first one requires the minimum number of robots, *i.e.*, five, and ensures exclusiveness. The robots use modifiable lights with only five states, and have a visibility range restricted to one. The second algorithm solves the problem with six robots and only three non-modifiable colors, still assuming visibility range one. The last algorithm requires seven identical robots without any light (*i.e.*, seven oblivious[1] anonymous robots) and ensures exclusiveness, yet assuming visibility range two. Our contributions are summarized below.

Visibility range	# of robots	# of colors	Modifiable colors?	Exclusiveness?
1	5 (opt)	5	Yes	Yes
1	6	3	No	No
2	7	1	N/A	Yes

In order to help the reader, animations, for each algorithm, are available online [6]. Our algorithms and their proofs of correctness can be found in a technical report online [5].

---

[1] *Oblivious* means that robots cannot remember the past.

3   Related Work

The model of robots with lights have been proposed by Peleg in [13,18]. In [8], the authors use robots with lights and compare the computational power of such robots with respect to the three main execution model: fully-synchronous, semi-synchronous, and asynchronous. Solutions for dedicated problems such as *weak gathering* or *mutual visibility* have been respectively investigated in [16] and [17].

Mobile robot computing in infinite environments has been first studied in the continuous two-dimensional Euclidean space. In this context, studied problems are mostly *terminating* tasks, such as *pattern formation* [11] and *gathering* [15], *i.e.*, problems where robots aim at eventually stopping in a particular configuration specified by their relative positions. A notable exception is the *flocking* problem [19], *i.e.*, the infinite task consisting of forming a desired pattern with the robots and make them moving together while maintaining that formation.

When considering a discrete environment, space is defined as a graph, where the nodes represent the possible locations that a robot can take and the edges the possibility for a robot to move from one location to another. In this setting, researchers have first considered finite graphs and two variants of the exploration problem, respectively called the *terminating* and *perpetual* exploration. The terminating exploration requires every possible location to be eventually visited by at least one robot, with the additional constraint that all robots stop moving after task completion. In contrast, the perpetual exploration requires each location to be visited infinitely often by all or a part of robots. In [9], authors solve terminating exploration of any finite grid using few asynchronous anonymous oblivious robots, yet assuming unbounded visibility range. The exclusive perpetual exploration of a finite grid is considered in the same model in [3].

Various terminating problems have been investigated in infinite grids such as *arbitrary pattern formation* [4], *mutual visibility* [1], and *gathering* [10,12]. The possibly closest related work is that of Emek *et al.* [14]. In this paper, authors consider a treasure search problem, which is roughly equivalent to the IGE problem, in an infinite grid. They consider robots that operate in two models: the semi-synchronous and synchronous ones. However, they do not impose the exclusivity at all since their robots can only sense the states of the robots located at the same node (in that sense, the visibility range is zero). The main difference with our settings is that they assume all robots agree on a *global compass*, *i.e.*, they all agree on the same directions North-South and East-West; while we only assume here a *common chirality*. This difference makes their model stronger, indeed they propose two algorithms that respectively need three synchronous and four asynchronous robots, while in our settings the IGE problem (even in its non-exclusive variant) requires at least five robots. They also exclude solutions for two robots. Brandt *et al.* [7] extend the impossibility result of Emek *et al.* Indeed, they show the impossibility of exploring an infinite grid with three semi-synchronous deterministic robots that agree on a common coordinate system. Although proven using similar techniques, this result is not correlated to ours. Indeed, the lower bound of Brandt *et al.* holds for robots that are weaker in terms of synchrony assumption (semi-synchronous *vs.* fully synchronous in our case), but stronger in terms of coordination capabilities (common coordinate system *vs.* self-inconsistent compass in our case). In other words, our impossibility results does not (even indirectly) follows from those of Brandt *et al.* since in our model difficulties arise from the lack of coordination capabilities and not the level asynchrony. As a matter of facts, based on the results of Emek *et al.* [14], four (asynchronous) robots are actually necessary and sufficient in their settings, while in our context five robots are required.

References

1. Adhikary, R., Bose, K., Kundu, M.K., Sau, B.: Mutual visibility by asynchronous robots on infinite grid. In: Gilbert, S., Hughes, D., Krishnamachari, B. (eds.) ALGOSENSORS 2018. LNCS, vol. 11410, pp. 83–101. Springer, Cham (2019). https://doi.org/10.1007/978-3-030-14094-6_6
2. Baldoni, R., Bonnet, F., Milani, A., Raynal, M.: Anonymous graph exploration without collision by mobile robots. Inf. Process. Lett. **109**(2), 98–103 (2008)
3. Bonnet, F., Milani, A., Potop-Butucaru, M., Tixeuil, S.: Asynchronous exclusive perpetual grid exploration without sense of direction. In: Fernàndez Anta, A., Lipari, G., Roy, M. (eds.) OPODIS 2011. LNCS, vol. 7109, pp. 251–265. Springer, Heidelberg (2011). https://doi.org/10.1007/978-3-642-25873-2_18
4. Bose, K., Adhikary, R., Kundu, M.K., Sau, B.: Arbitrary pattern formation on infinite grid by asynchronous oblivious robots. In: Das, G.K., Mandal, P.S., Mukhopadhyaya, K., Nakano, S. (eds.) WALCOM 2019. LNCS, vol. 11355, pp. 354–366. Springer, Cham (2019). https://doi.org/10.1007/978-3-030-10564-8_28
5. Bramas, Q., Devismes, S., Lafourcade, P.: Infinite grid exploration by disoriented robots. Technical report, arXiv, May 2019. https://arxiv.org/abs/1905.09271
6. Bramas, Q., Devismes, S., Lafourcade, P.: Infinite grid exploration by disoriented robots: animations, May 2019. https://doi.org/10.5281/zenodo.2625730
7. Brandt, S., Uitto, J., Wattenhofer, R.: A tight bound for semi-synchronous collaborative grid exploration. In: 32nd International Symposium on Distributed Computing (DISC), New Orleans, Louisiana, October 2018
8. Das, S., Flocchini, P., Prencipe, G., Santoro, N., Yamashita, M.: Autonomous mobile robots with lights. Theor. Comput. Sci. **609**(P1), 171–184 (2016)
9. Devismes, S., Lamani, A., Petit, F., Raymond, P., Tixeuil, S.: Optimal grid exploration by asynchronous oblivious robots. In: Richa, A.W., Scheideler, C. (eds.) SSS 2012. LNCS, vol. 7596, pp. 64–76. Springer, Heidelberg (2012). https://doi.org/10.1007/978-3-642-33536-5_7
10. Di Stefano, G., Navarra, A.: Gathering of oblivious robots on infinite grids with minimum traveled distance. Inf. Comput. **254**, 377–391 (2016)
11. Dieudonné, Y., Petit, F.: Circle formation of weak robots and Lyndon words. Inf. Process. Lett. **101**(4), 156–162 (2007)
12. Dutta, D., Dey, T., Chaudhuri, S.G.: Gathering multiple robots in a ring and an infinite grid. In: Krishnan, P., Radha Krishna, P., Parida, L. (eds.) ICDCIT 2017. LNCS, vol. 10109, pp. 15–26. Springer, Cham (2017). https://doi.org/10.1007/978-3-319-50472-8_2
13. Efrima, A., Peleg, D.: Distributed models and algorithms for mobile robot systems. In: van Leeuwen, J., Italiano, G.F., van der Hoek, W., Meinel, C., Sack, H., Plášil, F. (eds.) SOFSEM 2007. LNCS, vol. 4362, pp. 70–87. Springer, Heidelberg (2007). https://doi.org/10.1007/978-3-540-69507-3_5

14. Emek, Y., Langner, T., Stolz, D., Uitto, J., Wattenhofer, R.: How many ants does it take to find the food? Theor. Comput. Sci. **608**(P3), 255–267 (2015)
15. Flocchini, P., Prencipe, G., Santoro, N., Widmayer, P.: Gathering of asynchronous robots with limited visibility. Theor. Comput. Sci. **337**(1), 147–168 (2005)
16. Luna, G.A.D., Flocchini, P., Chaudhuri, S.G., Poloni, F., Santoro, N., Viglietta, G.: Mutual visibility by luminous robots without collisions. Inf. Comput. **254**, 392–418 (2017)
17. Ooshita, F., Datta, A.K.: Brief announcement: feasibility of weak gathering in connected-over-time dynamic rings. In: Izumi, T., Kuznetsov, P. (eds.) SSS 2018. LNCS, vol. 11201, pp. 393–397. Springer, Cham (2018). https://doi.org/10.1007/978-3-030-03232-6_27
18. Peleg, D.: Distributed coordination algorithms for mobile robot swarms: new directions and challenges. In: Pal, A., Kshemkalyani, A.D., Kumar, R., Gupta, A. (eds.) IWDC 2005. LNCS, vol. 3741, pp. 1–12. Springer, Heidelberg (2005). https://doi.org/10.1007/11603771_1
19. Yang, Y., Souissi, S., Défago, X., Takizawa, M.: Fault-tolerant flocking for a group of autonomous mobile robots. J. Syst. Softw. **84**(1), 29–36 (2011)

A Bounding Box Overlay for Competitive Routing in Hybrid Communication Networks

Jannik Castenow$^{(\boxtimes)}$, Christina Kolb, and Christian Scheideler

Heinz Nixdorf Institute and Computer Science Department,
Paderborn University, Paderborn, Germany
{jannik.castenow,christina.kolb,scheideler}@upb.de

Keywords: Wireless ad hoc networks · c-competitive routing · Bounding boxes

1   Introduction

We consider a real-life scenario of people using their smartphones in urban areas. A smartphone can be interpreted as an ad hoc device since it is able to communicate for instance via WiFi Direct or Bluetooth. Smartphones are different as they combine multiple communication modes in one device and hence are also able to use a cellular infrastructure to communicate. In metropolitan areas, the density of smartphones is sufficiently high such that the ad hoc network of smartphones is connected and, in principle, the entire data transmission between smartphones could be solely carried out via ad hoc links. The main challenge of data transmission in an ad hoc network of smartphones is routing. Human made obstacles like buildings interfere wireless communication leading to radio holes in the ad hoc network. Routing in an ad hoc network without knowledge about shapes and locations of radio holes potentially leads to very long detours [5]. To overcome this drawback, the locations and shapes of radio holes can be determined efficiently via the cellular infrastructure. For a fast computation of routing paths, it makes sense to not consider the exact shape of a hole but a more coarse-grained abstraction like a bounding box. In metropolitan areas, buildings are usually convex and rectangular shaped such that the bounding boxes of radio holes only rarely intersect, rendering the bounding box a practical hole abstraction in the described scenario.

1.1   Model

The model and definitions are close to those of [3]. We model the participants of the network as a set of nodes $V \subset \mathbb{R}^2$ in the Euclidean plane, where $|V| = n$.

This work was partially supported by the German Research Foundation (DFG) within the Collaborative Research Center 'On-The-Fly Computing' (SFB 901).

K. Censor-Hillel and M. Flammini (Eds.): SIROCCO 2019, LNCS 11639, pp. 345–348, 2019.
https://doi.org/10.1007/978-3-030-24922-9_26

Each node is associated with a unique ID (e.g., its phone number). For any given pair of nodes u, v, we denote the Euclidean distance between u and v by $\|uv\|$. The network is a hybrid directed graph $H = (V, E, E_{AH})$ where the node set V represents the set of cell phones, an edge (v, w) is in E whenever v knows the phone number (or simply ID) of w, and an edge $(v, w) \in E$ is also in the *ad hoc edge set* E_{AH} whenever v can send a message to w using its WiFi interface. For all edges $(v, w) \in E \setminus E_{AH}$, v can only use the cellular infrastructure to directly send a message to w. Since WiFi-communication can only be used over short distances, E_{AH} can only contain edges which are part of the Unit Disk Graph of V (UDG(V)). UDG (V), is a bi-directed graph that contains all edges (u, v) with $\|uv\| \leq 1$. Assume UDG (V) to be connected so that a message can be sent from every node to every other node in V by just using ad hoc edges.

While the potential ad hoc edges are fixed, the nodes can change E over time: If a node v knows the IDs of nodes w and w', then it can send the ID of w to w', which adds (w, w') to E. Alternatively, if v deletes the address of some node w with $(v, w) \in E$, then (v, w) is removed from E. There are no other means of changing E, i.e., a node v cannot learn about an ID of a node w unless w is in v's UDG-neighborhood or the ID of w is sent to v by some other node.

Moreover, we consider synchronous message passing in which time is divided into rounds. We assume that every message initiated in round i is delivered at the beginning of round $i + 1$.

1.2   Related Work

For a survey on geometric routing, we refer the reader to [1]. A severe problem of local routing strategies is the presence of sparse regions in the ad hoc network. We denote these regions as *radio holes*. In order to find efficient routing paths avoiding radio holes, Hybrid Communication Networks have been introduced [3]. Hybrid Communication Networks enrich the ad hoc network by a secondary global communication channel which can be used to exchange information about locations and shapes of radio holes fast. More precisely, in a Hybrid Communication Network, participants can communicate in their ad hoc range for free and also make use of long-range links to directly communicate with every other participant whose ID is known. These long-range links, called *Cellular Infrastructure*, are costly since a third party (e.g. a cell phone provider) is involved. To find efficient routing paths, the authors in [3] compute an Overlay Network via the Cellular Infrastructure in which holes are represented by their convex hulls. It is assumed that the convex hulls of the holes do not intersect. The storage requirements for some nodes are asymptotically in the size of the sum of all holes. In this work, we aim to reduce the storage requirements for these nodes and investigate also the challenging question of c-competitive routing through intersections of hole abstractions.

1.3   Our Contributions

We consider any hybrid graph $G = (V, E, E_{AH})$ where the Unit Disk Graph of V is connected. Let H be the set of radio holes in G and $P(h)$ denotes the length of the perimeter of a radio hole $h \in H$. For every radio hole, the nodes with maximal/minimal x- and y-coordinates are called *extreme points*. We say that a path p is c-competitive to a path p' if $\|p\| \leq c \cdot \|p'\|$ where $\|p\|$ denotes Euclidean length of p which is the sum of all Euclidean distances between pairs of consecutive vertices of p. Our main contribution is:

Theorem 1. *For any distribution of the nodes in V that ensures that UDG(V) is connected and of bounded degree, our algorithm computes an abstraction of UDG(V) in $\mathcal{O}(\log^2 n)$ communication rounds using only polylogarithmic communication work at each node. In case of non-intersecting bounding boxes, this algorithm finds 18.55-competitive paths between all source-destination pairs outside of bounding boxes. In case of pairwise overlapping bounding boxes, the paths are 28.83-competitive.*

The storage needed by the four extreme points of each radio hole is $\mathcal{O}(|H|)$. For every other node, the space requirement is constant.

For multiple bounding box intersections, we prove that in case we can find a c-competitive path between outer intersection points of bounding boxes, we can also find a $(10.68 + c \cdot 12.83)$-competitive path between all source-destination outside of bounding boxes. The full version of this paper is available online [4].

2   High Level Description

We assume that the edges of the ad hoc network form a 2-localized Delaunay Graph [6]. This graph is a 1.998-spanner of the Unit Disk Graph [7] and can be constructed in a constant number of communication rounds [6] given an initial connected Unit Disk Graph. This can be established in a short setup phase. To find c-competitive routing paths in this network, we make use of a Visibility Graph approach. A Visibility Graph is a geometric graph in which two vertices are connected if they have a direct line of sight. In this work, we consider two nodes to be connected in the Visibility Graph of the hole structure, if the direct line segment between the two nodes does not intersect any hole of the ad hoc network. Two nodes which share an edge in the Visibility Graph are called to be *visibile* from each other. The interesting relation about Visibility Graphs and our approach is that a shortest path between to nodes s and t in the Visibility Graph of the hole structure can be easily converted into a c-competitive path in the 2-localized Delaunay Graph. Therefore, we prove for visible vertices s and t in the 2-localized Delaunay Graph that every triangle which is intersected by the direct line segment $\overline{st}$ is also contained in the Delaunay Graph of the same point set. For Delaunay Graphs there are several c-competitive online routing strategies known. We make use of the latest, MixedChordArc [2], which finds 3.56-competitive paths. Combining both, the core idea is as follows: We compute a Visibility Graph of the hole structure of the 2-localized Delaunay Graph (in a distributed manner) and translate a shortest path p in the Visibility Graph into a c-competitive routing path in the ad hoc network by applying MixedChordArc along every edge $e \in p$.

The Visibility Graph can be very large since holes can have many nodes on their boundaries. We aim for a reduction of the number of considered nodes. In [3], convex hulls have been used. In this work, we reduce the number of considered node per hole to a constant by only taking the axis-parallel bounding box of each hole into account. We denote Visibility Graphs containing the axis-parallel bounding box of each hole *Bounding Box Visibility Graphs*. We show for non-intersecting and pairwise intersecting bounding boxes that Bounding Box Visibility Graphs contain c-competitive paths to usual Visibility Graphs. For multiple intersecting bounding boxes, we show how a slight modification of Bounding Box Visibility Graph can help to find c-competitive paths in the ad hoc network. Lastly, we also consider the distributed computation of Bounding Box Visibility Graphs and the embedding of these graphs in the network. The result is that every extreme point of a bounding box (a node having a maximal or minimal x/y-coordinate) stores a Bounding Box Visibility Graph of the network and moreover every node lying on the boundary of a hole knows the coordinates of the closest extreme points.

The actual routing works as follows: A source node s that wants to send data to a target node t starts sending its message via the MixedChordArc-algorithm in the ad hoc network. Either the message arrives at t or the message gets stuck at a radio hole. In this case, the message is redirected to the closest extreme point which computes a shortest path in the Bounding Box Visibility Graph and afterwards the message is sent via MixedChordArc along this path.

References

1. Ahmed, N., Kanhere, S.S., Jha, S.: The holes problem in wireless sensor networks: a survey. SIGMOBILE Mob. Comput. Commun. Rev. **9**(2), 4–18 (2005)
2. Bonichon, N., Bose, P., Carufel, J.D., Despré, V., Hill, D., Smid, M.H.M.: Improved routing on the delaunay triangulation. In: 26th Annual European Symposium on Algorithms, ESA 2018, pp. 22:1–22:13, Helsinki, Finland, 20–22 August 2018
3. Jung, D., Kolb, C., Scheideler, C., Sundermeier, J.: Competitive routing in hybrid communication networks. In: Gilbert, S., Hughes, D., Krishnamachari, B. (eds.) ALGOSENSORS 2018. LNCS, vol. 11410, pp. 15–31. Springer, Cham (2019). https://doi.org/10.1007/978-3-030-14094-6_2
4. Kolb, C., Scheideler, C., Sundermeier, J.: A Bounding Box Overlay for Competitive Routing in Hybrid Communication Networks (2018)
5. Kuhn, F., Wattenhofer, R., Zhang, Y., Zollinger, A.: Geometric ad-hoc routing: of theory and practice. In: Proceedings of the Twenty-Second ACM Symposium on Principles of Distributed Computing, PODC 2003, Boston, pp. 63–72, 13–16 July 2003
6. Li, X.-Y., Calinescu, G., Wan, P.-J.: Distributed construction of a planar spanner and routing for ad hoc wireless networks. In: Proceedings of the 21st Annual Joint Conference of the IEEE Computer and Communications Societies, vol. 3, pp. 1268–1277. IEEE Press, New York (2002)
7. Xia, G.: The stretch factor of the delaunay triangulation is less than 1.998. SIAM J. Comput. **42**(4), 1620–1659 (2013)

Mobile Robots with Uncertain Visibility Sensors

Adam Heriban$^{(\boxtimes)}$ and Sébastien Tixeuil

LIP6, Sorbonne Université, 75005 Paris, France
{adam.heriban,sebastien.tixeuil}@lip6.fr

1 Introduction

We consider a set of mobile robots, modelled as points, that move freely in a continuous 2-dimensional Euclidean space (the current terminology refers to this model as the $\mathcal{OBLOT}$ model [3]).

Two variants for the robots' visual sensors have been considered so far. With *complete* visibility, every other robot is viewed by the sensor and its position in the ego-centred coordinate system of the observer is returned. A weaker model that has been considered is the *limited* visibility sensor [1], where there exists a constant $\lambda > 0$ (generally unknown to the robots) such that every robot at distance less than λ from from observer is included in the returned view in a Look phase, while every robot further than λ from the observer is *not* included in the returned view.

Ever since the $\mathcal{OBLOT}$ model was introduced, its full visibility sensor was considered unrealistic by practitioners: since robot visual sensors have physical limitations (*e.g.* limited resolution for omnidirectional 3D cameras, this intrinsically yields a limitation of the visibility range. Limited visibility only partly addresses this issue. In the real world, the reliability of visual sensors generally decreases with distance, but not in a binary fashion, and may yield two types of incorrect outputs: *(i)* false positives: no robot exists at a position, but one is output by the vision sensor; *(ii)* false negatives: a robot exists at a position, but none is output by the vision sensor. False positives can be alleviated using known techniques (such as marker based detection) so we do not consider them in this work. False negatives, that are not addressed by the limited visibility model, are the focus of this paper. They are generally due to noise, quality of the sensors and environmental conditions such as smoke or robots unpredictably blending with the background. We define *uncertain* visibility sensors for mobile robots as sensors that satisfy the two following properties: *(i)* every robot closer that λ is output by the sensor; *(ii)* a subset of the robots further than λ is output by the sensor. Note that when the subset includes all such robots, we fall back to the complete visibility model, while the case where the subset is empty equates the limited visibility model.

Since we are interested in characterising the exact limits of models for the computability of tasks in the $\mathcal{OBLOT}$ model, we consider that the subset of robots beyond λ that remain visible is decided by an adversary. Also, a robot R

© Springer Nature Switzerland AG 2019
K. Censor-Hillel and M. Flammini (Eds.): SIROCCO 2019, LNCS 11639, pp. 349–352, 2019.
https://doi.org/10.1007/978-3-030-24922-9_27

that is at the same distance from two distinct robots A and B may be output by A's visibility sensor, but not by B's.

We unify and generalise the two previously studied visibility models: full visibility and limited visibility. In more details, we consider that robots further than λ away may not see each other, depending on the choice of an adversary. We consider two classes of adversaries, the k-random adversary and the k-enemy adversaries. The k-random adversary randomly select up to k visibility relations to be blocked, while the k-enemy adversary purposely selects those k visibility relations.

Then, we explore the impact of this new visibility model on the feasibility of benchmarking tasks in mobile robots computing: gathering, luminous rendezvous, and leader election. For each task, we determine the weakest visibility adversary that prevents task solvability, and the strongest adversary that enables task solvability. It turns out that for all three tasks, our characterisation is tight with respect to k, the parameter of the visibility adversary. Our works sheds new light on the impact of visibility sensors in the context of mobile robot computing, and paves the way for more realistic algorithms that can cope with uncertain visibility sensors.

Due to space constraints, proofs are omitted from this brief announcement.

2   Model

With the notable exception of restricted visibility sensors, our model matches the classical $\mathcal{OBLOT}$ model [3] model. Robots are modelled as points in a bidimensional Euclidean space, are anonymous and uniform (that is, they execute the same code and have no identifiers), and unless specified otherwise, cannot communicate explicitly (but can observe other robots positions in their ego-centred coordinate system) and are oblivious (that is, they cannot remember their past actions).

In this paper, our focus is on the *uncertain* part of the visibility model. As in the $\mathcal{OBLOT}$ model, we consider that $\lambda > 0$ in unknown to the robots. To introduce *selective* vision among robots, we model the Look phase as the sending and receiving of "visibility messages" between robots. We consider that correctly viewing a robot A is similar to correctly receiving a "visibility message" from A. Then, the adversary may simply block a subset of the visibility messages among robots when they are further than λ away from one another. The scope of this paper is limited to the FSYNC scheduler, as it more closely matches the synchronous setting of the paper by Santoro and Widmayer [5], which studied the impact of similar omission faults on consensus. So, in every synchronous step, every robot is scheduled for execution, and performs a complete Look-Compute-Move cycle. We introduce two classes of visibility messages adversaries:

Definition 1 (k-random). *k-random adversaries can make up to k visibility messages disappear in each synchronous round, but those k messages are chosen uniformly at random.*

Definition 2 (k-enemy). *k-enemy adversaries can make up to k visibility messages disappear in each synchronous round, and those k messages are chosen by the adversary.*

3 Results

From definitions of *uncertain* visibility, we can make the following observations, considering a FSYNC network of N robots where $N(N-1)$ visibility messages are sent each round: *(i)* 0-random and 0-enemy adversaries are identical, and equivalent to full visibility, *(ii)* $N(N-1)$-random and $N(N-1)$-enemy adversaries are identical, equivalent to limited visibility, *(iii)* The k-enemy adversary is stronger than the k-random adversary.

Gathering. We first consider the benchmarking problem of gathering N robots at the same location, not known beforehand, starting from any initial configuration. In general, the problem is impossible to solve deterministically in the SSYNC $\mathcal{OBLOT}$ model, but possible in the FSYNC $\mathcal{OBLOT}$ model when robots execute the "centre of gravity" algorithm.

Lemma 1. *In FSYNC, if N robots can achieve convergence under a k-enemy adversary using the* move to centre of gravity *algorithm, then they also achieve gathering.*

Theorem 1. *In FSYNC, deterministic gathering can only be achieved under a k-random adversary if $k \leq N(N-1) - 1$.*

Theorem 2. *In FSYNC, gathering can only be achieved under a k-enemy adversary if $k \leq 2N - 3$.*

Rendezvous. We then study the effect of uncertain visibility in the context of luminous robots [2,4] for the problem of gathering for two robots, or rendezvous, that is solvable in ASYNC.

Theorem 3. *Any luminous rendezvous algorithm that achieves rendezvous in the SSYNC full visibility model fails under the FSYNC 1-enemy adversary.*

Leader Election. Within the mobile robot literature, there were several proposed definitions of leader election specification. We define two different types of leader election. *Strict leader election* is based on the notion of *agreement*, while *soft leader election* is based around the partition in elected and non-elected states.

Definition 3 (Strict Leader Election). *A leader election process is* strict *if at any given time in the execution after leader election, every robot in the network knows which robot is the leader.*

Definition 4 (Soft Leader Election). *A leader election process is* soft *if at any given point in the execution after the leader has been elected, no other robot is elected, and the elected leader remains elected.*

Theorem 4. *In FSYNC, strict leader election is possible with a 0-random adversary and trivially impossible under the 1-random scheduler.*

Theorem 5. *In FSYNC, soft leader election based on positions is possible with a 0-random adversary and impossible under the 1-random scheduler.*

4   Conclusion

Overall, we introduced the notion of uncertain visibility for mobile robots. For a practical point of view, the proposed notion is the most realistic to date. From a theoretical point of view, uncertain visibility generalises visibility constraints studied in previous and yields interesting twists in the design of algorithms. We introduced two variants for uncertainty: a randomised version and an adversarial version. We presented lower bounds and matching algorithms in the context of gathering, rendezvous, and leader election protocols. For specifications that must maintain a global invariant (such as leader election), even the weakest non-trivial adversary (1-random) prevents the existence of a solution. For specifications with eventual safety properties (such as gathering and rendezvous), the results are contrasted: when the algorithm allows all robots to move at every round, strong adversaries can be handled, when there exists a synchronisation mechanism (as in rendezvous algorithms for luminous robots), even the weakest non-probabilistic adversary (1-enemy) precludes any deterministic solution.

References

1. Ando, H., Oasa, Y., Suzuki, I., Yamashita, M.: Distributed memoryless point convergence algorithm for mobile robots with limited visibility. IEEE Trans. Robot. Autom. **15**(5), 818–828 (1999). https://doi.org/10.1109/70.795787
2. Das, S., Flocchini, P., Prencipe, G., Santoro, N., Yamashita, M.: The power of lights: Synchronizing asynchronous robots using visible bits. In: 2012 IEEE 32nd International Conference on Distributed Computing Systems. IEEE Computer Society, Macau, pp. 506–515, 18–21 June 2012. https://doi.org/10.1109/ICDCS.2012.71
3. Flocchini, P., Prencipe, G., Santoro, N. (eds.): Distributed Computing by Mobile Entities-Current Research in Moving and Computing. LNCS, vol. 11340. Springer, Cham (2019). https://doi.org/10.1007/978-3-030-11072-7
4. Heriban, A., Défago, X., Tixeuil, S.: Optimally gathering two robots. In: Bellavista, P., Garg, V.K. (eds.) Proceedings of the 19th International Conference on Distributed Computing and Networking, ICDCN 2018, Varanasi, India, 4–7 January, pp. 3:1–3:10. ACM (2018). https://doi.org/10.1145/3154273.3154323
5. Santoro, N., Widmayer, P.: Time is not a healer. In: Monien, B., Cori, R. (eds.) STACS 1989. LNCS, vol. 349, pp. 304–313. Springer, Heidelberg (1989). https://doi.org/10.1007/BFb0028994

A Strongly-Stabilizing Protocol for Spanning Tree Construction Against a Mobile Byzantine Fault

Koki Inoue[1], Yuichi Sudo[1($\boxtimes$)], Hirotsugu Kakugawa[2], and Toshimitsu Masuzawa[1]

[1] Osaka University, Suita, Japan
{ko-inoue,y-sudou,masuzawa}@ist.osaka-u.ac.jp
[2] Ryukoku University, Kyoto, Japan
kakugawa@rins.ryukoku.ac.jp

1 Introduction

Self-stabilization [2] is a promising paradigm for designing distributed systems that are highly-tolerant of *transient* faults and adaptive to topology changes, since it guarantees that a system can recover its intended behavior even when its configuration (or global state) is arbitrarily changed by transient faults or topology changes. However, the recovery to the intended behavior requires a sufficiently long period of stable network environments (with no fault or topology changes). Self-stabilization guarantees nothing when the network has *permanent* faults or continuous topology changes. Thus, self-stabilization in the presence of permanent faults is a challenging and attractive issue.

Combination of tolerance to transient faults (or self-stabilization) and tolerance to permanent faults is practically important in distributed systems. It is common, as also in this paper, that permanent fault tolerance assumes some upper bound, say f, on the number of permanent faults: protocols work correctly only when the number of permanent faults never exceeds the bound f throughout their executions. However in an unexpected event like major power outage, the number of faults may temporarily exceed the bound. In such cases, permanent fault tolerance solely can guarantee nothing, but its combination with self-stabilization can guarantee eventual recovery to the intended behavior of the distributed systems.

Especially, self-stabilization in the presence of (permanent) *Byzantine faults* is a challenging issue. A self-stabilizing system has to respond to a corruption caused by transient faults to recover its intended behavior, but this makes it difficult that the system becomes stable in a legitimate configuration in spite of Byzantine faults: every time Byzantine faults corrupt the system configuration, the self-stabilizing system responds to the corruption, which may lead the whole system into permanently unstable behavior.

To circumvent the difficulty, Nesterenko and Arora [4] introduced the *strict stabilization* with the aim of containing the influence of repeated malicious actions of Byzantine faults within some distance (called *containment radius*)

© Springer Nature Switzerland AG 2019
K. Censor-Hillel and M. Flammini (Eds.): SIROCCO 2019, LNCS 11639, pp. 353–356, 2019.
https://doi.org/10.1007/978-3-030-24922-9_28

from Byzantine processes. It allows the processes within the containment radius from Byzantine processes to permanently deviate from the intended behavior (even though they are non-faulty). Unfortunately, the strict stabilization is attainable only for *local problems*, where the legitimacy of a configuration is locally checkable), such as vertex coloring and link coloring. To circumvent the impossibility, Dubois et al. [3] introduced the *strong stabilization* as an extension of the strict stabilization, and presented strongly stabilizing protocols for non-local problems such as the tree orientation problem and the spanning tree construction problem. The strong stabilization allows processes outside the containment radius from Byzantine processes to deviate from the intended behavior but only a *finite number of times* even when Byzantine processes make an infinite number of malicious actions.

This paper considers *mobile Byzantine faults*, which are modeled as malicious agents that move process to process and make the visited processes behave maliciously. After an agent leaves a process, the process regains its correct behavior (or its correct program code). A distributed system containing mobile Byzantine faults can be represented as a system having a bounded number of Byzantine processes at any time and their locations change over time. This model captures phenomena like virus infection over a network. Various models of a distributed system with mobile Byzantine faults have been proposed and studied.

2 Our Contribution

We consider strong-stabilizing spanning tree construction in the presence of both mobile Byzantine faults and transient faults. We adopt the model introduced by Buhrman et al. [1] where malicious agents can move only when messages are sent. We newly introduce the mechanism of *blocking links* in the model. When a process blocks a link, it refuses to receive messages and agents through the link. Blocking links has the pros and cons. The pros is that the blocked links can prevent a clean region of the system from being intruded and contaminated by malicious agents. The cons is that the blocked links also prevent processes from receiving useful information from their non-faulty neighbors. The introduction of blocking links to the model is necessary because otherwise a malicious agent can move to all processes infinitely often, thus we can never contain influence of the malicious agents in some area of a network.

First, we formalize the notion of strong stabilization for a system with *mobile* Byzantine faults. As mentioned above, strong stabilization for an *static* Byzantine faults requires that the processes outside the containment range deviate from its intended behavior only *finite* times. The containment range is defined as the set of processes whose distance from an Byzantine faulty process is within a parameter c. However, this definition does not suit to a system with mobile Byzantine faults because the containment range changes every time a Byzantine agent moves among processes. Instead, we define strong stabilization based on the number of legitimate processes. Specifically, we say that an protocol is (k, f)-strongly stabilizing protocol if, in every execution of the protocol with at most f

Byzantine agents, at most k processes deviates from the intended behavior after some point of the execution. Intuitively, this means that the Byzantine agents must stop harmful actions eventually or they are eventually contained in the set of at most k processes.

Next, we give the first strongly stabilizing protocol that tolerates mobile Byzantine faults. This protocol constructs a spanning tree rooted at a designated process r in a synchronous network. We consider a single Byzantine agent (or $f = 1$) under two distinct models depending on whether or not processes can *detect the destination* of the Byzantine agent when the agent leaves from them. The results are summarized in Table 1. If the processes have an ability to detect the destination of the agent, the proposed protocol is $(1, 1)$-strongly stabilizing for any 2-vertex-connected system. Otherwise, the proposed protocol is $(2, 1)$-strongly stabilizing for any 3-vertex-connected system.

Table 1. Properties of the proposed protocol

	With destination detection	Without destination detection
Property	$(1, 1)$-strongly stabilizing	$(2, 1)$-strongly stabilizing
Topology restriction	2-vertex-connected	3-vertex-connected

3 Problem Specification

We assume that each process has a tamper-proof memory where it safely stores the correct protocol code. When an agent leaves a process, the process can detect that it was infected in the previous round. Then, it recovers the correct protocol code from the tamper-proof memory. In this paper, we assume that exactly one (Byzantine) agent exists in the system. We call this agent *the agent* and denote the process that it stays on by v_B.

Each process v has an *output variable* $\mathtt{parent}_v \in \{0, 1, \ldots, \Delta_v\}$ to designate one of its neighbors as its parent. ($\mathtt{parent}_v = 0$ means v has no parent.) The goal of spanning tree construction is to set $\mathtt{parent}_v$ of every process v to form a spanning tree rooted by r. Process v_B can behave as if it is the root *forever*. Hence, the other processes cannot determine which is the true root, r or v_B. Thus, we accept a configuration with a spanning forest consisting of two trees, one is rooted by r and the other is rooted by v_B, as a legitimate configuration. We also assume that v_B never exists on r at the beginning of an execution because otherwise we cannot solve the problem anymore.

4 Key Idea of Proposed Protocol

The variables $\mathtt{parent}$ of all the processes form trees and cycles on the system. Among those trees, we call the tree rooted by r *the good tree* and the tree rooted by v_B *the bad tree*. We ignore the value of $\mathtt{parent}$ in v_B, thus v_B never belongs to the good tree. In a legitimate configuration, all processes join either the good tree or the bad tree. The key idea of this protocol is that the root always blocks its all ports and each process receives a message only from its parent and blocks all the other ports. This guarantees that the agent cannot move to any process in the good tree. Each process maintains a kind of its *reliability* as a non-negative integer, which corresponds to the number of the passed rounds since it changes its parent for the last time, and sends its reliability to all its neighbors at each round. Each process also maintains its *level* as a non-negative integer, which stores the distance from the process to r or v_B in the constructed tree(s), and sends its level to all its neighbors at each round. A non-root process memorizes the reliabilities of its neighbors. At each step, a non-root process receives a message from its parent, which contains the level and the reliability of the parent. If the non-root process finds any inconsistency between the message from its parent and the information it memorizes, it resets the reliabilities of itself and its parent to zero in its memory, and chooses a neighbor with the maximum reliability in its memory as a new parent. Otherwise, i.e., if it finds no inconsistency, it increments its reliability (and the reliabilities of all the neighbors) in its memory. When the agent leaves a non-root process, the process resets its reliability to zero and unblocks all its ports (except for the destination port in the detectable model) to receive messages from all its neighbors so that it can collect the reliabilities of every neighbor. In the next round, it choose a neighbor with the maximum reliability as its parent and blocks the ports for the other neighbors.

The processes of the good tree find no inconsistency because the agent cannot make any influence on them. Therefore, the reliability of the processes in the good tree increases unboundedly. Thanks to this unbounded increase, the good tree on any 2-vertex-connected graph in the detectable model (resp. any 3-vertex-connected graph in the non-detectable model) keeps on growing and its size will become $n - 1$ (resp. $n - 2$ or more) unless the agent stops harmful actions which prevent the system from reaching a legitimate configuration or deviate the system from a legitimate configuration.

Acknowledgment. This work was supported by JSPS KAKENHI Grant Numbers 17K19977, 18K18000, and 19K11826 and Japan Science and Technology Agency (JST) SICORP.

References

1. Buhrman, H., Garay, J.A., Hoepman, J.H.: Optimal resiliency against mobile faults. In: Twenty-Fifth International Symposium on Fault-Tolerant Computing. Digest of Papers, pp. 83–88. IEEE (1995)
2. Dijkstra, E.W.: Self-stabilizing systems in spite of distributed control. Commun. ACM **17**(11), 643–644 (1974)
3. Dubois, S., Masuzawa, T., Tixeuil, S.: Bounding the impact of unbounded attacks in stabilization. IEEE Trans. Parallel Distrib. Syst. **23**(3), 460–466 (2012)
4. Nesterenko, M., Arora, A.: Tolerance to unbounded byzantine faults. In: Proceedings of the 21st IEEE Symposium on Reliable Distributed Systems, pp. 22–29. IEEE (2002)

Explicit and Tight Bounds of the Convergence Time of Average-Based Population Protocols

Yves Mocquard[1], Bruno Sericola[2], and Emmanuelle Anceaume[3]([✉])

[1] University of Rennes, CNRS, Inria, IRISA, Rennes, France
yves.mocquard@irisa.fr
[2] Inria, University of Rennes, CNRS, IRISA, Rennes, France
bruno.sericola@inria.fr
[3] CNRS, University of Rennes, Inria, IRISA, Rennes, France
emmanuelle.anceaume@irisa.fr

Abstract. This paper focuses on the deep analysis of average-based problems in the population protocol model [1], a model in which agents are identically programmed, with no identity, and they progress in their computation through random pairwise interactions.

1 Introduction

This paper focuses on the deep analysis of average-based problems in the population protocol model [1], a model in which agents are identically programmed, with no identity, and they progress in their computation through random pairwise interactions. A considerable amount of work has been done so far to determine which properties can emerge from pairwise interactions between finite-state agents, together with the derivation of bounds on the time and space needed to reach such properties. In this work, we are primarily interested in problems that aim at quantifying properties on the system population, such as the proportion problem [2] or the counting problem [3]. Namely, we consider a set of n agents, interconnected by a complete graph, that asynchronously start their execution in one of two distinct states A (associated with some positive integer m) and B (associated with 0), and such that n_A (resp. n_B) is the number of agents whose initial state is A (resp. B). Such problems can be solved by relying on average-based population protocols [2–4]. Briefly, n agents starting independently from each other with an initial integer state, interact randomly by pairs, and at each interaction, keep the average of both states as their new state. Average-based protocols have also been used in gossip-based aggregation protocols as well as in consensus protocols [5].

2 Average-Based Population Protocols

Average-based population protocols use the average technique to compute the proportion of agents that started their execution in a given state A. The notion of

© Springer Nature Switzerland AG 2019
K. Censor-Hillel and M. Flammini (Eds.): SIROCCO 2019, LNCS 11639, pp. 357–360, 2019.
https://doi.org/10.1007/978-3-030-24922-9_29

time in population protocols refers to as the successive steps at which interactions occur, while the parallel time refers to as the successive number of steps each agent executes. Agents do not maintain nor use identifiers, however, for ease of presentation the agents are numbered $1, 2, \ldots, n$. We denote by $C_t^{(i)}$ the state of agent i at time t. The stochastic process $C = \{C_t, \ t \geq 0\}$, where $C_t = (C_t^{(1)}, \ldots, C_t^{(n)})$, represents the evolution of the population protocol. This means that, for every $i = 1, \ldots, n$, we have $C_0^{(i)} \in \{0, m\}$. At each discrete instant t, two distinct indices i and j are uniformly chosen among $1, \ldots, n$, that is with probability $1/(n(n-1))$. Once chosen, the couple (i, j) interacts, and both agents update their respective local state $C_t^{(i)}$ and $C_t^{(j)}$ by applying the transition function f, leading to state C_{t+1}, given by $f(C_t^{(i)}, C_t^{(j)}) = (C_{t+1}^{(i)}, C_{t+1}^{(j)})$ with

$$\left(C_{t+1}^{(i)}, C_{t+1}^{(j)}\right) = \left(\left\lfloor \frac{C_t^{(i)} + C_t^{(j)}}{2} \right\rfloor, \left\lceil \frac{C_t^{(i)} + C_t^{(j)}}{2} \right\rceil\right) \text{ and } C_{t+1}^{(r)} = C_t^{(r)} \text{ for } r \neq i, j.$$

We denote by ℓ the mean value of the sum of the entries of C_t and by L the row vector of $\mathbb{R}^n$ with all its entries equal to ℓ, that is $\ell = \sum_{i=1}^n C_t^{(i)}/n$ and $L = (\ell, \ldots, \ell)$. We denote by $\|.\|$ the Euclidean norm and by $\|.\|_\infty$ the infinite one. Let $\lambda = \min\{\ell - \lfloor \ell \rfloor, \lceil \ell \rceil - \ell\}$, which is the distance between ℓ and its nearest integer. It is easily checked that we have $0 \leq \lambda \leq 1/2$. In Theorem 4 of [2], we dealt with the case where $\ell - \lfloor \ell \rfloor = 1/2$. This case implies that n is even. In the following theorem, we generalize these results to the case where n is odd. We denote by $1_{\{A\}}$ the indicator function, which is equal to 1 if condition A is true and 0 otherwise.

Theorem 2.1. *For all $\delta \in (0, 1)$, if $\lambda = \left(n - 1_{\{n \ odd\}}\right)/(2n)$ and if there exists a constant K s.t. $\|C_0 - L\| \leq K$ then, for every $t \geq (n-1)(2 \ln K - \ln \delta - \ln 2)$, we have*

$$\mathbb{P}\left\{\|C_t - L\|_\infty > \frac{n + 1_{\{n \ odd\}}}{2n}\right\} = \mathbb{P}\left\{\max_{1 \leq i \leq n} C_t^{(i)} - \min_{1 \leq i \leq n} C_t^{(i)} > 1\right\} \leq \delta.$$

The Shadow Process. We introduce what we call a shadow process of the stochastic process $C = \{C_t, \ t \geq 0\}$. This shadow process is a stochastic process denoted by $D = \{D_t, \ t \geq 0\}$ and defined at time $t = 0$ by $D_0^{(i)} = C_0^{(i)} + 1_{\{i \in B_0\}}$, where B_0 is a non empty subset of b agents with $b \leq n - 1$, i.e. $B_0 \subset \{1, \ldots, n\}$ and $|B_0| = b$. For every $t \geq 1$, the shadow process D_t is defined as process C_t, that is, when the couple (i, j) is chosen to interact at time t, the vector D_{t+1} is given by

$$\left(D_{t+1}^{(i)}, D_{t+1}^{(j)}\right) = \left(\left\lfloor \frac{D_t^{(i)} + D_t^{(j)}}{2} \right\rfloor, \left\lceil \frac{D_t^{(i)} + D_t^{(j)}}{2} \right\rceil\right) \text{ and } D_{t+1}^{(r)} = D_t^{(r)} \text{ for } r \neq i, j.$$

Both stochastic processes C and D behave identically and evolve following the same interactions. The only difference lies their initial values. Note that process

C is a part of the protocol but not process D which is introduced only for the probabilistic analysis of C. As we did for process C, we denote by ℓ_D the mean value of the sum of the entries of D_t and by L_D the row vector of $\mathbb{R}^n$ with all its entries equal to ℓ_D. Lemma 2.2 shows that if at time $t = 0$, D_0 is in the shadow of C_0 then for any time $t \geq 0$, D_t remains in the shadow of C_t.

Lemma 2.2. *For all $t \geq 0$, there exists a non empty set B_t of b agents, i.e. $B_t \subset \{1, \ldots, n\}$ and $|B_t| = b$, such that for all $i \in \{1, 2, \ldots, n\}$, we have*

$$D_t^{(i)} = C_t^{(i)} + 1_{\{i \in B_t\}}.$$

Lemma 2.3. *For all $t \geq 0$, we have*

$$\left| \|D_t - L_D\|_\infty - \|C_t - L\|_\infty \right| \leq 1 - \frac{1}{n} \quad and \quad \left| \|D_t - L_D\| - \|C_t - L\| \right| < \sqrt{n}.$$

In Theorem 2.1, we obtained a first bound on the convergence time in the particular case where $\lambda = \left(n - 1_{\{n \text{ odd}\}} \right) / (2n)$. This result together with Lemmas 2.2 and 2.3 are used to obtain general results, i.e. results which apply for any value of λ.

Theorem 2.4. *For all $\delta \in (0, 1)$, if there exists a constant K such that $\|C_0 - L\| \leq K$, then, for every $t \geq (n - 1) \left(2 \ln \left(K + \sqrt{n} \right) - \ln \delta - \ln 2 \right)$, we have*

$$\mathbb{P} \left\{ \max_{1 \leq i \leq n} C_t^{(i)} - \min_{1 \leq i \leq n} C_t^{(i)} > 2 \right\} \leq \delta \quad and \quad \mathbb{P} \left\{ \|C_t - L\|_\infty \geq \frac{3}{2} \right\} \leq \delta.$$

3   Application: Solving the Proportion Problem

We apply our results to the proportion problem.

Definition 3.1. (Proportion Problem). *A population protocol solves the proportion problem with precision $\varepsilon \in (0, 1)$ and with probability at least $1 - \delta$, $\delta \in (0, 1)$, in $\tau(n, \varepsilon, \delta)$ interactions, if at any time $t \geq \tau(n, \varepsilon, \delta)$, any node is capable of computing n_A/n with an ε-precision without the knowledge of the population size n.*

We denote by γ_A the proportion of nodes starting with A, i.e. $\gamma_A = n_A/n$, where n_A is the number of nodes starting with A. The following theorem gives an evaluation of the first instant t from which the distance between $C_t^{(i)}/m$ and γ_A is less than a fixed ε with any high probability $1 - \delta$.

Theorem 3.2.   *For all $\delta \in (0, 1)$ and $\varepsilon \in (0, 1)$, by taking $m = \lceil 3/(2\varepsilon) \rceil$, we have, for all $t \geq (n - 1) \left[\ln n - \ln \delta + 2 \ln(2 + 1/\varepsilon) + \ln(9/32) \right]$,*

$$\mathbb{P} \left\{ |C_t^{(i)}/m - \gamma_A| < \varepsilon, \text{ for all } i = 1, \ldots, n \right\} \geq 1 - \delta.$$

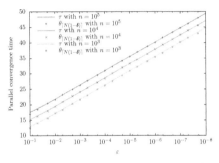
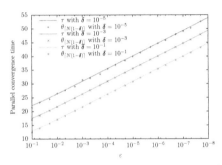

(a) $\theta_{\lceil N(1-\delta)\rceil}$ as a function of ε, with $\delta = 10^{-1}$ and $N = 10^4$. From top to the bottom, we have $n = 10^5$, $n = 10^4$ and $n = 10^3$ respectively.

(b) $\theta_{\lceil N(1-\delta)\rceil}$ as a function of ε, with $n = 10^3$ and $N = 10^6$. From top to the bottom, we have $\delta = 10^{-5}$, 10^{-3} and 10^{-1} respectively.

Fig. 1. Comparing the estimation $\theta_{\lceil N(1-\delta)\rceil}$ with the theoretical bound τ of the parallel convergence time (Theorem 3.2).

Experimental Results. We show how tight our bounds are by comparing the theoretical bound τ of the parallel convergence time to the results obtained via extensive simulations (see Fig. 1). We introduce the parallel convergence time θ defined by $\theta = \inf\{t \geq 0 \text{ s.t. for all } i,\ |C_t^{(i)}/m - \gamma_A| < \varepsilon\}/n$. We have run N independent simulations of θ and stored the N values of the parallel convergence times denoted by $\theta_1, \ldots, \theta_N$. The estimation of the first instant t such that $\mathbb{P}\{\theta < t\} \geq 1 - \delta$ is thus given by $\theta_{\lceil N(1-\delta)\rceil}$.

References

1. Angluin, D., Aspnes, J., Diamadi, Z., Fischer, M.J., Peralta, R.: Computation in networks of passively mobile finite-state sensors. Distrib. Comput. **18**(4), 235–253 (2006). https://doi.org/10.1007/s00446-005-0138-3
2. Mocquard, Y., Anceaume, E., Sericola, B.: Optimal proportion computation with population protocols. In: Proceedings of the IEEE NCA (2016)
3. Mocquard, Y., Anceaume, E., Aspnes, J., Busnel, Y., Sericola, B.: Counting with population protocols. In: Proceedings of the IEEE NCA (2015)
4. Sauerwald, T., Sun, H.: Tight bounds for randomized load balancing on arbitrary network topologies. In: Proceedings of the IEEE FOCS (2012)
5. Cordasco, G., Gargano, L.: Space-optimal proportion consensus with population protocols. In: Proceedings of the IEEE SSS (2017)

Visiting Infinitely Often the Unit Interval While Minimizing the Idle Time of High Priority Segments

Oscar Morales-Ponce[✉]

Department of Computer Engineering and Computer Science,
California State University Long Beach, Long Beach, CA, USA
oscar.moralesponce@csulb.edu

1 Introduction

Most of the previous works assume that all the points in the border have the same priority to be visited. However, the border may consist of sections with different priority of patrolling such as static guards protecting some sections of the border. However, these sections are required to be visited regularly to detect points of failure. On the other side, every high priority point is required to be visited as often as possible. What strategy must the robots follow to give the maximum protection to the points of the high priority sections while visiting infinitely often every point of the border? To answer this question, we consider the *idle time* to measure the efficiency of a strategy. Intuitively, the idle time of a given strategy with k robots measures the maximum period that any high priority point remains unvisited. In the remaining of the paper, we summarize the upper and lower bounds.

Model and Problem Statement. Without loss of generality, we model the border as a segment of unit length $C = [0, 1]$. The segment is partitioned with two subsets H and L where H represents the "high priority" sections, and L the "low priority" sections. We take H to be a finite union of closed intervals. Let n be the number of high priority segments.

We consider k identical robots with a maximum speed of one. Thus, a robot can traverse the unit interval in one unit of time. We assume that the acceleration is infinite. Hence, robots can change speed instantly. Each robot r_i follows a continuous function $f_i(t)$ that defines the position of the robot r_i in the unit interval C at time t. A *strategy* consists of k continuous functions. The idle time of the strategy is defined as the minimum maximum time that any point in H remains unvisited.

Given a partition (H, L) of the unit interval $C = [0, 1]$ and a set of k robots with the same maximum speed, the problem is to determine the optimal idle time that any strategy can attain, i.e., $I_k^* = \inf_{\forall A}(I_A)$.

Related Work. Different variations of the patrolling problem have been studied recently. A closely related version was studied in [2] where the border is divided

© Springer Nature Switzerland AG 2019
K. Censor-Hillel and M. Flammini (Eds.): SIROCCO 2019, LNCS 11639, pp. 361–365, 2019.
https://doi.org/10.1007/978-3-030-24922-9_30

into two different types of segments. Namely, the vital sections that are required to be visited with the minimum time and the neutral sections that are not required to be visited, but robots can traverse to reach vital sections. They provide an optimal strategy for the unit segment and the ring. The problem studied in this paper requires that all points must be visited infinitely often unlike [2].

Another closely related problem was recently studied in [1] where two robots are required to patrol a set of points on a unit segment. In that paper, points can be assigned different priorities. The problem asks to find strategies that guarantee that the maximum time that a point remains unvisited is at most the given priority. The authors provide a $\sqrt{3}$-approximation algorithm. A similar problem was studied in [6] where a single robot is required to visit a set of n points with priorities. These priorities are updating at a steady rate, and the problem asks to find a strategy that minimizes the maximum priority ever observed. The authors study two different variants of the model and provide upper bounds. Patrolling without priorities have been studied in different contexts. See for example, [3–5, 7].

2   Contributions

The lower and upper bounds are based on single lid covers and double lid covers. An l_k -lid cover [2] is a set $\{\ell_1, \ell_2 \ldots, \ell_k\}$ of k lids such that every high priority point in $p \in H$ is covered by at least one lid, i.e., for every point $p \in H$ there exists ℓ_i such that $p \in \ell_i$.

Throughout the paper we consider the minimum lid length, denote as λ_k, such that C admits a single λ_k-lid cover with k lids. Let us denote the order set of lids of such a cover as $\mathcal{W}_k$. A *block* in $\mathcal{W}_k$ is the tuple $B = \{\ell_{a(1)}, \ell_{a(2)}, \ldots \ell_{a(b)}\}$ of b lids such $Right(\ell_{a(i)}) = Left(\ell_{a(i+1)})$ for all $i \in [1, b-1]$ where $Left(\ell)$ and $Right(\ell)$ denote the leftmost and rightmost point of ℓ. Now we can define *critical blocks* that are of particular importance for the lower bound and upper bound proofs. A critical block B of $\mathcal{W}_k$ is a block such that $Left(B)$ is the leftmost point of a high priority segment and $Right(B)$ is the rightmost point of a high priority segment. We show that every single λ_k-lid cover has a critical block. To reduce the number of possible single l_k-lid covers, we shift to the right all lids of $\mathcal{W}_k$ as far as possible without uncover the high priority points. Let $\mathcal{W}^{\rightarrow}_k$ denote such a shifted single l_k-lid cover. Analogously, we shift to the left all lids of $\mathcal{W}_k$ as far as possible without uncover the high priority points. Let $\mathcal{W}^{\leftarrow}_k$ denote such a shifted single l_k-lid cover.

Lower-Bound. To show the lower bound, first we extend the definition of single l_k-lid cover to strong double l_k-lid cover. A strong double l_k-lid cover is the set $\{\ell_1, \ell_2 \ldots, \ell_k\}$ of k lids such that the unit segment C is fully covered and every high priority point in $p \in H$ is covered by at least two distinct lids, i.e., $\bigcup_{i=1}^{k} \ell_i = C$ and there exist ℓ_i and $\ell_{j \neq i}$ such that $p \in \ell_i \cap \ell_j$ for all $p \in H$. We refer as "strong" to emphasize that it covers the unit interval.

Throughout the paper we consider the minimum lid length, denoted as Λ_{2k}, such that C admits a strong double Λ_{2k}-lid cover with $2k$ lids. Let $\mathcal{S}_{2k}$ denote the order set of lids of such a cover. Similar as before we define $\mathcal{S}_{2k}^{\rightarrow}$ and $\mathcal{S}_{2k}^{\leftarrow}$ by shifting to the left or right the lids as far as we can without violating the strong double cover property. For the lower bound proof, it is essential to compare Λ_{2k} and λ_{k-1}. Indeed, when $\Lambda_{2k} < \lambda_{k-1}$, every lid of the strong double Λ_{2k}-lid cover covers at least one high priority point which means that the high priority segments must be uniformly distributed in the unit interval.

A *component* of a strong double Λ_{2k}-lid cover is a set $W = \{\ell_{a(1)}, \ell_{a(2)}, ..., \ell_{a(w)}\}$ of consecutive lids such that $\ell_{a(i)} \cap \ell_{a(i+1)} \cap H \neq \emptyset$. We say that W is a *maximal component* if it cannot be extended. A key property of any strong double Λ_{2k}-lid cover when $\Lambda_{2k} < \lambda_{k-1}$ is that every maximal component is of even length.

For the lower-bound we need to find $k + 1$ points at distance at least $\min(\Lambda_{2k}, \lambda_{k-1})$ where k points are in H. Thus, there is a time where one robot visits the low priority point and $k - 1$ points cannot visit k high priority point in time less than $2 \min(\Lambda_{2k}, \lambda_{k-1})$. We say that a set of points P of C is in *general position* if for all $a, b, c, d \in P$, with $a \leq b$ and $c \leq d$, if $b - a$ is a rational multiple of $d - c$, then $a = c$ and $b = d$. A high priority set H is in *general position* if the set of all endpoints of intervals in H are in general position. A key property for the lower bound is that both $\mathcal{S}_{2k}^{\rightarrow}$ and $\mathcal{S}_{2k}^{\leftarrow}$ have a critical block. Indeed, the critical block is common when $\Lambda_{2k} < \lambda_{k-1}$ and H is in general position. Using all previous results, we can find $k + 1$ points in $\mathcal{W}_k$ and $\mathcal{S}_{2k}$ where k points are in H and we can obtain the following theorem.

Theorem 1. If the priority set H is in general position, then $I_k^* \geq 2 \min(\Lambda_{2k}, \lambda_{k-1})$.

Upper Bound. We provide three different strategies for showing the upper bound. In Strategy 1, we consider $\mathcal{W}_{k-1} = \{\ell_1, \ell_2, ..., \ell_{k-1}\}$ with $k - 1$ lids of length λ_k. Let robot r_i cover (move back and forth) at maximum speed lid ℓ_i and robot r_k cover the unit segment. Observe that when r_k is at distance at least λ_{k-1} from any lid ℓ_i, it cannot help robot r_i. Therefore, the idle time of the strategy is $2\lambda_{k-1}$.

For the second and third strategy we consider $\mathcal{S}_{2k} = \ell_1, \ell_2, ..., \ell_{2k}$ with $2k$ lids of length Λ_{2k}. We define a cap c_i as the union of two consecutive lids ℓ_{2i-1}, ℓ_{2i}, i.e., $c_i = \{\ell_{2i-1}, \ell_{2i}\}$ for all $i \in [1, k]$. For each i let $c_i = \{\ell_{2i-1}, \ell_{2i}\}$ be the i-th cap, and let $c^* = \max_{\forall i}(\mathrm{len}(c_i))$, where $\mathrm{len}(c_i)$ is the length of the i-th cap. Furthermore, for every cap c_i, let x_i be the center of c_i. In Strategy 2, we place robot r_i at x_i and let it move back and forth at maximum speed in the segment $[x_i - c^*/2, x_i + c^*/2]$. Thus, the unit segment is split into k segments of length c^* not necessarily disjoint. Observe that this strategy requires that robots patrol their segments synchronously, i.e., robots reach the leftmost and rightmost point of their respectively segment concurrently. Therefore, although, the length of c^* can be up to $2\Lambda_{2k}$, the idle time of the second strategy is $3\Lambda_{2k}$ since robots help cover their neighboring segments.

We can improve the previous strategy with a more complex strategy where the segments of the robots are not necessarily of equal length, which means that robots cannot patrol synchronously. In Strategy 3, each robot r_i first overs lid ℓ_{2i-1} until (at time t_0) for each i lid ℓ_{2i-1} is periodically covered by r_i. Then, the first time after t_0 that r_1 reaches $Right(\ell_1)$, r_1 continues to cover ℓ_2. Once ℓ_2 is periodically covered and r_2 reaches $Right(\ell_3)$, r_2 continues to cover ℓ_4, and so on until for each i the lid ℓ_{2i} is periodically covered by r_i. Then the process is reversed, with r_k switching to covering ℓ_{2k-1}, and so on until finally r_1 switches to cover ℓ_1, and we repeat. Observe that the idle time of this strategy depends on the size of the lids. Thus, the idle time of the third strategy is $2\Lambda_{2k}$, Combining Strategy 1 and 3 we obtain the following theorem.

Theorem 2. There exists an algorithm that attain optimal idle time.

Computing Optimal Lid Covers. The previous strategies rely on a single λ_{k-1}-lid cover with $k-1$ lids and a strong double Λ_{2k}-lid cover with $2k$ lids. Therefore, it is of vital importance to computing them efficiently. To compute the values, we define a feasible solution where at least one block is critical, although not necessarily of optimal length. The idea is that given a length l, we can determine in $O(\max(k,n))$ time with a simple greedy algorithm whether the unit segment accepts a strong double cover with $2k$ lids of length l, respectively. Once we determine that C accepts a strong double l-lid cover, we can compute the closest smallest feasible solution. More precisely, if C admits a (strong double) l-lid cover, then we can obtain in $O(n)$ time a new lid length $l' \leq l$ such that C admits a feasible strong double l'-lid cover. Similarly, we can obtain a feasible single l'-lid cover. Now can find both Λ_{2k} and λ_{k-} such that C accepts a strong double Λ_{2k}-lid cover and a simple Λ_{2k}-lid cover in $O(\max(n,k)\log n)$ time using a binary search. Thus, we obtain the following theorem.

Theorem 3. There exists an algorithm that determines λ_{k-1} and Λ_{2k} in $O(\max(n,k)\log n)$ time.

References

1. Chuangpishit, H., Czyzowicz, J., Gąsieniec, L., Georgiou, K., Jurdziński, T., Kranakis, E.: Patrolling a path connecting a set of points with unbalanced frequencies of visits. In: Tjoa, A.M., Bellatreche, L., Biffl, S., van Leeuwen, J., Wiedermann, J. (eds.) SOFSEM 2018. LNCS, vol. 10706, pp. 367–380. Springer, Cham (2018). https://doi.org/10.1007/978-3-319-73117-9_26
2. Collins, A., et al.: Optimal patrolling of fragmented boundaries. In: Proceedings of the Twenty-Fifth Annual ACM Symposium on Parallelism in Algorithms and Architectures, pp. 241–250. ACM (2013)
3. Czyzowicz, J., Gąsieniec, L., Kosowski, A., Kranakis, E.: Boundary patrolling by mobile agents with distinct maximal speeds. In: Demetrescu, C., Halldórsson, M.M. (eds.) ESA 2011. LNCS, vol. 6942, pp. 701–712. Springer, Heidelberg (2011). https://doi.org/10.1007/978-3-642-23719-5_59

4. Czyzowicz, J., Gasieniec, L., Kosowski, A., Kranakis, E., Krizanc, D., Taleb, N.: When patrolmen become corrupted: monitoring a graph using faulty mobile robots. Algorithmica **79**(3), 925–940 (2017)
5. Czyzowicz, J., Kosowski, A., Kranakis, E., Taleb, N.: Patrolling trees with mobile robots. In: Cuppens, F., Wang, L., Cuppens-Boulahia, N., Tawbi, N., Garcia-Alfaro, J. (eds.) FPS 2016. LNCS, vol. 10128, pp. 331–344. Springer, Cham (2017). https://doi.org/10.1007/978-3-319-51966-1_22
6. Gąsieniec, L., Klasing, R., Levcopoulos, C., Lingas, A., Min, J., Radzik, T.: Bamboo garden trimming problem (perpetual maintenance of machines with different attendance urgency factors). In: Steffen, B., Baier, C., van den Brand, M., Eder, J., Hinchey, M., Margaria, T. (eds.) SOFSEM 2017. LNCS, vol. 10139, pp. 229–240. Springer, Cham (2017). https://doi.org/10.1007/978-3-319-51963-0_18
7. Kawamura, A., Kobayashi, Y.: Fence patrolling by mobile agents with distinct speeds. In: Chao, K.-M., Hsu, T., Lee, D.-T. (eds.) ISAAC 2012. LNCS, vol. 7676, pp. 598–608. Springer, Heidelberg (2012). https://doi.org/10.1007/978-3-642-35261-4_62

Author Index

Printed in the United States
By Bookmasters